MODERN OPHTHALMIC OPTICS

This book is a comprehensive account of the most recent developments in modern ophthalmic optics. It makes use of powerful matrix formalism to describe curvature and power, providing a unified view of the optical and geometrical properties of lenses. This unified approach is applicable to the design and properties of not only spectacle lenses, but also contact and intraocular lenses (IOL). The newest developments in lens design, manufacturing, and testing are discussed, with an emphasis on the description of free-form technology, which has surpassed traditional manufacturing methods and allows digital lenses to be specifically designed with the unique requirements of the user. Other important topics covered include modern lens materials, up-to-date lens measuring techniques, contact and intraocular lenses, progressive power lenses, low vision aids, ocular protection, and coatings. Providing a broad overview of recent developments in the field, it is ideal for researchers, manufacturers, and practitioners involved in ophthalmic optics.

JOSÉ ALONSO is head of R&D at Indizen Optical Technologies (IOT), the company he founded together with Dr. D. Crespo and Dr. J. A. Quiroga. He is also a professor in optics at the Complutense University of Madrid. His research interests are optical metrology and ophthalmic optics, with a focus on the design of advanced customized lenses.

JOSÉ A. GÓMEZ-PEDRERO is a professor at the College of Optics and Optometry at the Complutense University of Madrid, where he lectures on geometrical optics, ophthalmic optics, and digital imaging processing. His current research interests include ophthalmic and visual optics, image processing, computer vision, and optical design.

JUAN A. QUIROGA is co-head of R&D at IOT. He is also a professor in optics at the Complutense University of Madrid. His current principal areas of interest are digital image processing techniques applied to optical metrology, with a focus on ophthalmic technologies.

MODERN OPHTHALMIC OPTICS

JOSÉ ALONSO
Indizen Optical Technologies

JOSÉ A. GÓMEZ-PEDRERO
Complutense University of Madrid

JUAN A. QUIROGA
Indizen Optical Technologies

CAMBRIDGE
UNIVERSITY PRESS

CAMBRIDGE
UNIVERSITY PRESS

University Printing House, Cambridge CB2 8BS, United Kingdom

One Liberty Plaza, 20th Floor, New York, NY 10006, USA

477 Williamstown Road, Port Melbourne, VIC 3207, Australia

314–321, 3rd Floor, Plot 3, Splendor Forum, Jasola District Centre, New Delhi – 110025, India

79 Anson Road, #06–04/06, Singapore 079906

Cambridge University Press is part of the University of Cambridge.

It furthers the University's mission by disseminating knowledge in the pursuit of
education, learning, and research at the highest international levels of excellence.

www.cambridge.org
Information on this title: www.cambridge.org/9781107110748
DOI: 10.1017/9781316275474

First published 2019
Reprinted 2019

Printed in the United Kingdom by TJ International Ltd, Padstow Cornwall

A catalogue record for this publication is available from the British Library.

Library of Congress Cataloging-in-Publication Data
Names: Alonso, José, 1965– author. | Gómez-Pedrero, José A., 1971– author. |
Quiroga, J. Antonio (Juan Antonio) author.
Title: Modern ophthalmic optics / José Alonso, Indizen Optical Technologies, Madrid,
José A. Gómez-Pedrero, Universidad Complutense, Madrid,
Juan A. Quiroga, Indizen Optical Technologies, Madrid.
Description: Cambridge, United Kingdom ; New York, NY :
Cambridge University Press, 2019. | Includes bibliographical references and index.
Identifiers: LCCN 2018045074 | ISBN 9781107110748 (hardback)
Subjects: LCSH: Ophthalmic lenses. | Ophthalmology. | Eyeglasses. | BISAC: SCIENCE / Optics.
Classification: LCC RE976 .A46 2019 | DDC 617.7/52–dc23
LC record available at https://lccn.loc.gov/2018045074

ISBN 978-1-107-11074-8 Hardback

Contents

Foreword

A few years back I discovered a book by Theodore Obrig titled *Modern Ophthalmic Lenses and Optical Glass*. Published in 1935, its mouldering pages depicted the state of the spectacle lens art eight decades ago. I needed the book to make a point in a patent litigation case. Although its teachings comprised a useful historical perspective, they no longer provided guidance for spectacle lens designers in the present day. That same patent case caused me to seek a comparable text that could be used to define the state of the art today, but I could not find one. Patiently I assembled a dossier with hundreds of documents including patent applications, journal articles, advertisements and technical manuals. I gathered books on differential geometry, lens design and ophthalmic dispensing. It was a tedious task.

That is why I was so delighted to learn about the publication of this book. It will be a tremendous resource for both students and experts in ophthalmic optics. I have spent many pleasant hours in conversation and correspondence with Jose Alonso. My dossier included works by him and his coauthors and they demonstrate a complete mastery of the most relevant and *modern* subject matter. To their great credit, they have shared much of that mastery and it is all right here, in a single reference with blessedly consistent notation.

The preface rightly emphasizes the importance and utility of matrices for representation and calculation. Yet there is so much more. Let's start at the beginning: the breadth of information about ophthalmic materials is outstanding. Of course, the text teaches the fundamentals of refractive error, but the rigor of presentation is remarkable. One of my most fervent wishes is that our industry will learn to their methods to perform statistical analysis of dioptric power data. This could prevent many unfortunate errors, for example in reports of ophthalmic clinical trials. The rigorous and complete treatment of the lens-eye system and its analysis by wavefront or rays could have saved me years of work, had it been published sooner. With patience and concentration, we also can learn here how to design, fabricate, measure and evaluate lens surfaces including progressives. Speaking of progressive lenses, I was very happy to see the work of my former colleagues Scott Fisher and David Pope given rightful credit; I was even happier to read the discussion of Julie Preston's dissertation. Those of us privileged to work in R&D at SOLA were proud to support her research.

I encourage readers to pay careful attention to the content in the Appendices. The mathematical methods are powerful yet easy to implement in code or even spreadsheets. Learning them will open doors to careers and advancement in research, education and industry. Other books aim to teach the eyecare professional how to recommend, provide and dispense ophthalmic lenses. Such a clinical emphasis is necessary for practising that which is already known. But the concepts you can learn herein will enable you to make further discoveries that may lead us to the next phase of modernity.

Michael Alan Morris O.D.
Santa Rosa, California

Preface

This book attempts to be a comprehensive account of the most relevant topics in modern ophthalmic optics. It stands out from other books on the subject in the profuse use of the unifying and powerful matrix description of curvature to offer a unified view of the properties of ophthalmic compensation elements, from classic single-vision spherical lenses to the modern, customized, free-form progressive lenses. Throughout the book we discuss the newest developments in the paraxial properties of lenses, lens design, and manufacturing and testing, with special mention of free-form technology, which has surpassed traditional manufacturing methods. Additionally, we cover other important subjects such as lens materials, contact and intraocular lenses, low vision aids, and ocular protection and coatings.

Matrix methods are especially well suited for describing astigmatic optical systems, and visual optics is more *about* astigmatic powers and vergences than spherical ones. But matrix methods are not just a set of rules to get tidier formulas. Curvature is a tensor, and its natural representation in a Cartesian coordinate system is a matrix. When we abstractly think of the velocity of an object in 3D space we do not think about the individual polar or cartesian components; we rather think of the velocity as a whole vector, a single concept expressing speed and direction at the same time. Similarly, we should think of curvature as a single concept; it just happens to have three components. This kind of thinking proves to be an extremely powerful method to understand, work, and develop astigmatic optics. Along the last thirty years, matrix methods have been commonplace in many research papers in optometry, and visual and ophthalmic optics. However, a compilation of these techniques and at least the most representative results that have been achieved with them was lacking. For 25 years we have been teaching ophthalmic optics in both undergraduate and graduate courses at the University Complutense of Madrid. In 2008 we changed the standard mathematical approach and started using a matrix formalism as the *fabric* the subject was constructed upon. We initially feared that students could get jammed with the new tools, but experience showed otherwise. Not only were the matrix tools a tidier, more uniform, and easier to remember way to write down the geometrical and optical relationships, but they did not interfere with the understanding of the basic phenomena. We would rather say the unification of curvature and power that the matrix formalism makes so evident boosted this understanding. The notes we wrote back then for these courses transformed into the seed material for this book.

The book is primarily aimed at eye care professionals (ECP), researchers, and graduate students in the field that wish to expand their knowledge in ophthalmics optics. We think it will also be useful to optical engineers involved in the design of optometric and ophthalmic instruments and lenses. To get to such a wide audience we have kept a practical approach: the book is not clinical, but we provide hints and references on how clinicians can benefit from the formalism. Neither is it too technical: We have sacrificed the most abstract foundations of the formalism (for example, we avoid discussing symplecticity and the Hamiltonian origin of the matrix transfer formalism) to provide a problem-solving approach. Similarly, most of the contents are kept within the paraxial approximation. Deeper incursions in non-paraxial methods could have been of interest to some optical engineers, but they would have required a substantial raise on the mathematical complexity, making the book too harsh for a large community of vision professionals, which we preferred could enjoy the beauty and power of the matrix methods in visual and ophthalmic optics. Still, matrix paraxial and pseudo-paraxial approaches provide an excellent starting point to lens design, as we exemplify with the calculation of contact and intraocular lenses and with a new surface model for progressive surfaces. With some trimming of the most advanced material, the book can serve as a textbook for undergraduate courses in ophthalmic optics. Having this possible application in mind, we decided to include topics not directly related with the tensor nature of curvature, such as optical materials, lens manufacturing and measurement, frames, and protection filters. In this way the book has become more self-contained with respect to the standard curriculum in ophthalmic optics.

The book starts with a description of the most common optical materials used in the ophthalmic lens industry. Using a classic approach, the study is divided into optical glass and optical polymers, presenting their most relevant optical, mechanical, and chemical properties. We also include a simple representation technique that may help the ECP with the selection of ophthalmic materials.

The second chapter is devoted to the study of the curvature and other geometrical properties of the surfaces typically used in ophthalmic optics. The description is a bit more complete than the usual introduction offered in traditional textbooks; in return, it provides the interested reader with the tools and expressions for accurate thickness calculation and lens representation. The description of surfaces will also be useful for those attempting to proceed with exact ray-tracing techniques. Central to the book, in this chapter we establish the relation between curvature and the Hessian matrix.

Chapter 3 is the link between the description of surfaces – physical and wavefronts – and the ophthalmic lens theory and practice. In this chapter we describe the duality between wavefronts and rays. By identifying the vergence with the tensor curvature of wavefronts, and the rays with the vectors perpendicular to the same wavefronts, we provide our own view of the bridge between the ray picture widely developed by W. F. Harris, and the wavefront-based picture described, for example, by E. Acosta and R. Blendowske.

Once we have the key ingredients ready, we present a first description of ophthalmic lenses. Their geometrical properties follow from the matrix curvature of their surfaces, and the back vertex power follows from the propagation of matrix vergences through the lens. In this chapter we mainly deal with the back vertex power, the one we are interested in, though we also make a small incursion into the generalization for astigmatic systems of the traditional equivalent power. We also provide here matrix and scalar expressions and algorithms to compute lenses.

In Chapter 5 we model the lens-eye system to understand the way a compensating lens interacts with the eye. We define and describe refractive error in terms of second-order eye aberrations, and provide the links between the different ways to describe power: spherocylindrical prescription, power matrix, power vectors, and second order Zernike polynomials. To study prismatic effect and magnification with the greatest accuracy and generality, we introduce the 4×4 transfer matrix formalism using the tools developed in Chapter 3. We present a novel interpretation of the relation between Remole's dynamic magnification and Keating's magnification matrix. According to the problem-solving philosophy mentioned before, we avoid the introduction of the standard 5×5 augmented matrix formalism and study decentered lenses using the standard 4×4 transfer matrices and Harris' methods for managing decentered and tilted optical systems. In the end, the 5×5 augmented matrix formalism does not introduce new insight, it is just a tidier arrangement that will be useful for systems with many decentered or tilted surfaces (as an eye implanted with an IOL, for example) but it is not that necessary for a single decentered ophthalmic lens.

After describing the interaction between the ophthalmic lens and the eye within the paraxial approximation, we discuss the aberrations that mainly affect ophthalmic lenses and how they determine their design. First, we outline the classical third order theory of oblique astigmatism and mean power error leading to the Tscherning theory of ophthalmic lens design. We then describe how modern lenses are designed using lens-eye models beyond the scope of the classical Tscherning theory and show some examples of what free form technology allows.

The study of contact and intraocular lenses (IOL), even at the most basic level, requires separate books, as their optical properties amount to just a small fraction of the complex, nonoptical interaction they have with the eye. However, we wanted to include a small chapter, the seventh, just to show the reader how the matrix formalism presented in the previous chapters can be readily applied to the paraxial calculation of complex bitoric contact lenses and toric IOLs. Of course, we do not even mention other important topics on this type of lenses such as band design, oxygen permeability and material biocompatibility, contact lens adaptation, haptic design, and any medical or clinical issues.

Chapter 8 is devoted to multifocal lenses. We start with a brief summary of presbyopia, lens effectiveness when accommodation is in play, and the computation of the ranges of clear vision. Then we study bifocals, where once again we use the matrix formalism to obtain the basic power relations and to compute prismatic effects and image jump. Our approach to progressive power lenses (PPL) is somewhat unusual. Actual design and evaluation of PPLs requires massive amounts of numerical computation not adequate for a standard course in ophthalmic lenses. Analytical models that could allow the nonspecialized designer to understand, compute, and test their basic properties were lacking, and dissertations on progressive lenses have been limited to historically guided enumerations of PPL designs. In most cases, the only source of information about those designs comes from the designer, using a qualitative and in many cases vague language. Even worse, in recent years the advance of PPL seems to come from the adding of particular "technologies" applied here and there that are supposed to enhance the lens performance, but no model or theory is provided that could support why or when the new lens design behaves better. In fact, a complete theory or model does not exist yet, but we put forward a model for progressive surfaces that will help the reader to understand the nature of progressive variation of curvature, its restrictions, and their consequences for lens performance. It is well known that PPL design is about balancing pros and cons of each design feature: we show how to put numbers into these balances, at least those for which a simple analytical model can be laid out.

The next chapter is dedicated to low vision aids and high powered lenses describing the practical and dispensing issues related with them. Two main types of low vision aid, magnifiers and telescopes, are discussed.

Chapters 2–9 deal with the geometrical and refractive properties of ophthalmic lenses. The remaining two chapters address measurement, manufacturing, and coatings. In Chapter 10 we explain the general principles of lens manufacturing with special focus on freeform technology. We also describe different techniques for lens measurement beyond the classical focimeter, introducing the main technologies used by electronic lensmeters, lens mappers, and profilometers. We include a discussion on quality control for ophthalmic lenses, describing current ISO standards and presenting new methods for quality control of progressive power lenses.

The final chapter of the book is about the control of lens transmittance, either to filter out all or part of the light spectrum, or to enhance transmittance. We first include a study of the possible damages derived from the interaction between electromagnetic radiation and the ocular tissues, then we describe the different technologies used to control the transmittance and reflectance of spectacle lenses, the associated manufacturing processes, and some guidelines for filter prescription.

The topic on frames, frame adaptation, and lens centering has been consigned to the first appendix. On the one hand, full understanding of the rules used for lens centering requires a similarly good understanding of the optical properties of lenses. On the other hand, knowledge of frame coordinate and dimensioning systems lets us understand the constraints imposed by the frame in lens alignment. According to this, the first three sections of the appendix can be read at any time, probably best before starting Chapter 5. The fourth section can also be read at any time if the centering rules are considered axiomatically. To fully understand the reasons for these rules, Chapters 5, 6, and 8 should be first studied.

Appendices B and C are mathematical complements on matrix algebra and geometry that introduce and briefly describe the tools needed in the book. Appendix E contains a brief description of Seidel aberrations and their relation to Zernike polynomials. Finally, the last appendix contains Abelés' theory of multilayers, a supplement to those looking for a deeper understanding of the workings of interferential coatings.

Acknowledgments

We thank our families – with special mention to Elizabeth and Raquel – for their support while writing this book. We are also grateful for assistance from Indizen Optical Technologies (IOT); its co-founder and CEO Daniel Crespo; and Carolina Gago, head of IOT Spain. For José Alonso and Juan Antonio Quiroga – also co-founders of IOT – this entrepreneurial adventure has provided a profound appreciation for the challenges of applying research insights to the marketplace – an insight difficult to achieve from the academic world. We are also grateful to Madrid's Complutense University, alma mater and academic home to the three of us.

José Antonio Gómez-Pedrero acknowledges the continued backing of the Spanish Ministry of Economy and Competitiveness for his research in ophthalmic and visual optics.

Symbols and Acronyms

ACD	Anterior chamber depth
BFL	Back focal length
BFP	Back focal power
BVP	Back vertex power
C	Cylindrical component of astigmatic power
CTF	Contrast transfer function
D	Diameter, optical density
DOE	Diffractive optical element
DOF	Depth of field
DRP	Distance reference point
EFL	Effective focal length
FFL	Front focal length
FFP	Front focal power
FSL	Frame symmetry line
g	Shape factor (scalar)
$\mathbb{H}$	Hessian matrix of a surface
$\mathbb{G}$	Shape factor (matrix)
H, K	Mean and Gaussian curvatures/powers
HCL	Horizontal center line
HOA	High order aberrations
IOL	Intraocular lens
MTF	Modulation transfer function
NPD	Naso-pupillary distance
NRP	Near reference point
O_R	Remote point
O_P	Near point
OTF	Optical transfer function
$\mathbf{p}$	Prism (vector)
$\mathbb{P}$	Power matrix, general (refractive power, lens power)
P	Scalar power, general (refractive power, lens power)

P_B	Refractive power of the base curve of a toric surface
PPL	Progressive power lens
P_T	Refractive power of the cross curve of a toric surface
PRP	Prism reference point
P_S	Refractive power of a spherical surface (for spherocylindrical lenses)
PSF	Point spread function
P_V, P_E, P_{TL}	Back vertex power, equivalent power, thin lens power (scalar)
$\mathbb{P}_V, \mathbb{P}_E, \mathbb{P}_{TL}$	Back vertex power, equivalent power, thin lens power (matrix)
P_x, P_y, P_t	Components of the power matrix
$P_{1,2}$	Refractive powers of the two surfaces of a lens
P_-, P_+	Minimum and maximum principal powers
R	Radius (general), reflectance
$\mathbb{R}, R$	Refractive error (matrix/scalar)
r	Radial coordinate, radius of the circular contour of a round lens, radius of cap
RGP	Rigid gas permeable
S	Spherical component of astigmatic power
t	Thickness
t_e	Edge thickness
t_0	Center thickness
T, T_v	Transmittance, luminous (visible) transmittance
V	Vergence (scalar)
$\mathbb{V}$	Vergence (matrix)
$X_{1,2}$	Any magnitude referred to the front (1) or back (2) surface
λ	Wavelength of light, length along ray
n	Refractive index
κ	Scalar curvature
$\kappa_B \, \kappa_T$	Curvatures of the base and cross curves of toric surfaces
ν	Abbe number, constringence
p	Prism (modulus)
Φ_w	Light flux (radiant)
$[\ ]_\otimes$	Curvature/power matrix in the reference system of its main meridians

1

Ophthalmic Materials

1.1 Introduction

The physical, chemical, and optical properties of a modern ophthalmic lens are the result of the combination of three elements: the material the lens is made with, the geometry of the lens surfaces, and the coatings that may have been deposited on these surfaces. Therefore, it seems proper to start the study of ophthalmic lenses by summarizing the main characteristics of the ophthalmic materials. As customary, we will describe two families of ophthalmic materials: glasses and plastics (or polymers). Understanding the main properties and characteristics of ophthalmic materials is important for the practitioner in order to guide the users through the selection of the material that best fits their needs, and to instruct them on the proper care that should be provided to their lenses to achieve the best possible compensation for their visual problems.

We begin our study of materials with a brief review of glass and glass manufacturing history. We will review the development of the different types of glasses, from the common silica glass (the classical "crown" glass) to the more exotic rare-earth glasses that appeared as a result of the extensive research carried out in the years before the beginning of the Second World War and throughout the whole duration of this conflict. The main optical, mechanical, and chemical properties of glass will be reviewed in the next section, where important concepts such as dispersion, hardness, chromatic aberrations, etc. will be introduced. The next two sections of the chapter will be devoted to ophthalmic plastics, materials that have reached the preponderance in the ophthalmic industry. We will follow the same scheme, describing first the history of ophthalmic plastics and then their main properties. We will finish the chapter with some guidelines for the proper selection of ophthalmic materials according to the needs of the user.

1.2 History of Glass and Glass Manufacturing

Under the denomination of *glass,* we refer to a number of amorphous substances composed mainly of silica and other inorganic oxides which present the same properties as those of an undercooled liquid [1]. In nature, we can find glasses produced as a result of different phenomena. Most natural glasses are of volcanic origin and they are produced when the molten ingredients of the glass, after being mixed up in a volcanic chamber, are expelled to the outside as part of the magma, and then slowly cooled until glass is formed. This kind of natural glass, known as *obsidian*, has been employed by numerous civilizations in the manufacturing of jewelry and other objects such as the obsidian mirror, crafted by Aztec artisans, that can be found in the collection of Museum of the Americas in Madrid [2]. The impact of lightning on a beach or sandy surface may also give rise to a kind of natural glass known as *fulgurite*.

Although glass is produced in nature in the rather extreme conditions described earlier, it was possible to make glass with the technology available to mankind in the Bronze Age. Indeed, the first written account of glass discovery is due to Pliny the Elder [3] who attributed this invention to chance when a group of Phoenician natron merchants (or nitre according to Rasmussen [4]), moored on a beach, used this product in lieu of stones for supporting the cauldrons used to cook their food. When the merchants started the fire, the three fundamental ingredients of the glass, namely the silica coming from the sand and the sodium and calcium oxides obtained through the decomposition of the natron were mixed up and heated together, resulting in the appearance of streams of molten glass [3, 4]. In 1920 William L. Monroe attempted to prove the feasibility of a casual glass discovery in the way described by Pliny, with mixed results [4].

Legends aside, it is beyond doubt that glass was manufactured in the Near East throughout the Bronze Age. Glass objects such as beads have been discovered both in Syria and Egypt, dated around 2500 BCE [4], and a glass manufacturing industry flourished in Egypt around 1500 BCE. It is no coincidence that the origins of glass can be traced back to the era when the first metallurgic industry appeared. Indeed, current historiography suggests that glass may have been obtained as a byproduct of the metallurgical procedure employed to manufacture bronze objects [4]. An alternative hypothesis points out the origin of glass as an evolution of the techniques employed to produce glazing coatings for ceramic objects [4]. According to this theory, the origin of glass can be found in the type of ceramic (known as *faience*) that is obtained when a clay object is covered with sand before heating it in the kiln, so the resulting pottery appears coated by a glossy surface. The evolution of this technique, known by the Egyptians and Sumerians, may have given rise to the production of glass when the proper components [4] were melted for the first time in a furnace.

There is evidence of glass usage in the ancient World by the Roman, Indian, and Chinese [3, 5, 6] cultures. In China, glass production is documented in the fifth century BCE, presenting in its composition higher amounts of lead and barium than the glass produced in the Middle East [7]. In Roman culture, the technique of glass manufacturing was imported from the Hellenistic kingdoms around the late Republican era and it was highly developed both in output and quality in the early Roman Empire. Proof of the degree of sophistication in Roman glass-making technology is the production of transparent glass through a careful selection of sands in order to avoid glass impurities due to metallic oxides or, alternatively, by using bleachers such as antimony [8]. Transparent glass is a necessary step in order to produce optical glass but, unfortunately, most Roman glass-making knowledge was lost at the end of the Empire, so the appearance of optical glass in Europe had to wait more than a millennium. Roman glass manufacturing was not only focused on the inner market; indeed glass accounted for a significant part of Roman exports through the so called Silk Road, as evidenced by the discovery of Roman glass artifacts in Afghanistan, India, and China.

As stated before, at the end of the Roman empire glass production declined, as shown by the archeological evidence [9]. However, by the beginning of the first millennium CE, the new production technique of glass blowing appeared in Germany [5], leading to the development of the great glass manufacturing centers in Murano, an island of Venice, and in the region of Bohemia in the modern-day Czech Republic. At the end of the Middle Ages highly skilled glass manufacturing techniques were achieved, as is well demonstrated by the beauty of the magnificent windows that decorate the churches and cathedrals of this period. Through the middle ages, some differences in glass composition can be found between the glasses produced in Northern Europe, which usually present a higher amount of potash, than those produced in Southern Europe [10] made mainly from soda. And, finally, it is in the Middle Ages with the invention of spectacles that we find the first examples of optical glasses in history. Technically, an optical glass can be defined as a transparent glass with such a degree of homogeneity that it can be considered a homogeneous and isotropic optical medium [11] suitable to be employed in the production of lenses.

The early centuries of the modern age are marked by both technical and scientific discoveries in the field of optical glass making. Among the first, a new kind of glass known as "flint glass" or "flint crystal" was produced from the silica rich in lead oxides that comes from flint, hence its name. Flint glass presented a higher refractive index than the existing lime-soda or "crown" glass and it was highly appreciated in the manufacturing of vessels and other "crystal" objects following its discovery by Ravenscroft in 1672 [11]. By the same time, Newton formulated a theory on light dispersion [12], the fact that light of different wavelengths propagates at different speeds within the same medium. A consequence of light dispersion is the appearance of colored borders in the images produced by any optical system formed by refractive lenses, a phenomenon that is known as *chromatic aberration*. It was easily proved that flint glasses presented more dispersion than crown ones, and that this fact could be advantageously used to reduce chromatic aberration by a proper combination of a crown and a flint lens in the so-called achromatic doublet developed around 1750.

The coming of the Industrial Revolution at the beginning of nineteenth century brought with it a rupture with the traditional ways in which objects and materials were produced. This change, which affected glass making, led to the mass production of glass with higher quality than before. One of the first milestones in this process was the method devised by Guinand and Fraunhofer [13] around 1814 for removing bubbles by stirring the molten glass before cooling. Another crucial finding by Faraday in 1829 [14] was the use of platinum crucibles with better endurance against chemical attack by the molten glass. By mid-century the basic industrial glass-making process was definitively established and it can be summarized in three major operations: batching, melting, and forming [1]. In the first process, the glass components are selected and processed in order to avoid impurities, such as Fe_2O_3. Afterward, the ingredients are mixed and heated in a melting pot until molten glass is formed. Finally, the glass is cooled and a forming process (cutting, pressing, or casting) is carried out to obtain the final product [1].

Until 1880 there were only two available glass types: crown and flint. Crown glasses are low-index glasses with reduced dispersion, while flint glasses are just the opposite; they usually present a high index with great dispersion. It is in this year that Schott and Abbe [11] started a research program aimed at the production of high refractive index glasses with low dispersion in order to widen the range of available options for the optical designer. Their efforts culminated in the discovery of a whole family of barium-oxide glasses, which can be regarded as "high-index crowns," allowing the design of improved optical systems such as Celor or Dagor photographic lenses. This process of a scientific and systematic search of new glasses pioneered by Schott and Abbe was continued throughout the twentieth century, particularly in the interwar years by the effort of scientists like G. W. Morey and C. W. Frederik of Kodak Research Laboratories [11] who worked on the development of high-index and low-dispersion glasses using oxides of heavy atomic elements, especially rare earths such as lanthanum and thorium, embedded in a borate glass. The development of these rare-earth glasses was enhanced in the Second World War when more than 125,000 pounds (around 62,500 kilograms) of rare-earth glasses were produced in the United States [11]. With the use of rare-earth oxides, the refractive index can be as high as 1.9 while keeping the dispersion relatively low. Ultimately, both barium oxide and rare-earth glasses reached the ophthalmic industry by the last half of the twentieth century.

1.3 Glass Properties

As any other material, glass has a number of distinctive properties. Foremost of them are the optical ones, but there are also other important mechanical and chemical characteristics that must be known in order to understand properly the behavior of glass as an ophthalmic material. Thus, we are going to give first a brief summary of glass composition and, afterward, we will focus on the main optical properties: refractive index, dispersion, and transmittance. We will finish this section with a short description of relevant nonoptical properties.

1.3.1 Composition

Glass composition is not unique. In fact, different mixtures and proportions of materials lead to glasses with distinct properties. In general, glass is formed by the mixture of inorganic components [1]. For example, a typical crown glass (lime-sode glass) is formed [1] by 71.5% of silica (SiO_2) as the main ingredient, together with 14% of NaO_2, which comes from sodium carbonate Na_2CO_3 (soda), 13% of CaO (lime) and a lesser amount (about 1.5%) of Al_2O_3 (alumina). The role of each of these basic components is precisely defined. NaO_2 is employed to reduce the melting point of glass, while lime improves the chemical resistance and enhances glass formation, and alumina adds durability to the glass. Other additive widely employed in the glass industry is boric oxide, B_2O_3, which reduces viscosity and melting temperature, and forms the family of borate glasses [1].

Apart from the basic components, other substances can be added to the glass to alter its properties. For instance, it has been known since the discovery of flint glasses by Ravenscroft in the seventeenth century, that the addition of lead oxide, PbO, increases the refractive index and dispersion of the resulting glass. As mentioned previously, in the nineteenth century, Schott researched thoroughly in order to find new glass components such as barium oxide, BaO, obtaining in this way the barium-oxide glasses which present a higher refractive index than crown ones but lower dispersion than flints [1]. Another example of index changing substances are the rare-earth oxides, such as lanthanum or thorium, employed in high refractive index glasses [1, 11]. But not only can the refractive index be varied using additives, other properties such as color and transmittance can be altered in a significant way just by adding small amounts of metal oxides to the molten glass. This is why an accurate control of impurities is of paramount importance in the manufacturing process of glass. For more details of glass composition we refer to the specialized literature, particularly the works of Musikant [1] and Twyman [11].

1.3.2 Refractive Index and Dispersion

The glasses employed by the ophthalmic industry are homogeneous and isotropic, so their optical properties are the same everywhere within the medium and do not depend on the direction of light propagation. Thus, for a given wavelength, the refractive index, n, is a constant defined as the ratio between the speed of light a vacuum and within the material [15]. As the former is always larger than the latter, the refractive index is a number larger than one. Its value, for most optical glasses, is between $n = 1.5$ and $n = 2$.

In ophthalmic optics the importance of the refractive index is that, as we will see in the following chapters, this magnitude relates the optical properties of an ophthalmic lens, namely refractive and prismatic power, with the lens geometry, particularly curvature and thickness. An example of this relation is given by the following equation:

$$P = (n - 1)\left(\frac{1}{R_1} - \frac{1}{R_2}\right) \equiv (n - 1)K, \tag{1.1}$$

which relates the power of a thin lens, P, and the radius of curvature, R_1 and R_2, of the two lens surfaces through the refractive index, n. By combining the geometrical factor of equation (1.1) in a single magnitude K, we can easily see that different combinations of materials and geometries may result in the same power. Indeed, as a rule of thumb, the higher the refractive index, the flatter the lens surfaces, as shown in the following example.

Example 1 A convex-plane lens made in crown glass BK7 (refractive index $n_A = 1.523$) has a curvature radius for the first surface $R_{1A} = 85$ mm. Calculate the curvature radius of the anterior surface for another convex-plane lens with equivalent power made in an extra-dense flint glass ($n_B = 1.7506$) using equation (1.1).

As those lenses are convex-plane ones, the curvature radius of the second surface is $R_2 = \infty$ for both. Therefore, for the crown lens the geometrical factor is $K_A = 1000/85 - 1/\infty = 11.77$ D and, consequently, the power of this lens is $P_A = (1.523 - 1) \times 11.77 = 6.15$ D.

As we want the flint lens to have the same power, then its geometrical factor must be $K_B = P_B/(n_B - 1) = 6.15/(1.7506 - 1) = 8.20$ D and the resulting curvature radius for the first surface can be easily computed as $R_{1B} = 122$ mm. Thus, the high-index lens presents a flatter convex surface whilst maintaining the same power.

Dispersion is the dependence of the refractive index of a medium with the wavelength. The refraction index establishes the link between the optical and geometrical properties of a lens. As a consequence, the presence of chromatic dispersion will result in a blurring of the image and the appearance of rainbow-like fringes at the border of the objects located in the periphery of the lens field of view. This effect is known as chromatic aberration [15]. As the correction of chromatic aberration is not possible with a single lens, it constitutes an unavoidable feature of ophthalmic lenses. Another important consequence of dispersion is that the wavelength for which the refractive index has been measured should be specified when referring to this index. The usual convention in optics establishes that when a refractive index is given without explicit information about the wavelength, it is assumed that it corresponds to Fraunhofer's spectral d line ($\lambda_d = 587.56$ nm).

Figure 1.1 shows the dispersion curves of four common glasses, where we can see that refractive index varies smoothly with the wavelength. Several analytical functions such as Cauchy's or Selmeier's formulas have been devised that provide an analytical approximation to these dispersion plots. However, the overall dispersion of a given material is usually characterized by a single parameter known as the Abbe number, which is defined as:

$$v_d = \frac{n_d - 1}{n_F - n_C} \tag{1.2}$$

where n_d, n_C, and n_F are the values of the refractive index measured at the wavelengths corresponding to Fraunhofer's spectral lines d ($\lambda_d = 587.56$ nm), C ($\lambda_C = 656.28$ nm), and F ($\lambda_F = 486.13$ nm), respectively, being the difference $n_F - n_C$ the *principal dispersion* [1] of the glass. Notice that the lines C and F are located, respectively, at the red and blue extremes of the visible spectrum, while the d line is placed about in the middle. The bigger

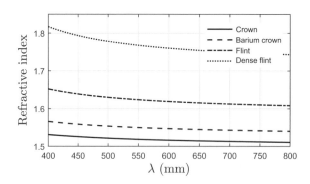

Figure 1.1 Variation of refractive index against wavelength for several glass types.

the value of the Abbe number, the smaller the material dispersion. Ideally, a material with no dispersion would have $n_F = n_C$ and then an Abbe number $\nu_d = \infty$, according to equation (1.2).

The value of the Abbe number allows to distinguish between "crown" and "flint" (i.e. low- and high-dispersive) glasses. In general, a crown (low-dispersive) glass would be a glass whose Abbe number is greater than 50 and, conversely, a flint glass would have an Abbe number lower than this value. Note that this classification applies only to the material dispersion, not to the composition, so a glass with intermediate properties such as a barium oxide can be flint or crown depending on the value of the Abbe number (see, for example, Figure 1.1). As a rule of thumb, the larger the refractive index n_d of a material, the larger the dispersion (smaller Abbe number) as is shown in the so-called Abbe diagram [16]. However, the reader must be cautious about this rule. It is neither a law nor a fact; it is just a tendency.

As a consequence of dispersion, the optical properties of a lens are wavelength-dependent through the refractive index. For example, by combining equation (1.1) with the definition of the Abbe number (1.2), we get the following expression for the lens power variation between the two extrema of the visible spectrum

$$P_F - P_C = \frac{P_d}{\nu_d},\qquad(1.3)$$

which characterizes the so-called longitudinal chromatic aberration of a lens. As it is stated in this equation, the difference of lens power between the blue and red wavelengths is inversely proportional to the Abbe number, so the bigger the dispersion the greater the chromatic aberration. Equation (1.3) summarizes one of the main drawbacks of using high refractive index materials: as they are usually more dispersive than low-index glasses, they have a bigger chromatic aberration. As a consequence, a user of lenses with a high refractive index would perceive a greater loss of sharpness through the whole image field. As the user looks through points closer to the periphery of the lens, the effect of the chromatic aberration is the appearance of colored borders in high-contrast objects. This manifestation of the chromatic aberration is called *transverse chromatic aberration* (TCA), and is given by

$$TCA = \frac{|yP_d|}{\nu_d},\qquad(1.4)$$

where y is the distance from the optical center to the point the user is looking through.

1.3.3 Transmittance

One of the most important optical properties of a glass is transmittance. The amount of energy per unit of time carried by light is known as radiant flux. The amount of radiant flux per wavelength is *spectral* radiant flux. Let us consider now a block of glass limited by two plane surfaces, and a certain amount of radiant energy incident on one of the planes. In these conditions, the ratio between the emergent and incident radiant flux is transmittance

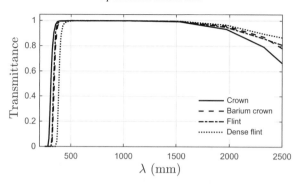

Figure 1.2 Spectral transmittance for several common glass types considering a glass thickness of 10 mm.

(see Chapter 11 for a more precise definition of transmittance). It is worth noting that transmittance should be specified for a given thickness of glass. For optical glasses an important magnitude is spectral transmittance, defined as

$$T(\lambda) = \frac{\Phi_t(\lambda)}{\Phi_i(\lambda)}, \tag{1.5}$$

where $\Phi_t(\lambda)$ is the transmitted and $\Phi_i(\lambda)$ the incident radiant flux. This magnitude is of interest not only for the visible region of the electromagnetic spectrum but also for the UV and infrared regions. Of particular interest is the behavior of spectral transmittance at the UV region. As the radiation from this part of the spectrum is potentially harmful to the eye, UV blocking is an important property of ophthalmic materials (see Chapter 11 for further details).

As can be seen in Figure 1.2, where we have plotted the spectral transmittance for the four characteristic glass types also represented in Figure 1.1, the spectral transmittance decreases abruptly in the UV region. This allows for the definition of a magnitude known as the UV cutoff wavelength λ_{UV} for which spectral transmittance reaches the value $T(\lambda_{UV}) = 0.05$ (which means that only 5% of the incident radiation is transmitted). For the glass types represented in Figure 1.2, this cutoff wavelength has a value comprised between 290 nm for the crown to 370 nm for the dense flint glasses. This means that most glasses block, at least, part of the UVA and UVB bands of ultraviolet radiation.

1.3.4 Mechanical Properties

Regarding the mechanical properties of ophthalmic glass, three main questions arise: would the material be safer enough against sudden impacts? Would it be rigid enough to withstand mechanical strain from the frame? Would it be hard enough to prevent the appearance of surface damage? These issues are addressed by manufacturing standards in which recommended values for the mechanical properties are given, as well as the tests that must be carried out to measure those properties.

In regard to the *impact resistance*, in the United States, the Title 21 of the Code of Federal Regulations requires that all lenses must be impact-resistant except when the practitioner rules otherwise attending to the specific requirements of a particular patient (chapter I, subchapter H, part 801, see [17]). According to this standard, the impact test consists of dropping a steel ball of 5/8 inch (15.87 mm) diameter and 0.56 ounce (15.87 g) of weight from a height of 50 inches (1.270 m) upon the horizontal surface of the lens, with additional provisions on the impact point location, lens support, etc. To pass the test the lens must not fracture, so the code specifies the particular conditions that hold when a lens is considered fractured [17]. In the European Union, according to the norm EN165 [18], the lenses classified as "individual protective equipment" (such as the sunglasses according to the European Directive 89/686/ CEE) must stand an impact test described in the Norm EN166 [19], which consists of throwing a ball of 22 mm of diameter and 43 g with a required speed of 5.1 m/s against the lens surface. Similar to the Code of Federal Regulations, Norm EN166 [19] specifies in which circumstances the lens has failed the test.

The international standard that specifies the mechanical properties for uncut finished ophthalmic lenses is the ISO 14889:1997 [20]. According to this standard, an uncut finished lens should be able to pass a strength test consisting of the application of a constant load of 100 ± 2 N for 10 ± 2 seconds with a speed not exceeding 400 mm per min. The lens must pass the test without breaking or showing permanent deformation. The European Union requires that all ophthalmic lenses must fulfill the provisions of ISO 14889:1997, while in the United States this is not mandatory, although the adherence to this standard is considered a "good practice" by the Food and Drug Administration. In the EU, sunglasses, as they are considered "individual protective equipment," must pass the more exhaustive strength test described in the European Norm EN166 [19], as it must be passed by any other eye-protection equipment such as laser protection goggles and so on. Similar provisions for safety lenses are required by the Occupational Safety and Health Administration (OSHA) of the Federal Government of the United States under Title 29 of the Code of Federal Regulations 1910.133, which requires that these lenses must comply with the standard ANSI Z87.1-2010 "USA Standard for Occupational and Educational Eye and Face Protection."

Mechanical *resistance to abrasion* by ophthalmic lens surfaces is regulated by the international standard ISO-8980-5:2005, which defines abrasion resistance as the ability of the lens surface to resist the appearing of surface defects such as scratches in the normal daily use. The standard consists in the application of 25 cycles of abrasion with a mechanical tool wrapped with a folded cheesecloth. The lenses pass the test when no apparent scratches appear to the naked eye using an illuminating system described in the norm.

Therefore, there are three mechanical properties of interest when studying glasses applied to ophthalmic lenses: impact resistance or fracture toughness, tensile strength, and hardness. Glasses naturally have high compressive strength, but not so good tensile strength. For example, fused silica glass has a tensile strength of 49.6 MPa, which means that it is susceptible to breaking under a sudden impact. To overcome this problem, glass can be thermally or chemically treated in order to increase its toughness in a process known as tempering. Another relevant fact regarding glass resistance is that, although glass can sustain relatively high amounts of compressive strength, on occasions it can break under a smaller load if there are surface defects (such as scratches). This fact must be taken into account in the mounting process, particularly when mounting glass lenses in metallic frames, which can exercise bigger loads than plastic ones. In order to avoid this fracture under compressive load, the practitioner must take care in polishing the lens edges before mounting and checking, using a polariscope, that no important tensions appear due to the pressure of the mount over the lens. Regarding hardness, silica glass has a Knoop hardness coefficient of around $500 \, kg/mm^2$, which is relatively high. For high-index glasses this magnitude is usually bigger, making them more prone to present surface defects.

Finally, another mechanical property to be considered is the density, which, generally, grows with the refractive index due to the use of heavy atoms such as lead or rare-earth elements in the composition of high-index glasses. For crown glasses, the density is roughly $2.2 \, g/cm^3$, while for glasses like barium oxide Ba_2O the density is near to $3.5 \, g/cm^3$, reaching the extreme values of $6.19 \, g/cm^3$ for some high-index glasses, such as the P-SF68 produced by Schott Company [16]. Although there is no exact correlation between density and mass because the volume of a lens is highly dependent on the curvature radius, in general, the usage of denser glasses tends to result in heavier lenses.

1.3.5 Chemical Properties of Glass

The chemical properties of interest in ophthalmic optics are climatic resistance, staining resistance, and resistance to acid or alkali attacks. The first property can be defined as the ability of the glass to conserve its transmittance when suffering extreme temperature and humidity changes. The exposure to high degrees of humidity and abrupt temperature changes may result in the deterioration of the transmissive properties of the glass surfaces, manifested as a loss of transmittance. In general, the glasses employed in the ophthalmic industry are among the most weather resistant, and the standard ISO 8980-3:2005 only specifies weather resistance test for lenses with coated surfaces.

Staining is an optical effect caused by the formation of a coating due to chemical reactions between glass components, particularly alkali or alkali earth oxides, and water or water vapor. As a consequence of this coating, an interference pattern is formed that can be visible in some orientations by the user or by external observers. The glass manufacturers classify their glasses according to their degree of resistance to staining by the implementation of a test that consists of the submersion of glass in a diluted acid solution and the measurement of the time necessary for the development of the interference pattern. Similar tests are employed for establishing resistance to acid and alkali solutions.

A chemical property that is actually specified in the standard ISO 14889:1997 is flammability. According to the norm, flammability is tested by placing a rod heated to $650 \pm 20° \, C$ on the surface of the lens to be tested for 5 s. The lens passes the test if it does not show signs of combustion after retiring the rod. Due to the chemical composition of glasses and their manufacturing process, flammability is not an actual issue for these substances.

1.4 History of the Optical Polymers Employed in the Ophthalmic Industry

With the term *optical polymers* we will designate a group of substances also known as optical plastics or organic materials. The "optical plastic" denomination refers to the mechanical properties, particularly that these substances can be permanently deformed (plasticity), while the name "organic materials" indicates the presence of molecules containing carbon in their composition. As this denomination is particularly misleading, because "organic" is a term usually employed when referring to living entities, we will restrict ourselves to speaking of optical polymers or optical plastics. Thus, we will define an optical polymer as any polymeric substance that is homogeneous, isotropic, transparent to visible light and with mechanical and chemical properties that make it suitable for the production of optical components.

The history of optical polymers is closely related to the development of the aeronautic and automotive industries in the first half of the twentieth century. The requirement of lightweight materials for its utilization in aircraft manufacturing, especially in the years preceding the Second World War and during the whole conflict, led to a great research effort centered on plastics. As a result of this activity, the first optical plastics appeared in the early thirties, such as Plexiglas (also known as Lucite, Acrylite, or Perspex) introduced in 1933 by the Röhm and Hass Company of Germany and immediately employed in forming the transparent panes of aircraft cockpits [21] due to its low density (between 1.17 and 1.20 g/cm^3) and impact resistance. Plexiglas is technically known as PMMA from the acronym for polymethyl methacrylate, and it has been employed in a multitude of applications, ranging from architectural (as in the Olympic Stadium of Munich [21]) to a number of products, such as computer screens. Other materials discovered before the Second World War were polyvinyl butyral or PVB, which has been employed in the production of laminated glass since the late thirties, and the nowadays ubiquitous polyvinyl chloride (PVC) and polystyrene.

Regardless of the previous discoveries, it was in the Second World War that the optical polymer *par excellence* was first synthesized: a resin named allyl diglycol carbonate, also known as CR-39. The complete story about the development of CR-39 can be found in reference [22] but it can be summarized as follows: in 1940 the Pittsburgh Plate Glass Company (PPG) started a new line of research in order to find a material that could be used in aircraft manufacturing. The actual research was done by a subcontracted company named Columbia Southern Chemical Company, which started to test a number of resins that were identified with a code composed by the initials "CR" (for Columbia Resin) and a number [22]. One of these substances, the resin coded CR-39, actually allyl diglycol carbonate, showed the exact properties that were expected. The Pittsburgh Plate Glass Company started its production, and it was first applied in the manufacturing of new lightweight fuel tanks for the Boeing B-17 bomber. This application was followed by other uses of CR-39 in aircraft manufacturing, which resulted in increasing production of the resin. When the war ended, PPG had a great stock of surplus resin, which led to an effort to find companies that would be interested in developing applications for CR-39. The result of this process was the use of CR-39 in the manufacturing of transparent sheets to be employed in a number of applications such as windows for welding masks, together with the first efforts to manufacture ophthalmic lenses from CR-39, carried out by Univis Lens Company [22]. This was not immediately successful, as Univis would ultimately retreat from its objective, but in the long term, ophthalmic lens manufacturers learned how to properly process CR-39 in the production of lenses, so, by the mid-seventies, the major industrial use of CR-39 was in the ophthalmic lens industry.

In 1957 another milestone in the development of optical polymers was achieved with the introduction of polycarbonate, an optical material with a high impact resistance that is now widely employed as ophthalmic lens material, particularly for lenses intended for users with higher risk of lens breaking, like children, or potentially dangerous activities such as adventure sports. Although polycarbonate was first obtained in 1898 by Alfred Einhorn in Germany, it was not produced industrially until the late fifties after some patent litigation by its two major manufacturers, General Electric and Bayer. The introduction of polycarbonate as an optical material was delayed because of its brownish appearance, a problem that was not solved until 1970, when the first transparent polycarbonate was produced. The main drawbacks of polycarbonate are its high dispersion (low Abbe number, $v_d = 31$) and the need to acquire special edgers designed for cutting lenses made of this material.

While current development on ophthalmic glasses seems to be stalled (of course, research and development keeps going on for glasses in general, as there are many other applications for them), the field is very active in ophthalmic plastics. The ideal properties of any material intended for ophthalmic lenses are high refractive index, high Abbe number, low density, and high impact resistance. The material should also withstand the processes required for AR (anti reflection) coating, and should be compatible with tinting, hard coating, and other common processes. It seems to be impossible to develop materials that excel in all those requirements, so each material has its own pros and cons. The path to high-index polymers has been led by the japanese Mitsui Chemical Company. They developed a thiourethane-based polymer called MR-6 in 1987, the first high-index plastic for the ophthalmic industry, with a refractive index $n \simeq 1.6$ [23]. In 1991 they launched MR-7, with index $n \simeq 1.67$, and later, improved versions of the two, MR-8 (1999) and MR-10 (1998). In 2000 Mitsui launched an ophthalmic plastic material with the highest refractive index yet, MR-174, with $n = 1.732$. Similarly to MR-6, these high-index polymers are derived from thiourethane, presenting good optical and mechanical properties. They can also be tinted and are compatible with both photochromic and AR coatings [23]. On the other hand, the American PPG, beyond CR-39, has focused on lightweight materials with impact resistance comparable to that of polycarbonate, but keeping dispersion under control. *Trivex* is such a material, a urethane-based pre-polymer. Originally developed for the military as visual armor, it was adapted to the ophthalmic industry by PPG. *Trilogy* is a similar material developed by Younger Optics. Recently, PPG has launched *Tribrid*, a hybrid material that keeps much of the impact resistance of Trivex, but with a higher refractive index and similar dispersion.

1.5 Properties of Optical Polymers

1.5.1 Composition and Manufacturing of Polymers

A polymer may be defined as a chemical substance whose molecules are formed by the aggregation of simpler constitutive groups usually known as monomers [24]. The chemical and physical process that transforms the monomers into a polymer is known as polymerization. The union of the monomers to form the polymer can be done in multiple ways, with the two more common being the formation of linear chains of monomers and the formation of three-dimensional structures. There are many physicochemical mechanisms to induce polymerization, but one of the most employed is *photopolymerization* in which short wavelength radiation is absorbed by photoinitiator molecules dissolved in the monomer that start a polymerization chain reaction. Another polymerization mechanism is temperature-mediated polymerization, where the transition from monomer to polymer is induced by the temperature. Other factors that must be taken into account when considering the structure and composition of polymers is that in addition to the monomer other substances can be added before polymerization in order to alter the properties of the final polymer. For example, in the ophthalmic application additives may be used to change optical properties such as the refractive index or Abbe's number or just to enhance the absorption of tint molecules.

We may distinguish between two kinds of polymeric substances: thermoplastic and thermoset polymers. Thermoplastic soften with heat and can even liquify. On the other hand, thermoset polymers do not soften and decompose when heated. An example of the former is polycarbonate, while CR-39 is an example of a typical thermoset polymer.

Thermoplastic lenses are made by injection molding: an extrusion process heats and compresses the raw material, liquifying and forcing it into a metallic mold. The lens is released from the mold after fast cooling. Pressure is exerted by the mold over the lens throughout the whole cooling process to compensate for shrinkage.[1] On the other hand, thermoset lenses can be polymerized (or *cured*) in glass molds. The most common polymerization process is thermally induced. The liquid monomer along with small amounts of thermal photoinitiators are inserted in a mold made of two glass parts separated by a flexible gasket. The molds are put into ovens and subjected to complex temperature cycles designed to minimize shrinkage. Typical polymerization times are 6 h with temperatures around 80° C [25].

It is worth noting that, in many cases, the product that is obtained after polymerization or injection molding is a semifinished lens or blank, i.e. a rather thick lens whose back

---

[1] Shrinkage is the reduction of volume of a thermoplastic polymer when solidifying from a liquid state, or the reduction in volume of a thermoset polymer during the polymerization process.

surface is to be reshaped by generating and polishing at a later stage. The curvature of the front surface, which will remain untouched in the final lens, is known as the *base curve* of the semifinished blank (see Chapter 10 for further information on lens manufacturing).

1.5.2 Optical Properties of Polymers

Before starting to describe the optical properties of polymers, it is important to point out the role played by additives in defining these properties - including the optical ones – of these materials. An additive can be defined as any substance mixed with the monomer to modify its original properties. Blank manufactures may use a generic monomer (for example, allyl diglicol carbonate), adding to it its own additives to produce a slightly modified polymer that will commercialize under their own brand. Here we will only deal with the properties of the generic monomer originally developed, so we refer the reader to the catalogs of the different manufacturers where the optical properties of their branded materials are specified for a further insight on this point.

The refractive index of the polymers available for the ophthalmic industry ranges from $n = 1.498$ for CR-39 to $n = 1.74$ for high-index polymers, the usual choices being $n = 1.5$ (CR-39), $n = 1.58$ (polycarbonate), $n = 1.6$, and $n = 1.67$. Comparing to glass we find a similar range of options, although high indexes such as $n = 1.8$ or $n = 1.9$ are not available for plastics. The role of high-index polymers is exactly the same as that of high-index glasses: the same power can be achieved with flatter curvatures and lenses. However, for polymers it is no longer true that high index implies greater density. For example, the refractive index of polycarbonate is $n = 1.58$, bigger than that of CR-39; but their densities are $1.20 \, \text{g/cm}^3$ for polycarbonate and $1.30 \, \text{g/cm}^3$ for CR-39.

Chromatic dispersion tends to increase with refractive index, ranging from $v_d = 59.3$ for CR-39 to $v_d = 32$ for MR-174. Again, polycarbonate constitutes an exception, its Abbe number being $v_d = 30$, considerably lower than the usual value of $v_d \simeq 40$ for resins with equivalent index $n \simeq 1.6$.

Regarding light transmission and, more especially, the UV cutoff wavelength, it ranges from $\lambda_{UV} = 350 \, \text{nm}$ for CR-39 to $\lambda_{UV} = 380 \, \text{nm}$ for polycarbonate. Therefore, many plastic lenses are transparent to the UVA band, and this makes the usage of a proper UV blocking filter advisable (even mandatory for tinted lenses according to the particular regulations of each country).

1.5.3 Mechanical Properties of Polymers

The first mechanical property we will consider in this section is density. Due to its chemical structure and composition, polymers are considerably lighter than glasses. We advanced in the previous section that the density of CR-39 is $1.30 \, \text{g/cm}^3$, while the density of crown glass is $2.54 \, \text{g/cm}^3$. If we compare two lenses made of these materials having the same radii and thickness (and roughly the same power, as the refractive indexes are quite similar), the crown glass lens will weigh double the one made with CR-39. This difference will become even larger as we move to higher refractive indexes, because the density of glasses presents a clear growth with index, while this variation of density is much smaller in plastics. The density of ophthalmic glasses ranges between 2.5 and 4.3 g/cm^3. For plastics, density ranges from 1.11 g/cm^3 for Trivex to 1.46 g/cm^3 for MR-174.

In general, impact resistance is greater in polymers than in glasses. In particular, polycarbonate exhibits excellent impact resistance, which makes this material appropriate for safety eyeglasses and one of the choices when prescribing lenses for children or athletes. Trivex is another good option for these users due to its impact resistance properties. The eye care professional (ECP) must be aware of the difference between an occupational safety eyeglass intended for reducing the risk of injury in risky activities and ophthalmic lenses for everyday use. The two uses are subject to different legal requirements and, therefore, they must meet different standards as it was previously mentioned. Concerning other plastics, both CR-39 and high-index resins [23] present good impact resistance, lower than Trivex and polycarbonate, but still fulfilling the FDA tests.

Optical polymers are more prone to elastic and thermal deformation than glasses. For example, Young's modulus for polycarbonate is 2.43 GPa with a thermal expansion coefficient of $6.75 \times 10^{-5}° C^{-1}$ compared to the values of 73.7 GPa and $0.57 \times 10^{-6}° C^{-1}$ of the soda-lime glass [1]. This means the pressure exerted by the frame rim, particularly metallic ones, may cause significant deformations in plastic lenses [26, 27]. Stronger deformations can even be expected if the lenses are heated while subjected to compressive stress, as may happen when leaving the eyeglasses on a car dashboard directly exposed to sunlight on a hot day. The stress exerted by the frame on the lens induces a phenomenon called *birefringence*: The refractive index of the stressed material will depend on both the direction of propagation and the direction of polarization of light, that is, the stress field will change the material from isotropic to anisotropic. The effect can be observed using a polariscope, an instrument that reveals birefringence as a field of colored fringes. An excess of fringes would be a symptom of an over-tight lens. Birefringence may also be frozen in lenses due to an incorrect cooling process (in the case of injection molding) or curing cycles (for thermoset materials).

Surface hardness is considerably lower in plastic materials than in glasses, particularly for polycarbonate [1]. This implies a greater probability of getting cosmetic defects on the lens surfaces, particularly scratches, both during manufacturing and during the standard use of the lens. A solution to this drawback, besides gentler handling of the lenses, is the application of scratch-resistant coatings to the lens surfaces. Polycarbonate is so soft that these types of coatings are compulsory. See Chapter 11 for further details.

1.5.4 Chemical and Biological Properties of Polymers

The relevant chemical properties expected for the polymers used in ophthalmic lenses are, as with glasses, resistance to chemical attacks and to climatic conditions. It is also important that the material be resistant to staining and fogging. Biomedical compatibility is another property that must be taken into account when considering contact or intraocular lenses. Bio-compatibility means that the material should not be harmful in any way to the human body, which not only rules out toxic materials but also those materials that may cause allergic reactions. Bio-compatibility also implies that the material should perform as intended (for example an artificial bone must present the same mechanical properties as the natural ones) and must not interfere with the functioning of any other body part. For ophthalmic materials, a major breakthrough in the field of bio-compatibility was made by Sir Harold Ridley [28], the surgeon who performed the first intraocular lens implant in 1949, who discovered, in the years of the Second World War, that the eye injuries caused by PMMA healed faster and with fewer complications because PMMA behaved as an inert material for eye tissues. Subsequent research led to the discovery of other bio-compatible ophthalmic materials such as silicone or acrylic polymers [28].

1.6 Summary of Ophthalmic Materials

The aim of this section is to summarize the optical and mechanical properties of the most common materials in the ophthalmic industry. In Table 1.1 we have compiled the main specifications (refractive index, Abbe number, and density) of several common ophthalmic materials.

In order to compare the overall performance of the materials listed in Table 1.1, we are going to compute three relevant characteristics of any ophthalmic lens: the thickness at the center or the edge, whichever is bigger, the transverse chromatic aberration, and the lens weight. Of course, we will study all this in detail in subsequent chapters, but we will briefly anticipate here how to compute these characteristics for any given spherical lens power.

Table 1.1 *Summary of the main specifications of a selection of ophthalmic materials widely used by the ophthalmic industry. Data obtained from [1, 16, 23]*

Name	Refractive index n_d	Abbe number v_d	Density $\rho\left(g/cm^3\right)$	Type
Crown	1.523	58.6	2.54	Glass
1.6/41	1.601	41.5	2.63	Glass
1.7/42	1.700	41.6	3.21	Glass
1.8/35	1.802	34.6	3.65	Glass
1.9/32	1.885	31.9	3.99	Glass
PMMA	1.490	58	1.18	Plastic
CR-39	1.498	58	1.30	Plastic
Trivex	1.527	44	1.11	Plastic
Polycarbonate	1.586	34	1.22	Plastic
MR-8	1.592	41	1.30	Plastic
Tribrid	1.607	41	1.23	Plastic
MR-10	1.661	32	1.37	Plastic
MR-174	1.732	33	1.47	Plastic

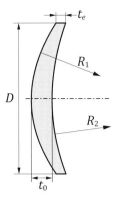

Figure 1.3 Profile of a typical ophthalmic lens with the main dimensions indicated.

Let us consider the spherical ophthalmic lens depicted in Figure 1.3. The different curvature radius of each surface, R_1 and R_2, confers the lens with an optical power P, as given by the lens maker equation (1.1) The different radii also cause the thickness to change from point to point. In particular, we are interested in the thickness at the center, t_0, and at the edge, t_e. For positive lenses we must have $R_1 < R_2$ and $t_0 > t_e$, while for negative lenses the opposite holds true, $R_1 > R_2$ and $t_0 < t_e$. In fact, for any lens with spherical surfaces, the center and edge thickness are related by (see Chapter 4)

$$t_0 + R_2 \left[1 - \sqrt{1 - (D/2R_2)^2} \right] = t_e + R_1 \left[1 - \sqrt{1 - (D/2R_1)^2} \right], \qquad (1.6)$$

where D is the diameter of the lens, as shown in Figure 1.3. In general, we fix the smaller thickness and compute the bigger one as a function of the lens radii, which in turn are related by the lens power equation. We may use $t_e = 1$ mm for positive lenses and $t_0 = 2$ mm for negative ones, these values being quite standard in actual lens manufacturing. The radii of the front surface are not arbitrary, they are selected from a handful of possible values. Large front-side radii are selected for negative lenses and relatively small radii for positive ones. Then, the procedure is as follows: For a given lens power, we select a value for R_1 (we well make the choice considered as standard in the industry), then from the refractive index of the lens material we compute R_2 by means of the lens marker equation (1.1).

Regarding the transverse chromatic aberration (TCA) and besides the previous definition of the same, suffice to say that it is an undesirable effect of material dispersion, and we will evaluate it at the edge of the lens, using equation (1.4),

$$TCA = \frac{|P_d| D}{2v_d}, \qquad (1.7)$$

where D is the lens diameter, P_d the lens power for the Fraunhofer's *d-line* of the spectrum, and v_d the Abbe number. According to equation (1.7), TCA diminishes when the Abbe number is increased. Transverse chromatic aberration is a dimensionless quantity, but in optometry and ophthalmic optics, the product of power (in diopters) times a distance (in centimeters) gets the name *prism diopter*, and the symbol Δ is used to represent $1\,\mathrm{D} \times \mathrm{cm}$.

Finally, the weight of a lens can be computed from the product of the density of the lens material times its volume. For spherical lenses the equation giving the weight is

$$w = \rho \left[\frac{\pi}{4} D^2 t_e + \frac{\pi}{3} z_1^2 (3R_1 - z_1) - \frac{\pi}{3} z_2^2 (3R_2 - z_2) \right], \qquad (1.8)$$

where ρ is the density and z_1 and z_2 are the sags of the front and back surfaces given by the following equation

$$z_i = R_i \left[1 - \sqrt{1 - (D/2R_i)^2} \right], \quad i = 1, 2. \qquad (1.9)$$

Thus, the weight of a lens does not only directly depend on the material density but it also depends, indirectly, on the refractive index through the relationship between power, curvature radius, and refractive index (see equation (1.1)).

Now, we can use the previous expressions to test the influence of the material on these three magnitudes, maximum thickness, TCA, and weight; the results are shown in Figure 1.4. We have selected a $-8\,\mathrm{D}$ (for high myopia), but very similar results are obtained for moderate power levels or positive lenses (though for positive lenses it is the center thickness instead of the edge thickness what we should evaluate). The lens is assumed be round with a diameter of 50 mm. The plot at the top represents edge thickness *versus* chromatic aberration. Each material is represented by a hollow circle whose diameter is proportional to the weight of the lens in said material; each label includes the weight in grams. The ideal material would be a small circle sitting at the bottom left corner of the plot: small edge thickness and low chromatic aberration, as a consequence of a large refractive index and a large Abbe number, respectively. Unfortunately, the tendency shows otherwise, as we already know. A high refractive index is related to small Abbe numbers, as the dispersion of the circles confirms. Although the difference in weight between glasses and plastics is notorious, within each group it is difficult to tell the circles apart by their sizes. This we can see better in the bottom graph. Here we show lens volume and weight as a function of refractive index. Glasses are indicated by star symbols and plastics by circles. The solid ones indicate weight, while the hollow ones stand for volume. We see that lens volume is a smooth and decreasing function of refractive index. The higher the index, the smaller the lens volume (as a consequence of the fact that higher refractive indexes yield the same surface power with larger curvature radii). Of course, for a negative lens of a given power, there is also a smooth relation between volume and edge thickness. The relation between volume/thickness and refractive index is not affected by the material being glass or plastic. They uniformly mix together in the curve volume *versus* index. The curve weight *versus* index is quite different; there are two completely different regimes for glasses and plastics. First, we see that glasses weigh about twice as much as plastics. Second, there is a different tendency in the behavior of weight with respect to index in each group. Though volume is always a decreasing function of index, for glasses this reduction

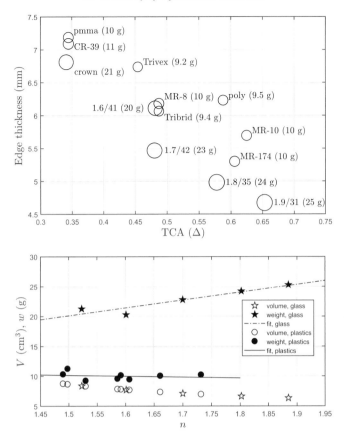

Figure 1.4 Performance of different lens materials regarding thickness, transverse chromatic aberration (TCA), and lens weight. The top graph shows edge thickness versus transverse chromatic aberration for a -8 D lens. Each circle represents a material, the diameter being proportional to the weight of the lens. The bottom graph shows volume and weight of the same lens as a function of refractive index, each point corresponding to a different material.

does not compensate the increase in density, so the tendency for lens weight is to grow as we select higher indexes. For plastics there is virtually no tendency. On average, the reduction in volume is compensated by a small increase in density, though if we look closer we see there are different subgroups within the plastics. Polycarbonate, Trivex, and Tribrid are lighter, Trivex being the lightest of all materials. Its exceptionally low density is not compensated by the reduction in volume of any other higher-index plastic lens. The thiourethane-based resins MR-8, MR-10, and MR-174 all have a very similar weight, about 10% higher that that of Trivex. Finally, the lens in CR-39 is the heaviest of all the plastic group.

The selection of the material should take into account all of the characteristics we have described so far. If the thickness of the lens is the absolute main concern of the user, 1.8 or 1.9 glasses should be considered, though selecting smaller frames, lenses with optimized diameter (for positive prescriptions), and even the use of strong aspherical surfaces available with modern free-form technology may provide better thickness saving than the material. If weight is a concern, no doubt plastic lenses are the choice, and among them, Trivex, Trilogy, Tribrid, or polycarbonate, though the difference with respect to the MR family is small. If price is not a concern, polycarbonate should be ranked below MR-8 or Tribrid, as its TCA is clearly worse. With respect to Trivex or Trilogy, the refractive index of polycarbonate is slightly higher, but the advantage probably does not pay off for the much higher chromatic aberration. If impact resistance is an important issue, the same four just listed are the best possible choice. Optimum combination between thickness and weight would be provided by the MR-174 resin, but TCA would be among the highest. In general, the ECP will have to guide the user through the different options, weighing the user preferences against the properties of the available materials (price included) to make the best possible selection.

2

Surfaces in Ophthalmic Lenses

2.1 Introduction

Optics is about controlling the propagation of light. This control can be achieved either by the use of surfaces separating two different media on which light refracts or reflects, or by propagating the light through media in which the refractive index changes from point to point. The latest approach is based on *GRadient refractive INdex* media (GRIN materials), and although there are some interesting applications using them (some types of optical fibers, scanner imagers, even the eye crystalline lens) most optical components and systems use materials with constant refractive index, and the surfaces separating different materials allow for the intended control of light. Most ophthalmic lenses are made of a piece of optical material bounded by two well-defined, polished surfaces, and the imaging properties of these lenses rely on the geometry of their surfaces. Some lenses (bifocals and trifocals) may have three or four optically functional surfaces. An important part of the optics of the eye is determined by the external surface of the cornea. Its geometry determines, to a large extent, the astigmatism and higher-order aberrations of the eye.

The geometry of the surfaces not only determines the optical properties of the system. It also determines the shape of the lenses, their thickness, and, along with the density of the materials used, the weight (see Section 1.6). It is quite clear that any degree of understanding of an optical system requires some understanding of the surfaces it contains. Most textbooks devoted to ophthalmic lenses deal with the properties of surfaces as they study the properties of lenses. In this way, the lenses are separated according to their symmetry or function: spherical lenses, cylindrical lenses, spherotoric lenses, etc. Before the arrival of special lenses for presbyopia, lenses could be distributed into two main categories: spherical and astigmatic. Spherical lenses had two spherical surfaces, whereas astigmatic lenses typically had a spherical surface and a cylindrical or toric surface. This approach was perfect when the majority of lenses could be labeled into these few categories, but over time, the typology of ophthalmic lenses grew more complex. Nowadays, with free-form technology, it would be impossible to accommodate the lenses in the market in just these few categories. For example, a bi-toric lens (that is, a lens in which the two surfaces are toric) could be nonastigmatic. A lens with two power-progressive surfaces can be made to be single vision. We therefore think that a modern approach to ophthalmic lenses should deal first with the properties of surfaces. Once we learn how to compute the general curvature and other properties of a surface, we will easily derive its refractive properties (Chapter 3). Then, lens properties, both optical and geometrical, will also easily emerge by compounding two surfaces, as we will see in Chapters 4 and 8.

Nowadays, an ever greater proportion of RX-lenses[1] have a free-form surface. That means a big proportion of surfaces in ophthalmic lenses lack any particular symmetry. Nevertheless, the classical surfaces are still fundamental to the optical industry: spheres and tori are still widely used. Cylindrical surfaces were abandoned for ophthalmic lenses almost 80 years ago, but pure astigmatic lenses in trial lens sets are still made as plano-cylindrical lenses. In summary, the study of standard, highly symmetric surfaces is still mandatory in a book about ophthalmic lenses, either because they are still used or because a good understanding of their simpler properties will help the reader progress toward more complex surfaces. So, this chapter is divided into three sections: First, we will study the properties of surfaces with revolution symmetry, the simpler of which is the sphere. Then we will study surfaces without revolution symmetry, but still with axial symmetry, such as toric, cylindrical, atoric, and biconic surfaces. Finally, we will present an introduction to the more complex free-form surfaces.

In order to properly follow this chapter, the reader will need some knowledge about the geometry of surfaces. A brief review of concepts and formulae is presented in Appendix C, along with a bibliography for further delving into the subject. However, we have considered it useful to include here an even briefer summary of fundamental definitions, formulas, and tools.

Surface. We have the intuitive concept of a surface as the boundary separating two different media. The mathematical concept is more subtle: a two-dimensional set of points embedded in the three-dimensional normal space. To describe a surface, we select a coordinate system XYZ in the three-dimensional space, and state the distance z from the XY plane to the surface by means of a function $z(x, y)$. This distance, usually named *sagitta* or just *sag* (or sag height), is measured along the direction perpendicular to the XY plane and passing through the point with coordinates (x, y) (see Figure 2.1). The one-dimensional functions $z(x, 0)$ and $z(0, y)$ represent plane curves embedded on the surface, the first one along the X axis (dashed line in Figure 2.1), the second one along the Y axis (dot-dashed line in Figure 2.1). Throughout this book, we will mostly choose the reference system so that the surface is tangent to the XY plane at the origin. We will call this the *standard representation* of a surface.

Curvature. Loosely speaking, curvature is a property of curves and surfaces indicating the *degree of bending*. In a curve, the curvature at each point is uniquely determined by

---

[1] According to the way they are processed, Ophthalmic lenses are classified as *RX-lenses* or *stock lenses*. The former are made by resurfacing the back side of stocked semi-finished lenses or blanks, and only after the order from the ECP has been placed. The latter are made with their final shape in advance, only for common prescriptions. *RX-labs* is the name used for installations where RX-lenses are made.

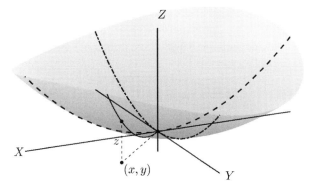

Figure 2.1 General surface and its relation to the coordinate axes X,Y, and Z.

a single number. In a surface it is a little bit more complex: As there are infinitely many directions passing through any point on the surface, we could think there are infinitely many different values of curvature at each point on the surface. It turns out that those values are not independent, and surface curvature at any point can be fully described by three numbers.

Surface Hessian matrix. This is a symmetric 2×2 matrix formed with the three distinct second-order derivatives of the surface function $z(x,y)$. If we name the elements of the Hessian matrix h_{ij}, and we use the short notation for the partial derivatives used in Appendix C, then $h_{11} = z_{xx}$, $h_{12} = h_{21} = z_{xy}$ and $h_{22} = z_{yy}$. When the function $z(x,y)$ describes a standard representation of a surface, its Hessian evaluated at the origin provides an exact representation of its curvature at that point. Because of this, we will also call *curvature matrix* to the Hessian matrix. For shallow surfaces,[2] h_{11} and h_{22} are good approximations to the curvature along the X and Y axes, respectively. Similarly, h_{12} quantifies the so-called *torsion* of the surface (see Appendix C and [29]). Approximations to the mean and Gaussian curvatures are given by $H = (h_{11} + h_{22})/2$ and $K = h_{11}h_{22} - h_{12}^2$, respectively. These expressions become exact at the vertex of the surface when $z(x,y)$ describes a standard representation. Fortunately, all the paraxial properties of ophthalmic lenses rely on the curvature of their surfaces at their vertexes, and in most cases the surfaces are tangent to the XY plane of their own reference systems. So the Hessian can be used in most applications to study, without error, the paraxial properties of the lenses.

Principal curvatures. At any point on the surface there are two perpendicular directions along which the curvature gets minimum and maximum values. These are named principal curvatures, and for shallow surfaces are given by $\kappa_- = H - |\sqrt{H^2 - K}|$ and $\kappa_+ = H + |\sqrt{H^2 - K}|$. The angle formed by the direction with curvature

---

[2] More precisely, for surfaces in which the first-order derivatives, z_x and z_y, are much smaller than 1.

κ_- and the X axis is $\theta_- = \arctan\left[(\kappa_- - h_{11})/h_{12}\right]$, and because of the perpendicularity between these principal directions, $\theta_+ = \theta_- \pm 90°$. When the principal directions coincide with the X and Y axes, then $h_{12} = 0$ and the Hessian matrix gets the simple form $\mathbb{H} = \begin{bmatrix} \kappa_+ & 0 \\ 0 & \kappa_- \end{bmatrix}$ if the bigger curvature lies along the horizontal direction, or $\mathbb{H} = \begin{bmatrix} \kappa_- & 0 \\ 0 & \kappa_+ \end{bmatrix}$ if the bigger curvature lies vertical.

Surface normal. This is the three-dimensional vector perpendicular to the surface at any point (x, y), and can be approximately computed by $\mathbf{N} = (-h_{11}x - h_{12}y, -h_{12}x - h_{22}y, 1)$. Particularly useful is the two-dimensional vector formed by the two first coordinates of $\mathbf{N}$. We call it $\mathbf{n}$ and it is the projection of $\mathbf{N}$ over the XY plane. In matrix notation, $\mathbf{n} = -\mathbb{H}\mathbf{r}$, where $\mathbf{r}$ is the column vector formed by the coordinates (x, y).

Concavity and convexity. There is always some confusion about these two terms. The reader will remember from elementary math courses that the concave/convex character of a plane curve is related to the sign of its second-order derivative, but it is easy to forget which type belongs to each sign. In this book we follow a simpler and easier to remember criterion: think of a surface as a very thin curved shell. The side of the shell that can hold a liquid will be the concave side. The other side will be the convex one. In the case of a lens, surfaces are not shells but the boundaries between the lens material and the surrounding media (normally air). The same criteria hold here. If one side of the lens can hold a liquid, then it will be a concave side. Those sides that cannot hold a liquid will be convex.

2.2 Surfaces with Revolution Symmetry

2.2.1 Sphere

Spheres are the most widely used surfaces in optics. A sphere can be defined in many ways; for example, as the set of points that are equidistant from a common center point. It can also be defined as the unique surface in which all the points have identical, constant, and nonzero main curvatures. A sphere with radius R has a constant curvature along any direction given by $\kappa = 1/R$. The implicit equation of a sphere with its center at the origin of coordinates and radius R is well known from elementary geometry: $x^2 + y^2 + z^2 = R^2$. In case the center is located at a different point, say (x_0, y_0, z_0), the equation would be $(x - x_0)^2 + (y - y_0)^2 + (z - z_0)^2 = R^2$. In general, the form of the equation will depend on the location and orientation of the reference system with respect to the surface. In optics, and whenever it is possible, we select a reference system in which the origin is located at the vertex of the surface. We can achieve this situation by locating the center of the sphere at point $(0, 0, R)$. If we substitute this center point on the previous equation and solve for z, then we get $z = R \pm \sqrt{R^2 - (x^2 + y^2)}$, the plus and minus signs being used to describe the upper and lower portions of the sphere, respectively (see Figure 2.2). Of course, we do not

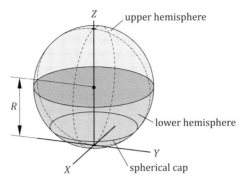

Figure 2.2 Sphere with center located at $(0, 0, R)$. It is described by $z = R \pm \sqrt{R^2 - (x^2 + y^2)}$, where the plus sign defines the upper hemisphere while the minus sign defines the lower one. The part of the sphere that will effectively be used as a lens surface is the cap shown at the bottom of the lower hemisphere.

need the whole sphere, only a spherical cap that will make any of the lens surfaces, so we discard the plus sign and limit the range of the x and y coordinates so that $x^2 + y^2 < R^2$. The final, more useful explicit equation for the sphere is then

$$z(x, y) = R \left(1 - \sqrt{1 - \frac{x^2 + y^2}{R^2}} \right). \tag{2.1}$$

This expression has some advantages for optical applications. First, the sphere is tangent to the XY plane at the origin, that is $z(0, 0) = 0$ and the normal vector at the origin is parallel to the Z axis. Second, the resulting surface is single valued; by discarding the plus sign in the previous equations we remove the upper hemisphere. Finally, the factorized radius can have negative values that would yield negative z-values. In optics, a positive-valued surface has its concave side toward the positive Z axis, while a negative-valued surface has its concave side toward the negative Z axis.

Note that x and y appears in (2.1) in the particular combination $x^2 + y^2$. This is the square of the distance r from the point on the plane XY for which we are evaluating the surface sag to the origin of the coordinate system. This means the surface sag does not depend on x and y independently, but depends on the coordinates through the radial distance r. In other words, the surface sag will obtain the same values for all the points (x, y) for which r is constant. These points are circumferences centered at the origin. This is to say that our sphere cap has *revolution symmetry* around the Z axis:[3] the surface can be obtained by rotating any section of the same (for example, the section on the XZ plane) around the Z axis. Because of this symmetry, we can write the expression for the surface sag as

$$z(x, y) = z(r) = R \left(1 - \sqrt{1 - (r/R)^2} \right).$$

---

[3] Indeed the complete sphere has more symmetries, but our cap is restricted to revolution symmetry.

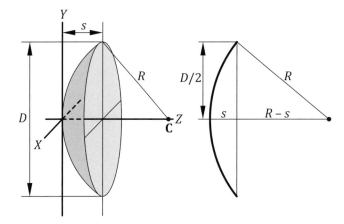

Figure 2.3 Spherical cap. To the right, the YZ section of the spherical cap, illustrating the relation between radius, maximum sag, and cap diameter.

It is important that the reader gets used to the relation between curvature, radius, and diameter of a spherical cap. The diameter D would be twice the maximum value of r, and is illustrated in Figure 2.3. From the section shown to the right and applying Pythagoras theorem, $(D/2)^2 = R^2 + (R - s)^2$, and solving for the radius

$$R = \frac{(D/2)^2 + s^2}{2s}, \tag{2.2}$$

which is a standard equation for circular sections of a sphere shown in most textbooks on ophthalmic lenses. The distance s is the maximum sag for a given cap diameter, so $s = z(D/2, 0) = z(0, D/2)$. Using equation (2.1), we get

$$s = R\left(1 - \sqrt{1 - \frac{D^2}{4R^2}}\right).$$

Example 2 The maximum sag for a spherical cap of diameter 65 mm is 5.25 mm. Compute the curvature radius.

Substituting the problem data in equation (2.2) we have

$$R = \frac{32.5^2 + 5.25^2}{10.5} = 103.2 \, \text{mm},$$

which is the requested radius.

Expression (2.1) is simple and fast to evaluate with computers. This is an important feature when exact ray tracing or wavefront computation has to be done to assess lens performance, as in a typical computation the surface sag has to be evaluated millions of times. But on many occasions we also prefer an even simpler formula that can be easily manipulated to obtain closed paraxial formulas on lens power, lens thickness, etc., as we will see in the following chapters. We can obtain such a formula by means of a linear approximation to the square root obtained from its Taylor polynomial expansion. Effectively, consider the Taylor expansion of $\sqrt{1+w}$:

$$\sqrt{1+w} = 1 + \frac{w}{2} - \frac{w^2}{8} + \frac{w^3}{16} - \frac{5w^4}{128} + \cdots .$$

When $w \ll 1$ we may neglect the second and higher-order terms, so $\sqrt{1+w} \simeq 1 + w/2$. Assuming then that $x^2 + y^2 \ll R^2$, we can apply this approximation to equation (2.1) to obtain

$$z(x, y) \simeq \frac{x^2 + y^2}{2R}, \tag{2.3}$$

which is known as the parabolic approximation to the sphere. Of course, it preserves the rotation symmetry of the sphere and can be written in terms of the radial distance, $z(r) = r^2/(2R)$.

We may proceed now to compute the Hessian of the sphere. If we use the exact expression (2.1), then we get

$$\mathbb{H} = \frac{1}{R^3 \left(1 - \frac{x^2+y^2}{R^2}\right)^{3/2}} \begin{bmatrix} R^2 - y^2 & xy \\ xy & R^2 - x^2 \end{bmatrix}.$$

We know that, by definition, the curvature of a sphere is constant all over the surface, but the Hessian is not. Indeed, to compute the exact curvatures of any surface we have to proceed as in Example 2 of Appendix C, using the first and second fundamental forms. However, the evaluation of the Hessian at the vertex of the surface, at $(x, y) = (0, 0)$ gives

$$\mathbb{H}(0, 0) = \begin{bmatrix} 1/R & 0 \\ 0 & 1/R \end{bmatrix} = \begin{bmatrix} \kappa & 0 \\ 0 & \kappa \end{bmatrix},$$

and at this point it gives the correct curvature of the sphere. This is because at the origin the sphere is tangent to the XY plane, and hence the first-order derivatives of the sag function vanish.

We could also use the approximate expression for the sphere sag function (equation 2.3). In that case, the approximate Hessian matrix turns out to be constant and equal to its exact value at the vertex. The reader has to be cautious here: the Hessian of a surface gives us an approximation to the curvature of this surface at any point, but the Hessian of a parabolic approximation of a sphere is the right curvature of that sphere at any point. This coincidence only happens with spherical surfaces having constant curvature. It is interesting to check the sag error made when computing the sag of the sphere with the parabolic approximation (2.3) and the curvature error made when computing curvature with the Hessian. Both are plotted in Figure 2.4. To the left we have plotted the difference between the actual curvature of the sphere, $\kappa = 1/R$, and the mean curvature obtained from the Hessian. They have been plotted for three different radii and against the relative radial distance, up to a maximum of 35 mm. For $R = 50$ mm ($\kappa = 20$ D), the relative radial distance reaches 0.7 and the maximum error grows beyond 15 D, which is close to 100% of error. However, for $R = 100$ mm ($\kappa = 10$ D), the maximum relative radial distance reaches 0.35 and the

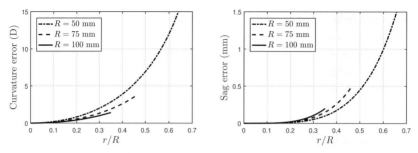

Figure 2.4 Sag and curvature errors made when computing the sag of a spherical surface with the parabolic approximation (to the left) and when computing curvature with the Hessian (to the right).

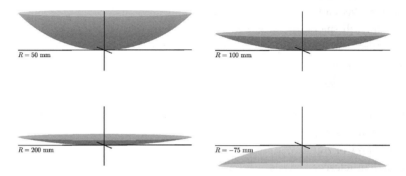

Figure 2.5 Appearance of spherical caps with 70 mm diameter and with different radii.

maximum error is about 1.5 D. As expected, the bigger the radius, the flatter the surface and the smaller the error made when computing curvature with the Hessian matrix. Something similar can be said about the error made when computing the surface sag with the parabolic approximation. At a radial distance of 35 mm, the error is about 1.5 mm for $R = 50$ mm and about 0.2 mm for $R = 100$ mm. There is a difference between curvature and sag errors. The first is proportional to $1/R$, whereas the second is proportional to R. Then, for a given relative radial system, the curvature error becomes smaller with bigger radii, but the sag error becomes bigger. Had we computed relative errors instead of differences, the curves would had been independent of the radius. Finally, before proceeding to study the next type of surfaces, we direct the reader to pay attention to the appearance of various spherical caps with the same diameter but different radii. By checking these caps with the results shown in Figure 2.4, the reader may understand what we call *shallow* or *step* surfaces, or how flat a surface should be in order to use the typical approximations within some accuracy level.

2.2.2 Conicoids

Conicoids are the surfaces obtained when a plane conic curve is rotated around its symmetry axis [30]. The conic curves (also known as conic sections) are the ellipse, the circumference, the parabola, and the hyperbola. When we rotate an ellipse, we get an ellipsoid. A parabola generates a paraboloid, the circumference generates a sphere, and the hyperbola generates a hyperboloid. Despite the sphere being a particular class of conicoid, we have studied it separately because it is the most widespread surface in optics, and this is because its higher symmetry allows a much easier manufacturing process. Conic sections were studied by ancient Greek mathematicians and many of the more distinguished mathematicians in history have made different contributions to the field. Our approach to conics and conicoids is going to be quite simple: ellipses, parabolas, and hyperbolas have some particular curvature at their vertexes and, contrary to the circumference that keeps this curvature constant at any point, the remaining conics have increasing or decreasing curvature as we consider points away from their vertexes. Suppose we need to surface one side of a lens with a central curvature κ. We can do this with *any* type of conicoid. If we use a sphere, this curvature will be constant all over the surface. With another conicoid the curvature will smoothly increase or decrease from the center value as we move away from the vertex, and this fact has interesting applications in optics. In particular, and within the field of ophthalmic optics, some oblique aberrations affecting the lens-eye system can be better reduced by using the adequate conicoids. We will study these aberrations in Chapter 6.

Spheres have been the default choice in optics since the beginning of lens manufacturing and because of this, the surfaces departing from the spherical shape have been generically named *aspheres* or *aspheric* surfaces. A well-known example was the primary mirror of the Newtonian telescope, which needed to be shaped as a paraboloid to avoid spherical aberration. Conicoids were also the first aspheric surfaces used in ophthalmic optics. Some classical examples are

Katral lens. Developed by Gullstrand and commercialized by Zeiss in 1912. It was a lens for high hyperopia or for aphakic patients. Its concave surface (closer to the eye) was spherical, but the convex surface was an ellipsoid to achieve improved reduction of oblique aberrations.

Merté's lenses. Anastigmatic lenses patented in 1950 by W. Merté [31]. This is a family of lenses characterized by combining a plane and a conicoidal surface. Positive lenses had the conicoid on the convex side, as opposed to negative lenses.

Hyperbolic Jalie's lenses. In 1981 M. Jalie disclosed a type of lenses combining spherical and hyperboloidal surfaces [32]. The surface with the bigger center curvature was the one with the hyperbolidal shape. The objective was to reduce the oblique aberrations keeping the lens thin and flat.

There are many different ways to mathematically describe a conicoid. We will use a formula that closely resembles the general equation for the sphere (2.1). The sag of a general conicoid with curvature $\kappa = 1/R$ at its vertex can be obtained with the expression

$$z(x, y) = \frac{R}{p}\left(1 - \sqrt{1 - p\frac{x^2 + y^2}{R^2}}\right),$$ (2.4)

Table 2.1 *Types of conicoids according to
the value of the asphericity parameter*

p	Type of conicoid
$p < 0$	Hyperboloid
$p = 0$	Paraboloid
$0 < p < 1$	Prolate ellipsoid
$p = 1$	Sphere
$p > 1$	Oblate ellipsoid

where p is the dimensionless asphericity parameter. Indeed, the only difference between the general equation for the sphere and the conicoids is the presence of the asphericity parameter p. In fact, the sphere is a conicoid with asphericity parameter $p = 1$. It is clear that the type of conicoid will depend on the value, or range of values, of p, as shown in Table 2.1. Equation (2.4) describes a conicoid tangent to the XY plane and with the vertex on the origin of the coordinate system. The symmetry axis of a conicoid with such a position and orientation coincides with the Z axis. As we saw for the sphere, the revolution symmetry of the conicoid allows writing the sag formula in terms of the radial coordinate, $z(r) = (R/p)\left(1 - \sqrt{1 - pr^2/R^2}\right)$.

The effect of the asphericity parameter can be appreciated in Figure 2.6. In each plot we represent the XZ sections of four conicoids with the same center curvature. The first conclusion is that the asphericity parameter doesn't affect the behavior of the conicoids at the vertex, all of them having the same curvature $1/R$ at $(x, y) = 0$. As we move away from the vertex, the sphere, with $p = 1$, keeps constant curvature. The conicoids with $p > 1$ become more curved, and the conicoids with $p < 1$ become flatter. In other words, oblate ellipsoids, with the form of a lentil, become more curved than spheres toward the edge of the conicoidal cap, whereas prolate ellipsoids, paraboloids, and hyperboloids become flatter than the sphere. In general, if we compare two conicoids $z_1(x, y)$ and $z_2(x, y)$ with the same curvature at the vertex and with asphericity parameters p_1 and p_2, and we name their respective curvatures $\kappa_1(x, y)$ and $\kappa_2(x, y)$, we can state that if $p_1 > p_2$, then $\kappa_1(x, y) > \kappa_2(x, y)$ at any point other than the vertex. Another interesting fact is that the effect of the asphericity parameter is much stronger in conicoids with small vertex radius than in shallower (flatter) conicoids. For example, the sag difference between the sphere and a $p = -10$ hyperboloid, 40 mm away from the vertex, is about 11.4 mm for $R = 50$ mm and just 0.810 mm for $R = 150$ mm.

The maximum diameter of a conicoidal cap depends both on R and p. This maximum diameter is determined by the sign of the radicand in equation (2.4). For a given value of r, the surface will exist if the inequality $1 \geq pr^2/R^2$ holds. It is obviously satisfied for any r when $p \leq 0$, but for $p > 0$ the surface can exist only for radial distances satisfying $r \leq R/\sqrt{p}$. For example, the maximum diameter for the conicoid with $R = 50$ mm and $p = 5$ (as the one shown in the leftmost plot in Figure 2.6) is $r = 22.4$ mm, and this is why

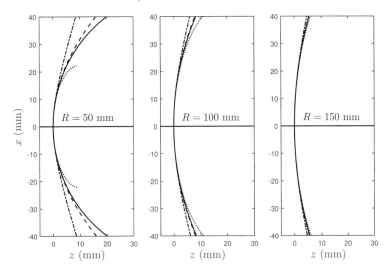

Figure 2.6 Effect of the asphericity parameter for different values of the center radius. The continuous line corresponds to a sphere, with $p = 1$. For the other curves: dashed, paraboloid ($p = 1$); dotted, oblate ellipsoid ($p = 5$); dash-dot, hyperboloid ($p = -10$).

its XZ section gets terminated at this value of the x coordinate, whose absolute value, on this section, equals the distance to the vertex.

If we fix the vertex curvature, there is just one sphere and one paraboloid with that precise vertex curvature. However, there are an infinite number of ellipsoids and hyperboloids, as their asphericity parameter may vary continuously in the intervals $(0, 1) \cup (1, \infty)$ for ellipsoids and $(-\infty, 0)$ for hyperboloids. Indeed, the uniqueness of the paraboloid takes us to a technical problem: If we substitute $p = 0$ in equation (2.4), we get an indeterminate result of type $0/0$. We can overcome this problem by computing the limit

$$\lim_{p \to 0} \frac{R}{p} \left(1 - \sqrt{1 - p \frac{r^2}{R^2}} \right) = \frac{r^2}{2R} = \frac{\kappa}{2} r^2.$$

The result we obtain is exactly the same as the approximation to the sphere obtained before. We can explore this result even further if we apply the Taylor expansion of the square root to the general conicoid expression stated in equation (2.4),

$$\frac{R}{p} \left(1 - \sqrt{1 - p \frac{r^2}{R^2}} \right) = \frac{r^2}{2R} + p \frac{r^4}{8R^3} + p^2 \frac{r^6}{16R^5} + \cdots . \tag{2.5}$$

The parabolic approximation to any type of conicoid with vertex radius R is the paraboloid with the same radius. This second-order approximation describes well the surface in a small region around the origin where the curvature of all these conicoids is the same. Also, we conclude that if we apply a parabolic approximation to a conicoid, we lose the information on its asphericity coefficient. This information only appears if we extend the approximation to the fourth or higher order. The same reasoning can be used to show that the Hessian matrix of a conicoid with vertex curvature κ computed at this very vertex is

$$\mathbb{H}_{conicoid}(0,0) = \begin{bmatrix} \kappa & 0 \\ 0 & \kappa \end{bmatrix}, \tag{2.6}$$

identical to the Hessian of a sphere. That is, the Hessian of a conicoid evaluated at its vertex does not depend on the asphericity parameter, and it just reflects that the curvature at the vertex is the same along any direction, as expected in a revolution symmetry surface.

Exact Curvatures of a Conicoidal Surface

If we know the form of the function $z(x, y)$, we can always compute exact curvatures, as the formulas presented in Appendix C only involve the computation of derivatives. However, with just a few exceptions, the formulas grow too large to be of practical use. In the case of conicoids they are still quite manageable and we can show them here to explore the behavior of the curvatures and the accuracy of the approximated expressions. To make the expressions lighter, we will use the quotient of the radial coordinate to the vertex radius of the conicoid, or relative radial coordinate, $\rho = r/R$. The exact Gaussian curvature of a conicoid is

$$K = \frac{1}{R^2} \frac{1}{\left[1 - (p-1)\rho^2\right]^2}, \tag{2.7}$$

whereas the exact mean curvature is given by

$$H = \frac{1}{2R} \frac{2 - (p-1)\rho^2}{\left[1 - (p-1)\rho^2\right]^{3/2}}. \tag{2.8}$$

Exact principal curvatures are given by $H \pm \sqrt{H^2 - K}$, but we have to be cautious about how we name them. The numerator of $H^2 - K$ turns out to be a perfect square whose sign depends on the value of p, so selection of either sign for the square root does not guarantee we get the smaller, κ_-, or the bigger, κ_+, curvatures. Instead, some algebra leads us to the conclusion that the main curvature we get by selecting the plus sign is oriented along the radial direction. We will call it tangential curvature,

$$\kappa_t = H + \sqrt{H^2 - K} = \frac{\kappa}{\left[1 - (p-1)\rho^2\right]^{3/2}}, \tag{2.9}$$

and its related principal direction is oriented at an angle $\theta_t = \arctan(y/x)$, which is the polar angle of the point with coordinates (x, y). Selection of the minus sign gives the sagittal curvature,

$$\kappa_s = H - \sqrt{H^2 - K} = \frac{\kappa}{\left[1 - (p-1)\rho^2\right]^{1/2}}, \tag{2.10}$$

whose principal direction is perpendicular to the radial direction, $\theta_s = \arctan(-x/y)$. For surfaces with revolution symmetry, as with conicoids, it is customary to define the

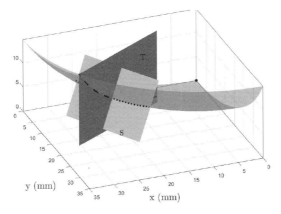

Figure 2.7 Tangential and sagittal planes and corresponding sections at the point (21.7, 12.5) mm for a conicoid with parameters $R = 100$ mm, $p = 2.5$. Tangential plane (T) is in darker gray than the sagittal plane (S), whereas the surface is plotted with some transparency degree.

tangential and sagittal planes at any point in the surface as those containing the surface normal at that point and whose sections generate curves with tangential and sagittal curvatures (see Figure 2.7). These names and curvatures are important when exploring the performance of ophthalmic lenses made with at least one conicoidal surface, as we will see in Chapter 6.

Which of the two curvatures κ_t and κ_s is bigger depends on the value of the asphericity parameter. For oblate ellipsoids, $p > 1$ and $\kappa_t \geq \kappa_s$, that is, the curvature along the radial direction is bigger than the curvature along the perpendicular direction. For other conicoids with $p < 1$, $\kappa_t \leq \kappa_s$. These relations hold for every r at which the surface is defined, the equality being reached only at the origin for all conicoids. In terms of the minimum and maximum curvature,

$$\kappa_- = \begin{cases} \kappa_s, & p > 1 \\ \kappa_t, & p < 1 \end{cases},$$
(2.11)

and the opposite

$$\kappa_+ = \begin{cases} \kappa_t, & p > 1 \\ \kappa_s, & p < 1 \end{cases}.$$
(2.12)

Of course, for the spherical surface, $p = 1$, the curvature equations reduce to $H = \kappa_- = \kappa_+ = 1/R$, and $K = 1/R^2$. The behavior of the main curvatures for the general case is shown in Figure 2.8. We can see two surfaces, each one corresponding with the tangential and sagittal curvatures. They have been computed for $R = 100$ mm that correspond with a vertex curvature $\kappa = 10$ D. For small values of the relative radial coordinate, the two main curvatures are almost identical, and the two surfaces in Figure 2.8 stick together at the plot resolution. They also keep identical for $p = 1$, where the two sheets intercept each other.

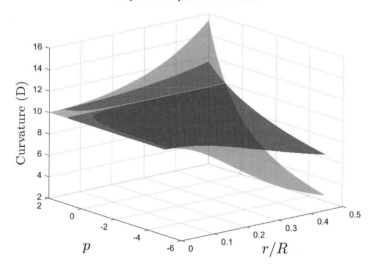

Figure 2.8 Tangential (lighter gray) and sagittal curvatures of the general conicoid as a function of the relative radial coordinate $\rho = r/R$ and the asphericity parameter p.

Finally, the reader may find useful another expression for a conicoid tangent to the XY plane and whose symmetry axis coincides with the Z axis,

$$z(x, y) = \frac{\kappa \left(x^2 + y^2\right)}{1 + \sqrt{1 - p\kappa^2 \left(x^2 + y^2\right)}}. \tag{2.13}$$

This expression provides identical results to those of expression (2.4), but it is still valid for $p = 0$. However, it requires some more computational effort than (2.4).

Example 3 Compute the sag corresponding to a point located 25 mm away from the vertex for: (a) a paraboloid, (b) a sphere, and (c) an ellipsoid with $p = 2.75$ if the curvature radius at the vertex for all the three surfaces is $R = 75$ mm.

The surface curvature is $\kappa = 75^{-1} = 0.0133\,\mathrm{mm}^{-1}$. The distance of a point of coordinates (x, y) from the vertex is $r = \sqrt{x^2 + y^2}$. Thus, according to the problem statement, we have $x^2 + y^2 = 25^2 \equiv 625\,\mathrm{mm}^2$. Substitution of these values in equation (2.13) with $p = 0$ (paraboloid), $p = 1$ (sphere), and $p = 2.75$ (ellipsoid) gives $z = 4.17$ mm, $z = 4.29$ mm and $z = 4.54$ mm, respectively. Notice that the sagitta of the conicoid is a growing function of the asphericity parameter.

2.2.3 Other Aspherics with Revolution Symmetry

Any surface that can be written as a function of the radial coordinate $z(x, y) = z(r)$ has revolution symmetry. It can be imagined as the surface generated by the one-dimensional function $z(r)$ as its curve is rotated around the Z axis. Useful surfaces in optics have to be smooth at the vertex. That means the first derivative of the function at the origin must be zero, $z'(0) = 0$. We also demand that the curve itself is zero at the origin, $z(0) = 0$. These conditions ensure the surface is tangent to the XY plane at the vertex and the curvature is continuous. The simplest curve we can imagine is polynomial, which has to start with the second order to meet the previous conditions. The case $z(r) = cr^2$ is the paraboloid already studied, with vertex curvature $\kappa = 2c$. Adding more terms to the quadratic one doesn't change the curvature at the vertex, as it only depends on the second derivative, but we add more flexibility to control surface behavior (curvature included) away from the vertex. Although we could add terms with odd powers equal or greater than 3, only polynomials with even powers are used in optics. So the typical polynomial surface would be

$$z(r) = \sum_{n=1}^{N} c_n r^{2n}, \tag{2.14}$$

where the first coefficient is half the vertex curvature, $c_1 = \kappa/2$. As we can deduce from the discussion on exact curvatures in the previous section that tangential and sagittal curvatures are main curvatures in a surface with revolution symmetry, we only do not know, a priori, which one is the biggest and which one the smallest. There are analytical expressions for directly computing tangential and sagittal curvatures in revolution symmetry surfaces without the need for previously computing mean and Gaussian curvatures. These are

$$\kappa_t = \frac{z''}{(1 + z'^2)^{3/2}}, \tag{2.15}$$

$$\kappa_s = \frac{z'}{r\sqrt{1 + z'^2}}. \tag{2.16}$$

When the slope given by the derivative $z'(r)$ is small with respect to 1 (which either happens at any point on the surface if the overall curvature is small or at points close enough to the origin when the curvature is large; remember that $z'(0) = 0$), we may approximate the previous expressions by the handier

$$\kappa_t \simeq z'' \left(1 - \frac{3}{2}z'^2\right),$$

$$\kappa_s \simeq \frac{z'}{r} \left(1 - \frac{1}{2}z'^2\right).$$

Example 4 Using the general expressions for the tangential and sagital curvature of a revolution symmetry surface, determine these curvatures for an arbitrary conicoid up to second order on the radial coordinate.

To obtain this second-order approximation, we may use the polynomial expansion of the conicoid up to fourth order, that is, $z = (\kappa/2)r^2 + p(\kappa^3/8)r^4$. Computing its first and second derivatives, substituting in the previous expressions and neglecting third and higher-order powers of the radial coordinate, we get

$$\kappa_t = \kappa \left[1 + \frac{3}{2}(p-1)\kappa^2 r^2 \right],$$

$$\kappa_s = \kappa \left[1 + \frac{1}{2}(p-1)\kappa^2 r^2 \right].$$

Polynomial surfaces are seldom used. Much more popular are so-called generalized conicoids, which are defined as the sum of a conicoid and a polynomial of even powers,

$$z(r) = \frac{R}{p}\left(1 - \sqrt{1 - p\frac{r^2}{R^2}} \right) + \sum_{n=2}^{N} c_n r^{2n}. \tag{2.17}$$

The smaller power in the polynomial is set to 4, so that the curvature of the surface at its vertex is only defined by the conicoid, $\kappa = 1/R$. These combinations are used because some types of behavior cannot be practically achieved just by using polynomials. For example, the hyperboloid gets flatter as we move away from the vertex. Asymptotically, it tends to a cone, its tangential curvature approaching zero as we move away from the vertex. A polynomial surface cannot do that, unless we use an impractically high number of terms. On the other side, the conicoids present a rigid structure: Given the vertex curvature, there is only one parameter controlling their behavior far from the vertex. The polynomial part of the generalized conicoid may provide such a flexibility.

Many sets of aspherical base curves[4] used in the ophthalmic industry are described by generalized conicoids.

2.3 Surfaces with Axial Symmetry

A surface with axial symmetry is identical to either side of an axis passing through its vertex. The surface may have a unique symmetry axis (such as the old, symmetric progressive surfaces that we will study in Chapter 8) or, more usually, it will have two orthogonal symmetry axes, as most surfaces used for astigmatic lenses have (cylinders, but mainly tori being the more extensively used). The main characteristics of a surface with axial symmetry that has lost revolution symmetry is that the curvature at its vertex now depends on the direction considered, that is, the surface is now *astigmatic*. We will use this term to refer to surfaces in which curvature at the vertex depends on the direction considered and, hence, has two different main curvatures. From the mathematical point of view, the Hessian matrix of such a surface is no longer a multiple of the identity matrix but a general symmetric 2×2 matrix. Astigmatic surfaces are not very common in the optics of imaging instruments, where all lenses usually have revolution symmetry. However, the eye is a biological system and its optical surfaces are more prone to error than those in precision imaging instruments. Indeed it is quite rare for the cornea or the crystalline lens to have revolution symmetry, and

---

[4] Base curves is the name the ophthalmic sector gives to a set of semi-finished lenses made by casting (or injection molding in the case of polycarbonate). These lenses have the front surface finished with a predefined geometry. Their back surface will be reshaped, upon order from the optical shop, to confer the lens with the power required by the prescription.

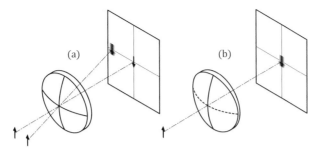

Figure 2.9 Difference between oblique astigmatism and axial astigmatism. (a) Imaging system: its lenses have revolution symmetry. When the object is on-axis, the image is free from astigmatism. For off-axis objects oblique astigmatism may be present. In this case, even though the surfaces have revolution symmetry, the oblique incidence of the rays refracting through them may cause astigmatism on the output wavefront. The amount and orientation of this oblique astigmatism also has revolution symmetry, and it can be reduced by adequate selection and/or combination of surfaces with revolution symmetry. (b) Astigmatic system. In this case, the optical system lacks revolution symmetry and even the image of on-axis objects will present astigmatism. The lack of symmetry manifests in the different curvature along the vertical and horizontal meridians, and the variable edge thickness of the lens. It has to be compensated with astigmatic lenses with surfaces having axial symmetry.

for a high percentage of eyes the astigmatism of the corneal surface is not negligible at all. It is because of this reason that we usually need astigmatic lenses to compensate for eye astigmatism. The reader should be aware of the difference between oblique astigmatism that may be present for off-axis objects in any imaging system and astigmatism produced on-axis by an optical system due to the lack of revolution symmetry of any of its surfaces (see Figure 2.9). The compensation of the first error will require adequate selection of revolution surfaces with the right tangential and sagittal curvatures. The compensation of the second will require the use of at least one astigmatic surface without revolution symmetry, but with axial symmetry. Finally, let's observe that the astigmatic eye compensated with astigmatic lenses will also be affected by oblique astigmatism, as we will see in Chapter 6.

In this section we will present the properties of the most common surfaces with axial symmetry used in the ophthalmic sector: the cylinder, the tori, and the different types of aspherical tori, also called atoric surfaces.

2.3.1 Cylinder

The cylinder is a ruled surface generated by the rotation of a straight line called the generatrix around a revolution axis parallel to it. The direction defined by the generatrix or the rotation axis is called the *cylinder axis*. The cylindrical surface is invariant under translation along the axis direction. For example, if the cylinder axis lies along the X axis, then the function describing the cylindrical surface won't depend on the variable *x*. There is another way to construct the cylinder as a translation surface that will be useful in the next section; we can draw a circumference and parallel translate it along a

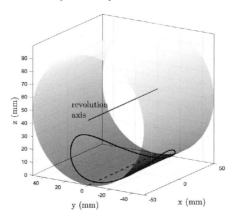

Figure 2.10 Complete cylinder generated by the parallel translation of a circle with radius $R = 50$ mm, located on the YZ plane, with center at $(0, 0, 50)$ mm, and translated along the X axis direction. The revolution axis of the cylinder is the line described by its center. The curved solid line represents the contour of a cylindrical cap that could be used as a lens surface. The dashed line is parallel to the revolution axis, but located on the surface. This line will be considered as the *cylinder axis*, not from the geometrical point of view, but from the optical point of view.

straight line. As the circumference moves it generates the cylinder, and its center generates a straight line that will become the revolution axis of the surface. For example, consider a circumference of radius R lying on the YZ plane, with its center at point $(0, 0, R)$. If we translate this circumference along the X axis direction, then we generate the cylinder shown in Figure 2.10. This cylinder is tangent to the XY plane. A cylindrical cap that could be used for optical purposes is also shown in the figure. We have shown two axes in the figure: The actual revolution axis, as a solid line, and another dashed line, parallel to it, on the cylindrical cap tangent to the XY plane. We will only use the latter, which will be referred to as the cylinder axis, and it provides the orientation of the direction along which the cylinder is invariant.

From what we learned in the section on spherical surfaces, it is not difficult to deduce that the equation of the cylindrical cap shown in Figure 2.10 is $z(x, y) = R\left(1 - \sqrt{1 - y^2/R^2}\right)$, and as expected, it does not depend on the variable x. This equation is of little help, as cylindrical surfaces used in the ophthalmic sector may have any orientation, in particular that required by the astigmatism of the eye. Hence, we need a cylinder description in which the axis may have any orientation in the XY plane. We can obtain this by the rotation of the coordinate system. Let us assume the cylinder axis forms an angle θ with the X axis. If we create an orthogonal coordinate system UV such that the U axis coincides with the cylinder axis, then the equation of the cylinder in the UV system will be the same as previously written: $z(u, v) = R\left(1 - \sqrt{1 - v^2/R^2}\right)$; it won't depend on the u coordinate. But now, the coordinates in the UV system are obtained from the coordinates in the XY system by a rotation around the Z axis, the relation between both pairs of coordinates being

$$u = x\cos\theta + y\sin\theta$$
$$v = -x\sin\theta + y\cos\theta. \tag{2.18}$$

We just have to substitute the value of v in the original equation

$$z(x, y) = R\left(1 - \sqrt{1 - \frac{(-x\sin\theta + y\cos\theta)^2}{R^2}}\right), \tag{2.19}$$

to obtain the equation of a cylinder whose orientation (its axis) lies along a direction making an angle θ with the X axis.

Intuitively, it is pretty clear that the main curvatures of cylindrical surfaces are zero along the axis and $\kappa = 1/R$ along the generating circumference. Which one is the smallest depends on the sign of R. The curvature matrix at the origin can be obtained from direct computation of the second-order derivatives of the expression (2.19). The result is

$$\mathbb{H}(\theta) = \begin{bmatrix} \kappa\sin^2\theta & -\kappa\sin\theta\cos\theta \\ -\kappa\sin\theta\cos\theta & \kappa\cos^2\theta \end{bmatrix}. \tag{2.20}$$

As we would expect, the curvature matrix of a rotated cylinder is not diagonal, as its principal directions do not coincide with the X and Y axes. For those particular cases when $\theta = 0°$ or $\theta = 90°$, the curvature matrix simplifies to

$$\mathbb{H}(0) = \begin{bmatrix} 0 & 0 \\ 0 & \kappa \end{bmatrix}, \quad \mathbb{H}(90) = \begin{bmatrix} \kappa & 0 \\ 0 & 0 \end{bmatrix}.$$

In the first case, the curvatures along the X and Y axes are 0 and κ, as shown in Figure 2.11(a). In the second case, the cylinder axis lies along the Y axis, and the curvatures get interchanged. Let us introduce a notation that will prove to be useful in the chapters that follow. The elements of a Hessian matrix depend on the orientation of the reference system used. If we choose the X axis to lie along the axis of a cylinder, its Hessian matrix will take the form $\mathbb{H}(0)$, even though the axis is tilted with respect to the horizontal axis. We will use the subindex $\otimes$ attached to the parenthesis of the matrix to indicate that it is referred to the main meridians of the surface it describes. According to this, the Hessians of the rotated cylinder would be

$$\mathbb{H}(\theta) = \begin{bmatrix} \kappa\sin^2\theta & -\kappa\sin\theta\cos\theta \\ -\kappa\sin\theta\cos\theta & \kappa\cos^2\theta \end{bmatrix} = \begin{bmatrix} 0 & 0 \\ 0 & \kappa \end{bmatrix}_{\otimes}. \tag{2.21}$$

The first matrix is referred to the general (horizontal/vertical) reference system, while the second matrix is referred to a rotated system whose X axis lies along the cylinder axis.

From equation (2.19) it follows that in the general case, the curvature along the X axis is given by $\kappa_x = \kappa\sin^2\theta$. This result is presented in a slightly different way in many classical textbooks. Instead of using a Cartesian reference system to describe the general cylinder curvature in its matrix form, we may consider a cylinder that is not necessarily related to a particular reference system, and we can ask about its curvature along a meridian making an angle θ with the cylinder axis, whatever its direction may be. Then we can always select

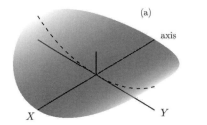
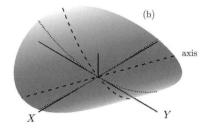

Figure 2.11 (a) Cylindrical cap oriented at $\theta = 0°$. The main sections (dashed lines) lie along the X and Y axes. (b) Cylindrical cap oriented at $\theta = 160°$. The X and Y sections (dotted lines) are no longer main sections, and the curvature matrix of the surface is not diagonal.

a reference system in which the X axis lies along the meridian that we want to know the curvature of. The cylinder axis would then make an angle θ with the X axis, and therefore the curvature on this meridian would be

$$\kappa(\theta) = \kappa \sin^2 \theta.$$

Addition of Cylinders

In ophthalmic practice it is usually said that the superposition of two identical cylindrical lenses with axes perpendicular to each other is equivalent to a sphere with the same radius as the cylinders. Let us explore the geometry behind this claim with the tools we are presenting in this chapter. Consider a surface function that can be written as the sum of two other functions, $z(x, y) = z_1(x, y) + z_2(x, y)$. Because of the linearity of the derivatives, it is clear that the Hessian matrices of these functions satisfy $\mathbb{H} = \mathbb{H}_1 + \mathbb{H}_2$. Notice that this is not true for the main curvatures, as they may lie along different directions and then cannot be directly added together, but it holds true for the Hessian or curvature matrix. Let us consider the surface given as the sum of two cylinders with identical maximum curvature κ and with perpendicular axis, $z(x, y) = z_\theta(x, y) + z_{\theta+90°}(x, y)$, where the sub-index denotes the orientation of the cylinder axis. The matrix curvature of the composed surface is given by

$$
\begin{aligned}
\mathbb{H} &= \mathbb{H}(\theta) + \mathbb{H}(\theta + 90°) \\
&= \begin{bmatrix} \kappa \sin^2 \theta & -\kappa \sin \theta \cos \theta \\ -\kappa \sin \theta \cos \theta & \kappa \cos^2 \theta \end{bmatrix} + \begin{bmatrix} \kappa \cos^2 \theta & \kappa \cos \theta \sin \theta \\ \kappa \cos \theta \sin \theta & \kappa \sin^2 \theta \end{bmatrix} \\
&= \begin{bmatrix} \kappa & 0 \\ 0 & \kappa \end{bmatrix},
\end{aligned}
$$

where we have used $\cos(\theta + 90°) = -\sin\theta$ and $\sin(\theta + 90°) = \cos\theta$. The important conclusion is that the sum of two cylindrical surfaces with the same maximum curvature and axes perpendicular to each other is equivalent to a surface with revolution symmetry and with the same curvature at the vertex. Of course, this result is exact only at the vertex. As we move away from this point, the equivalence becomes less approximate. We can get

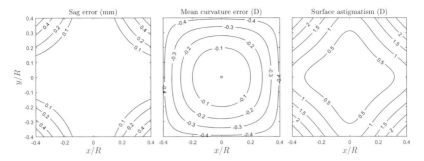

Figure 2.12 Differences in sag and curvature between a perfect sphere and the sum of two cylinders with perpendicular axes. The three surfaces have $R = 100$ mm. Although sag errors seem to be quite small (smaller than 0.1 mm inside a circle with a diameter of about 50 mm), curvature error, and especially surface astigmatism, becomes quite large toward the periphery of the lens.

an idea of the quality of the approximation by taking a look at Figure 2.12. The contour map to the left shows the sag-difference between a sphere with radius $R = 100$ mm and the sum of two cylinders with the same radius and axis oriented at $0°$ and $90°$. For radial distances up to a quarter of the curvature radius, the difference in sag is smaller than 0.1 mm. We also see that the sum of the two cylinders does not have real revolution symmetry, as the error lacks this symmetry. The map at the center shows the difference between the mean curvature of the sum of cylinders and 10 D, which is the mean curvature of the spherical surface. At a distance of 40 mm from the center, the curvature of the sum of cylinders is 0.4 D smaller than the curvature of the spherical surface. Finally, the map on the right shows the differences between the main curvatures of the composed surface, which we call surface astigmatism. As expected, it is zero at the center, but grows above 2 D at about 40 mm from the center.

There are two important remarks in relation to the previous example. At this point in the book the reader may be wondering about the relation between the superposition of two cylindrical lenses and our example about creating a surface by adding two different ones. Although we are going to deeply delve into this question in the next chapters, a quick explanation here may help the reader in case the previous question arises.

Surfaces are important because they determine the properties of lenses. Let us name the front and back surfaces of a lens $z_f(x, y)$ and $z_b(x, y)$ (we are assuming light goes from the front to the back surface). We will see that the main optical property of a lens is its power, which is nothing but curvature[5] and as such can be described by a 2×2 symmetrical matrix $\mathbb{P}$ given by the expression $\mathbb{P} = (n - 1)\left(\mathbb{H}_f - \mathbb{H}_b\right)$, where n is the refractive index of the lens and $\mathbb{H}_f$ and $\mathbb{H}_b$ the Hessian matrices of the front and back surfaces. Now let us assume we locate two lenses, one just after the other. The power of the set is, once again, additive, $\mathbb{P} = \mathbb{P}_1 + \mathbb{P}_2$ so

---

[5] The curvature of the wavefront emerging from the lens when a collimated beam of rays impinges on it.

$$\mathbb{P} = (n-1)\left(\mathbb{H}_{f1} - \mathbb{H}_{b1}\right) + (n-1)\left(\mathbb{H}_{f2} - \mathbb{H}_{b2}\right)$$
$$= (n-1)\left[\left(\mathbb{H}_{f1} + \mathbb{H}_{f2}\right) - \left(\mathbb{H}_{b1} + \mathbb{H}_{b2}\right)\right].$$

That is, the combined power of the two lenses has the same power as a single lens whose front and back surfaces can be expressed as the sum of the front and back surfaces of the two lenses. In this way, we may establish parallelism between the addition of two cylindrical surfaces and the superposition of two cylindrical lenses, but the concept is fully general and can be applied to any type of surface. We have advanced many concepts to bring this demonstration up, but we think it will be useful for the reader and will reinforce future material. It is also important to keep in mind that astigmatic power can only be added as shown in the example when power and curvature are represented by their matrix description.

The second remark is about the concept of the cylinder and how it permeates the nomenclature used in visual optics. Let us consider an arbitrary surface $z(x, y)$ with main curvatures κ_- and κ_+. Because of the additive principle just exposed, we could split our surface into the sum of a sphere $z_{sph}(x, y)$ and a cylinder $z_{cyl}(x, y)$ so that the (only) curvature of z_{sph} is κ_- and the maximum curvature of the cylinder is $(\kappa_+ - \kappa_-)$. Thanks to the revolution symmetry of the sphere and its curvature being the same along any meridian, the main curvatures of these two surfaces can be added without the need for adding the matrix forms. The cylinder has zero curvature along its axis, so the only curvature of the composed surface is that from the sphere, κ_-. Along the perpendicular direction, the sum of the two curvatures gives $\kappa_- + (\kappa_+ - \kappa_-) = \kappa_+$. So, we recover the main curvatures of the original surface. It is because of this possibility that the difference between the main curvatures of a surface is known as its *cylinder* or *surface astigmatism*. The same nomenclature is used for lenses and prescriptions, even when there are no actual cylindrical surfaces in any of them. Although cylinders are no longer used to make spectacle lenses, they are still widely used for measuring the refractive state of the eye, both in sets of trial lenses or as components inside phoropters. Once again, we will develop these ideas further in future chapters.

2.3.2 Toric Surfaces

Tori are principal surfaces for ophthalmic optics, as most standard lenses designed to compensate for eye astigmatism have a back toric surface. From the geometrical point of view, a torus is a surface generated by a circumference rotating around an axis contained in the same plane. This surface is shown in Figure 2.13, in particular, the well-known doughnut or tire-shaped ring torus. For this representation, we have chosen the rotation axis to be vertical, and the generating circumference is contained in a plane that also contains the rotation axis. The point in the circumference furthest from the axis is Q, and upon rotation around the axis it generates the equator of the toric surface, a maximum diameter circumference in this example. Because of the symmetry, it is easy to understand that for any point on the equator, the principal curvatures are precisely those of the equator and the generating circumference. Tori have considerable importance in some branches of

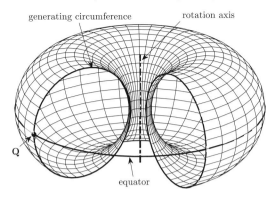

Figure 2.13 Toric surface with ring geometry. The generating circle revolves around a vertical rotation axis. The surface has been cut in order to better appreciate its geometry.

mathematics, especially topology. In optics, we do not use entire toric surfaces but patches of them, just because of their astigmatism. The point of using toric patches in optics is that, because of their geometry, they are easily generated by milling machines known as toric generators (see Section 10.2.2).

According to the relative sign of the main curvatures and the position of the generating circumference with respect to the rotating axis, there are four different types of toric patches that may have different optical properties when used as diopters to refract light. They are depicted in Figure 2.14. Let us call the radius of the generating circumference and the distance from its center to the rotation axis R_g and d. Let us also assume the Z axis points from the surface toward the revolution axis. As usual, if the patch curves toward the positive direction of the Z axis the curvature will be positive, and negative otherwise. When $d > R_g$, the generating circumference does not intersect the revolution axis, the generated surface is a ring torus with the typical doughnut shape, and there are two possible patches:

Ring patch. Is the one located at the outer part of the ring torus, shown in subplot (a) in Figure 2.14. It has main curvatures $\kappa_+ = 1/R_g$ and $\kappa_- = 1/(R_g + d)$. Both are positive, and the smaller one corresponds with the equator of the torus.

Capstan patch. Also called "pulley" for obvious reasons, is located at the inner part of the ring torus. It is shown in subplot (b). It has principal curvatures $\kappa_- = -1/R_g$ and $\kappa_+ = 1/(d - R_g)$ and they have opposite signs. It is very rarely used as an optical surface. In particular, it would produce rather large aberrations if used as a surface in astigmatic ophthalmic lenses.

When the generating circumference intersects the revolution axis, $d < R_g$, the complete torus does not present its typical hole. Instead, a self-intersecting spindle torus is formed with an inner sheet completely inside the outer one. Depending on which sheet we get the patch from, we have once again two types:

Ring patch. Shown in subplot (c). In optical applications, it has the same name as the outer patch of ring-type tori, because the curvatures behave pretty much the same

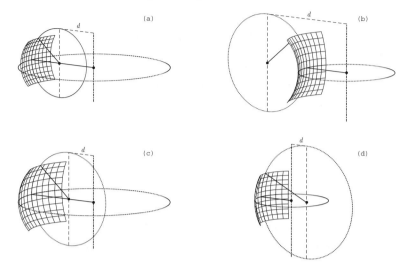

Figure 2.14 Different patches of toric surfaces according to the position of the generating circumference in relation to the rotating axis and the sign of the main curvatures. In the upper row, patches from a ring torus. In (a), a ring patch from the outer part of the ring torus. In (b), the capstan (inner) patch from the same type of torus, characterized for having curvatures with opposite signs. In the lower row, patches from a spindle torus. The subplot (c) shows the patch from the outer sheet, also named the ring-type patch as its geometry is basically identical to the one in (a). At (d), the patch from the inner sheet of the spindle torus, which is known in optics as a barrel patch. Main meridians are always circumference arcs, shown here with dotted circumferences. The revolution axis is vertical in all four cases and represented with a dash-dot line. For ring patches, the flatter meridian is that of the equator. However, the equator is the more curved meridian in the barrel patch.

all over the patch, as we will see next. This patch on the outer surface has principal curvatures $\kappa_+ = 1/R_g$ and $\kappa_- = 1/(R_g+d)$, the same as the ring-type patch from a ring torus. The only difference now is that $R_g+d < 2R_g$, whereas in the previously seen ring type $R_g + d > 2R_g$. For ring-type patches, the equator is the flattest of the two principal meridians, regardless if whether they come from ring-type or spindle-type tori.

Barrel patch. Finally, from the inner surface of a spindle torus, we get a toric patch with principal curvatures $\kappa_+ = 1/(R_g - d)$ and $\kappa_- = 1/R_g$. Both are also positive but now there is a significant geometrical difference: the shallower (flatter) principal meridian is the generating circumference, whereas the meridian with maximum curvature is the equator. This change will produce some differences in the distribution of curvatures on the surface, as we will see next. This last type of toric patch is known as "barrel" type and is shown in subplot (d) in Figure 2.14.

For ring and barrel patches, where the principal curvatures have the same sign, it is customary in ophthalmic optics to name the flatter principal meridian the *base curve* and the most curved principal meridian the *cross curve*. For ring-type toric patches, the base curve is the equator of the torus, whereas the cross curve is the generating circumference. For barrel-type toric patches, the opposite holds; the base curve is the generating circumference and the cross curve is the equator.

Ring and Barrel Types of Toric Surfaces

Proper design of maximum quality astigmatic ophthalmic lenses should take into account which type of toric surface better behaves from the optical point of view. Some prescriptions will require a ring-type surface; some others will have smaller aberrations with a barrel type. With modern free-form technology, more sophisticated surfaces can be sought, but to compute these, a good starting point is the right type of toric surface. At the center of the lens, the ring and barrel types will behave the same. It is when we consider points away from the vertex that the two surfaces will differ. Let's see this with an example. Assume we want to create an astigmatic toric surface with principal radii of 125.0 mm and 83.33 mm. Principal curvatures will be $\kappa_+ = 1/0.08333 = 12\,\mathrm{D}$ and $\kappa_- = 1/0.125 = 8\,\mathrm{D}$. We also want the flattest meridian (that with curvature κ_-) to lie horizontal and the other one vertical. According to the nomenclature just described, we want the base curve to lie horizontal and the cross curve vertical. To create a ring-type toric patch with these curvatures and horizontal base curve, we select a generating circumference with a radius $R_g = 83.33$ mm, located in any plane containing the Z axis, for example, the XZ plane. The rotating axis will be the Z axis. To generate an equator with a radius of 125 mm, we need a distance $d = 41.67$ mm from the center of the generating circumference to the rotating axis. In this example $R_g > d$, so the axis will intersect the generating circumference and the torus will be spindle-type. The outer sheet of this torus, its principal meridians, and the round contour of the desired patch are shown in Figure 2.15. Alternatively, we may create a barrel-type toric patch with the same curvatures at the vertex by selecting a generating circumference with a radius $R_g = 125$ mm. As this has to lie horizontal, the rotation axis has to be horizontal also, now coinciding with the Y-axis. We may first imagine the generating circumference on the XY plane and then rotate it around the Y axis. The distance from the axis to the center of the generating circumference is once again $d = 41.67$ mm, but now the arc closer to the Y axis is the one generating the toric patch. The construction is also depicted in the lower plot in Figure 2.15.

So far we have generated two toric patches having the very same curvatures along the horizontal and vertical directions, but with entirely different construction strategies. The way this affect the curvatures all over the patch is shown in Figure 2.16. We have represented here the curvature maps of the two patches, ring-type in the top row and barrel-type in the bottom row. Maps for the smaller principal curvature, that of the base curve, are to the left. Maps for the bigger curvature, that of the cross curve, are to the right. We see that the horizontal principal curvature is not constant all over the surface of a ring-type patch. It rather decreases when moving upward or downward from the equator. On the other side, the vertical principal curvature is not constant in the barrel-type surface, as it gets bigger as we move left or right from the equator of the torus (which is vertical now). Cross curvature in the ring-type and base curvature in the barrel-type are constant all over the surfaces, as

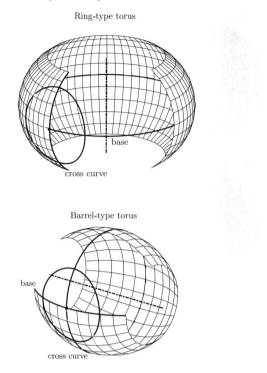

Figure 2.15 Ring-type (top) and barrel-type (bottom) toric surfaces to generate a toric patch with horizontal base curve with a curvature of 8 D and a vertical cross curve with a curvature of 12 D. Both tori are plotted to the same scale.

the corresponding principal meridian is always the generating circumference. In summary, our two toric patches have the same principal curvatures at their vertex, 8 and 12 D, but the distribution of principal curvatures at any other point on the patches will depend on the type of toric patch selected.

Sag Equations for Toric Patches

A torus with base and cross curves radii R_B and R_T is best described by parametric equations

$$x = [(R_B - R_T) + R_T \cos v] \cos u$$
$$y = [(R_B - R_T) + R_T \cos v] \sin u \tag{2.22}$$
$$z = R_T \sin v,$$

where u and v are angular parameters, both ranging from 0 to 2π to generate the entire surface. These equations are easy to handle and they have been used to generate the plots in Figures 2.13, 2.14, and 2.15, but they are not in the standard form we use to describe surface patches with their vertex tangent to the XY plane. To obtain the explicit equation

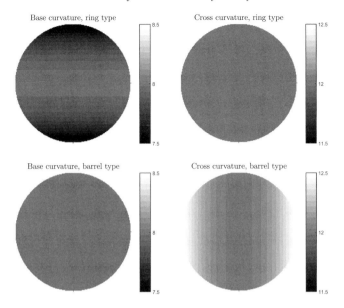

Figure 2.16 Curvature maps of two toric patches with the same main curvatures at the vertex. Top row, ring-type toric patch. Bottom, barrel-type toric patch.

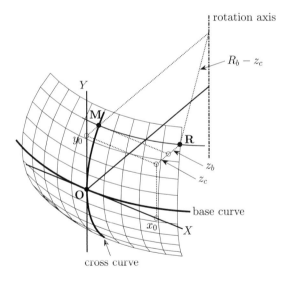

Figure 2.17 Sag height of a toric surface computed as the sum of the sags z_T and z_B of two circumference arcs.

for a ring-type toric patch with its base curve parallel to the X axis, let us take a look at Figure 2.17. We have still plotted the revolution axis as vertical, for easier visualization, but the reader should observe that this rotation axis is parallel to the Y axis, and that we are choosing the XY plane to be tangent to the toric patch at its vertex **O**. Let **M** be a point on the cross curve passing through the vertex $\mathbf{O}$, such that its y-coordinate is y_0. As the cross curve is an arc of circumference with radius R_T, its sag at point $\mathbf{M}$ is given by

$$z_T = R_T \left(1 - \sqrt{1 - y_0^2/R_T^2} \right).$$

Now let us consider the rotation of point $\mathbf{M}$, as part of the whole cross curve, around the rotation axis of the torus up to the position of point $\mathbf{R}$. As $\mathbf{M}$ rotates around the rotation axis, it describes another circumference arc with radius $R_B - z_T$ (take into account that only the vertex $\mathbf{O}$ would rotate with the base curve radius, R_B). The sag corresponding to the new rotation is given by

$$z_B = (R_B - z_T) \left[1 - \sqrt{1 - \frac{x_0^2}{(R_B - z_T)^2}} \right],$$

where x_0 is the x-coordinate of the point $\mathbf{R}$. The sag of the ring-type toric patch at $\mathbf{R}$ is then given by the sum of the two sags, $z = z_T + z_B$. More explicitly, the sag of a ring-type toric patch with its base curve aligned with the X axis, at any point with coordinates (x, y) is given by

$$z(x, y) = z_T(y) + \left[R_B - z_T(y) \right] \left\{ 1 - \sqrt{1 - \frac{x^2}{\left[R_B - z_T(y) \right]^2}} \right\}, \qquad (2.23)$$

where $z_T(y) = R_T \left(1 - \sqrt{1 - y^2/R_T^2} \right)$. For a barrel-type toric patch, we just need to interchange variables $x \leftrightarrow y$ and $R_T \leftrightarrow R_B$. In case the base curve were lying at any other direction making an angle θ with the X axis, we use the same trick that led to the equation (2.19) for the cylinder with arbitrary orientation. We just have to substitute x by $x \cos \theta + y \sin \theta$ and y by $-x \sin \theta + y \cos \theta$. For the ring-type patch we would have

$$z(x, y) = z_T(x, y) + (R_B - z_T(x, y)) \left\{ 1 - \sqrt{1 - \frac{(x \cos \theta + y \sin \theta)^2}{\left[R_B - z_T(x, y) \right]^2}} \right\}, \qquad (2.24)$$

where

$$z_T(x, y) = R_T \left[1 - \sqrt{1 - (-x \sin \theta + y \cos \theta)^2/R_T^2} \right]. \qquad (2.25)$$

As before, we just interchange variables $x \leftrightarrow y$ and $R_T \leftrightarrow R_B$ to obtain the equation of a general barrel-type toric patch with the base curve at an angle θ with the X axis.

Example 5 A barrel-type torus has a base radius of 125 mm and a cross one of 105 mm, being the base curve oriented at an angle of 30°. Compute the elevation of this surface at the point with coordinates $(15, -20)$ mm.

As we are dealing with a barrel-type torus, we have to compute first the sagitta along the base curve, that is

$$z_B = 125 \times \left(1 - \sqrt{1 - (20 \times \sin 30 + 15 \times \cos 30)^2 / 125^2} \right) = 2.13 \text{ mm}.$$

Notice that we have interchanged the variables x and y in equation (2.25) as indicated. Having computed the base curve sag, the resulting torus elevation is

$$z = 2.13 + (105 - 2.13)\left[1 - \sqrt{1 - \frac{(15 \times \cos 30 - 20 \times \sin 30)^2}{(105 - 2.13)^2}}\right] = 2.17\,\text{mm}.$$

If we had computed the sag at the same point for a ring-type torus with the same parameters, we would have obtained a different value $z_{ring} = 3.01$ mm.

Curvature Matrix of Toric Surfaces

Once we have the standard representation of a toric surface with arbitrary orientation, we can obtain its Hessian matrix at the vertex by computing the second-order derivatives and evaluating them at the origin. The computation, although straightforward, is tedious, as there are chained radicals in the expression for $z(x, y)$. We will skip the algebra and will directly go to the result. We will give the names κ_B and κ_T to the curvatures along the base and cross curves, respectively. As before, θ will be the angle made by the base curve with the X-axis, and $\phi = \theta + \pi/2$ the corresponding angle for the cross curve. The result is

$$\mathbb{H} = \begin{bmatrix} \kappa_B \cos^2\theta + \kappa_T \sin^2\theta & -(\kappa_T - \kappa_B)\sin\theta\cos\theta \\ -(\kappa_T - \kappa_B)\sin\theta\cos\theta & \kappa_B \sin^2\theta + \kappa_B \cos^2\theta \end{bmatrix}, \tag{2.26}$$

which can be written as

$$\mathbb{H} = \begin{bmatrix} \kappa_B \cos^2\theta & \kappa_B \sin\theta\cos\theta \\ \kappa_B \sin\theta\cos\theta & \kappa_B \sin^2\theta \end{bmatrix} + \begin{bmatrix} \kappa_T \sin^2\theta & -\kappa_T \sin\theta\cos\theta \\ -\kappa_T \sin\theta\cos\theta & \kappa_T \cos^2\theta \end{bmatrix}. \tag{2.27}$$

We already found the second matrix in the previous section: It is the Hessian of a cylinder with maximum curvature κ_T and cylinder axis making an angle θ with the X axis (equation (2.19)). The first matrix can be written in the same way if we take into account that $\cos\phi = -\sin\theta$ and $\sin\phi = \cos\theta$,

$$\begin{bmatrix} \kappa_B \cos^2\theta & \kappa_B \sin\theta\cos\theta \\ \kappa_B \sin\theta\cos\theta & \kappa_B \sin^2\theta \end{bmatrix} = \begin{bmatrix} \kappa_B \sin^2\phi & -\kappa_B \sin\phi\cos\phi \\ -\kappa_B \sin\phi\cos\phi & \kappa_B \cos^2\phi \end{bmatrix},$$

so the general Hessian of a torus with its base curve making an angle θ with the X-axis is the sum of the Hessians of two cylinders, one with maximum curvature κ_T and axis direction θ, the other with maximum curvature κ_B and axis direction $\phi = \theta + \pi/2$, perpendicular to the first one. There is another way to write the Hessian of a toric surface. Consider the upper left component of the Hessian, the curvature along the X-direction in (2.26), $\kappa_x = \kappa_B \cos^2\theta + \kappa_T \sin^2\theta$. If we add and subtract $\kappa_B \sin^2\theta$ to κ_x, we do not modify it, but we can rearrange terms getting to $\kappa_x = \kappa_B + (\kappa_T - \kappa_B)\sin^2\theta$. Using the same trick for the curvature along the Y-direction, we get the expression

$$\mathbb{H} = \begin{bmatrix} \kappa_B + (\kappa_T - \kappa_B)\sin^2\theta & -(\kappa_T - \kappa_B)\sin\theta\cos\theta \\ -(\kappa_T - \kappa_B)\sin\theta\cos\theta & \kappa_B + (\kappa_T - \kappa_T)\cos^2\theta \end{bmatrix}, \tag{2.28}$$

which can also be split as

$$\mathbb{H} = \begin{bmatrix} \kappa_B & 0 \\ 0 & \kappa_B \end{bmatrix} + \begin{bmatrix} (\kappa_T - \kappa_B) \sin^2 \theta & -(\kappa_T - \kappa_B) \sin \theta \cos \theta \\ -(\kappa_T - \kappa_B) \sin \theta \cos \theta & (\kappa_T - \kappa_B) \cos^2 \theta \end{bmatrix}.$$

This means the Hessian of the toric surface can also be computed as the sum of the Hessian of a sphere with curvature κ_B and that of a pure cylinder with curvature $\kappa_T - \kappa_B$ and axis θ. We can also repeat the same procedure, but isolating κ_T instead of κ_B in the diagonal terms of (2.26). We would then get the alternative expression

$$\mathbb{H} = \begin{bmatrix} \kappa_T + (\kappa_B - \kappa_T) \sin^2 \phi & -(\kappa_B - \kappa_T) \sin \phi \cos \phi \\ -(\kappa_B - \kappa_T) \sin \phi \cos \phi & \kappa_T + (\kappa_B - \kappa_T) \cos^2 \phi \end{bmatrix}, \tag{2.29}$$

so the Hessian of the toric surface can also be obtained as the sum of the Hessian of a sphere with curvature κ_T and the Hessian of a pure cylinder with curvature $\kappa_B - \kappa_T$ and axis $\theta + \pi/2$.

We should note that these Hessians only depend on three parameters: κ_B, κ_T, and θ. There is no way to tell if a torus patch is ring-type or barrel-type from its Hessian matrix. Or in other words, the Hessian does not depend on the type of toric surface we are considering. Of course, this is because we are only considering the Hessian matrix at the surface vertex. Only the principal curvatures and their directions determine the Hessian, and at the vertex, ring-type and barrel-type tori with the same base and cross curves and the same orientation angle are essentially the same. Out of the vertex, the Hessian matrices of each type of surface would differ, as their principal curvatures do. The farther from the vertex, the larger the difference. But as we have said before, many fundamental properties of lenses only depend on the properties of their surfaces at their vertexes, so studying them just requires the Hessian matrices shown here. The study of aberrations will require us to analyze points other than the vertex and reconsider the ring or barrel nature of the toric surfaces.

Let us finish this section by giving the explicit expressions of the Hessian of a toric surface for three particular values of the orientation angle: $\theta = 0$, $\pi/4$ and $\pi/2$. Respectively, the curvature matrices corresponding to each of them are

$$\mathbb{H}(0) = \begin{bmatrix} \kappa_B & 0 \\ 0 & \kappa_T \end{bmatrix}, \quad \mathbb{H}(\pi/4) = \frac{1}{2} \begin{bmatrix} \kappa_B + \kappa_T & \kappa_B - \kappa_T \\ \kappa_B - \kappa_T & \kappa_B + \kappa_T \end{bmatrix}, \quad \mathbb{H}(\pi/2) = \begin{bmatrix} \kappa_T & 0 \\ 0 & \kappa_B \end{bmatrix}.$$

If we recall the discussion preceding equation (2.21), we can apply the same idea to toric surfaces as well. If we write the general Hessian of a toric surface (2.28) in a rotated reference system in which the X axis is parallel to the base curve, then it takes the form $\mathbb{H}(0)$, that is,

$$\begin{bmatrix} \kappa_B + (\kappa_T - \kappa_B) \sin^2 \theta & -(\kappa_T - \kappa_B) \sin \theta \cos \theta \\ -(\kappa_T - \kappa_B) \sin \theta \cos \theta & \kappa_B + (\kappa_T - \kappa_B) \cos^2 \theta \end{bmatrix} = \begin{bmatrix} \kappa_B & 0 \\ 0 & \kappa_T \end{bmatrix}_\otimes.$$

The matrices just obtained tell us nothing about the toric nature of the surfaces they represent, except for the principal curvatures and their orientations. This means other astigmatic surfaces with the same principal curvatures and directions would have the very same curvature matrices (see next section.) When the orientation angle of the base curve is 0, the base curve is parallel to the X axis, the cross curve parallel to the Y axis. The principal directions are then parallel to the coordinate axis; henceforth the Hessian matrix is diagonal. The curvature along the X direction is κ_B and the curvature along the Y direction κ_T. When the orientation angle of the base curve is $\pi/2 = 90°$, the principal directions are still parallel to the coordinate axis, so the Hessian is still diagonal. The difference is that the curvature along the X axis is now κ_T and the curvature along the Y axis is κ_B. Finally, if the orientation angle of the base curve is $\pi/4 = 45°$, the principal directions are not parallel to the coordinate axis, and the Hessian matrix is no longer diagonal. As the X and Y axes are now angle bisectors of the principal directions of the surface, the curvature along them is the arithmetic mean of κ_B and κ_T. The torsion term (nondiagonal) is $(\kappa_B - \kappa_T)/2$.

Finally, we may use the curvature matrices to easily compute the curvature of a toric surface along any oblique meridian, making an arbitrary angle α with the base curve, or an arbitrary angle β with the X axis. Let us assume the base curve of the toric surface makes an angle θ with the X axis (see Figure 2.18).

The curvatures of this toric surface along the horizontal and vertical meridians are given by the components of the Hessian matrix on the diagonal. The curvature on the horizontal curve is $\kappa_X = \kappa_B + \kappa_T \sin^2 \theta$, whereas the curvature on the vertical meridian is $\kappa_Y = \kappa_B + \kappa_T \cos^2 \theta$. To obtain the curvature on the meridian making an angle α with the base curve, we may rotate the coordinate system so that the X axis will lie along this direction. The base curve of the surface would then make an angle $-\alpha$ with the new X axis, and then $\kappa_\alpha = \kappa_B + \kappa_T \sin^2 \alpha$. The minus sign has no effect as the sine function is squared.

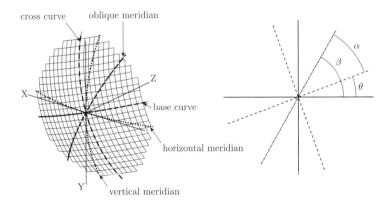

Figure 2.18 To the left, 3D image of a toric surface. The grid orientation is that of the principal meridians, the base and the cross curves, which are drawn with thick dashed lines. Horizontal and vertical meridians are plotted with dotted lines. They are the intersections of the ZX and ZY planes with the surface, respectively. Finally, an oblique meridian is also shown on the surface with a dash-dot line. The X,Y and Z axis are shown. To the right, the surface and the aforementioned meridians as seen from a point in the Z-axis. This is the representation that we will mainly use through the book. Horizontal and vertical meridians are superimposed here to the coordinate axis X and Y.

Finally, if the meridian orientation is given with respect to the X axis, that is, we are given the angle β, then the curvature will obviously be $\kappa_\beta = \kappa_B + \kappa_T \sin^2 (\beta - \theta)$.

2.3.3 Astigmatic Surfaces with Aspherical Sections

The principal meridians of a toric surface, regardless whether it is barrel-type or ring-type, are circumference arcs. The curvature along these meridians is thus constant. The same arguments and facts that made conicoids worth studying apply for toric surfaces. We may look for astigmatic surfaces in which curvature increases or decreases as we move away from the vertex along the principal meridians. There are different ways to create generalizations of toric surfaces in which both the base curve and the cross curve are conic curves, with arbitrary asphericity parameters. For example, we could choose the base curve to be elliptical, with increasing curvature toward the periphery of the surface, and the cross curve to be hyperbolic, with decreasing curvature toward the periphery. In general, we may call p_b and p_c to the asphericity parameters for the base and cross curves, respectively.

The reader may try some of these generalizations by starting from equation (2.23). The idea is to change the nested circumference sags $R(1 - \sqrt{1 - u^2/R^2})$ into conicoid sags $(R/p)(1 - \sqrt{1 - pu^2/R^2})$, u standing for either x or y, using the adequate curvature radius for each principal meridian. However, we won't bother the reader with these generalizations as there are other families of astigmatic surfaces with aspherical principal meridians that are easier to describe, such as *biconic* surfaces. When their principal meridians are parallel to the X and Y axes, a biconic surface with vertex curvatures κ_B and κ_T along its flattest and steepest meridians is given by

$$z(x, y) = \frac{\kappa_B x^2 + \kappa_T y^2}{1 + \sqrt{1 - p_B \kappa_B x^2 - p_T \kappa_T y^2}}. \tag{2.30}$$

For the general case in which the base curve makes an angle θ with the X axis, it suffices to apply the already known changes $x \rightarrow x\cos\theta + y\sin\theta$ and $y \rightarrow -x\sin\theta + y\cos\theta$. If we set $p_B = p_T$ and $\kappa_B = \kappa_T$ in the biconic equation, we recover equation (2.13) for conicoid surfaces with revolution symmetry. Also, it is worth noting that if we set $p_B = p_T = 1$ in the biconic equation, but we keep $\kappa_B \neq \kappa_T$, we get an astigmatic surface with circular principal meridians, but such a surface is not a torus, as it cannot be obtained by rotation of a circumference around an axis. The biconic is not a revolution surface for any value of its asphericity parameters.

The Hessian of a biconic surface can be obtained, as usual, by computing the second derivatives of the surface at the origin. In this case, the computation is not as lengthy as for toric surfaces, but still we leave the reader to check that the Hessian of a biconic surface does not depend on the asphericity parameters p_B and p_T. Indeed, the Hessian of the biconic is identical to that of a torus with the same principal curvatures and orientations, the one in equation (2.28). We can say the same for any other type of atoric surfaces. The Hessian of an astigmatic surface at its origin only depends on the values of its second derivatives also at the origin. From the optical point of view, the consequence of all these aspherical surfaces having the same Hessian is that they all behave the same in the paraxial regime, that is, for rays impinging close to the vertex of the lens. Indeed, were we only interested in the paraxial region, we could select the asphericity parameters for which the astigmatic surface gets its simpler form: $p_B = p_T = 0$, which gives us an astigmatic paraboloid with its base curve parallel to the X axis

$$z(x, y) = \frac{1}{2} \left(\kappa_B x^2 + \kappa_T y^2 \right). \tag{2.31}$$

It is a good exercise to work out the expression for the astigmatic paraboloid with arbitrary orientation by using the standard substitutions $x \rightarrow x \cos \theta + y \sin \theta$ and $y \rightarrow -x \sin \theta + y \cos \theta$ in the previous equation. We readily get

$$z(x, y) = \frac{\kappa_B \cos^2 \theta + \kappa_T \sin^2 \theta}{2} x^2 + \frac{\kappa_B \sin^2 \theta + \kappa_T \cos^2 \theta}{2} y^2$$
$$-(\kappa_T - \kappa_B) \sin \theta \cos \theta \, xy. \tag{2.32}$$

We immediately recognize the coefficients of $x^2/2$, $y^2/2$, and xy as the elements of the toric Hessian (2.26). In effect, the second derivatives of a general parabolic surface $z(x, y) = \frac{1}{2} \left(ax^2 + by^2 + 2cxy \right)$ are $z_{xx} = a$, $z_{yy} = b$ and $z_{xy} = c$, so the Hessian of such a surface is

$$\mathbb{H} = \begin{bmatrix} a & c \\ c & b \end{bmatrix},$$

and now we know how the principal curvatures and orientations relate to those coefficients

$$a = \kappa_B \cos^2 \theta + \kappa_T \sin^2 \theta = \kappa_B + (\kappa_T - \kappa_B) \sin^2 \theta,$$
$$b = \kappa_B \sin^2 \theta + \kappa_T \cos^2 \theta = \kappa_B + (\kappa_T - \kappa_B) \cos^2 \theta, \tag{2.33}$$
$$c = (\kappa_T - \kappa_B) \sin \theta \cos \theta.$$

Any astigmatic surface with principal curvatures κ_B and κ_T and base orientation θ can be approximated in the neighborhood of its vertex by a paraboloid with the same curvatures and orientation, with the paraboloid being given by equation (2.32), and they all share the same curvature matrix, whose components are proportional to the coefficients of the paraboloid.

Example 6 Compute the approximate sagitta of the torus described in the previous example, i.e. base curve radius 125 mm, cross curve radius 105 mm, and base curve orientation 30° at a point of coordinates $(15, -20)$ mm.

Let us compute first the Hessian matrix of the torus. The base and cross curvatures are $\kappa_B = 125^{-1} = 0.008 \, \text{mm}^{-1}$ and $\kappa_T = 105^{-1} = 0.00952 \, \text{mm}^{-1}$, respectively. From them and equation (2.28) we compute the Hessian matrix, and from it the sagitta,

$$z = \frac{10^{-3}}{2} \begin{bmatrix} 15 & -20 \end{bmatrix} \begin{bmatrix} 8.38 & -0.658 \\ -0.658 & 9.14 \end{bmatrix} \begin{bmatrix} 15 \\ -20 \end{bmatrix} = 2.96 \, \text{mm}.$$

Notice that this is an intermediate value between the exact ones obtained, at the same point, for a barrel-type torus, $z_{barrel} = 2.17$ mm and a ring-type torus, $z_{ring} = 3.01$ mm, both with the same curvatures and orientation as the one considered in this example.

2.4 Surfaces with Many Degrees of Freedom

So far, we have studied surfaces described by relatively simple analytical equations. These equations incorporate a few parameters that determine how the surface behaves. For example, the equation for a spherical surface (2.1) depends on a single parameter, the curvature radius, R, or its inverse, the curvature κ. The paraboloid has two parameters, κ and p; the first now specifies the curvature only at the vertex, whereas the second determines how this curvature changes away from the vertex. Although p can change continuously, each value of the same determines a rigid way in which curvature changes along the radial direction. If we wanted to change this behavior, we would need to add more parameters, as in the generalized paraboloids. The most complex function we have studied from the point of view of the number of parameters is the biconic (2.30). If we use the rotated coordinates to allow for any orientation of the surface, five parameters are required: κ_B, κ_T, p_B, p_T, and θ. In this family of surfaces there are two main curvatures at the vertex; they change along the main meridians according to the values of p_B and p_T, and the orientation is determined by θ. Yet, those changes in curvature are rigid; they depend on the functional form of (2.30).

The number of parameters needed to describe a family of surfaces is known as the *number of degrees of freedom*. The more degrees of freedom, the more ways the curvature of the surface can change from one point to an other. On many occasions we need surfaces in which the curvature changes from one point to another in ways not allowed by any of the families studied up to now. The best example for these surfaces can be found in the progressive lenses used to compensate for presbyopia that we will study in Chapter 8. These lenses typically have two reference points, very close to the Y axis and separated 10 to 20 mm along the vertical direction. These are called the near and far vision points, as the lens provides the right power for viewing at distance or near objects when looking through the corresponding points on the lens. One of the lens surfaces has a progressive nature, the other one typically being a sphere or a torus. The progressive surface may have two regions around each of the far and near points in which it is mainly a portion of a sphere. The important fact is that the curvature around the far point is different than the curvature around the near point. None of the surfaces studied here have this property.

To generate, describe, and study complex surfaces (such as those of progressive lenses) we need many more degrees of freedom and we also need tools and techniques that can describe *any* desired surface. Mastering this type of surfaces requires a knowledge of geometry and numerical methods beyond the scope of this book. Also, the handling of these types of surfaces requires a lot of programming as most techniques for evaluation of sags and curvatures rely on heavy numerical computing. Nevertheless, we think the brief review we present next may help the reader to get a better understanding of the contents of Chapter 8, and it can serve as an introduction to those who want to dive deeper in the subject.

2.4.1 Mesh-Defined Surfaces

An arbitrary surface can be approximately described by the coordinates of a finite subset of its points. Let us set an example with a planar curve: the function $z(x) = x^2$ describes the infinite set of points making the curve represented by the function, in this case a simple parabola. Let us now select a set of five values for the X coordinate: $\{x_n\}_{n=1,...,5} = \{-1, -0.5, 0, 0.5, 1\}$. The sag heights corresponding to these points are given by $z_n = x_n^2$, so $\{z_n\} = \{1, 0.25, 0, 0.25, 1\}$. The set $\{P_n\} = \{(x_n, z_n)\}$, $n = 1, \ldots, 5$ contains five points belonging to the curve $z(x) = x^2$, as shown in Figure 2.19. If we did not know the expression for the curve, but only the set of points on it, we still may compute the sag heights at any other points by using interpolation methods, which are algorithms to guess the values of unknown points from the values of neighboring, known points. For example, if we use linear interpolation, we may compute the unknown points between any pair of known points by assuming the former lie in the straight line joining the later. With quadratic interpolation, we take three consecutive known points, compute the unique quadratic curve passing through them, and use this parabola to compute the sag height of any point in between. Using the same idea, we may use arbitrarily high-order interpolation. In practice, linear to cubic interpolation algorithms are mainly used.

For the two-dimensional case, that is, for surfaces, the procedure is quite similar. Instead of having a function $z(x, y)$ that allows the computation of the sag height at any point (x, y), we may have a grid of points on the XY plane, $\{(x_n, y_n)\}_{n=1,...,N}$, for which we know the sag of the surface, $\{z_n\}_{n=1,...,N}$. Two-dimensional interpolation is then used to compute the sag height of any point not in the grid. The grid is usually a set of uniformly spaced points, that is, $x_{n+1} = x_n + \Delta x$ and $y_{n+1} = y_n + \Delta y$, where Δx and Δy are the grid intervals along the X and Y directions. Interpolation is an approximate technique, and its accuracy depends

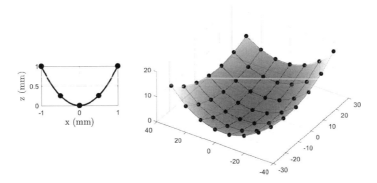

Figure 2.19 To the left: parabolic curve and a subset of 5 points on it. To the right, surface and a subset of 49 points on it. The XY coordinates of these points are arranged in a 7×7 Cartesian grid.

on the grid intervals and the interpolation algorithms. If the grid intervals are smaller than the lateral size of any bump the surface could have, we say that the surface is well sampled by the set of points. In that case, cubic interpolation produces, in general, better results than quadratic interpolation, which in turn is more accurate than linear interpolation. Higher-order interpolation algorithms can be used, but usually require more computation, so the normal practice is to use a grid dense enough so that quadratic or cubic interpolation provides the desired degree of approximation.

For the type of surfaces and range of curvatures mainly used in the making of ophthalmic lenses, and assuming we use a grid interval of 1 mm for both the X and Y directions, cubic interpolation will allow us to compute the sag height at any point with better accuracy than any manufacturing method (for ophthalmic lenses) could achieve. In terms of curvature, the accuracy of the main curvatures computed from a cubic interpolation algorithm applied to a grid with 1 mm spacing is around 0.01 D, more than enough for ophthalmic lenses. Of course, these results will not apply to surfaces with bumps or features with a lateral size smaller than a few millimeters. These surfaces would have very large curvatures (hundreds or thousands of diopters) and we would need much denser grids to describe them properly.

Mesh-defined surfaces are nowadays constantly used in lens manufacturing labs using *free-form* technology (see Chapter 10). Free-form generators and polishers can produce arbitrary optical surfaces and the process goes like this: When an order from the optical shop arrives at the lens manufacturing lab, a software named LDS (lens design software) is called to compute the lens. This software selects a semifinished lens with a previously defined front surface (usually a spherical one) and computes the optimum back surface for the given prescription and user parameters. In general, this surface lacks any type of symmetry and requires many degrees of freedom. The LDS then computes the surface at the locations of a predefined grid and writes the resulting set of points to a file. This file is sent to the generator and polisher, which, by means of interpolations algorithms, will cut the surface on the rear side of the semifinished lens, so producing the final lens. The files containing the surface mesh are simple text files. An example of a standard file format, the *.HMF (Height Map File), is shown next:

```
Aspherical base file File Version=1.2 General height-map
properties [Properties] Count=81 Interval=1.000000 The
height-map dat begins here [Data] 9.572899, 9.329654,
9.092923, 8.862678, 8.638891, 8.421534, 8.210581,
8.006008, 7.807789, 7.615902, 7.430324, 7.251034,
7.078009, 6.911231, 6.750680, 6.596337, 6.448185,
6.306207, 6.170386, ......... 7.251034, 7.430324,
7.615902, 7.807789, 8.006008, 8.210581, 8.421534,
8.638891, 8.862678, 9.092923, 9.329654, 9.572899
```

The file has a single header with basic information about the grid: The variable `Count` is a variable storing the number of points in the grid along both the X and Y directions.

`Interval` stores the grid interval (which is the same for both axes). The mesh then describes the surface inside a squared region with a side size of 80 mm. Just behind the label `[Data]` starts the list of sag heights in mm. The first 81 numbers correspond to the first row of the grid, the second 81 numbers to the second row, and so on. The total number of points amounts to $81 \times 81 = 6561$. The corresponding x and y coordinates are easily deduced: as the interval is 1 mm, both coordinates range from -40 to 40 mm, in steps of 1 mm. For example, the sag height at $(-40, -40)$ is 9.572899 mm, at $(-39, -40)$ is 9.329654 mm, etc.

2.4.2 Surfaces Defined by Polynomials

In the previous section we introduced the software that *designs* the lenses. This means looking for the surface that best matches the expected optical properties. In general there is not a direct and unique solution for that quest. The lens with perfect imaging for all sight directions is not achievable; there are optical aberrations that cannot be fully eliminated and in most cases the improvement of one property worsens others. The designer then weighs the importance of the different properties and looks for the best possible lens according to those weights. This process is called optimization. The degrees of freedom of the surface are used as variables for this optimization algorithm. Mesh-defined surfaces, as they are used for manufacturing, have a very large number of degrees of freedom: each sag height in the grid is a degree of freedom. The optimization of a surface defined by a standard HMF file would require the handling of 6561 variables. A problem of this size can be handled with appropriate computing strategies. However, there are preferable surface descriptions with a smaller number of degrees of freedom but still with large enough flexibility to produce any surface we would need.

Once again, mathematics offers a very large range of methods to tackle this problem, but probably those most employed in optics and engineering use polynomials. Any one-dimensional function can be approximated by a polynomial, $f(x) \simeq a_0 + a_1 x + a_2 x^2 + \ldots + a_N x^N$. A well-known example is the Taylor polynomial expansion. In a similar way, a two-dimensional function, such as those required to describe surfaces, can be approximated by two-dimensional polynomials, $f(x, y) \simeq a_{00} + a_{10}x + a_{01}y + a_{20}x^2 + a_{02}y^2 + a_{11}xy + \ldots + a_{NN}x^N y^N$. Let us look at the problem the other way around. Instead of considering the polynomial approximation of a given function, why do not we consider the polynomial *as the* function? Each coefficient of the polynomial is a degree of freedom. If we define our surface as

$$z(x, y) = \sum_{n,m=1}^{N} a_{nm} x^n y^m, \qquad (2.34)$$

then it depends on $(N+1)^2$ coefficients, which are the degrees of freedom of the surface. If the polynomial is to be a standard representation of the surface, then $z(0, 0)$ should be zero, as well as the first derivatives computed at the origin. This leads to $a_{00} = a_{10} = a_{01} = 0$. The second-order coefficients of the polynomial-defined surface (2.34) are the components of the curvature matrix at the origin, $\mathbb{H} = \begin{bmatrix} 2a_{20} & a_{11} \\ a_{11} & 2a_{02} \end{bmatrix}$. Finally, the behavior of the surface (and its curvatures) as we move away from the vertex is determined by the higher-order coefficients. A polynomial with $N = 15$ has $16 \times 16 = 256$ degrees of freedom and can properly describe most continuous surfaces that could eventually be used in ophthalmic lenses, from spheres to the most complex progressive surfaces. Be aware that the finite polynomial in (2.34) cannot be an *exact* representation of a spherical surface, or any of the analytical surfaces explained before (except for paraboloids, which are nothing but polynomials), but it can get very close to any of them by proper selection of the coefficients and if N is large enough.

These polynomials made from simple monomials (which are the expressions $a_{nm}x^n y^m$) have some drawbacks related to their mathematical behavior. For example, the high powers of the x and y coordinates receive enormous values as these variables get bigger. For the surface to make any sense, the corresponding coefficients have to take very small values, yet they are very important in determining the surface shape and geometry. A tiny change in any a_{nm} with n, m large will produce significant changes in the surface. This fact poses big challenges to the optimization algorithms that look for the optimum surface. Also, there are no direct, simple relations between the higher-order coefficients and the "visible" properties of the surface away from the vertex such as the sag heights, slopes, or curvatures. As already mentioned, the latter are determined by the former, but the relation is obscure, all the higher-order coefficients cooperate to determine the surface in a nonintuitive way.

There are other approaches based on so-called *sets of orthogonal polynomials*. A proper introduction to these sets would take us on a large mathematical detour, so we address the interested reader to the specialized literature. See, for example, Doman [33] for a thorough, but not too technical, general introduction to the matter. Dai [34] is also an excellent source for studying the orthogonal polynomials used in visual optics. We have added a very brief description in Appendix E, where a set particularly relevant for optical applications, Zernike polynomials, are better defined. By now, let us just assume there are sets of polynomials so that each of them can be assigned one or more integer labels or indexes. They can be one-dimensional with just one index, $\{P_n(x)\}$, two-dimensional with two indexes, $\{P_{nm}(x, y)\}$, or even two-dimensional with one index, $\{P_n(x, y)\}$. For example, we show in Table 2.2 the first four polynomials of the sets known as Laguerre

Table 2.2 *Laguerre and Legendre polynomials up to order 3*

	Laguerre polynomials	Legendre polynomials
0	1	1
1	$1 - x$	x
2	$\frac{1}{2}(x^2 - 4x + 2)$	$\frac{1}{2}(3x^2 - 1)$
3	$\frac{1}{6}(-x^3 + 9x^2 - 18x + 6)$	$\frac{1}{2}(5x^3 - 3x)$

and Legendre polynomials. Both sets contain an infinite number of polynomials, as n runs from 0 to infinity.

These two sets of polynomials are one-dimensional, but we easily construct the two-dimensional ones by direct multiplication of the polynomials evaluated on the x and y coordinates: $\{P_{nm}(x, y)\} = \{P_n(x)P_m(y)\}$. These sets of polynomials were found as part of the solutions of important differential equations, with many applications in physics and engineering. Our current application requires two-dimensional polynomials and, for simplicity, they will be rearranged to be single-indexed. Function approximation is achieved with the linear combination

$$f(x, y) \simeq \sum_{n=0}^{N} c_n P_n(x, y), \tag{2.35}$$

where a finite number of polynomials, N, is used. The coefficients will depend on the set of polynomials selected, for example they could be Laguerre's or Legendre's (there are many more of them). The fact that $f(x, y)$ is computed as a linear combination of polynomials reminds us of vectors in 2D or 3D space, which can be expressed as linear combinations of orthogonal (perpendicular) vectors. Orthogonality in vectors is established through the dot product: two vectors $\mathbf{s}$ and $\mathbf{t}$ are said to be orthogonal if their dot product is zero, $\mathbf{s} \cdot \mathbf{t} = 0$. In a similar way, orthogonal polynomials can be considered "vectors" in a space of infinite dimensions, where the dot product is typically defined as

$$P_n(x, y) \cdot P_m(x, y) = \iint_I P_n(x, y)P_m(x, y)dxdy,$$

and the integral is extended to the region I. For one-dimensional polynomials the integral would also be one-dimensional, and I would be an interval of numbers. For two-dimensional polynomials the region I is a two-dimensional area, typically a rectangle or a circle centered on the coordinate system. Two polynomials are said to be orthogonal when their dot product is zero. A set of polynomials is orthogonal in their specified region if

$$P_n(x, y) \cdot P_m(x, y) = \begin{cases} 0 & \text{if } n \neq m \\ A_n & \text{if } n = m \end{cases},$$

where the A_n are constants to be determined. If $A_n = 1$ for all n, the polynomials are said to be *normalized*, and the set is said to be *orthonormal*.

Orthogonality confers a lot of good mathematical properties to the set of polynomials, some theoretical, some practical (see [33]). From our point of view, the advantages of using orthogonal polynomials are listed next.

Coefficient determination. The coefficients of an approximation using orthogonal polynomials are simply the dot product of our function and the corresponding polynomial, $c_n = f \cdot P_n$, that is, $c_n = \iint_I f(x, y)P_n(x, y)dxdy$.

Coefficient uniqueness. The value of each coefficient does not depend on the number of polynomials used in the linear combination (2.35). If we first approximate our function with, let us say, five polynomials, compute c_1 to c_5, and then add extra polynomials to improve accuracy, the coefficients c_1 to c_5 will not change after the new addition. A fitting of the function with nonorthogonal polynomials behaves differently: If we change the number of polynomials used, the whole set of coefficients will change.

Minimum fitting variance. Each term $c_n P_n(x, y)$ gives the best possible fit to the function that can be achieved with the nth polynomial. Adding more terms improves the fitting, but any combination (even with noncorrelative polynomials) will provide optimum fitting. In contrast, the Taylor expansion up to order N only makes sense if all the lower-order terms $n < N$ are included.

Mathematical stability. Orthonormality makes the computation of expansions such as (2.35) less prone to numerical instabilities when we face practical cases [35].

Orthogonality region. As said before, each set of polynomials has its own orthogonality region. For example, Laguerre and Legendre polynomials are orthogonal in the interval $[-1, 1]$, and their two-dimensional counterparts in the square defined by $x, y \in [-1, 1]$. Zernike polynomials $Z_n(x, y)$ are orthogonal in the unit circle defined by $x^2 + y^2 \leq 1$. This circumstance does not limit the region in which we can approximate a function. For example, if our function $f(x, y)$ extends to a circle of size r mm, being $r \neq 1$, the expansion is made as follows

$$z(x, y) = \sum_{n=0}^{N} c_n Z_n \left(\frac{x}{r}, \frac{y}{r} \right).$$

In this way the polynomials are evaluated within their range of orthogonality and all the coefficients have the same dimension as z.

Sampled functions. We finally may wonder how can we compute the coefficients of (2.35) if the function $f(x, y)$ is unknown in the first place. In practice there are two main applications leading to the function $f(x, y)$. In the first, the function does not exist and we want to create it according to some criteria. For example, the function may describe the surface of a lens, and we want to find the optimal one according to a set of optical criteria. An optimization algorithm is then performed that will find the coefficients c_n that generate this optimal surface. In the second application the function represents something that can be measured, for example the surface of a lens or the wavefront of an eye. The measurement provides sampled values of the function f_i at some specific measurement points (x_i, y_i). We can then compute the coefficients of the expansion (2.35) by approximating the integrals with sums

$$c_n = \sum_{i=1}^{M} f_i Z_n \left(\frac{x_i}{r}, \frac{y_i}{r} \right).$$

Depending on the density of the sampling, this approximation may require adjusting the polynomials to keep their orthogonality even when the dot product is computed on sampled polynomials, a process known as Gram-Schimdt orthogonalization. The reader may find further detail in the references cited above.

The number of polynomials required for a good fitting depends on the application. For example, to describe a wavefront sampled at M points, the number of polynomials (or degrees of freedom) to be used must typically remain below M to avoid computational issues (and because adding extra polynomials will not improve the accuracy of the description, as the information on the function is limited by the number of sampling points). To describe complex lens surfaces, such as those used in progressive lenses, between 150 and 250 polynomials will be needed, depending on the size of the surface and its geometrical subtleties.

3

Wavefronts and Rays

3.1 Introduction

In the previous chapter we studied the basic properties of some surfaces that are useful to ophthalmic optics. We were especially interested in two aspects: the computation of sagittas, which is necessary for computing lens thickness, and the understanding, handling, and calculation of curvatures, which will determine the power of the lenses. In the present chapter, we are going to apply these geometrical tools to the description and handling of wavefronts, the surfaces describing how light propagates. We will also learn how wavefronts are transformed when light refracts through a surface separating two different media.

If we picture the development of this book as the building of a house, the last chapter dealt with the materials needed to build the house, such as bricks, concrete, etc. In this chapter, we will construct the foundations and walls of the house, while leaving the fine details necessary for completing the house to the remaining chapters. In summary, we will present the relevant concepts of geometrical optics necessary for the study of ophthalmic lenses. Following the usual trend in ophthalmic and physiological optics, we will make extensive use of the concept of vergence and the relationship between rays and wavefronts.

3.2 Vergence and Wavefront

In geometrical optics, light is modeled as bundles of rays along which energy is transported. When the medium in which light propagates is homogeneous and has a constant refractive index, the rays are straight lines; they only change their propagation direction when they pass through surfaces separating media with different refractive indexes. An object to be imaged is described as an infinite set of points, and from each of these points, rays are generated in all directions. The imaging of a complex object is then reduced to the imaging of each of the points composing it. In geometrical optics, the behavior of optical systems is analyzed by studying the refraction of rays coming from each of the object points (or sources) that form a given object.

3.2.1 Vergence of Spherical Wavefronts

Given a bundle of rays, we define its wavefronts as the surfaces that are perpendicular to all the rays in the bundle. Alternatively, it is also possible to define the wavefronts as the surfaces formed by the points that are reached by the light at the same time. The time of flight is proportional to the *optical path*, which is the product of the distance traveled by the ray (or along the ray) times the refractive index of the medium in which the ray travels. If we consider two wavefronts of the same light bundle, the optical path between them is the same along any ray connecting them. If a ray passes through different media, its optical path is the sum of the distances traversed in each medium multiplied by the corresponding refractive index.

The wavefront of geometrical optics cannot be defined for bundles of rays coming from different points of a complex object. We then restrict the wavefront definition to the rays coming from the same source point or converging to the same image point. In Figure 3.1, we show a point source and the beam of rays coming from it within a single homogeneous medium. As all rays come from the source point F, the surfaces perpendicular to these rays must be spheres centered on F. We will define the *vergence* of the ray beam at a point A as the curvature of the wavefront surface at this point. From this definition, the vergence of a spherical wavefront is the inverse of the distance from its apex to the point where its associate rays converge Mathematically, the vergence of a spherical wavefront at point A is then

$$V_A = \kappa_A = 1/R_A, \tag{3.1}$$

where R_A is the curvature radius of the sphere with center at F that contains point A. Notice that, according to this definition, for a diverging wavefront (such as the one depicted in Figure 3.1) vergence is negative. However, if a beam converges, so that all the beam rays aim at a focus point, then its vergence is positive.

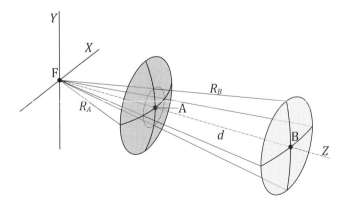

Figure 3.1 Wavefronts at two points (A and B) along the direction Z separated by a distance d. Vergence at each point is the curvature of the wavefront at the same point.

Vergence changes when we move along a ray beam. If we take into account the sign in the value of vergence, the longer the beam propagates, the greater is its vergence, with independence of whether the beam is divergent or convergent. For convergent beams, vergence raises when we are closer to the source. For diverging wavefronts, although the absolute value of vergence diminishes as we move away from the source, it becomes greater algebraically, as it is a negative quantity. If we consider the absolute value of vergence instead, the closer the point to the source, the greater the absolute value of the vergence at this point.

If we know the vergence of a wavefront at point A, it is possible to compute the vergence at any other point B if we know the distance z between them:

$$V_B = \frac{1}{R_B} = \frac{1}{R_A - z}.$$

If we multiply both numerator and denominator of the right side of the former equation by V_A we get

$$V_B = \frac{V_A}{1 - zV_A}, \tag{3.2}$$

which is the traditional formula for spherical vergence propagation. We will sometimes find a linear approximation to this propagation formula useful. For this we need the Taylor expansion of $(1 - x)^{-1}$, which is

$$\frac{1}{1 - x} = 1 + x - x^2 + x^3 \dots.$$

We can apply this expansion to (3.2) by setting $zV_A \to x$ and retaining just the linear term in z. The result is

$$V_B \simeq V_A + zV_A^2, \tag{3.3}$$

accurate whenever $|zV_A| \ll 1$, which happens in many situations in practice. From either the exact or the approximate expressions, we can see that vergence always grows when propagating along the positive z direction. If the beam is divergent with negative vergence, its absolute value becomes smaller with propagation across positive z, but taking the sign into account, it still grows.

Example 7 The vergence of two spherical beams at point A is (a) $V_A = 2$ D, and (b) $V_A = -20$ D. Determine the vergence at point B located 3 mm to the right of point A.

(a) As B is located to the right of A, z is positive so vergence will grow:

$$V_B = \frac{2}{1 - 0.003 \times 2} = 2.012 \, \text{D}.$$

The change is very small, approximately 0.60%. This makes sense, given that point A is located 500 mm to the left of the focal point, while B is only 3 mm closer, so the curvature change is small.

(b) For the negative vergence (divergent beam),

$$V_B = \frac{-20}{1 + 0.003 \times 20} = -18.87\,\text{D}.$$

Now, the relative change has grown to 6.4%. The absolute value of the vergence gets smaller, as the divergent wavefront becomes flatter upon propagation.

The imaging properties of lenses, that is, their capability of focusing the rays coming from an object point toward an image point, can be seen as a change of the vergence of the light beam refracting through them. Let us illustrate this with another example.

Example 8 A thin lens is imaging a source point located 25 cm to its left. The image is formed 35 cm away to the right of the lens. Compute the change of vergence at the lens plane.

The vergence of the incident beam at the plane of the lens is $V = 1/-0.25 = -4$ D. After refraction, the new vergence is $V' = 1/0.35 = 2.86$ D. The change of vergence is thus

$$\Delta V = V' - V = 6.86\,\text{D}.$$

We will see later that the lens power is closely related to this change of vergence.

3.2.2 *Vergence of Astigmatic Wavefronts*

The wavefronts that come from a real point source are necessarily spherical, so their geometry is completely specified by a single number: vergence or curvature. However, when a light beam is refracted by an astigmatic surface, it loses its revolution symmetry and becomes an astigmatic beam. Given that the vergence of a spherical beam is the curvature of its wavefront, the natural generalization for astigmatic beams seems to be the association of vergence with the Hessian matrix. We will then define the vergence of an astigmatic wavefront at point A as its Hessian matrix. We will use the letter $\mathbb{V}$ to refer to this curvature matrix,

$$\mathbb{V}_A = \begin{bmatrix} V_x & V_t \\ V_t & V_y \end{bmatrix},$$

where we will use the lighter notation V_x, V_t, and V_y to denote the second derivatives of the wavefront. The matrix formalism is also valid for spherical wavefronts. We just multiply the scalar vergence by the identity matrix, $\mathbb{V} = V\mathbb{I}$.

The same properties studied for surfaces apply for wavefronts. Suppose that we set a coordinate system with the Z axis parallel to the optical axis, and we put the XY plane at the point A (so it becomes the origin of our reference system). Then the sagitta or elevation of the wavefront with respect to the transverse plane at A is

$$z(x, y) = \frac{1}{2}\mathbf{r}\mathbb{V}_A\mathbf{r} = \frac{1}{2}V_x x^2 + V_t xy + \frac{1}{2}V_y y^2.$$

On the other hand, the vector normal to the wavefront at a point of coordinates (x, y) is given by

$$\mathbf{n} = -\mathbb{V}\mathbf{r}. \tag{3.4}$$

This equation is of particular importance, as the ray directions are, precisely, perpendicular to the wavefront. If we know the normal vector at any point of the wavefront we will know the direction vector of the ray that passes by this point, and we can, therefore, evaluate its trajectory.

3.3 Ray Propagation

Let us consider a light beam in a homogeneous medium, its propagation direction matching the Z axis of the reference system. Consider the wavefront of this ray at $z = z_0$, and the ray passing through the point $\mathbf{q}_0 = (x_0, y_0, z_0)$. Let us also assume we know the direction vector of the ray (u, v, w). The coordinates of any other point along the ray path, $\mathbf{q} = (x, y, z)$, can be computed by means of the vector equation of a straight line, $\mathbf{q} = \mathbf{q}_0 + \lambda(u, v, w)$. Now, the ray is perpendicular to the wavefront, so its direction vector coincides with the normal to the wavefront at $\mathbf{q}_0$ (see figure 3.2). Also, we know that if the wavefront is not too curved, the third component of the normal can be assumed to be 1, and the first two components, $\mathbf{n}_0 = (u, v)$, can be approximately computed from the curvature matrix of the wavefront, that is, from its vergence at z_0,

$$\mathbf{n}_0 = -\mathbb{V}\begin{pmatrix} x_0 \\ y_0 \end{pmatrix}.$$

We can now split the three-dimensional equation of the ray into a two-dimensional equation and a scalar equation. Let us define $\mathbf{r}_0 = (x_0, y_0)$ and $\mathbf{r} = (x, y)$. Then it is clear that the ray equation is equivalent to the set

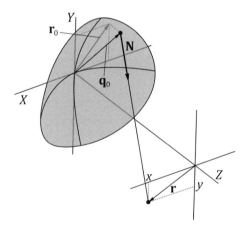

Figure 3.2 Propagation of the ray that passes by point $\mathbf{q}_0 = (x_0, y_0, z_0)$.

$$\mathbf{r} = \mathbf{r}_0 + \lambda \mathbf{n}_0,$$

$$z = z_0 + \lambda.$$

Basically, we are assuming that if the wavefront is not too curved, the parameter λ is just the propagation distance, $z - z_0$ (as the rays will make a small angle with the Z axis). Under this approximation, the change of the transversal coordinates of the ray is related to the vergence of the beam

$$\mathbf{r} - \mathbf{r}_0 = -(z - z_0)\mathbb{V}\mathbf{r}_0. \tag{3.5}$$

The previous equation can be used to explore the focal properties of a wavefront. Spherical beams have a well-defined focus point on the Z axis to which the rays converge. To find it we just have to set $\mathbf{r} = 0$ in the scalar form of equation (3.5), that is, $-\mathbf{r}_0 = -(z - z_0)\mathbb{V}\mathbf{r}_0$, which yields $z - z_0 = 1/V$. As expected, the propagation distance from z_0 to the focal point is just the inverse of the vergence of the beam at z_0. However, what would happen for an astigmatic beam with a nondiagonal vergence matrix? To answer this question, we will use the same approach, trying to find the propagation distance λ at which the ray intercepts the Z axis.

$$\mathbf{r}_0 - \lambda\mathbb{V}\mathbf{r}_0 = 0, \tag{3.6}$$

which may be written as

$$(\mathbb{I} - \lambda\mathbb{V})\,\mathbf{r}_0 = 0. \tag{3.7}$$

This is a linear system on the coordinates (x_0, y_0). If the determinant of its matrix is nonzero, then the only solution would be $\mathbf{r}_0 = 0$ (the ray matching the Z axis). Nontrivial solutions require the determinant of the matrix $\mathbb{I} - \lambda\mathbb{V}$ to be zero. After some algebra we get to the equation

$$1 - \lambda 2H + \lambda^2 K = 0, \tag{3.8}$$

where $H = (V_x + V_y)/2$ is the mean vergence (half the trace of the matrix) and $K = V_x V_y - V_t^2$, its Gaussian curvature (the determinant of the matrix). The two solutions of equation (3.8) are $\lambda_- = 1/V_-$ and $\lambda_+ = 1/V_+$, where $V_\pm = H \pm \sqrt{H^2 - K}$ are the main curvatures, or vergences, of the vergence matrix. Finally, after substitution of the values $\lambda_\pm$ into equation (3.7), we may compute the vectors $\mathbf{r}_{0\pm}$ for which the propagated rays will intercept the Z axis. The condition is $(\mathbb{I} - \lambda_\pm\mathbb{V})\,\mathbf{r}_{0\pm} = 0$, and yields $\mathbf{r}_{0\pm} = (\cos\theta_\pm, \sin\theta_\pm)$, where $\theta_\pm = (V_\pm - V_x)/V_t$ (see equations C.46). The vectors $\mathbf{r}_{0\pm}$ are perpendicular to each other and defined except for a constant, so the points at which rays are passing that will intercept the Z axis lie in perpendicular meridians. The rays passing by points $\mathbf{r}_{0+}(\mathbf{r}_{0-})$ will intercept at a distance $1/V_+(1/V_-)$. These rays are shown with dashed lines in Figure 3.3. A ray passing by any other point or meridian (shown with a continuous line in the same figure) *will not intercept* the Z axis.

Let us consider now the propagation of a set of rays passing through points $(\rho\cos\theta, \rho\sin\theta, z_0)$. They form a circumference lying parallel to XY plane with its center at

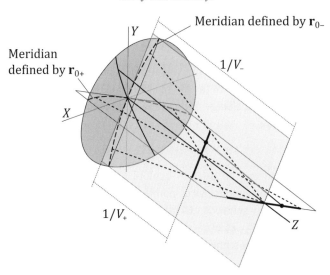

Figure 3.3 Rays contained within the main sections of an astigmatic wavefront (dashed lines) that intercept the Z axis. The ray outside these sections (solid line) will not intercept the Z axis.

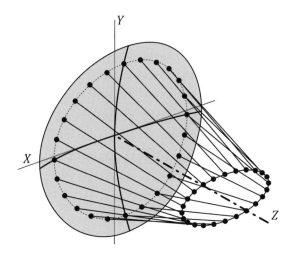

Figure 3.4 Transverse elliptical section of an astigmatic beam after propagating before the focal lines.

$(0, 0, z_0)$. After propagating these rays a distance z, the points defined by the interception of the rays with the plane perpendicular to the optical axis now form an ellipse (see Figure 3.4). In principle, we could study this ellipse using equation (3.7) with any vergence matrix $\mathbb{V}$. However, the calculus is simplified if we choose a reference system with X and Y axes parallel to the main directions of the wavefront. Let us call these axes, and the coordinates along them, ξ and η. This selection does not change the output of the analysis; it only makes it simpler. Let us recall from Chapter 2 that we use the notation $[\ ]_\otimes$ to denote any curvature matrix expressed in the reference system formed by its main directions. Then we can write the vergence matrix

$$\mathbb{V} = \begin{bmatrix} V_- & 0 \\ 0 & V_+ \end{bmatrix}_{\otimes}.$$

Without loosing generality, we can also set the origin of the Z axis at the vertex of the wavefront, that is, we can set $z_0 = 0$. From equation (3.7) we get the following expression

$$\begin{aligned} \xi(z, \theta) &= \rho(1 - zV_-) \cos \theta, \\ \eta(z, \theta) &= \rho(1 - zV_+) \sin \theta. \end{aligned} \tag{3.9}$$

These equations allow us to determine the way in which an astigmatic beam evolves. First, we realize that equation (3.9) defines an ellipse with semi-axes $|\xi(z, 0)| = |\rho(1 - zV_-)|$ and $|\eta(z, \pi/2)| = |\rho(1 - zV_+)|$, as we can see in Figure 3.4. Let us now consider the properties of this astigmatic beam when it propagates along the Z axis.

Focus generated by V_-. When $\xi(z, 0)$ becomes zero, the ellipse collapses into a line par-
allel to the η axis. This happens when $1 - zV_- = 0$, that is at $z_- = 1/V_-$. If we substitute this axial coordinate into the equation for η, we obtain $\eta(z_-, \pi/2) = \rho(1 - V_+/V_-)$, which can be positive or negative. The length of the focal line associated to the main meridian along the ξ coordinate is then twice the absolute value of the η coordinate, $T_- = 2\rho |(V_- - V_+)/V_-|$.

Focus generated by V_+. At $z_+ = 1/V_+$, the semi-axis along the η axis becomes zero, and the ellipse transforms into a focal line along the ξ coordinate. The value of the ξ coordinate at the focal position is now equal to $\xi(z_+, 0) = \rho(1 - V_-/V_+)$, and the length of the focal line $T_+ = 2\rho |(V_+ - V_-)/V_+|$.

Circle of least confusion. This is formed at the axial coordinate z_c for which the two semi-axis of the ellipse are equal, so the ellipse transforms into a circle. The mathematical condition is $|\xi(z_c, 0)| = |\eta(z_c, \pi/2)|$, that is,

$$\rho(1 - z_c V_-) = \pm\rho(1 - z_c V_+).$$

The plus sign leads to the solution $z_c = 0$, which means the beam already had a circular section at the plane where we started its propagation. The minus sign returns the solution $z_c = 2/(V_- + V_+)$, which is the position of the circle of least confusion. The forced selection of the minus sign means that at the circle of least confusion the coordinates $\xi(z_c, 0)$ and $\eta(z_c, \pi/2)$ have opposite signs. In turn, this means that at the circle of least confusion one of the rays at either $\theta = 0$ or $\theta = \pi/2$ has already passed through a focus, while the other has not. In other words, the circle of least confusion is located between the two focal lines of the astigmatic beam. If we plug the value of z_c into any of the factors $\rho(1 - zV_\pm)$, we get the size of the circle of least confusion,

$$T_c = 2\rho \left| \frac{V_- - V_+}{V_- + V_+} \right|.$$

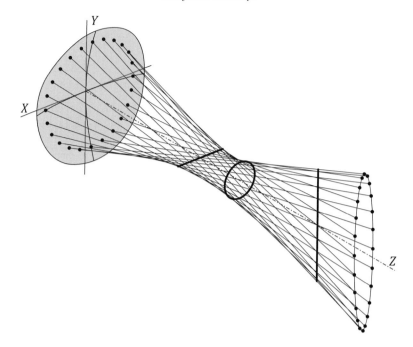

Figure 3.5 Sturm's conoid described by the rays coming from an astigmatic wavefront. In this particular case, the main sections of the wavefront are parallel to the X and Y directions. If this is not the case, the focal line will rotate in the same way as the principal meridians do.

> Finally, let us point out that there is an exception to the previous discussion. When the main vergences of the beam have equal absolute value but different sign, $V_- = -V_+$, then the equations for z_c and T_c break down. The only plane at which the beam has a circular section is the plane we started from, the focal lines lying before and after it.

The structure formed by the two focal lines and the circle of least confusion of an astigmatic beam is known as *Sturm's conoid*, and it is represented in Figure 3.5.

The computation presented here requires that we know the main vergences $V_\pm$, as we were working on a reference system with axes parallel to the main meridians. The general procedure and formulae are summarized in Table 3.1. From the three elements of the vergence matrix V_x, V_y, and V_t, we compute the mean and Gaussian vergences H and K, respectively. From them we get the main vergences V_- and V_+ from which we can compute the position and size of both focal lines and the circle of least confusion using the former equations. It is worth noticing that the focal line associated with a main curvature is always perpendicular to the corresponding main direction. The reason is simple: All the rays refracted by a main section cross at the center of curvature (focus) of the corresponding meridian. This means that the transversal section of the beam is null in the direction of this principal meridian. The rays that are not contained in the main section of the wavefront

Table 3.1 *Characteristics of the Sturm's conoid of an astigmatic beam*

Vergence	H, K	Main vergences	Orientation
$\begin{bmatrix} V_x & V_t \\ V_t & V_y \end{bmatrix}$	$H = (V_x + V_y)/2$ $K = V_x V_y - V_t^2$	$V_- = H - \sqrt{H^2 - K}$ $V_+ = H + \sqrt{H^2 - K}$	$\theta_- = \arctan\left[(V_- - V_x)/V_t\right]$ $\theta_+ = \theta_- \pm 90°$

—

Characteristic	Location	Size	Observations
Focal line generated by V_-	$z_- = 1/V_-$	$T_- = 2\rho \left\| \frac{V_+ - V_-}{V_-} \right\|$	Oriented along θ_+
Focal line generated by V_+	$z_+ = 1/V_+$	$T_+ = 2\rho \left\| \frac{V_+ - V_-}{V_+} \right\|$	Oriented along θ_-
Circle of least confusion	$z_c = 2/(V_- + V_+)$	$T_c = 2\rho \left\| \frac{V_+ - V_-}{V_+ + V_-} \right\|$	

do not intercept the optical axis at the center of curvature of this section (this can only happen for a spherical wavefront that forms a point focus). These rays form a focal line perpendicular to the principal meridian of the astigmatic wavefront. The whole Sturm's conoid generated by an astigmatic wavefront is depicted in Figure 3.5.

3.3.1 Propagation of Astigmatic Vergences

We will now study the problem of vergence propagation for astigmatic wavefronts. At the beginning of this chapter we analyzed the propagation of vergence for a spherical wavefront. Let us suppose now that $\mathbb{V}_A$ is the vergence of the wavefront at a point with axial coordinate z_A. As usual, we take the direction of propagation of the beam as the Z axis of our coordinate system. Our problem is to compute the vergence $\mathbb{V}_B$ corresponding to the wavefront propagated through a distance $z = z_B - z_A$.

Let us consider now a ray with transversal coordinates $\mathbf{r}_A = (x, y)$ at z_A. After propagation through the distance z using equation (3.5), the new coordinates for the ray will be

$$\mathbf{r}_B = \mathbf{r}_A - z\mathbb{V}_A \mathbf{r}_A. \tag{3.10}$$

The direction of the ray at a point defined by the position vector $\mathbf{r}_B$ is $\mathbf{k} = -\mathbb{V}_B \mathbf{r}_B$. As rays do not change direction when propagated in homogeneous media, we have

$$-\mathbb{V}_B \mathbf{r}_B = -\mathbb{V}_A \mathbf{r}_A,$$

therefore

$$\mathbb{V}_B \left(\mathbb{I} - z\mathbb{V}_A\right) \mathbf{r}_A = \mathbb{V}_A \mathbf{r}_A.$$

As this equation must hold for any value of $\mathbf{r}_A$, we conclude that the matrices at both sides of the latter equation must be equal, so

$$\mathbb{V}_B \left(\mathbb{I} - z\mathbb{V}_A\right) = \mathbb{V}_A,$$

and multiplying both sides of this equation by the inverse of $(\mathbb{I} - z\mathbb{V}_A)$, we get

$$\mathbb{V}_B = \mathbb{V}_A \, (\mathbb{I} - z\mathbb{V}_A)^{-1}. \tag{3.11}$$

Because of the symmetry properties of the matrices involved $(\mathbb{I} - z\mathbb{V}_A)$ and its inverse commute with $\mathbb{V}_A$ we can write the matrix product in any order. This equation is formally identical to the scalar one (3.2) we derived for spherical wavefronts. There will be more instances in this book in which we will be able to write general expressions for astigmatic systems that are formally identical to the corresponding scalar equations for systems with revolution symmetry.

Using the symmetry of the curvature matrices, it is interesting to note that the propagation formula can be written in the following way

$$\mathbb{V}_B = \frac{1}{1 - 2Hz + Kz^2} \, (\mathbb{V}_A - zK\mathbb{I}), \tag{3.12}$$

where H and K are the mean and Gaussian curvatures of vergence $\mathbb{V}_A$. Proof of this expression follows from direct use of the inverse matrix formula as given in Appendix B, and we leave it to the reader.

If the propagation distance z is small, the following approximation holds

$$\frac{1}{1 - 2Hz + Kz^2} \simeq 1 + 2Hz.$$

If we plug it into the last expression for $\mathbb{V}_B$ and keep neglecting second-order terms in z we get to

$$\mathbb{V}_B \simeq (1 + 2Hz) \, (\mathbb{V}_A - zK\mathbb{I})$$
$$\simeq \mathbb{V}_A + z \, (2H\mathbb{V}_A - K\mathbb{I}).$$

Now, it can be shown that the relation $2H\mathbb{V} - K\mathbb{I} = \mathbb{V}^2$ holds for any 2×2 symmetrical matrix $\mathbb{V}$, so we finally have

$$\mathbb{V}_B \simeq \mathbb{V}_A + z\mathbb{V}_A^2, \tag{3.13}$$

which is, once again, formally identical to the approximate propagation formula for spherical beams.

Example 9 An astigmatic beam propagating along the Z axis has, at its origin, a diameter of 10 mm, and a vergence

$$\mathbb{V} = \begin{bmatrix} 4 & -1 \\ -1 & 6 \end{bmatrix} \mathrm{D}.$$

Determine the position, size, and orientation of the focal lines and the circle of least confusion. Also determine the vergence of the beam after propagating 20 mm along the Z axis.

The mean and Gaussian curvatures of the wavefront are $H = 5\,\mathrm{D}$ and $K = 23\,\mathrm{D}^2$, respectively. The main vergences

$$V_- = H - \sqrt{H^2 - K} = 3.59\,\mathrm{D},$$

$$V_+ = H + \sqrt{H^2 - K} = 6.41\,\mathrm{D},$$

and the main directions

$$\theta_- = \tan^{-1}\left[(V_- - V_x)/V_t\right] = \tan^{-1}\left(\frac{3.59 - 4}{-1}\right) = 22.3^\circ.$$

The positions of the Sturm's focal lines measured from the origin of the wavefront are

$$z_- = 1/V_- = 278.5\,\mathrm{mm}, \quad z_+ = 1/V_+ = 156\,\mathrm{mm}.$$

According to Table 3.1, the focal line corresponding to the main vergence V_- has a length of

$$T_m = 2\rho\frac{|V_+ - V_-|}{|V_-|} = 15.8\,\mathrm{mm},$$

and it is oriented at an angle $\theta_+ = \theta_- \pm 90^\circ = 102.3^\circ$. For the other focal line we have

$$T_+ = 2\rho\frac{|V_+ - V_-|}{|V_+|} = 8.8\,\mathrm{mm},$$

and its orientation is $\theta_m = 22.3^\circ$. Finally, the position of the circle of least confusion is

$$z_c = \frac{2}{V_- + V_+} = 200\,\mathrm{mm},$$

while its size is

$$T_c = 2\rho\frac{|V_+ - V_-|}{|V_+ + V_-|} = 5.66\,\mathrm{mm}.$$

The vergence of the wavefront, after propagation through 20 mm along the optical axis is

$$\mathbb{V}_B = \frac{1}{0.8092}\begin{bmatrix} 4 - 0.02 \times 23 & -1 \\ -1 & 6 - 0.02 \times 23 \end{bmatrix} = \begin{bmatrix} 4.37 & -1.24 \\ -1.24 & 6.85 \end{bmatrix}\mathrm{D}.$$

Notice the increment of the vergences in the main diagonal, which indicates that we have a converging wavefront. We could have arrived at this result using equation (3.11) instead.

3.4 Refraction

In this section we will study how the vergence of a light beam changes when the beam refracts through a surface separating two different media. The reader will probably be familiarized with the image formation properties of spherical surfaces. Consider one such surface with curvature radius R, separating two media with refractive index n and n'. Let us take any straight line passing by the center of curvature of the sphere as the optical axis and consider an object point O located on this axis at a distance s from the vertex of the sphere. If the rays coming from the object are not too tilted with respect to the optical axis, and if the radius of curvature of the sphere is large with respect to s (the conditions of the *paraxial approximation*), the rays from the object will converge, after refraction, on the image O' located at a distance s' from the surface. The quantity $n(1/s - 1/R)$ is equal at either side of the surface, that is, $n(1/s - 1/R) = n'(1/s' - 1/R)$, and it is known as the Abbe's invariant. Its invariance can also be written as

$$\frac{n'}{s'} - \frac{n}{R} = \frac{n' - n}{R}. \tag{3.14}$$

The inverse distances $1/s$ and $1/s'$ are precisely the vergences, at the surface vertex, of the wavefronts propagating from the object and toward the image. In term of vergences, equation (3.14) may be written as

$$n'V' - nV = P, \tag{3.15}$$

where $P = (n' - n)/R$ is known as the refractive power of the surface. This equation is meaningful, and can be read in the following way: The change of vergence (times refractive index[1]) that undergoes a beam refracting through a spherical surface that separate two media is the refractive power of said surface. Another way of looking at it is that the refractive power of a surface is its capacity to change the vergence (times refractive index) of the light beams that refract through it.

As long as we restrict the parameters of the refraction process to the paraxial environment, the previous description is valid not just for spherical surfaces but for any surface with revolution symmetry. In this case, the optical axis must be the symmetry axis of the surface and R its radius of curvature at the vertex. The ray analysis of the refraction of astigmatic wavefronts by astigmatic surfaces turns out to be quite tangled, particularly so when the main directions of the incident wavefront and that of the astigmatic surface do not have the same orientation. This fact limits the capability of analyzing astigmatic systems by ray tracing, even within the paraxial approximation. However, by using the matrix formalism that we have introduced so far, we will be able to generalize the Abbe's invariant and equation (3.15), deriving expressions that will allow us to deal with astigmatic systems in the same way that we do with those having revolution symmetry.

3.4.1 Snell's Law in Space

Basic geometrical optics is usually based on a two-dimensional version of Snell's law because in systems with revolution symmetry, rays which are initially coplanar with the optical axis always stay coplanar after refraction. In this case, Snell's law reads

$$n \sin \theta = n' \sin \theta', \tag{3.16}$$

where θ and θ' are the angles formed by the incident and refracted ray with the surface normal. From this 2D Snell's law we are going to derive a 3D version. For this, we need

---

[1] Some authors call the product nV *reduced vergence*.

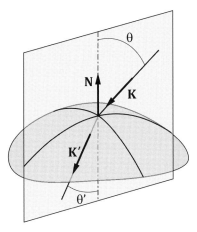

Figure 3.6 Relationship between the direction vectors of the incident and the refracted rays, $\mathbf{K}$ and $\mathbf{K}'$, and the normal vector to the surface at the incidence point. The plane highlighted by a rectangular boundary is the incidence plane.

to characterize rays in space with their direction vectors, $\mathbf{K}$ for the incident ray and $\mathbf{K}'$ for the refracted one. We also need the vector normal to the surface at the incidence point, $\mathbf{N}$. These three vectors are coplanar, so they can be expressed as a combination of the vector normal $\mathbf{N}$ and a vector $\mathbf{T}$ tangent to the surface and contained in the incidence plane. From Figure 3.6, we derive this relationship between them

$$\mathbf{K} = \cos\theta\mathbf{T} + \sin\theta\mathbf{N},$$
$$\mathbf{K}' = \cos\theta'\mathbf{T} + \sin\theta'\mathbf{N}.$$

By applying 2D Snell's law to the second equation, we get

$$\mathbf{K} = \cos\theta\mathbf{T} + \sin\theta\mathbf{N},$$
$$\mathbf{K}' = \cos\theta'\mathbf{T} + \left(n/n'\right)\sin\theta\mathbf{N},$$

and, after multiplying the first equation by $-n/n'$,

$$-\left(n/n'\right)\mathbf{K} = -\left(n/n'\right)\cos\theta\mathbf{T} - \left(n/n'\right)\sin\theta\mathbf{N},$$
$$\mathbf{K}' = \cos\theta'\mathbf{T} + \left(n/n'\right)\sin\theta\mathbf{N}.$$

Finally, if we add both left and right terms of the previous equation, we obtain, after some rearrangement,

$$n'\mathbf{K}' - n\mathbf{K} = \left(n'\cos\theta' - n\sin\theta\right)\mathbf{N}. \tag{3.17}$$

This equation is known as the vector form of Snell's law and it is a necessary tool to perform exact ray tracing or paraxial ray tracing in astigmatic systems. If the incidence angle θ is small enough, we can reduce both equations (3.16) and (3.17) to their paraxial form,

$$n\theta = n'\theta'$$

$$n'\mathbf{K}' - n\mathbf{K} = (n' - n)\,\mathbf{N}. \tag{3.18}$$

3.4.2 Abbe's Invariant for Astigmatic Systems

Let us consider now the refraction of an astigmatic wavefront on an arbitrary surface, which may also be astigmatic. Just before refraction, the vergence of the incident beam is $\mathbb{V}$. Immediately after refraction, the curvatures of the wavefront have changed and the vergence becomes $\mathbb{V}'$. In turn, the refracting surface is characterized, within the parabolic approximation, by a Hessian matrix $\mathbb{H}$. We will also suppose that the surface and the incident and refracted beams share a common Z axis, so the central ray of the beam impinges perpendicularly on the refracting surface. Finally, as vectors $\mathbf{K}$, $\mathbf{K}'$, and $\mathbf{N}$ make small angles with the Z axis, the value of their third component is very close to 1 and we only have to track their first two components (see Appendix B).

In these paraxial environments, let us consider any ray belonging to the incident beam and passing through the point $\mathbf{r} = (x_0, y_0)$. Its direction vector is normal to the incident wavefront at $\mathbf{r}$, so $\mathbf{k} = -\mathbb{V}\mathbf{r}$. Similarly, the direction vector of the refracted ray at the same point is $\mathbf{k}' = -\mathbb{V}'\mathbf{r}$. In addition, the normal to the refracting surface at $\mathbf{r}$ is given by $\mathbf{n} = -\mathbb{H}\mathbf{r}$. Substituting these expressions into the vector form of the paraxial Snell's law, equation (3.18), we derive

$$-n'\mathbb{V}'\mathbf{r} + n\mathbb{V}\mathbf{r} = -(n' - n)\,\mathbb{H}\mathbf{r}.$$

This equation must hold for any incident point $\mathbf{r}$, as we have not made any assumption about it. So the next matrix equation follows

$$n'\mathbb{V}' = (n' - n)\,\mathbb{H} + n\mathbb{V}. \tag{3.19}$$

Equation (3.19) is formally identical to its scalar counterpart (3.15), but generalized for astigmatic surfaces and wavefronts with arbitrary orientations. Obviously, the Abbe's invariant also has a matrix counterpart

$$n'\left(\mathbb{V}' - \mathbb{H}\right) = n\left(\mathbb{V} - \mathbb{H}\right). \tag{3.20}$$

Finally, we can also define the astigmatic *refractive power* as $\mathbb{P} = (n' - n)\mathbb{H}$, so the refraction equation reads

$$n'\mathbb{V}' - n\mathbb{V} = \mathbb{P}. \tag{3.21}$$

Refractive power depends on the surface curvature and the change of refractive index across the surface. The change of vergence (always multiplied by the index) is, precisely, the refracting power of the surface, or, in other words, the refractive power of a diopter measures the ability of a surface for changing the vergence of a given beam after refracting on this surface. Just as the surface has main curvatures $\kappa_\pm$, the refractive power matrix has main refractive powers given by $P_\pm = (n' - n)\kappa_\pm$.

Let us end this chapter with a pair of examples illustrating the use of the matrix refraction equation. First, we will analyze the formation of a Sturm's conoid by an astigmatic surface.

Example 10 A toric surface is given by a Hessian

$$\mathbb{H} = \begin{bmatrix} 6.75 & -3\sqrt{3}/4 \\ -3\sqrt{3}/4 & 8.25 \end{bmatrix} D,$$

which separates two media of refractive index 1 and 1.5, respectively. A ray beam, coming from an object located at infinity, incides parallel to the Z axis. Determine its vergence after refraction and the location of its focii.

The vergence of the incoming beam is the null matrix, as it is a beam of parallel rays coming from infinite. The refracting power of the surface is

$$\mathbb{P} = (n' - n)\mathbb{H} = \begin{bmatrix} 3.375 & -3\sqrt{3}/8 \\ -3\sqrt{3}/8 & 4.125 \end{bmatrix} D$$

and the image vergence

$$\mathbb{V}' = \frac{1}{n'}\mathbb{P} = \begin{bmatrix} 2.25 & -\sqrt{3}/4 \\ -\sqrt{3}/4 & 2.75 \end{bmatrix} D.$$

To compute the position of the main focii we need the main vergences. We readily find $H = 2.5$ D and $K = 6\,\mathrm{D}^2$, so

$$V'_- = 2.5 - \sqrt{6.25 - 6} = 2\,\mathrm{D},$$
$$V'_+ = 2.5 + \sqrt{6.25 - 6} = 3\,\mathrm{D}.$$

Finally, the location of the focal lines is

$$z_- = 1/V'_- = 500\,\mathrm{mm},$$
$$z_+ = 1/V'_+ = 333\,\mathrm{mm}.$$

The second example shows the combined use of astigmatic refraction and vergence propagation for describing a complex astigmatic system such as an aphakic eye corrected with an intraocular lens.

Example 11 An astigmatic cornea with refractive index 1.33 has the following curvature matrix

$$\mathbb{H}_c = \begin{bmatrix} 126.67 & -2.887 \\ -2.887 & 130 \end{bmatrix} D.$$

We will neglect the back surface of the cornea, assuming that the refractive indexes of the cornea and the aqueous humor at the anterior chamber are equal. A thin convex-plane intra-ocular lens (IOL), is placed at a distance of 9 mm from the back surface of the cornea. The retina is located 15 mm away from the IOL, and we want, for a distant object point, a sharp focus to be formed at the retina. Assuming that the refractive index of the vitreous humor is the same of that of the aqueous humor, compute the orientation and main refractive powers of the IOL.

We want to form at the retina a sharp image of a distant object, so the vergence of the incident beam at the cornea will be null. The refractive power of the cornea is

$$\mathbb{P}_c = (1.33 - 1)\mathbb{H}_c = \begin{bmatrix} 41.8 & -0.953 \\ -0.953 & 42.9 \end{bmatrix} D,$$

so the vergence refracted by the cornea will be

$$\mathbb{V}'_c = \frac{1}{n'}\mathbb{P}_c = \frac{1}{1.33}\begin{bmatrix} 41.8 & -0.953 \\ -0.953 & 42.9 \end{bmatrix} = \begin{bmatrix} 31.43 & -0.716 \\ -0.716 & 32.26 \end{bmatrix} D.$$

Next, we have to compute the vergence arriving at the front surface of the IOL, so we have to propagate the refracted vergence at the cornea through a distance of $z = 9$ mm,

$$\mathbb{V}_1 = \mathbb{V}'_c \left(\mathbb{I} - z\mathbb{V}'_c\right)^{-1} = \begin{bmatrix} 43.84 & -1.407 \\ -1.407 & 45.46 \end{bmatrix} D.$$

After refracting through the first surface of the lens whose refractive power is $\mathbb{P}_1$, the new vergence must comply with the following equation

$$n_L\mathbb{V}'_1 - n\mathbb{V}_1 = \mathbb{P}_{1L},$$

where n is the refractive index of the aqueous humor (and vitreous in our approximation), n_L that of the lens, and $\mathbb{P}_{1L}$ is the refractive power of the front surface of the lens. We will focus now on the refraction at the back surface of the lens. As we consider the lens to be thin, the incident wavefront at the back surface of the lens is $\mathbb{V}_2 = \mathbb{V}'_1$. On the other hand, as the back surface is planar, both its Hessian and its refractive power are null so $\mathbb{P}_{2L} = \mathbb{O}$, where $\mathbb{O}$ is the null matrix. Therefore, the expression of Abbe's invariant for refraction at the back surface of the lens would be

$$n\mathbb{V}'_2 = \mathbb{P}_{2L} + n_L\mathbb{V}_2 \equiv n_L\mathbb{V}_2.$$

As we want the eye to focus a distant object at the retina, placed 15 mm to the right of the IOL, the exit vergence at the back surface of the IOL would be

$$\mathbb{V}'_2 = \begin{bmatrix} 1/0.015 & 0 \\ 0 & 1/0.015 \end{bmatrix} = \begin{bmatrix} 66.67 & 0 \\ 0 & 66.67 \end{bmatrix} D.$$

Therefore, the incident vergence at the back surface of the lens will be

$$\mathbb{V}_2 = \frac{n}{n_L}\mathbb{V}'_2 = \frac{1.33}{1.49}\begin{bmatrix} 66.67 & 0 \\ 0 & 66.67 \end{bmatrix} D = \begin{bmatrix} 59.51 & 0 \\ 0 & 59.51 \end{bmatrix} D.$$

As described before, as we consider the lens to be thin, $\mathbb{V}'_1 = \mathbb{V}_2$, so the refractive power of the front surface of the IOL is

$$\mathbb{P}_{1L} = 1.49\begin{bmatrix} 59.51 & 0 \\ 0 & 59.51 \end{bmatrix} - 1.33\begin{bmatrix} 43.84 & -1.407 \\ -1.407 & 45.46 \end{bmatrix} = \begin{bmatrix} 30.36 & 1.87 \\ 1.87 & 28.21 \end{bmatrix} D.$$

Thus, the Hessian of the front surface of the IOL would be

$$\mathbb{H}_{1L} \equiv \frac{\mathbb{P}_{1L}}{(n_L - n)} = \begin{bmatrix} 189.75 & 11.69 \\ 11.69 & 176.25 \end{bmatrix} D.$$

Finally, let us compute the main curvatures of the convex surface of the IOL. The mean and Gaussian curvatures are $H = 183.0\,\mathrm{D}$ and $K = 33310\,\mathrm{D}^2$, so

$$\kappa_- = 183 - \sqrt{183^2 - 3331} = 183.5 - 13.5 = 170\,\mathrm{D},$$

$$\kappa_+ = 183 + \sqrt{183^2 - 33310} = 197\,\mathrm{D}.$$

The corresponding radii of curvature are $R_- = \kappa_m^{-1} = 5.88$ mm and $R_+ = \kappa_+^{-1} = 5.08$ mm, and the orientation of the meridian with curvature κ_-

$$\theta_- = \arctan\left(\frac{170 - 197}{11.69}\right) \approx 30°.$$

In Chapter 7 we will see more examples of IOL calculation using the matrix formalism.

4

Single Vision Lenses

4.1 Introduction

In Scientific problems that are amenable to be solved mathematically, the output (unknown) variables are usually a complex function of the input (known) variables. If we limit the scope of our problem to small variations of the input variables around some fixed state, the complex function can be approximated by a simpler, linear one. For example, let us assume that our problem depends on two variables, x and y, that can represent any magnitude. Let us also assume that there are two output variables, u and v, that similarly represent arbitrary magnitudes. The relationship established by the math describing the problem can be expressed as

$$u = f(x, y), \quad v = g(x, y).$$

When the functions f and g are complex and/or difficult to handle, a trick that eases the problem is *linearization*, that is, finding linear substitutes for f and g that are a good approximation to both of them. This usually works well if the variables x and y stay close to some initial state x_0, y_0. After linearization, the functions f and g will always have the form

$$u = a(x - x_0) + b(y - y_0)$$
$$v = c(x - x_0) + d(y - y_0),$$

where a, b, c, and d are constants that may depend on the initial state. After linearization, the problem can always be written in matrix form,

$$\begin{bmatrix} u \\ v \end{bmatrix} = \begin{bmatrix} a & b \\ c & d \end{bmatrix} \begin{bmatrix} x \\ y \end{bmatrix} - \begin{bmatrix} a & b \\ c & d \end{bmatrix} \begin{bmatrix} x_0 \\ y_0 \end{bmatrix},$$

and whenever the initial state can be set to zero, we only need the first term on the right side of this matrix equation. We have seen in Chapter 2 that when we approximate a surface by a paraboloid, the problem of calculating its normal becomes linearized, the matrix involved being the curvature of the surface. We have also seen in Chapter 3 that when we use linearized equations for computing normals along with the linearized Snell's law, the refraction of vergences through surfaces turns into another linear problem, this time mediated by the refractive power of the surface. Lenses are made by the composition of two surfaces, so the problem of vergence/ray propagation through them keeps linear and can be easily described with matrices.

The linearization of the propagation of rays through optical systems is due to Gauss, but the introduction of a matrix formalism that could deal with astigmatic systems was laid out in an elegant way by Luneburg in 1964 [36]. Luneburg's approach was, however, abstract, and some of the properties and applications of the matrices he introduced, called ABCD matrices, were not known. From Luneburg's book to the mid-eighties many authors developed applications for ABCD matrices in many different fields of optics [37, 38, 39, 40, 41, 42], but it was Fick [43], Long [44], and Keating [29, 45] who were the pioneers of introducing matrix methods in ophthalmic and visual optics and optometry. We must also cite here the extensive work by Harris in disseminating the applications of matrices in visual optics, advancing the comprehension of their properties and proposing new ideas, definitions, and techniques. Harris [46] provides a summary of astigmatic systems and contains most of his most relevant references.

4.1.1 Matrices and Lens Power

Once we have gone through the necessary ideas and tools for handling surfaces and their geometry, and once we have learned how to describe light beams by means of the curvature of their wavefronts, and how this curvature changes when the wavefronts refract through surfaces, we are in a position to deal with our main subject of interest: the ophthalmic lens. A lens may be defined as a combination of two surfaces enclosing material of homogeneous refractive index. Ophthalmic lenses are usually immersed in air, but there are exceptions, such as graduated lenses for swimming, or the lenses used in diving masks. Contact and intraocular lenses are also in contact with media other than air. In this chapter, we will study the main geometrical and optical properties of single vision (or monofocal) lenses. Subsequent chapters will deal with other aspects of ophthalmic lenses, such as the eye–lens interaction, aberrations and design of ophthalmic lenses, and so on.

So far, we have associated a curvature matrix to single surfaces. For the physical surfaces of a lens, their Hessian matrices are the mathematical objects that describe their curvatures at their vertex. For a wavefront, its Hessian matrix is, once again, its curvature (or vergence, as it is called for wavefronts). We have also described the optical properties of the surface separating two media with different refractive index by means of a symmetrical matrix called refractive power. It is given by the change of refractive index across the surface times its actual curvature, $\Delta n\mathbb{H}$. If the medium to the right has $n = 1$ and the the vergence of the input beam is zero, then the refractive power would exactly match the curvature of the refracted wavefront. What can we expect with respect to lenses? Would it be possible to assign a curvature-related matrix to them? We will see that indeed this is the case. When the lens is in air and a plane wavefront impinges on it, the curvature of the output wavefront will be its *back vertex power,* the type of power mostly used in ophthalmic optics. We will see that a lens can be associated with other power-related matrices that, in general, do not represent curvature and will be necessary to compute accurate deviations of rays refracting through a lens.

4.2 Geometrical Aspects

As we have already stated in the introduction to Chapter 1, modern ophthalmic lenses combine three different aspects with their particular technologies:

Geometry. By lens geometry we understand everything related to the shape of the surfaces, the relative position between them, and all the optical, ergonomic, and esthetic properties that are derived from those. Lens geometry effectively determines the image-forming properties of the lens: lens power, magnification, field, prismatic effect, and monochromatic aberrations. In multifocal lenses the geometry also determines how these properties are distributed across the lens, and how the user can access them. Surface shape and relative position determine the thickness and weight of the lens, hence the ergonomic aspects for adapting and wearing it. And of course, all this affects the esthetic of the lens, a main factor in user satisfaction with the lens.

Material. Determines the physical and chemical properties of the lens, particularly weight, mechanical resistance, hardness, and material dispersion, which sets the chromatic aberration of the lens.

Coatings. Modify the physical and chemical properties of the lens surface, enhancing or changing light transmission, adding photochromic or polarization capabilities, improving scratch resistance, etc.

Traditionally, the practitioner advised the patient on the geometry, the material, and the coatings that were best suited for each case. Nowadays, it is fairly common to find "labels" or trademarks that wrap a particular selection on all three categories under a commercial brand. Thus, the "X" lens made by manufacturer "Y" is characterized by a given geometry (spherical, aspherical, multifocal, etc.), material, and set of coatings (anti-reflective, hydrophobic, etc.). From a marketing point of view, these commercial concepts might be useful as they may help to improve sales. Unfortunately, they reduce the flexibility of choice available to the prescribing professional. As we have already said, we are going to focus, in this chapter, on the first of these properties: the geometry of the lens, which, as we will see, determines the image formation properties, the thickness, and the volume of the lens.

4.2.1 Classification Based on the Curvature Radius of the Lens Surfaces

The radius of curvature is measured from the surface apex to the local center of curvature. Throughout this book, we will follow the usual sign convention of geometrical optics [47]: light will propagate along the Z axis, from left to right, as shown in Figure 4.1. The direction of propagation of light then coincides with the positive direction of the Z axis. According to this convention, surfaces with positive curvature radii have their center or

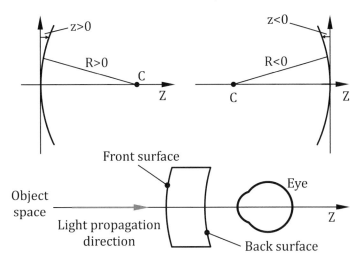

Figure 4.1 Top: Surfaces with curvature radius positive (left) and negative (right). Bottom: Location of the front and back surface of an ophthalmic lens.

curvature to the right of the surface, and their sagittas are also positive. On the contrary, surfaces with negative radii have their centers located to the left of the surface, and their sagittas are negative too, see Figure 4.1. In the case of spectacle lenses, the surface to the left, which is the one encountered first by the incident light, is known as the *front surface*, while the right surface, the one closest to the eye, is known as the *back surface*. Other terminology used for the front surface includes "first," "outermost," "anterior," or "object," the last because it is the surface facing the objects we see. For the back surface we may also find the terms "second," "inner," "posterior," and "eye." "Object" and "eye" surfaces are commonly employed in patents on ophthalmic lenses, but they are seldom used in academic environments. We will adhere to the adjectives *front* and *back*, as indicated in Figure 4.1, and will use the subindex "1" and "2" to mathematically address the two surfaces.

The naming of the refractive index is a little tricky if we want to avoid cluttering the formulas with subindexes. Generically, the indexes to the right and left of a given surface are typically named n and n'. Δn will be the change of refractive index across the surface as it is traversed from left to right, $\Delta n = n' - n$. If we apply this rule to a lens, its material would be to the right of the first surface, and should be named n', and to the left of the second surface, so it should be named n. We can solve this problem by attaching an index to the refractive index of the lens, for example n_L. Then, if the lens in immersed in air, for the first surface we would have $n = 1$ and $n' = n_L$, while for the second surface $n = n_L$ and $n' = 1$. If various lenses are to be analyzed in tandem, each refraction requires a change in the name of the refractive index. As we will try not to overuse subindexing, we advise the reader to get familiar with the required changes in notation.

Concavity and convexity are qualitative indicators of the side to which a curve or surface bends. The quality of concavity or convexity is related to the sign of the second derivative in plane curves but is more complex in surfaces. Even with curves there is no universal agreement about labeling a curve with, say, positive second derivative, as concave or convex, as you could look to the curve from above or from below. In surfaces, concavity depends on the sign of the two main curvatures. If both are positive, the surface "curves up." If both are negative, we would say the surface "curves down." If the main curvatures at the vertex of the surface have different signs, the vertex is a "saddle point": some meridians are concave, others are convex. The main curvatures of both surfaces of a typical ophthalmic lens are positive. From the mathematical point of view, both surfaces "curve up" and they should have the same type of "convexity." However, it is totally standard in the ophthalmic industry to term the front surface convex and the back one concave. This is because the surfaces of a lens have "physical reality" and there is matter in between. The substance of the lens is in the direction of curving of the front surface and opposite to the direction of curving of the back surface. To settle the issue, consider a spherical shell. If our surface curves similarly to the outer part of the shell, we will say it is convex. If it curves like the inner part of the shell, we will say it is concave. An even more practical definition is the next one: If you can hold some liquid on top of the lens surface, then it is concave. If you cannot hold any liquid because it runs down the surface, it is convex. Ophthalmic lenses usually have a convex front surface and a concave back surface, what is known as *meniscus shape*.

We have represented in Figure 4.2 the four possible cases that may arise when combining the sign of the radii and the location of the surfaces. The surfaces of the upper row are both external, so $\Delta n > 0$. The external surface to the right has a positive radius, it is convex, and its refractive power is positive. Conversely, the surface to the right of the upper row has a negative radius, it is concave, and its refractive power is negative. For the bottom row, both surfaces are internal, and both have $\Delta n < 0$. The curvature radius of the one located to the left is positive, this surface being concave, and its refractive power is negative. The surface bottom-right has a negative radius, it is convex, and its refractive power is positive. According to our definition, a convex surface will always have positive refractive errors,

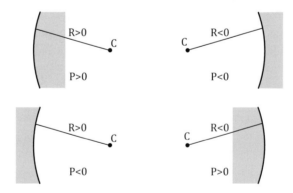

Figure 4.2 The sign of the refractive power of a lens diopter depends on the sign of the radius of curvature of the diopter and its location at the front or back of the lens. The shadowed areas indicate the lens material.

while concave surfaces will have negative ones (assuming the lens is immersed in a medium with refractive index smaller than that of the lens).

Table 4.1 contains a complete lens classification regarding the lens format. All lenses whose format is biconvex, convex-plane, or plane-convex are positive, while all biconcave,

Table 4.1 *Possible shapes for a lens immersed in air.*

Type	R_1	P_1	R_2	P_1	Shape
Biconvex	$R_1 > 0$	$P_1 > 0$	$R_2 < 0$	$P_2 > 0$	
Convex-plane	$R_1 > 0$	$P_1 > 0$	$R_2 = \infty$	$P_2 = 0$	
Convex-concave	$R_1 > 0$	$P_1 > 0$	$R_2 > 0$	$P_2 < 0$	
Plane-convex	$R_1 = \infty$	$P_1 = 0$	$R_2 < 0$	$P_2 > 0$	
Plane-concave	$R_1 = \infty$	$P_1 = 0$	$R_2 > 0$	$P_2 < 0$	
Concave-convex	$R_1 < 0$	$P_1 < 0$	$R_2 > 0$	$P_2 < 0$	
Concave-plane	$R_1 < 0$	$P_1 < 0$	$R_2 = \infty$	$P_2 = 0$	
Biconcave	$R_1 < 0$	$P_1 < 0$	$R_2 < 0$	$P_2 > 0$	

plane-concave, and concave-plane ones are negative. Convex-concave or concave-convex lenses are known as meniscus-shaped lenses. Both of them may be positive or negative (examples of both are show in Table 4.1). Only five lens formats out of the the nine shown in Table 4.1 are really physically different. If we turn around a concave-convex lens (by rotating the lens 180° around the X or Y axes), the sign of the lens radii changes and the lens becomes a convex-concave one. In other words, the front and back surfaces swap each other. The same happens with plane-convex and the plane-concave lenses, which, after a 180° rotation, become convex-plane and concave-plane, respectively. But the reader has to take into account that performance-wise, lens orientation may make a big difference. For example, if we want to focus a beam of parallel rays, we can use a plane-convex lens with either the convex or the plane surface facing the incident beam. However, it turns out that the convex surface facing the incident beam produces smaller aberrations and a sharper focus. Most ophthalmic lenses are convex-concave, as it is the lens format that provides the best performance for compensating refractive errors. Contact lenses are always convex-concave, in this case imposed by the high curvature of the cornea (see Chapter 7). Other formats are possible for intraocular lenses, depending on the ocular parameters and lens design, but the biconvex format is probably the one mostly used.

Semifinished Blanks and Lens Manufacturing

The manufacturing of ophthalmic lenses poses a limitation to the number of different geometries that can be used for the front surface of ophthalmic lenses. Lens manufacturing and measurement are covered in Chapter 10, but we need now to briefly advance some definitions and concepts that are necessary for the chapters to follow. According to the processing method, ophthalmic lenses can be classified into stock and RX-lenses. When the eye care professional (ECP) asks the lens supplier for a pair of lenses, she provides the supplier with the patient's prescription and possibly other custom parameters. The ECP also selects the material and the coatings for the lenses. Let us recall that prescriptions are measured in steps of 0.25 D both for the sphere and cylinder, so there is a finite number of possible prescriptions. The lens supplier may have a stock of finished lenses with the more prevalent prescriptions and with the more popular materials and coatings. If the order from the ECP matches some of the lenses in the stock, no manufacturing is needed. Of course, the number of possible combinations that prescriptions, materials, tints, and coatings can be put into is extremely high, so stock lenses can only be used for a very small set of these combinations.

When no stock lens is available, an RX-lens is produced at the RX-lab, which is the optical workshop at which lenses are produced. The lab then selects a semifinished lens, also known as blanks. These are thick lenses that are going to be reshaped to produce the final lenses with the right prescriptions and parameters. For more than 40 years, the reshaping process has become standard all over the world. The front surface of the blank is left unchanged, and only the back surface is reshaped. As the prescription can be split among the two surfaces almost arbitrarily, only a reduced number of front surfaces are needed. A compromise between cost, lens quality, and esthetics can be achieved if about

8 to 12 different front surfaces are used for each material and/or lens type. Most of these front surfaces are spherical or at least have revolution symmetry, and their vertex refractive power is named *base curve*.

This brief summary is enough to introduce the concepts of RX-lenses and labs, semifinished lenses, and above all, base curves, that we will need in this and subsequent chapters. The interested reader may take a look at Chapter 10 for a deeper review of these matters.

4.2.2 Classification Based on the Type of Surfaces

An alternative way of classifying lenses, parallel to the former one, is based on the type of surfaces that form the lens, regardless of the sign of their curvatures. Up to the first half of the twentieth century, only a few surface types were routinely used for making ophthalmic lenses, basically spheres, cylinders, and tori. Eventually, some types of aspherical surfaces were added, conicoids, generalized conicoids, atoric, and finally progressive surfaces. Before the arrival of free-form technology around 2000, RX-labs could only cut spheres and tori as back surfaces, the more exotic aspherical surfaces being already cast at the front of some blanks. Free-form technology has enabled the RX-lab with the possibility of cutting arbitrary surfaces[1] at the back of the semifinished blanks, so the spectrum of available surfaces has dramatically grown during recent years. According to this, the next classification is not intended to be exhaustive, it cannot be, but just a broad classification of lenses according to the types of surfaces most likely to be used.

Spherical lenses. The most simple type of lenses. Both surfaces are spherical, and the lens has symmetry of revolution around the optical axis, which is defined as the line that joins the two centers of curvature. In some cases, the adjective "spherical" is used as opposed to "astigmatic," meaning that the lens does not have a prescription cylinder. With this interpretation, any lens with revolution symmetry, having spherical or aspherical surfaces, could be considered "spherical." This interpretation is more clinical than technical. Asphericity is used to correct aberrations, but the refractive error we want to compensate is ultimately composed of a spherical term and a cylindrical term. If the cylindrical term is zero, the refractive error is said to be purely spherical, and the natural extension for the clinician is to consider "spherical" all the lenses intended to correct "spherical" errors. We will, however, keep the technical interpretation. Any lens with rotation symmetry, either with spherical or aspherical surfaces, is intended to correct spherical errors, but its optical properties (specially the thickness and amount of oblique aberrations) may depend to a great extent on the particular selection of surface geometry.

Astigmatic lenses. These lenses have, at least, one astigmatic surface. They do not have symmetry of revolution, and they are intended for compensating ocular astigmatism.

---

[1] With the only limitation of maximum local curvature between 15 and 20 D.

Aspheric lenses. This is a generic name for those lenses that have at least one aspherical surface. In the ophthalmic industry, the concept "aspheric surface" is somewhat ambiguous. For example, a progressive surface is, necessarily, aspherical. However, we sometimes find a distinction between aspherical and spherical progressive surfaces. The latter have two wide regions with constant spherical power, so these regions may be considered almost spherical. Currently, the concept of aspherical lenses is associated with lenses manufactured from semi-finished blanks whose front surface is aspherical, typically a generalized conicoid. The back surface is, usually, a sphere or torus. Free-form lenses used to have an external spherical surface, and an internal surface that can be arbitrary with just a limitation to the maximum local curvature. It is also possible to find bi-aspheric lenses for which both front and back surfaces are generalized conicoids.

Spherocylindrical lenses. These lenses combine a spherical surface with a cylindrical one. They were the first geometry used for compensating ocular astigmatism. Although a cylinder is, obviously, not a sphere, these lenses are not considered aspherical, only astigmatic. The cylindrical surface could be the front or the back. In the former case the cylinder was convex and positive, while in the latter, the cylinder was concave and negative. Cylinders are no longer in use for spectacle lenses, but they are standard in trial lenses and inside optometric instruments.

Spherotoric lenses. These are lenses made from a sphere and a torus. Standard modern astigmatic lenses are spherotoric, the torus being the back surface of the lens. Spherotoric lenses with external torus were also manufactured in the past, but very few specialized optical workshops would make them today. Spherotoric lenses, similarly to spherocylindrical ones, are not considered aspherical, despite being astigmatic, because the torus is one of the simplest astigmatic surfaces, as it is formed by a rotating circle.

Atoric lenses. These lenses have a spherical front surface and an atoric back surface. We could consider them part of the "aspheric" group, but the addition of "toric" appendix simply tells us the lens is intended to correct astigmatic refractive errors.

4.2.3 Lens Thickness

Thickness does not play an important role in the optical properties of ophthalmic lenses, which from the optical point of view are fairly thin, but it has outstanding practical, ergonomic, and esthetic importance. First, center thickness is critical to the impact resistance of the lens. It is also important to take into account the edge thickness for proper glazing. Having the edge thickness too large or too small may cause difficulties for glazing and mounting the lens, especially for rimless and semi-rimless frames. Second, lens weight is almost proportional to center thickness. As weight makes an important impact on comfortable wear, so does lens thickness. Finally, the appearance of medium-to high-powered lenses is usually very important to their wearers, and a thickness too

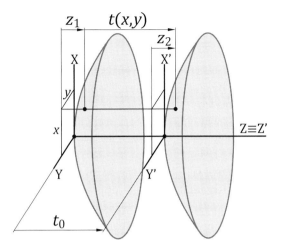

Figure 4.3 Definitions of lens thickness at an arbitrary point along the Z axis.

large, either at the edge in negative lenses or at the center in positive lenses, is considered unsightly. No wonder that lens designers take great care considering thickness issues, even if the effect on optical properties is very limited.

A precise definition of lens thickness requires a precise definition of the lens itself. In Chapter 2 we described a surface by means of a function $z(x, y)$ that yields the value of the sagitta z corresponding to a point (x, y) on the XY plane. Besides, we asked for *standard representations of surfaces*, that is, functions $z(x, y)$ such that $z(0, 0) = 0$, and also $z_x(0, 0) = z_y(0, 0) = 0$. This requirement was needed to ensure the Hessian matrix properly described the surface in a region around the origin of coordinates. A lens is defined by two such functions, $z_1(x, y)$ for the front surface and $z_2(x, y)$ for the back, as shown in Figure 4.3. To avoid the two surfaces touching each other at the origin, we use for $z_2(x, y)$ a reference system X'Y'Z' displaced with respect to XYZ a distance t_0 along the Z axis. This displacement is precisely the center thickness of the lens: the distance, along the Z axis, between the vertexes of the two surfaces that define the lens. Now consider points with coordinates (x, y) on planes XY and X'Y'. The sagittas from these two points to the respective surfaces are $z_1(x, y)$ and $z_2(x, y)$. The local thickness of the lens at point (x, y) will be defined as the distance between the two surfaces along the Z axis, at point (x, y). A look at Figure 4.3 immediately reveals that

$$t(x, y) = t_0 + z_2(x, y) - z_1(x, y). \tag{4.1}$$

There are some important considerations relative to the previous discussion and the following equation (4.1).

- We have been using so far *standard representations* of surfaces, but for equation (4.1) to work we only need that $z_1(0, 0) = z_2(0, 0) = 0$. The surfaces could be tilted around the vertex and still the thickness equation would work. Of course, if a surface is not tangent to the XY plane, the Hessian will not approximate the vertex curvature well, and the parabolic approximation C.33 will not be valid, as long as we use exact functions for the sagittas that do not affect the thickness.

- When the two functions z_1 and z_2 are both standard representations, then the common Z axis intercepts the two surfaces at right angles. For basic single vision lenses, the Z axis will be the optical axis of the lens (see Section 4.3.1). For more advanced lenses that may use surfaces without rotation or axial symmetry, the Z axis may not be an optical axis. In any case, this fact does not affect the validity of the thickness equation.

- Equation (4.1) determines thickness *along* the Z axis. This is the most common approach to determine and control thickness, and the one used when the lens is designed, but other definitions are possible. A situation in which a different definition is more convenient arises when glazing lenses for wraparound frames. Typically, these are frames for sunglasses with the frame strongly curving to wrap around the face. If powered lenses are to be fitted into these frames, they have to be produced with a highly curved front surface. Ideally, to achieve better esthetics, the curvature of the base curve should match the curvature of the frame. The geometry of frames and lenses is shown in Figure 4.4. In subplot (a) we can see a top view of the frame in which negative lenses are mounted. In subplot (b) we see two sections representing two possible shapes for this negative lens. To the left a shape with a flat base curve, possibly the selection in place for standard, flat frames. To the right, a shape with a steeper base curve that matches the curvature of the wrap around frame. The geometry of the frame forces the lenses to be tilted when mounted. These lenses should be glazed in such a way that the edge does not run parallel to the Z axis, but perpendicular to the front surface. As we can see in subplot (a), this type of glazing would produce a smaller edge thickness, and a far more appealing esthetics. If we want to predict the thickness of lenses glazed for this kind of frames, for example as a function of the refractive index, we should compute it as the distance AB, as shown in Figure 4.4(c). In general, these calculations require the use of computers and numerical approaches, as closed formulae such as equation (4.1) do not generally exist. The detailed procedure is outside the scope of this book, but at least it is important for the ECP to be aware of the occasional necessity of assessing thickness in ways other than along the Z axis.

Let us get back to equation (4.1) and explore it for some particular cases. For example, if the lens has spherical surfaces, the thickness at a point of coordinates (x, y), which is then placed at a distance $r = \sqrt{x^2 + y^2}$ from the optical center is

$$t(r) = t_0 + R_2 \left[1 - \sqrt{1 - (r/R_2)^2} \right] - R_1 \left[1 - \sqrt{1 - (r/R_1)^2} \right]. \tag{4.2}$$

If the lens has a circular contour centered at $(0, 0)$ with diameter D, the edge thickness, $t_e = t(D/2)$, is

$$t_e = t_0 + R_2 \left[1 - \sqrt{1 - (D/2R_2)^2} \right] - R_1 \left[1 - \sqrt{1 - (D/2R_1)^2} \right], \tag{4.3}$$

which is constant all around the lens perimeter.

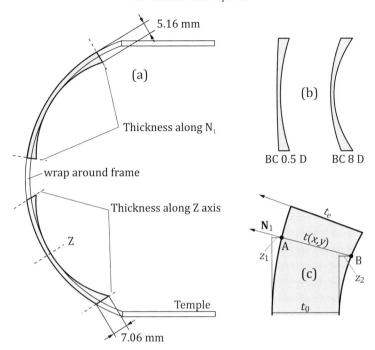

Figure 4.4 Wraparound frame in which a negative lens is mounted. (a) Top view of the wraparound frame. Right and left lenses have been glazed differently, the right one along the direction normal to the front surface and the left one along the Z axis. The edge thickness difference is almost 2 mm, favoring the right lens. (b) Sections of the negative lens made with two different base curves, 0.5 and 8 D. The one on the left is used for the wraparound frame. (c) Thickness along the direction normal to the front surface.

Lens designers would employ the exact expressions for the sagittas of both surfaces to compute lens thickness. However, on many occasions it is more practical to compute thickness using approximate formulas for the sagittas using the Hessian matrix. We will see later that this approximation will allow us to establish a direct relation between thickness and power. Let us recall that as long as z is a standard representation of a surface, then $z(x, y) = \frac{1}{2}\mathbf{r}^T H\mathbf{r}$, where $\mathbf{r}$ is in column form and its transpose in row form. Applying this parabolic approximation to equation (4.1) we get

$$t\,(\mathbf{r}) = t_0 + \frac{1}{2}\mathbf{r}^T\,(\mathbb{H}_2 - \mathbb{H}_1)\,\mathbf{r}, \qquad (4.4)$$

where $\mathbb{H}_1$ and $\mathbb{H}_2$ are the Hessians of the front and back surfaces of the lens, respectively. Let us see a couple of particular cases of the former equation. A spherical lens has Hessian matrices

$$\mathbb{H}_1 = \kappa_1\mathbb{I}, \ \mathbb{H}_2 = \kappa_2\mathbb{I},$$

which lead to the simple expression

$$t(r) = t_0 - \frac{1}{2}(\kappa_1 - \kappa_2) r^2.$$ (4.5)

The approximate edge thickness of a spherical lens with circular contour of diameter D would be $t_e = t_0 - (\kappa_1 - \kappa_2) D^2/8$.

Let us consider now a spherotoric lens, with a front spherical surface having curvature κ_1 and a back toric surface with main curvatures κ_{2B} and κ_{2T} (we shall drop sub-index 2 as we know the toric surface is on the back). Let us see first the case in which the base meridian is aligned with the X axis. The corresponding Hessians are

$$\mathbb{H}_1 = \kappa_1 \mathbb{I} \,, \mathbb{H}_2 = \begin{bmatrix} \kappa_B & 0 \\ 0 & \kappa_T \end{bmatrix},$$

and upon substitution into equation (4.4),

$$t(x, y) = t_0 - \frac{1}{2}(\kappa_1 - \kappa_B) x^2 - \frac{1}{2}(\kappa_1 - \kappa_T) y^2.$$

Finally, let us consider the general case, with the base of the toric surface making an angle θ with the X axis. If we call the curvature difference $\Delta\kappa = \kappa_T - \kappa_B$, the curvature matrix of the back surface is

$$\mathbb{H}_2 = \begin{bmatrix} \kappa_B & 0 \\ 0 & \kappa_B \end{bmatrix} + \begin{bmatrix} \Delta\kappa \sin^2\theta & -\Delta\kappa \sin\theta \cos\theta \\ -\Delta\kappa \sin\theta \cos\theta & \Delta\kappa \cos^2\theta \end{bmatrix},$$

and thickness

$$t(x, y) = t_0 - \frac{1}{2}(\kappa_1 - \kappa_B) r^2 +$$
$$+ \frac{1}{2}\left(x^2 \Delta\kappa \sin^2\theta + y^2 \Delta\kappa \cos^2\theta - 2xy\Delta\kappa \sin\theta \cos\theta\right).$$ (4.6)

Notice that a spherotoric lens with circular shape does not have constant edge thickness along the lens perimeter. This can be easily tested by substituting in the previous equations the coordinates $(D\cos\phi, D\sin\phi)$ that would generate the points on the lens contour as ϕ runs from 0 to 2π. The maximum and minimum values of edge thickness are obtained by taking the derivative of the resulting thickness function $t(\phi)$, equating it to zero, and solving for the angular coordinate ϕ. The reader can check that if the main curvatures κ_1, κ_B, and κ_T are all positive (as they should be in a convex-concave ophthalmic lens) the maximum and minimum values of the edge thickness are

$$t_e(max) = t_0 - \frac{1}{2}(\kappa_1 - \kappa_T)(D/2)^2,$$ (4.7)

$$t_e(min) = t_0 - \frac{1}{2}(\kappa_1 - \kappa_B)(D/2)^2.$$

4.3 Paraxial Optical Properties

In this section we are going to define and study the optical properties of thick lenses within the framework of paraxial approximation. For some of them we will provide little more than a definition. Others, more relevant to ophthalmic lenses, will be treated in more detail.

4.3.1 Optical Axis and Optical Center

Let start by considering lenses whose surfaces have revolution symmetry, for example spheres, conicoids, or generalized conicoids. These surfaces have a well-defined vertex at which the revolution axis crosses the surface perpendicular to it. The vertex is umbilical, that is, all meridians passing through the vertex have identical curvature. The center of curvature of the vertex is also located on the revolution axis. A centered lens should have its two surfaces sharing the same revolution axis. The optical axis of the lens is, precisely, the shared revolution axis. For lenses with spherical surfaces the definition is more relaxed, as any line perpendicular to a spherical surface is a revolution axis. We can see sections of lenses with spherical surfaces and with aspherical surfaces with rotation symmetry in Figure 4.5 (a) and (b). For both types, the line connecting the centers of curvature of the two surfaces is the optical axis. In subplot (c), the back spherical surface has been displaced downward, but the line connecting the two centers of curvature still crosses the two surfaces at a perpendicular angle, and is still the optical axis of the lens. The displacement of the surface produces a tilt of the optical axis and a displacement of the vertexes of both surfaces. The situation is entirely different for aspheric surfaces, as shown in subplot (d). As the back surface is displaced downward, the line connecting the centers of curvature is no longer

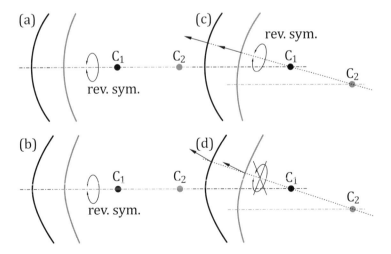

Figure 4.5 Symmetry relations and optical axis in lenses with surfaces having revolution symmetry. (a) Spherical surfaces. (b) Aspherical surfaces, aligned. (c) Spherical surfaces, the back one displaced. Rotation symmetry still exists around the axis joining the two centers of curvature. (d) Aspherical surfaces, the back one displaced. The lens loses both revolution and axial symmetry.

perpendicular to the surfaces at the intersection points (as the normal vectors indicate in the plot). This connecting line is no longer a revolution axis, neither is it an optical axis. Depending on the asphericities and curvatures of the surfaces, it may be possible that an axis crossing perpendicular to both surfaces does not even exist. Similarly, the paraxial cardinal points we are defining next do not exist or only exist in an approximate way. This fact does not imply we cannot make ophthalmic lenses like that. We can, and we can compute local power and prismatic deviation. We can even spot the point that should be aligned with the eye for optimum performance, but the lens may still not have a well-defined optical axis as it is traditionally defined.

Another element that is generally referred to but not precisely defined is the optical center. In many textbooks and even research papers on ophthalmic lenses, the optical center is assumed to be the point at which the optical axis intersects *the lens*. Of course that is a loose definition based on considering the lens a thin element. The property that the optical center is expected to have is that any ray passing through it will not deviate after refraction through the lens. As with the optical axis, a precise definition of the optical center can be made, but not for all types of lenses.

Let us consider a lens with revolution symmetry that must have surfaces with revolution symmetry *sharing* the same revolution axis. This axis, as we have just discussed, will be the optical axis. Two of these lenses are depicted in Figure 4.6. The principal points H and H′ are defined as the conjugate points on the optical axis for which magnification is unity. That means if an object is located in the plane transversal to the optical axis at H, its image will be formed, with the same size and orientation, at the plane transversal to the optical axis at H′. Similarly, nodal points are defined as a pair of conjugate points on the optical axis with the next property: If a ray enters the lens pointing to the nodal point N, the refracted ray exits the lens pointing from the conjugate nodal point N′ and with the same inclination as the incident ray. When the refractive index is the same at both sides of the lens, the principal and nodal points coincide.

Now, is there a single point such that any ray passing through it will not deviate after refraction through the lens? Indeed, this point exists and can be found using any of the next two techniques, both illustrated in Figure 4.6.

1. Let ρ be a ray passing from the nodal point N = H and intersecting the front surface of the lens at C. The refracted ray points from N′ = H′ and intersects the back surface at D. The line connecting C and D will cross the optical axis at O, which is the optical center of the lens. Any ray passing through O will not change its direction after refraction.
2. Let us draw a line from an arbitrary point A on the front surface to its center of curvature C_1. Now let us draw a line parallel to the first one, from the back surface, at B, to its center of curvature C_2. The line connecting points A and B will also cross the optical axis at O, the optical center of the lens.

An example of both constructions is shown in Figure 4.6. In the upper plot, (a), a positive meniscus-shaped lens is shown. For these types of lenses, the first principal

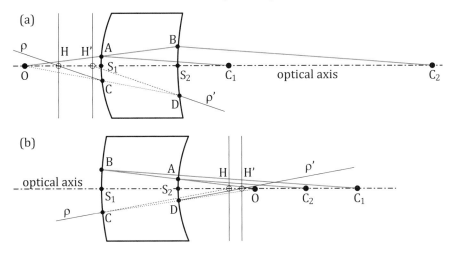

Figure 4.6 Optical axis, optical center, and principal planes of a thick, meniscus-shaped lens. (a) Positive lens, with $P_1 > |P_2|$. (b) Negative lens, with $P_1 < |P_2|$.

point is located before the lens, whereas the second principal point can be before or after the front surface, depending on the amount of lens bending. The optical center is always located before the first principal plane. In subplot (b) a negative lens is depicted, meniscus-shaped as well. For these lenses the second principal plane is located after the lens, whereas the first principal plane can be inside or after the lens, depending, once again, on the amount of bending. Some trigonometry can be used with any of the two rules above to get an expression relating the position of the actual optical center with respect to the vertex of the first surface, S_1, the lens thickness and the refractive power of its surfaces,

$$S_1O = \frac{t_0 P_2}{P_1 + P_2}. \tag{4.8}$$

The optical center is always located after the second principal plane. The reader should take into account that the thickness of the lenses depicted in Figure 4.6 is greatly exaggerated with respect to the real thickness of the typical spectacle lens. The curvature radii used in the figure for both lenses are 50 and 100 mm long (interchanging front and back surfaces for either lens), and for clarity, the thickness has been chosen to be 30 mm. In a real ophthalmic lens the thickness will be a few millimeters at most, the principal planes and optical centers will be much closer to the lens, and the three of them more closely packed. For such a thin lens, we may assume the optical center is on any of the lens surfaces. Though usually no distinctions are made, we would make the best approximation if we locate principal planes and optical center at the front surface of positive lenses, and on the back surface of negative lenses. Bi-convex and bi-concave lenses will have the principal planes within the lens and the optical center between the principal planes, but these lens forms are seldom used in ophthalmic optics. All these assertions can be deduced from equation (4.8). For example, for thin positive lenses t_0 and P_2 are small so $S_1O \ll t_0$ and the optical center is very close to S_1. For negative lenses P_1 is small and $S_1O \approx t_0$, so the optical center is very close to S_2. In general, and as long as $P_1 + P_2 \neq 0$, the smaller the thickness, the closer O gets to S_1, and the closer S_2 gets to S_1 as well. For plano lenses $(P_1 \approx -P_2)$ the optical center is not well defined; the lens does not deviate rays and any point on the optical axis acts as an optical center.

The next question is how do we extend these concepts to lenses without revolution symmetry, for example astigmatic lenses with axial symmetry and even progressive or free-form personalized lenses without any type of symmetry. For astigmatic lenses with axial symmetry and with no tilts and/or displacement between surfaces, that is, the two functions z_1 and z_2 being standard representations, the optical axis is well defined, and, as expected, it coincides with the shared Z axis. However, principal planes and optical center are not well defined in astigmatic lenses. The location of these depends on the lens power and curvatures on each surface. As astigmatic surfaces and lenses have two main meridians with different curvatures and powers, the location of principal points and optical center would be different for each mean meridian, and their properties could only be applied to rays coplanar with the optical axis and the corresponding main meridian. For any other nonprincipal meridian they would not even exist because the optical axis, the meridian, the incident ray, and the refracted ray cannot be coplanar. However, as long as the lens can be considered thin, the differences between the position of principal planes and optical centers of the two main meridians will be negligible, and we can say that an approximate optical center exists at the point the optical axis crosses either the front or the back surfaces, depending on whether the lens is positive or negative.

Finally, progressive and/or personalized lenses will not have any particular symmetry. The optical axis, as traditionally defined for optics with revolution symmetry, does not exist. However, designers usually define a point at which the normals to the two surfaces point in the same direction, and the line crossing the surfaces at this point, perpendicular to them, is the closest concept to an optical axis.

4.3.2 Back Vertex Power

Let us consider a lens with refractive index n, with two arbitrary surfaces whose geometry near the vertex is given by the respective curvature matrices $\mathbb{H}_1$ and $\mathbb{H}_2$. The corresponding refractive powers are $\mathbb{P}_1 = (n-1)\,\mathbb{H}_1$ and $\mathbb{P}_2 = (1-n)\,\mathbb{H}_2$. For the following analysis to be correct, we just need a common Z axis perpendicular to both surfaces. The thickness of the lens along this axis is t_0 and as it will frequently appear divided by the refractive index, we will use the lighter notation $\tau = t_0/n$. Some authors call this quotient *reduced thickness*, but we prefer not to see τ as a new concept, just a simplification of the nomenclature. A planar wavefront defining a beam of rays parallel to the optical axis impinges on the front surface of the lens with null vergence. We will define the back vertex power, $\mathbb{P}_V$, as the vergence of the beam after refraction through the lens, measured at the vertex of the back surface. The computation of back vertex power is straightforward: We just have to refract the incident vergence at the first surface, propagate the refracted vergence to the back surface, and refract again to get the final vergence at the exit of the lens.

Let us call $\mathbb{V}_i$, and $\mathbb{V}'_i$ the incident and refracted vergence at surface i. As the incoming beam is planar, then $\mathbb{V}_1 = \mathbb{O}$. The refraction of the beam through the lens has three steps:

Refraction at the front surface. Applying equation (3.19) to the front surface, $n\mathbb{V}'_1 = \mathbb{P}_1 + \mathbb{V}_1$, taking into account that the incoming beam is planar, then

$$\mathbb{V}'_1 = \mathbb{P}_1/n.$$

Propagation to the back surface. The refracted vergence $\mathbb{V}'_1$ has to be propagated a distance $z = t_0$ up to the back surface, so the incident vergence at the back surface is

$$\mathbb{V}_2 = \left(\mathbb{I} - t\mathbb{V}'_1\right)^{-1} \mathbb{V}'_1 = (\mathbb{I} - \tau\mathbb{P}_1)^{-1} \mathbb{P}_1/n.$$

Refraction at the back surface. Finally, refraction at the back surface requires equation (3.19) again,

$$\mathbb{V}'_2 = \mathbb{P}_2 + n\mathbb{V}_2 = \mathbb{P}_2 + (\mathbb{I} - \tau\mathbb{P}_1)^{-1} \mathbb{P}_1.$$

By definition, this exit vergence, corresponding to a planar incident wavefront, measured at the vertex of the back surface of the lens is the back vertex power, so

$$\mathbb{P}_V = (\mathbb{I} - \tau\mathbb{P}_1)^{-1} \mathbb{P}_1 + \mathbb{P}_2. \tag{4.9}$$

The matrix $(\mathbb{I} - \tau\mathbb{P}_1)^{-1}$ is a generalization of the *shape factor* typically defined for spherical front surfaces as

$$g = \frac{1}{(1 - \tau P_1)}. \tag{4.10}$$

As we allow the front surface to be arbitrary, the shape factor turns into a matrix that we will designate $\mathbb{G}$ and in terms of which the back vertex power reads

$$\mathbb{P}_V = \mathbb{G}\mathbb{P}_1 + \mathbb{P}_2. \tag{4.11}$$

The matrices $\mathbb{G}$ and $\mathbb{P}$ commute (see Appendix 3), so we may either write $\mathbb{G}\mathbb{P}_1$ or $\mathbb{P}_1\mathbb{G}$. When the front surface has revolution symmetry its refractive power is a multiple of the identity matrix $\mathbb{P}_1 = P_1\mathbb{I}$. In this case, the matrix shape factor is also a multiple of the identity matrix, the factor being the classical (and scalar) quantity. In this case, the back vertex power can be written $\mathbb{P}_V = gP_1\mathbb{I} + \mathbb{P}_2$.

With independence of whether the scalar or matrix form is used, back vertex power is the sum of the refractive power of the back surface $\mathbb{P}_2$, and the "modified version" of $\mathbb{P}_1$, which results from multiplying it by the shape factor $\mathbb{G}$.

From the physical point of view, the back vertex power is just a vergence, that is, a curvature, and so any of the descriptions we have presented for curvatures can be applied to it. We can define the mean (back vertex) power, the "Gaussian power" (the Gaussian curvature of the wavefront refracted through the lens when the incident wavefront is plane), and of course, the main powers. The matrix $\mathbb{P}_V$ describes curvature and its components as the second partial derivatives of the function that would describe the output wavefront, but as with vergence matrices, we will use the shorter single-subscript notation

$$\mathbb{P} = \begin{bmatrix} P_x & P_t \\ P_t & P_y \end{bmatrix}, \tag{4.12}$$

where P_x and P_y are the powers along the X and Y axis, respectively, and P_t is the torsional power component. For the mean and Gaussian powers we will use the usual letters, $H = (P_x + P_y)/2$ and $K = P_x P_y - P_t^2$. The main powers, as always, are given by

$$P_- = H - \sqrt{H^2 - K},$$
$$P_+ = H + \sqrt{H^2 - K}, \tag{4.13}$$

P_- being the smallest (in the algebraic sense, not necessarily the one with lower absolute value) and P_+ the bigger principal power. In turn, the orientation of the meridian with power P_- is

$$\theta_- = \arctan\left(\frac{P_- - P_x}{P_t}\right), \tag{4.14}$$

while the other meridian is oriented along the perpendicular direction, $\theta_+ = \theta_- \pm 90°$.

We may have a better grasp of the interpretation of the components of the back vertex power matrix if we study the propagation of rays refracted by the lens. For this we use equation (3.10) that we derived in the previous chapter. If a ray parallel to the Z axis refracts through the lens and exits at a point $\mathbf{r}_0$ on the back surface, its transverse coordinates $\mathbf{r}$ any distance z down the Z axis will be given by $\mathbf{r} = (\mathbb{I} - z\mathbb{P}_V)\mathbf{r}_0$. Expanding this equation into components,

$$x = x_0 (1 - zP_x) - y_0 z P_t,$$
$$y = -x_0 z P_t + y_0 (1 - zP_y).$$

Now let us assume the incident ray is contained on the plane XZ, so that it incides on the X axis, with coordinates $(x_0, 0)$. Its transverse coordinates after propagation through the same distance z will be

$$x = x_0 (1 - zP_x),$$
$$y = -x_0 z P_t.$$

If we look for the distance at which the ray intersects the Z axis, we should equate x and y to zero. The first equation yields $z = 1/P_x$, but the y coordinate cannot be zero unless $P_t = 0$. If the torsion is not null, the ray exits the lens making some angle with the plane XZ, and it will never intersect the Z axis. When the z coordinate is $1/P_x$ the ray is crossing the axis (its x coordinate will be zero), but with a y coordinate $y = -x_0 P_t / P_x$. This situation is shown in Figure 4.7. A lens with positive main powers is sketched with the vertex of its back surface located at $z = 0$. Rays are refracted through the lens and propagated up

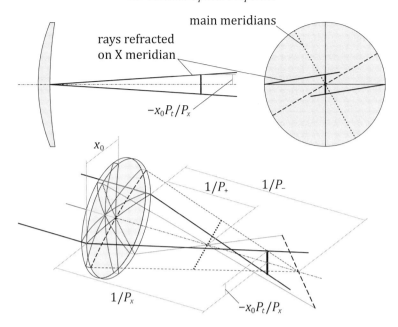

Figure 4.7 Sketch of a lens and the refraction of three pairs of rays, one pair on the X axis (solid thicker lines) and two pairs on the main meridians (dotted and dashed lines). Top-left is a side view of the rays propagating after refraction; top right is a frontal view; and the plot at the bottom is a projected view.

to the second Sturm's focus. Top-left is a lateral view of the sketch of rays, top-right is a frontal view, and at the bottom a projected view. Pairs of rays refract at opposite sites on the lens circumference, one pair for the X axis and two more for the principal directions. The six rays fall upon the lens along the direction of the Z axis, so when refracted, they are perpendicular to the wavefront whose vergence is the back vertex power of the lens. Rays refracted at the main meridians are drawn with dotted and dashed lines, as are the focal lines they make, while rays refracted at the X axis are represented with solid lines. The vertical thick segment joining these rays is located at $z = 1/P_x$ and it is vertical because at that axial coordinate $x = 0$ for the rays refracted at the X axis. In the projected view we can see the plane XZ, one of the solid rays above it, the other one below it. It is important to notice that the thick vertical segment at $z = 1/P_x$ is not a focal line; at that point the section of the whole beam would be elliptical.

We can use this information to provide an insightful interpretation of the torsion component of the power. The angle formed by either of the rays refracted on the X axis with the XZ plane is $\alpha = \arctan(y/z)$, where y is its vertical coordinate at a given propagation distance z. If we choose $z = 1/P_x$, then $y = -x_0 P_t/P_x$ and so

$$\alpha \simeq \tan\alpha = -x_0 P_t.$$

In other words, $-x_0 P_t$ determines the deviation angle outside of the XZ plane for a ray initially embedded in the XZ plane that incides on the lens at a distance x_0 from its center. If P_t is zero, either the X axis is a principal meridian or the lens has revolution symmetry, and the ray inciding from the XZ plane stays on the same plane. Exactly the same reasoning can be applied to rays inciding on the lens from the YZ plane.

Example 12 An aspheric lens has a front conicoid surface, with a refractive power 4 D and asphericity coefficient $p = -6$. The back surface is toric, with the base meridian oriented at $30°$ and a curvature radius of $R_B = 250$ mm. The radius of the cross curve is $R_T = 200$ mm. The center thickness of the lens is $t_0 = 3.5$ mm, and its refractive index $n = 1.5$. Determine the back vertex power of the lens.

Let us determine first the refractive powers of the lens surfaces. The front face has revolution symmetry and, despite being aspheric, its refractive power is independent of the surface asphericity, so $\mathbb{P}_1 = \begin{bmatrix} 4 & 0 \\ 0 & 4 \end{bmatrix}$ D. The curvatures of the toric surface are $\kappa_B = 1/0.25 = 4$ D and $\kappa_T = 1/0.2 = 5$ D, so $\Delta\kappa = \kappa_T - \kappa_B = 1$ D. The refractive power of the back surface is then

$$\mathbb{P}_2 = (1-n) \begin{bmatrix} 4 + \sin^2 30 & -\sin 30 \cos 30 \\ -\sin 30 \cos 30 & 4 + \cos^2 30 \end{bmatrix} = \begin{bmatrix} -2.125 & 0.2165 \\ 0.2165 & -2.375 \end{bmatrix} \text{D}.$$

The scalar shape factor is

$$g = 1/(1 - 0.0035 \times 4/1.5) = 1.0094,$$

and the back vertex power of the lens is

$$\mathbb{P}_V = g\mathbb{P}_1 \mathbb{I} + \mathbb{P}_2 = 1.0094 \begin{bmatrix} 4 & 0 \\ 0 & 4 \end{bmatrix} + \begin{bmatrix} -2.125 & 0.2165 \\ 0.2165 & -2.375 \end{bmatrix} = \begin{bmatrix} 1.913 & 0.217 \\ 0.217 & 1.663 \end{bmatrix} \text{D}.$$

From the power matrix, it is possible to get the main powers, being, in this case, $P_- = 1.64$ D, and $P_+ = 1.93$ D, with P_- oriented at $\theta_- = 30°$.

4.3.3 Thin Lens Approximation

When the thickness of a lens is small enough that the components of $\tau\mathbb{P}_1$ are negligible against 1, the shape factor becomes very close to the identity matrix, and the approximation $\mathbb{P}_V \simeq \mathbb{P}_1 + \mathbb{P}_2$ can be made. In fact, we *define* the thin lens power of a lens as the sum of its refractive powers

$$\mathbb{P}_{TL} = \mathbb{P}_1 + \mathbb{P}_2. \tag{4.15}$$

As $\mathbb{P}_1$ and $\mathbb{P}_2$ are symmetric matrices, so is $\mathbb{P}_{TL}$, henceforth it can be considered the Hessian of some wavefront. For lenses with revolution symmetry, refractive powers, shape factor, and back vertex power are scalar magnitudes. The front surface of an ophthalmic lens is convex almost without exception, so $P_1 > 1$. This implies $g > 1$, so the back vertex power is greater than the thin lens one, $P_V > P_{TL}$.

Another way to state the first phrase of this paragraph is that when the thickness is small enough, the back vertex power and the thin lens power get very close to each other, and the latter is a good approximation to the former. The mathematical condition for this approximation to be good is that $tP_{1x}, tP_{1y}, tP_{1t} \ll 1$. From an optical point of view, a lens is thin if we may neglect the change of vergence that takes place when the wavefront propagates from the front surface of the lens to the back one. This happens when the beam refracted at the first surface of the lens is very plane (which happens when the elements of $\mathbb{P}_1$ are small), when the propagation distance (the center thickness of the lens) is small, or when both conditions are met.

In Figure 4.8, we show the values of the (scalar) shape factor and the difference between back vertex and thin lens power as a function of the center thickness and the refractive power of the front surface, assuming it has revolution symmetry. For virtually all ophthalmic lenses, the scalar shape factor would be in the range $[1, 1.1]$. With respect to the error that we would make if thin lens power is used as a substitute for back vertex power, this would stay below 0.2 D for most ophthalmic lenses. Negative ophthalmic lenses have a center thickness in the range $[1.5, 2]$ mm. Less than 1.5 mm would render the impact resistance of the lens below the typically required levels, and above 2 mm the lens would be unnecessarily thick and heavy. Also, negative lenses are made with relatively flat front

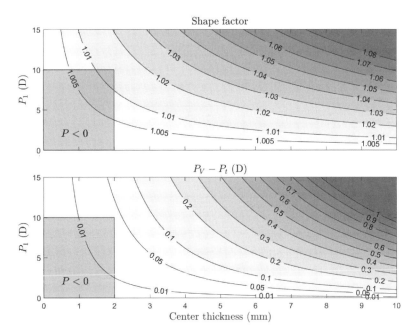

Figure 4.8 Thin lens approximation. In the top plot the shape factor for spherical lenses is plotted as a function of center thickness and refractive power of the front surface. The bottom plot shows the error made when using thin lens approximation as a function of the same parameters. Most ophthalmic negative lenses would lie on the gray rectangle at the bottom left of both figures, as their thickness is very rarely greater than 2 mm, and negative lenses employ relatively flat front surfaces.

surfaces, so P_1 (or any of the components of $\mathbb{P}_1$, if the surface is astigmatic) will be small, typically below 4 D and only above that for lenses intended for special frames, as we saw in Section 4.2.3. The gray rectangle at the bottom-left of the two plots in Figure 4.8 virtually contains all the negative lenses that could potentially be used for glasses. We see then that the thin lens approximation is indeed good for negative lenses, the shape factor being smaller than 1.01 and the error made when using thin lens power instead of the back vertex power typically smaller than 0.01 D. If we take into account that ISO and ANSI standards establish a tolerance for manufacturing power errors of 0.12 D for the spherical component of low- to medium-powered lenses, then we see the manufacturer could well use thin lens approximation to manufacture negative lenses, at least to give the right paraxial power at the optical or geometrical axis. For positive lenses the picture is different. Low-powered lenses will have base curves in the range of 4 to 8 D with center thickness between 3 and 6 mm.[2] This means the error of the thin lens approximation will be, for many positive lenses, 0.1 D or smaller, but it can grow above 0.3 D for lenses with base 8 D and center thickness above 6 mm. High-plus lenses for high hyperopia or aphakia (see following chapters) will require base curves above 10 D, even above 15 D, with center thickness that can go near to 10 mm. These lenses cannot be considered thin any longer and the error of thin lens approximation may grow over 1 D, which amounts to around 10% of lens power. Manufacturers cannot compute their plus lenses using the thin lens approximation, but still it is very useful for academic purposes.

When the product $t\mathbb{P}_1$ is small, but not completely negligible, we may approximate the shape factor as follows:

$$\mathbb{G} \approx \mathbb{I} + \tau \mathbb{P}_1,$$

so the back vertex power would be, in this case,

$$\mathbb{P} \approx (\mathbb{I} + \tau \mathbb{P}_1)\,\mathbb{P}_1 + \mathbb{P}_2$$
$$= \mathbb{P}_1 + \mathbb{P}_2 + \tau \mathbb{P}_1^2 \equiv \mathbb{P}_{TL} + \tau \mathbb{P}_1^2. \tag{4.16}$$

Finally, we will say that thin lenses are used in optics not only as an approximation to back vertex power, but as ideal components that can help us to analyze individual lenses and more complex systems, such as the eye, without necessarily introducing added approximations. For example, any thick lens can be replaced by a thin lens located at the vertex of the back surface of the former, the latter having a thin lens power identical to the back vertex power of the thick lens. The two lenses would behave just the same when a plane wavefront incides onto them; they would produce the same output vergence. Their performance would not be identical for a non-plane incident wavefront, and depending on how they are used, they could have slightly different magnification. In the next section we will use this approach to split the power of astigmatic lenses into spherical and cylindrical components.

---

[2] These ranges are typical and by no means fixed. There is no way to be sure about the base curve a manufacturer will decide to use, and the center thickness will depend, to a great extent, on the diameter of the lens to be made.

Example 13 Let us consider two ophthalmic lenses with refractive index $n = 1.5$. Both of them have the same back toric surface with base and cross radius $R_B = 200$ mm and $R_T = 100$ mm, oriented at $90°$ and $180°$, respectively. The first lens has a front spherical surface with radius $R_1 = 250$ mm and a center thickness of 2 mm. The second lens also has a spherical front surface with radius $R_1 = 62.5$ mm and a thickness of 8 mm. Determine the back vertex and thin lens power matrices for both lenses, along with the difference (expressed as a percentage) between the two types of power.

The common back surface has refractive power

$$\mathbb{P}_2 = (1-n) \begin{bmatrix} 1/R_T & 0 \\ 0 & 1/R_B \end{bmatrix} = \begin{bmatrix} -5 & 0 \\ 0 & -2.5 \end{bmatrix} D.$$

For the lens with the flatter front surface, $P_1 = (n-1)/R_1 = 2$ D, and $g = 1/(1 - \tau P_1) = 1.0026$. Its back vertex power is

$$\mathbb{P}_V = \begin{bmatrix} 1.0026 \times 2 & 0 \\ 0 & 1.0026 \times 2 \end{bmatrix} + \begin{bmatrix} -5 & 0 \\ 0 & -2.5 \end{bmatrix} = \begin{bmatrix} -2.995 & 0 \\ 0 & -0.495 \end{bmatrix} D,$$

while the thin lens power is

$$\mathbb{P}_{TL} = \begin{bmatrix} 2 & 0 \\ 0 & 2 \end{bmatrix} + \begin{bmatrix} -5 & 0 \\ 0 & -2.5 \end{bmatrix} = \begin{bmatrix} -3 & 0 \\ 0 & -0.5 \end{bmatrix} D.$$

The absolute error is the same for both meridians (as it should be, the first surface being spherical and the shape factor scalar): 0.05 D. The error percentages are $\Delta P_B = 1.01\%$ and $\Delta P_T = 0.18\%$.

For the lens with the steeper front curve we have $P_1 = 8$ D and $g = 1.044$, so

$$\mathbb{P}_V = \begin{bmatrix} 1.044 \times 8 & 0 \\ 0 & 1.044 \times 8 \end{bmatrix} + \begin{bmatrix} -5 & 0 \\ 0 & -2.5 \end{bmatrix} = \begin{bmatrix} 3.356 & 0 \\ 0 & 5.856 \end{bmatrix} D,$$

and

$$\mathbb{P}_{TL} = \begin{bmatrix} 8 & 0 \\ 0 & 8 \end{bmatrix} + \begin{bmatrix} -5 & 0 \\ 0 & -2.5 \end{bmatrix} = \begin{bmatrix} 3 & 0 \\ 0 & 5.5 \end{bmatrix} D.$$

Now, the absolute error is 0.36 D in both meridians. The relative errors are now $\Delta P_T = 11.9\%$ and $\Delta P_B = 6.5\%$, respectively.

4.3.4 Forms of Astigmatic Power: Spherocylindrical and Crossed-Cylinder

Regardless of the actual geometry of a lens, its back vertex power is nothing but curvature: that of the refracted wavefront corresponding to a plane incident one. As such, all the relationships among principal curvatures and meridians, and the components of the curvature matrix that we presented in Section 2.3.2 hold for power matrices. In particular, the three ways we split the Hessian of a toric surface, (2.26), (2.28), and (2.29), can be used for power matrices as well. Let us start by collecting here these three different ways in which the main powers and meridians relate with the components of the power matrix, using the notation $\Delta P = P_+ - P_-$ to make the many formulas more readable. The power along the X axis is,

$$
\begin{aligned}
P_x &= P_- + \Delta P \sin^2 \theta_- \\
&= P_+ - \Delta P \sin^2 \theta_- \\
&= P_- \sin^2 \theta_+ + P_+ \sin^2 \theta_-.
\end{aligned} \tag{4.17}
$$

For the power along the Y axis,

$$
\begin{aligned}
P_y &= P_- + \Delta P \cos^2 \theta_- \\
&= P_+ - \Delta P \cos^2 \theta_- \\
&= P_- \cos^2 \theta_+ + P_+ \cos^2 \theta_-.
\end{aligned} \tag{4.18}
$$

And finally, for the torsional component of the power,

$$
\begin{aligned}
P_t &= -\Delta P \sin \theta_- \cos \theta_- \\
&= \Delta P \sin \theta_+ \cos \theta_+ \\
&= -\Delta P \sin \theta_+ \cos \theta_+ - \Delta P \sin \theta_- \cos \theta_-.
\end{aligned} \tag{4.19}
$$

Now, each of the three lines on each equation lead to a different splitting of the back vertex power.

Spherical and positive cylindrical components. For this split, let us rename P_- as S, the positive difference as C, and θ_- as α. Then, the power matrix can be rewritten as

$$
\mathbb{P}_V = \begin{pmatrix} S & 0 \\ 0 & S \end{pmatrix} + \begin{pmatrix} C \sin^2 \alpha & -C \sin \alpha \cos \alpha \\ -C \sin \alpha \cos \alpha & C \cos \alpha \end{pmatrix}, \tag{4.20}
$$

the first matrix corresponding to a spherical vergence of curvature S and the second to a cylindrical one, with maximum curvature C and its axis lying at direction α. The meaning of this split is that our original lens is equivalent to a thin spherical lens with power $S = P_-$ superimposed with a thin cylindrical lens with power $C = \Delta P$ and axis oriented at θ_-. The two lenses are thin and located at the vertex of the back surface of the actual thick lens. A plane wavefront refracting through our lens or through this combination of thin lenses will get exactly the same refracted vergence $\mathbb{P}_V$. We will use the notation $[S, C \times \alpha]$ to collect the three new parameters, which will be named *spherocylindrical representation* of the power $\mathbb{P}_V$ *with positive cylinder*.

Spherical and negative cylindrical components. Now we will use the second line of equations (4.17) to (4.19) and the renaming $S' = P_+$, $C' = -\Delta P$ and $\alpha' = \theta_+$ to rewrite

$$
\mathbb{P}_V = \begin{pmatrix} S' & 0 \\ 0 & S' \end{pmatrix} + \begin{pmatrix} C' \sin^2 \alpha' & -C' \sin \alpha' \cos \alpha' \\ -C' \sin \alpha' \cos \alpha' & C' \cos \alpha' \end{pmatrix}. \tag{4.21}
$$

The interpretation is the same as before; the only differences are that we assign the spherical power to the bigger main power of the lens, the cylinder changes sign, and the cylinder axis turns out to be θ_+. The three new components are written $[S', C' \times \alpha']$, which are known as the *spherocylindrical representation* of the power $\mathbb{P}_V$ *with negative cylinder*.

Two cylindrical components. Finally, we use the last line of equations (4.17) to (4.19) to get the final way to write the back vertex power,

$$\mathbb{P}_V = \begin{pmatrix} P_- \sin^2 \theta_+ & -P_- \sin \theta_+ \cos \theta_+ \\ -P_- \sin \theta_+ \cos \theta_+ & P_- \cos^2 \theta_+ \end{pmatrix}$$
$$+ \begin{pmatrix} P_+ \sin^2 \theta_- & -P_+ \sin \theta_- \cos \theta_- \\ -P_+ \sin \theta_- \cos \theta_- & P_+ \cos^2 \theta_- \end{pmatrix}. \quad (4.22)$$

Each of the matrices corresponds to the vergence of a cylindrical wavefront, with maximum curvatures P_- and P_+ and respective axes θ_+ and θ_-. The lens effect can then be split as the effect of two crossed thin cylindrical lenses (with perpendicular axes), which we will represent with the notation $(P_- \times \theta_+)(P_+ \times \theta_-)$ and call *crossed-cylinder representation* of the power $\mathbb{P}_V$. Let us recall that the curvature of a pure cylinder along its axis is zero, so the vergence $(C \times \alpha)$ would have curvature C along $\alpha + \pi/2$ and zero along α. This is why the angles are swapped when passing from main meridians to the crossed-cylinder form.

The interpretation of both the spherocylindrical and crossed-cylinder representations in terms of thin lenses is not an approximation to vertex power. The writing of the power as a matrix or as any of the representations shown above does not change the fact that all are back frontal power. We can change from one to another without any approximation. They are all different forms for expressing back vertex power. If we had started using $\mathbb{P}_{TL}$ instead of $\mathbb{P}_V$ to generate the spherocylindrical and crossed-cylinder representations, then they would have been approximate back vertex powers. In ophthalmic optics, the transformation from one form of representing power to another is called *transposition*.

It is quite obvious that the spherocylindrical and crossed-cylinder representations are different ways of arranging the information provided by the principal powers and meridians. The four forms require knowledge of the orientation of the main meridians and they just differ on the naming of the same concepts (although there are interesting practical applications that are sensitive to the chosen representation of astigmatic power.) From the conceptual point of view, the reason we can rearrange power in so many different ways is that, within the paraxial approximation, curvature is additive. The focusing effect of our thick lens with main powers P_- and P_+ at respective angles θ_- and θ_+ is exactly the same as the effect of a thin spherical lens with power P_- superimposed with a thin cylindrical lens with power ΔP and axis θ_-. The first will turn the initial plane wavefront into a spherical one with curvature P_-, and the second will further change this curvature, adding ΔP diopters of curvature along the direction θ_+ and leaving the perpendicular direction, its axis, untouched. In this way, the two idealized components perform the same function as the original lens (at least in terms of focusing). However, we have to be cautious about curvature additivity. It is true that curvature is additive *on the same meridian*, but we simply cannot directly add curvatures from different meridians. If we had two superimposed thin cylindrical lenses whose axes are neither parallel nor perpendicular, we cannot simply add their main curvatures. The combined effect will have a spherical component and a cylindrical component, and the axis of the latter will not match with any of the axes of the original cylinders.

One of the advantage of the matrix representation of curvature (and power) is that its components are referred to a fixed XY reference system, and this gives the representation real additivity. If two thin lenses are represented by their matrix powers, their joint effect would just be the sum of the two matrices, regardless of the orientation of the main meridians for each lens. And if the lenses are separated by a nonnegligible distance, we can still handle it by propagating the astigmatic vergences between them.

Why should we worry about keeping five different, yet equivalent, power representations? The answer to this question is practical. From the point of view of a lens designer, the matrix representation will be, most of the time, the preferred one. For the lens manufacturer, probably the main powers and meridians would be the preferred choice, with occasional use of the spherocylindrical and matrix representations. For optometrists and other ECPs, the spherocylindrical (and occasionally the crossed-cylinder) forms will be more favored, one of the reasons being that many optometric tests are based upon the split of astigmatic power into spherical and cylindrical power.

All the expressions needed to transpose between different representations of astigmatic power have already been written down. However, the reader may find handy the summary presented in Table 4.2 to quickly swap between different forms.

Example 14 The back vertex power of a lens is $\mathbb{P}_V = \begin{bmatrix} 2.75 & -1.299 \\ -1.299 & 4.25 \end{bmatrix}$ D. Determine the spherocylindrical and crossed-cylinder forms of the lens power.

We first need the main power and meridians. Mean and Gaussian powers are $H = 3.5$ D and $K = 10\,\mathrm{D}^2$. The main powers are then

$$P_- = 3.5 - \sqrt{3.5^2 - 10} = 2\,\mathrm{D},$$

$$P_+ = 3.5 + \sqrt{3.5^2 - 10} = 5\,\mathrm{D},$$

$$\theta_- = \arctan\left(\frac{2 - 2.75}{-1.299}\right) = 30°.$$

To obtain the spherocylindrical form with positive cylinders, we set the smaller main power as sphere, $S = 3.5$ D, the difference between the bigger and the smaller main powers as cylinder, $C = 3$ D, and its orientation axis will be that of the main meridian with smallest power (the one set as sphere). Then, $[2, 3 \times 30°]$ is the spherocylindrical form with positive cylinder. To get the other spherocylindrical form we set the bigger main power as sphere, $S' = 5$ D, the difference between the smaller and the bigger as cylinder, $C = -3$ D, and

Table 4.2 *Table of conversion between the different forms of expressing the astigmatic power.*

	Matrix	Main powers	Spherocyl. $C > 0$	Spherocyl. $C' < 0$	crossed-cylinder
$H = (P_x + P_y)/2$ $K = P_x P_y - P_t^2$ $\begin{bmatrix} P_x & P_t \\ P_t & P_y \end{bmatrix}$		$P_- = H - \sqrt{H^2 - K}$ $P_+ = H + \sqrt{H^2 - K}$ $\theta_- = \arctan\left(\frac{P - P_x}{P_t}\right)$	$S = H - \sqrt{H^2 - K}$ $C = 2\sqrt{H^2 - K}$ $\alpha = \arctan\left(\frac{S - P_x}{P_t}\right)$	$S' = H + \sqrt{H^2 - K}$ $C' = -2\sqrt{H^2 - K}$ $\alpha' = \arctan\left(\frac{S' - P_x}{P_t}\right)$	$(P_- \times \theta_+)$ $(P_+ \times \theta_-)$ $P_- = H - \sqrt{H^2 - K}$ $P_+ = H + \sqrt{H^2 - K}$ $\theta_- = \arctan\left(\frac{P - P_x}{P_t}\right)$
(P_-, P_+, θ_-) $\Delta P = P_+ + P_-$	$P_x = P_- + \Delta P \sin^2 \theta_-$ $P_y = P_- + \Delta P \cos^2 \theta_-$ $P_t = -\Delta P \sin\theta_- \cos\theta_-$		$S = P_-$ $C = P_+ - P_-$ $\alpha = \theta_-$	$S' = P_+$ $C' = P_+ - P_-$ $\alpha' = \theta_- \pm \pi/2$	
$[S, C \times \alpha]$	$P_x = S + C \sin^2 \alpha$ $P_y = S + C \cos^2 \alpha$ $P_t = -C \sin\alpha \cos\alpha$	$P_- = S$ $P_+ = S + C$ $\theta_- = \alpha$		$S' = S + C$ $C' = -C$ $\alpha' = \alpha \pm \pi/2$	$P_- = S$ $P_+ = S + C$ $\theta_- = \alpha$
$[S', C' \times \alpha']$	$P_x = S' + C' \sin^2 \alpha'$ $P_y = S' + C' \cos^2 \alpha'$ $P_t = -C' \sin\alpha' \cos\alpha'$	$P_- = S' + C'$ $P_+ = S'$ $\theta_- = \alpha' \pm \pi/2$	$S = S' + C'$ $C = -C'$ $\alpha = \alpha' \pm \pi/2$		$P_- = S' + C'$ $P_+ = S'$ $\theta_+ = \alpha'$
$(P_- \times \theta_+)$ $(P_+ \times \theta_-)$	$P_x = P_- \sin^2 \theta_+ + P_+ \sin^2 \theta_-$ $P_y = P_- \cos^2 \theta_+ + P_+ \cos^2 \theta_-$ $P_t = -P_+ \sin\theta_- \cos\theta_- - P_- \sin\theta_+ \cos\theta_+$		$S = P_-$ $C = \Delta P$ $\alpha = \theta_-$	$S' = P_+$ $C' = -\Delta P$ $\alpha' = \theta_+$	

the orientation of the meridian with the bigger power as the cylinder axis, that is [5, −3 × 120°]. Finally, the crossed-cylinder form uses the main powers, but the axes orientations are swapped with respect to the main meridians, (2 × 120°)(5 × 30°).

4.3.5 *Power Referred to Principal Planes*

Let us recall that when a lens has revolution symmetry, it is possible to define the principal planes and the nodal points of the lens. The principal planes are traversed conjugated with a magnification equal to one. This means that if we could place an object at the object (or front) principal plane, designated usually by H, its image would be formed at the image (or back) principal plane, H', and this image would have the same size and orientation as those of the object. In geometrical optics, the image focal distance is the distance between the image principal plane and the image focus. Its inverse is the power referred to principal planes, or equivalent power, that we will call P_e. The relation between refractive powers, thickness, and equivalent power is known after Gullstrand and can be retrieved from any basic textbook in geometrical optics:

$$P_E = P_1 + P_2 - \tau P_1 P_2, \tag{4.23}$$

where we are using the same notation as in previous sections. Similarly, we can find the formulas giving the location of the principal points with respect to vertexes of the lens, S_1 and S_2,

$$\overline{S_2 H'} = -\tau \frac{P_1}{P_E},$$
$$\overline{S_1 H} = \tau \frac{P_2}{P_E}. \tag{4.24}$$

We may wonder whether the concept of principal planes could be extrapolated to astigmatic systems. We will see next that a generalization to equation (4.23) can be obtained through matrix optics, but in general it is not possible to define the principal planes for astigmatic systems.

Let us consider now a paraxial ray tracing through a lens with revolution symmetry, as shown in Figure 4.9. An incident ray refracted at point A on the front surface, propagates to point B on the back surface, where it is refracted again. The inclination angles (over the optical axis) of the incident and refracted rays are σ and σ', respectively. The height of the incident ray over the optical axis at point A is u, while the incident height measured at the object principal plane is y. To proceed with the ray tracing we need an auxiliary ray, parallel to the former one, passing through the object focus. According to the laws of geometrical optics, as both incident rays are parallel, they will intercept the back focal plane at the same point. Moreover, the auxiliary ray will exit the lens parallel to the optical axis, as it goes through the object focus.

From the figure, and considering the paraxial approximation is effective, we get $\sigma = h/f$ and $\sigma' = -(y - h)/f'$, where f and f' are respectively the object and image

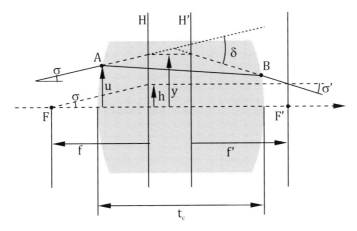

Figure 4.9 Ray tracing through a lens with symmetry of revolution for which the principal planes are well defined.

focal distances measured from the principal planes. If the lens is immersed in air, $f = -f'$, and we obtain

$$\delta = \sigma' - \sigma = -\frac{y}{f'} = -P_E y. \tag{4.25}$$

That is, the ray deviation is proportional to the equivalent power and the ray height between principal planes. If we relate the ray deviation with the height of the incident ray at the front surface, u, the result is

$$\delta = \sigma' - \sigma = -P_E u - \tau \sigma P_2. \tag{4.26}$$

Comparing the two previous equations, we deduce that when the height of the incident ray is given at the principal plane, the ray deviation does not depend on the angle of the incident ray, while if the height of the incident ray is given at the front surface, the ray deviation depends on τ, σ, and P_2. Of course, if the lens is thin enough, $\tau \sigma P_2$ will be negligible and $y \simeq u$, so $\delta \simeq -P_E u$.

Let us generalize this analysis to lenses with astigmatic surfaces. A beam of vergence $\mathbb{V}_1$ incides over the lens. Any ray of this beam will be perpendicular to the wavefront defined by $\mathbb{V}_1$, so the direction vector of a ray that incides over the front surface at a point $\mathbf{r} = (x, y)$ is given by $\mathbf{k} = -\mathbb{V}_1 \mathbf{r}$. The vergence of the refracted beam is $\mathbb{V}_1' = (\mathbb{P}_1 + \mathbb{V}_1)/n$, so the direction of the refracted ray is given by $\mathbf{k}' = -\mathbb{V}_1' \mathbf{r}$. This ray propagates to the second surface, intercepting it at the point with coordinates $\mathbf{r}' = (\mathbb{I} - t\mathbb{V}_1') \mathbf{r}$, as seen in the last chapter. Finally, the direction vector of the ray refracted through the second surface at point $\mathbf{r}'$ is given by $\mathbf{k}_2' = -\mathbb{V}_2' \mathbf{r}'$, so we need to compute the vergence of the output beam at the second surface. The procedure is identical to the one we used to derive the back vertex power, except for $\mathbb{V}_1$ not being zero now. The vergence of the wavefront reaching the back surface is obtained by propagation of $\mathbb{V}_1'$,

$$V_2 = \left(\mathbb{I} - t\mathbb{V}'_1\right)^{-1} \mathbb{V}'_1 = \frac{1}{n} \left[\mathbb{I} - \tau\left(\mathbb{P}_1 + \mathbb{V}_1\right)\right]^{-1} \left(\mathbb{P}_1 + \mathbb{V}_1\right),$$

and $\mathbb{V}'_2$ is obtained by refraction,

$$\mathbb{V}'_2 = \mathbb{P}_2 + n\mathbb{V}_2 = \left[\mathbb{I} - \tau\left(\mathbb{P}_1 + \mathbb{V}_1\right)\right]^{-1} \left(\mathbb{P}_1 + \mathbb{V}_1\right) + \mathbb{P}_2.$$

By substituting the corresponding vergences in the formulas for the direction vectors we find the following expression

$$
\begin{aligned}
\mathbf{k}'_2 &= -\left\{\left[\mathbb{I} - \tau\left(\mathbb{P}_1 + \mathbb{V}_1\right)\right]^{-1} \left(\mathbb{P}_1 + \mathbb{V}_1\right) + \mathbb{P}_2\right\} \left[\mathbb{I} - \tau\left(\mathbb{P}_1 + \mathbb{V}_1\right)\right] \mathbf{r} \\
&= -\left\{\mathbb{P}_2 \left[\mathbb{I} - \tau\left(\mathbb{P}_1 + \mathbb{V}_1\right)\right] + \left(\mathbb{P}_1 + \mathbb{V}_1\right)\right\} \mathbf{r} \\
&= -\left(\mathbb{P}_1 + \mathbb{P}_2 - \tau\mathbb{P}_2\mathbb{P}_1\right) \mathbf{r} + \tau\mathbb{P}_2\mathbb{V}_1\mathbf{r} - \mathbb{V}_1\mathbf{r}.
\end{aligned}
$$

The ray deviation is now a vector magnitude (as there can be horizontal and vertical deviations) given by the difference between the direction vectors $\mathbf{k}'_2$ and $\mathbf{k}_1$. If we define the matrix equivalent power as $\mathbb{P}_E = \mathbb{P}_1 + \mathbb{P}_2 - \tau\mathbb{P}_2\mathbb{P}_1$, then

$$\delta = \mathbf{k}'_2 - \mathbf{k}_1 = -\mathbb{P}_E\mathbf{r} - \tau\mathbb{P}_2\mathbf{k}_1, \tag{4.27}$$

which is formally identical to the scalar equation (4.26), but valid for lenses with arbitrary astigmatic surfaces. If the incident ray is parallel to the optical axis, $\mathbf{k}_1 = 0$ and $\delta = -\mathbb{P}_E\mathbf{r}$.

Similar considerations that applied for equation (4.26) do apply now for the general case described by equation (4.27):

- A direct relationship between equivalent power and back vertex power that the reader can easily check is $\mathbb{P}_V = \mathbb{P}_E\mathbb{G}$. The order in the matrix multiplication has to be observed, otherwise the result would not be correct.
- In general, the refractive powers $\mathbb{P}_1$ and $\mathbb{P}_2$ do not necessarily commute, so we have to be careful with the order in which we make the matrix product. $\mathbb{P}_2$ must left-multiply $\mathbb{P}_1$ always.
- The refractive powers commute in two cases: (1) when at least one of the surfaces has revolution symmetry, and its corresponding matrix refractive power is a scalar times the 2×2 identity matrix, and (2) when both surfaces are astigmatic having parallel main meridians.
- When $\mathbb{P}_1$ and $\mathbb{P}_2$ do not commute, $\mathbb{P}_E$ is not a symmetric matrix, and it cannot be interpreted as the vergence of a given wavefront, as wavefronts are always described by symmetrical vergence matrices.
- When $\mathbb{P}_1$ and $\mathbb{P}_2$ commute $\mathbb{P}_E$ is symmetrical, but it cannot be interpreted as the vergence of the wavefront measured at the principal planes, as we might think by analogy with the scalar case. This happens because the application of equations (4.24) would result in a different axial location of the principal planes for each of the principal meridians of the lens. We may check this assertion by *reductio ad absurdum*: let us assume that a plane wavefront is entering the lens and that $\mathbb{P}_E$ is indeed the vergence of the output wavefront at some distance d from the back surface. We know the output vergence at

the back surface is $\mathbb{P}_V$. Then we should have $(\mathbb{I} - d\mathbb{P}_V)^{-1}\mathbb{P}_V = \mathbb{P}_E$. Left-multiplying by $(\mathbb{I} - d\mathbb{P}_V)$, we have $\mathbb{P}_V = (\mathbb{I} - d\mathbb{P}_V)\,\mathbb{P}_E$, that is, $d\mathbb{P}_V\mathbb{P}_E = \mathbb{P}_E - \mathbb{P}_V$. But using once again $\mathbb{P}_V = \mathbb{P}_E\mathbb{G}$, we have $d = \mathbb{P}_E^{-1}\left(\mathbb{G}^{-1} - \mathbb{I}\right)$, that is, $d = -\tau\mathbb{P}_E^{-1}\mathbb{P}_1$. Formally, this equation is identical to the first one of (4.24), but here we encounter a problem: A scalar cannot be equal to a matrix, unless $\mathbb{P}_E^{-1}\mathbb{P}_1$ happens to be a multiple of the identity matrix. It is not difficult to show that in general this is not possible for astigmatic lenses, even if the second surface has revolution symmetry and $\mathbb{P}_E$ and $\mathbb{P}_1$ have the same principal meridians. So we conclude d is not a number, and our assumption was false. In general, there is no position in space at which the output wavefront has vergence $\mathbb{P}_E$.[3]

- When the thickness is small enough we can use thin lens approximation, $\mathbb{P}_V \simeq \mathbb{P}_E \simeq \mathbb{P}_{TL}$. In this case the difference between the nondiagonal terms in $\mathbb{P}_E$ will be negligible, and the ray deviation can be written as $\boldsymbol{\delta} = -\mathbb{P}_{TL}\mathbf{r}$, regardless of the direction of the incident ray. This is the matrix formulation of the so-called *Prentice's rule* [44] that we will study in Chapter 5.

4.3.6 Paraxial Computations for Ophthalmic Lenses

We have presented matrix relationships between different types of power and the geometrical parameters defining the lens, and from them we should be able to compute any lens property within the paraxial approximation. In some cases, most of them academic, we will have information on the geometry of the surfaces forming the lens, its center thickness, and its refractive index. From them we can compute power or any other optical property. The manufacturer and/or the lens designer will usually face a different problem: The desired lens power is known and it is for them to decide the material and geometry for the front surface. Then the geometry of the back surface has to be determined so that the final lens will have the required power and possibly other optical properties. In some cases the equations can be analytically solved, but there are some situations where the use of numerical methods is necessary. Let us now look at the most relevant problems that may arise when computing lenses, and how to solve them.

Power calculation from surface radii, refractive index, and thickness

This is the most direct problem we may find. From surface radii and refractive index we can directly compute $\mathbb{P}_1$, $\mathbb{G}$, and $\mathbb{P}_2$. From thickness and refractive index we get τ, and finally, the different types of power are computed with the expressions presented so far, $\mathbb{P}_V = \mathbb{G}\mathbb{P}_1 + \mathbb{P}_2$ for the back vertex power, $\mathbb{P}_{TL} = \mathbb{P}_1 + \mathbb{P}_2$ for the thin lens power, and $\mathbb{P}_E = \mathbb{P}_1 + \mathbb{P}_2 - \tau\mathbb{P}_2\mathbb{P}_1$ for the equivalent power. For lenses with revolution symmetry we may also compute the position of the principal planes from equations (4.24).

---

[3] There are particular cases in which $\mathbb{P}_E$ is the vergence of the output beam at a plane, when the astigmatism and the shape factor balance to allow for the existence of a principal plane, but they are of little practical interest. They require $\mathbb{P}_E$ to be proportional to $\mathbb{P}_1$ and $\mathbb{P}_2$ to be proportional to $\mathbb{P}_1\mathbb{G}$.

Power from index, thickness, front surface data, and any of the other types of power

In this case we need to obtain $\mathbb{P}_2$ from the given lens power. If we know the back vertex power, $\mathbb{P}_2 = \mathbb{P}_V - \mathbb{G}\mathbb{P}_1$. If we know the thin lens power, then $\mathbb{P}_2 = \mathbb{P}_{TL} - \mathbb{P}_1$. Finally, if we are given the equivalent power, then $\mathbb{P}_2 = (\mathbb{P}_E - \mathbb{P}_1)\,\mathbb{G}$ (order observed for the product $\mathbb{P}_E\mathbb{G}$). Once we have $\mathbb{P}_2$ we can directly compute the remaining types of lens power.

$\mathbb{P}_1$ from lens power, index, thickness, and back surface data

If we know the thin lens power, getting $\mathbb{P}_1$ from the remaining data is trivial, $\mathbb{P}_1 = \mathbb{P}_{TL} - \mathbb{P}_2$. If back vertex power is given, then $\mathbb{P}_1 = [\mathbb{I} + \tau\,(\mathbb{P}_V - \mathbb{P}_2)]^{-1}\,(\mathbb{P}_V - \mathbb{P}_2)$, the multiplying matrices commuting. Finally, if equivalent power is given, $\mathbb{P}_1 = (\mathbb{I} - \tau\mathbb{P}_2)^{-1}\,(\mathbb{P}_E - \mathbb{P}_2)$, order observed.

Lens maker problem

When the ECP places an order for a lens, the lens manufacturer faces a somewhat different problem than the ones previously described. The manufacturer knows the final back vertex power of the lens, the refractive index of the lens material, the refractive power of the front surface (base curve), and the diameter of the finished lens. From these data the manufacturer has to compute two parameters: (1) the center thickness, and (2) the curvatures of the back surface (the one to be reshaped) in order to produce a lens with the right vertex power. The problem is quite different depending on the sign of the main vertex powers. If both are negative, that is $P_-, P_+ < 0$, then we will find the minimum lens thickness at the center. The manufacturer then chooses the value of this center thickness attending mainly to safety reasons.[4] Once the center thickness has been set, the refractive power of the back surface is given by $\mathbb{P}_2 = \mathbb{P} - (\mathbb{I} - \tau\mathbb{P}_1)^{-1}\,\mathbb{P}_1$, and from it we get the main curvatures and directions. Negative lenses are then quite straightforward to compute. If any of the two main meridians is positive, then the minimum thickness will be located at some point of the lens contour, and this is the one that will be set. Now, vertex power does not depend on edge thickness but on center thickness, so our unknowns are t_0 and $\mathbb{P}_2$, and our equations are the thickness equation (4.1) and the vertex power equation (4.11), along with the relations between refractive power and curvatures for $\mathbb{P}_2$.[5] Explicitly, the system of equations we have to solve is

$$\mathbb{P}_2 = \mathbb{P} - (\mathbb{I} - \tau\mathbb{P}_1)^{-1}\,\mathbb{P}_1,$$

$$t = t_{min} + z_1\,(x_m, y_m) - z_2\,(x_m, y_m),$$

where (x_m, y_m) are the coordinates of the edge point where the thickness is minimal. This system has no analytical solution, even for the simplest case of a lens with spherical

---

[4] The manufacturer selects the minimum thickness that ensures the requirements of mechanical resistance according to the relevant standards. In practice, this translates into a center thickness between 1.2 and 1.5 mm for plastic negative lenses, and between 1.5 and 1.8 for glass lenses.

[5] $\mathbb{P}_2$ may have more than one unknown, but the matrix equation for vertex power splits into the same number of scalar equations, so unknowns and equations are always matched.

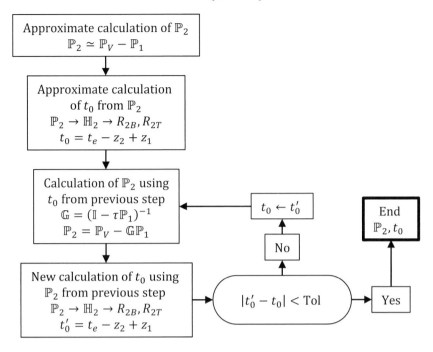

Figure 4.10 Flux diagram of a typical algorithm used for determining both the back surface of an ophthalmic lens and its center thickness, given the base curve (refractive power of the front surface), the minimum edge thickness, and the objective vertex power. In the first step we compute the approximate value of $\mathbb{P}_2$ using the thin lens approximation. With the obtained $\mathbb{P}_2$ we calculate the center thickness t_0 from the surface sagittas. The next step is similar, but now we use the shape factor computed with the center thickness obtained in the previous step to obtain a new value for $\mathbb{P}_2$ and with it a new value of the center thickness, t_0'. The next step is identical, but first we set $t_0 = t_0'$ and then the step generates a new t_0'. In each iteration we compare the two values of the center thickness, and when the difference between them is smaller than certain tolerance, the algorithm is stopped. In fact, the algorithm just needs two or three steps to compute the center thickness with micron-level accuracy.

surfaces. To solve it we must resort to a numerical method, such as the recursive algorithm illustrated in Figure 4.10.

Example 15 The curvature radius of the front surface of a spherical lens is 200 mm, while the radius of the back surface is 400 mm. The refractive index is $n = 1.6$ and the center thickness 3 mm. Determine the back vertex, thin lens, and equivalent powers, and the position of the principal planes of the lens.

As the lens has rotation symmetry, we do not need matrices to do the computations, the shape factor, and all powers being scalars. The refractive powers are

$$P_1 = \frac{n-1}{R_1} = \frac{0.6}{0.2} = 3\,\mathrm{D}, \; P_2 = \frac{1-n}{R_2} = \frac{-0.6}{0.4} = -1.5\,\mathrm{D}.$$

The thin lens power is $P_{TL} = P_1 + P_2 = \mathbf{1.5}$ D and the shape factor

$$g = \frac{1}{(1 - \tau P_1)} = \frac{1}{\left(1 - \frac{0.003 \times 3}{1.6}\right)} = 1.0057.$$

The back vertex power,

$$P_V = gP_1 + P_2 = 1.0057 \times 3 - 1.5 = 1.517\,\text{D},$$

while the equivalent power

$$P_E = g^{-1}P = \frac{1.517}{1.0057} = 1.508\,\text{D}$$

Finally, the location of the principal planes with respect to the lens vertexes,

$$S_2 H' = -\tau \frac{P_1}{P_E} = -3.7\,\text{mm}, \quad S_1 H = \tau \frac{P_2}{P_E} = -1.9\,\text{D}.$$

The next example illustrates the computation of the refractive power of the front surface from the back vertex power, refractive index, thickness, and curvature of the back surface.

Example 16 A bitoric lens is made with front and back surfaces having nonparallel main meridians. The front torus has main curvatures $\kappa_{1B} = 14$ D and $\kappa_{1T} = 19$ D, the base oriented along the X axis. The back torus has curvatures $\kappa_{2B} = 6$ D and $\kappa_{2T} = 8$ D, the base oriented at 120°. The center thickness of the lens is 6 mm, and its refractive index $n = 1.5$. Determine the back vertex and equivalent power of the lens.

We have all the relevant data to compute the paraxial properties of this lens. Although both surfaces are astigmatic and nonaligned, the computation is rather straightforward with matrix techniques. The Hessians of the surfaces are

$$\mathbb{H}_1 = \begin{bmatrix} 14 & 0 \\ 0 & 19 \end{bmatrix} \text{D}, \quad \mathbb{H}_2 = \begin{bmatrix} 6 + 2\sin^2 120 & -2\sin 120\cos 120 \\ -2\sin 120\cos 120 & 6 + 2\cos^2 120 \end{bmatrix} \text{D}.$$

The refractive powers,

$$\mathbb{P}_1 = \begin{bmatrix} 7 & 0 \\ 0 & 9.5 \end{bmatrix} \text{D}, \quad \mathbb{P}_2 = \begin{bmatrix} -3.75 & -0.433 \\ -0.433 & -3.25 \end{bmatrix} \text{D}.$$

The factor $\tau = t_0/n$ is 4 mm, and the shape factor

$$\mathbb{G} = (\mathbb{I} - \tau\mathbb{P}_1)^{-1} = \begin{bmatrix} 1.0288 & 0 \\ 0 & 1.0395 \end{bmatrix},$$

So the vertex power is

$$\mathbb{P}_V = \mathbb{G}\mathbb{P}_1 + \mathbb{P}_2 = \begin{bmatrix} 3.452 & -0.433 \\ -0.433 & 6.625 \end{bmatrix} \text{D},$$

and the equivalent power

$$\mathbb{P}_E = \mathbb{P}_V\mathbb{G}^{-1} = \begin{bmatrix} 3.355 & -0.417 \\ -0.417 & 6.374 \end{bmatrix} \text{D}.$$

The principal back vertex powers can be readily computed from the matrix form, $P_- = 3.39$ D and $P_+ = 6.68$ D, the orientation of P_- being $\theta_- = 7°$.

Finally, let us look at an example of the manufacturer's computation for a negative lens.

Example 17 We want to manufacture a spherotoric lens with back vertex power $[-4, -2 \times 10°]$, made with polycarbonate ($n = 1.59$). The base curve is 3 D and the center thickness 1.8 mm. Determine the main curvature radius and directions of the back surface of the lens.

When talking about a lens, base curve always refers to the front surface, so $P_1 = 3$ D. The back vertex power matrix is

$$\mathbb{P}_V = \begin{bmatrix} -4 - 2\sin^2 10 & 2\sin 10 \cos 10 \\ 2\sin 10 \cos 10 & -4 - 2\cos^2 10 \end{bmatrix} = \begin{bmatrix} -4.06 & 0.342 \\ 0.342 & -5.94 \end{bmatrix} D,$$

and the product $gP_1 = 3/\left[1 - 0.0018 \times 3/1.59\right] = 3.0102$. Thus, the refractive power of the back surface is

$$\mathbb{P}_2 = \mathbb{P}_V - gP_1\mathbb{I} = \begin{bmatrix} -4.06 & 0.342 \\ 0.342 & -5.94 \end{bmatrix} - \begin{bmatrix} 3.01 & 0 \\ 0 & 3.01 \end{bmatrix} = \begin{bmatrix} -7.07 & 0.342 \\ 0.342 & -8.95 \end{bmatrix} D,$$

whose main refractive powers are

$$P_{2B} = -7.01 \text{ D at } 10°, \quad P_{2T} = -9.01 \text{ D at } 100°.$$

Notice that the refractive power difference between the principal meridians of the back surface of the lens is just the prescription cylinder. This is a consequence of the lens being spherotoric with a toric back surface. The curvature radii of the back surface are

$$R_{2B} = \frac{1-n}{P_{2B}} = \frac{-0.59}{-7.01} \times 1000 = 84.2 \text{ mm at } 10°,$$

$$R_{2T} = \frac{1-n}{P_{2T}} = \frac{-0.59}{-9.01} \times 1000 = 65.5 \text{ mm at } 100°.$$

4.3.7 Spherotoric Lenses: Computing Power by Meridians

The most common way of manufacturing astigmatic lenses is the combination of a spherical and a toric surface. Spherotoric lenses may be computed using the general matrix formulas described in the preceding sections, as these equations are valid for any lens geometry. However, as one of the lens surfaces is spherical, it is also possible to compute the power for each meridian separately. This allows the computation of lens power without using the standard components of curvature matrices, making the calculations lighter, though not easier. Computing lens power by principal meridians is the traditional way of calculating astigmatic lenses, and it is included in all textbooks on ophthalmic optics. We will add now this way of computing power for the sake of completeness, and because a combined study of astigmatic systems using main meridians and curvature matrices will provide the best insight into the matter. Anyway, let us insist that the computation can be carried out on main meridians only when the main meridians of the two surfaces are aligned, or one of the surfaces has revolution symmetry. Otherwise, the use of standard curvature matrices referred to a fixed XY reference system is the only sensible approach.[6]

Even though we will compute the surfaces using the main meridians, we will still keep the matrix notation using a rotated reference system and the notation $[]_\otimes$, which means the matrix is expressed in the rotated reference system $\xi\eta$ aligned with the lens principal meridians. In this reference system the nondiagonal terms are zero, and the diagonal terms will be the main curvature/powers of the surface/lens our matrix represents. This way we handle the two principal curvature/powers at the same time, with a more ordered and easy-to-follow notation. Let us take a look to this notation before going into the subject matter. We are going to compute the geometry for astigmatic lenses with a back vertex power $\mathbb{P}_V$ and with a positive-cylinder spherocylindrical representation $[S, C \times \alpha]$. We can write the matrix power as $\mathbb{P}_V = S\mathbb{I} + \mathbb{C}$. In the XY reference system this equation reads

$$\begin{bmatrix} P_x & P_t \\ P_t & P_y \end{bmatrix} = \begin{bmatrix} S & 0 \\ 0 & S \end{bmatrix} + \begin{bmatrix} C\sin^2\alpha & -C\sin\alpha\cos\alpha \\ -C\sin\alpha\cos\alpha & C\cos^2\alpha \end{bmatrix}, \qquad (4.28)$$

whereas in a rotated reference system $\xi\eta$ in which the rotated X axis, ξ, aligns with the axis of the cylinder, the previous matrix equation is

$$\begin{bmatrix} P_- & 0 \\ 0 & P_+ \end{bmatrix}_\otimes = \begin{bmatrix} S & 0 \\ 0 & S \end{bmatrix} + \begin{bmatrix} 0 & 0 \\ 0 & C \end{bmatrix}_\otimes,$$

where we do not need the subindex $\otimes$ in the matrices of spherical components as they are always diagonal in *any* reference system. As we have chosen the positive-cylinder representation, the meridian P_- must align with the cylinder axis, and then $P_- = S$, $P_+ = S + C$. If we had chosen the negative-cylinder representation $[S', C' \times \alpha']$, the matrix equation would be

$$\begin{bmatrix} P_+ & 0 \\ 0 & P_- \end{bmatrix}_\otimes = \begin{bmatrix} S' & 0 \\ 0 & S' \end{bmatrix} + \begin{bmatrix} 0 & 0 \\ 0 & C' \end{bmatrix}_\otimes.$$

The reader must then be aware that the notation $[]_\otimes$ always means the reference system is rotated so that the original X axis now lies along the cylinder axis of whatever spherocylindrical representation we have selected, and henceforth the top-left component of the cylinder matrix in the rotated system will always be zero.

External Torus

Let us consider a spherotoric lens formed by a toric front surface and a spherical back surface. Let $[S, C \times \alpha]$ be the positive-cylinder spherocylindrical representation of its back vertex power. The matrix relation between lens power and surface refractive powers is $S\mathbb{I} + \mathbb{C} = \mathbb{G}\mathbb{P}_1 + P_2\mathbb{I}$. We will assume that the front surface will be convex, so the main

---

[6] There are other ways to express curvature that can handle the combination of astigmatic surfaces or wavefronts additively, but their components are linear combinations of the Hessian components. See Chapter 5 and references [48, 49].

refractive powers of the torus are positive, and hence $P_T > P_B$. The lens equation displayed in the $\xi\eta$ system is then

$$\begin{bmatrix} S & 0 \\ 0 & S \end{bmatrix} + \begin{bmatrix} 0 & 0 \\ 0 & C \end{bmatrix}_\otimes = \begin{bmatrix} g_B P_B & 0 \\ 0 & g_T P_T \end{bmatrix}_\otimes + \begin{bmatrix} P_2 & 0 \\ 0 & P_2 \end{bmatrix}.$$

We knew the base curve of the torus should be aligned with the cylinder axis of the spherocylindrical representation because we chose $C > 0$ and the torus to be convex, so $P_T > P_B > 0$. This means the bigger principal power of the lens, P_+, has to be split either as $S + C$ or $g_T P_T + P_2$. Now, we can split the previous matrix equation into two scalar equations and solve for both P_B and P_T, but we already solved this equation in matrix form in Section 4.3.6, with the result

$$\mathbb{P}_1 = [\mathbb{I} + \tau\,(\mathbb{P}_V - \mathbb{P}_2)]^{-1}\,(\mathbb{P}_V - \mathbb{P}_2),$$

where now $\mathbb{P}_V = S\mathbb{I} + \mathbb{C}$. As this equation is valid on any reference system, we can apply it to our problem,

$$\begin{bmatrix} P_B & 0 \\ 0 & P_T \end{bmatrix}_\otimes = \begin{bmatrix} S - P_2 & 0 \\ 0 & S + C - P_2 \end{bmatrix}_\otimes \begin{bmatrix} \frac{1}{1+\tau(S-P_2)} & 0 \\ 0 & \frac{1}{1+\tau(S+C-P_2)} \end{bmatrix}_\otimes, \qquad (4.29)$$

where we are using the fact that the inverse of a diagonal matrix is another diagonal matrix with its components inverted. In terms of the scalar components of the main diagonal,

$$P_B = \frac{S - P_2}{1 + \tau(S - P_2)},$$

$$P_T = \frac{S + C - P_2}{1 + \tau(S + C - P_2)}. \qquad (4.30)$$

Example 18 We want to manufacture a spherotoric lens, made in CR-39 ($n = 1.5$), with back vertex power $[6, -2 \times 10°]$, using a toric front surface with base curve $P_B = 6.5$ D. The center thickness is $t = 5.5$ mm. Compute the cross curve of the torus and the refractive power of the back surface.

This is a mixed problem, in which we know one main meridian of the torus, and do not know the back sphere. First we need to transpose the back vertex power into the positive-cylinder form, which is $[4, 2 \times 100°]$. Next we use the direct equation $S = g_B P_B + P_2$ to obtain P_2. The form factor for the base meridian is $g_B = 1/(1 - \tau P_B) = 1.0244$, so $P_2 = S - g_B P_B = -2.66$ D. Now we can use the second equation in (4.30) to get the cross meridian of the torus

$$P_T = \frac{S + C - P_2}{1 + \tau(S + C - P_2)} = 8.39\,\text{D}.$$

The orientation of the base curve of the torus is that of the (positive) cylinder axis, so $\theta_B = 100°$.

If we had used the thin lens approximation, the results would be

$$P_2 = S - P_B = -2.5\,\text{D},$$
$$P_T = S + C - P_2 = 8.5\,\text{D}.$$

Internal Torus

Now the front surface is spherical and the toric surface is the back one. The relation between vertex power and surface refractive powers is $S\mathbb{I} + \mathbb{C} = gP_1\mathbb{I} + \mathbb{P}_2$, so $\mathbb{P}_2 = S\mathbb{I} + \mathbb{C} - gP_1\mathbb{I}$. Now, if we keep using the positive-cylinder spherocylindrical representation, the cross curve of the toric surface will be aligned with the axis of the cylinder, as $P_T < P_B < 0$ for convex toric surfaces. In the rotated reference system we then have

$$\begin{bmatrix} P_T & 0 \\ 0 & P_B \end{bmatrix}_\otimes = \begin{bmatrix} S & 0 \\ 0 & S+C \end{bmatrix}_\otimes - \begin{bmatrix} \frac{P_1}{1+\tau P_1} & 0 \\ 0 & \frac{P_1}{1+\tau P_1} \end{bmatrix}_\otimes. \tag{4.31}$$

If we had used the negative-cylinder spherocylindrical, $[S', C' \times \alpha']$, the base curve of the toric surface would align with the axis of the cylinder,

$$\begin{bmatrix} P_B & 0 \\ 0 & P_T \end{bmatrix}_\otimes = \begin{bmatrix} S' & 0 \\ 0 & S'+C' \end{bmatrix}_\otimes - \begin{bmatrix} \frac{P_1}{1+\tau P_1} & 0 \\ 0 & \frac{P_1}{1+\tau P_1} \end{bmatrix}_\otimes.$$

Both equations are indeed identical, as $S' = S + C$, $S' + C' = S$, and the cylinder axes of both representations are interchanged. In terms of the scalar components we finally use the positive-cylinder representation,

$$P_B = S + C - \frac{P_1}{1 - \tau P_1},$$
$$P_T = S - \frac{P_1}{1 - \tau P_1}. \tag{4.32}$$

The reader may be confused by the need of a "a priori" selection of the meridian of the toric surface that has to be aligned with the axis of whatever spherocylindrical representation is used. This is not a loose end of the matrix formulation; it is simply a consequence of having given arbitrary names to the main meridians of the torus. Calling the flatter meridian the "base curve" is just an arbitrary convention, and if we want to keep track of what direction this flatter meridian lies along, we have to pay attention to the numerical values of the lens main powers and take into account whether the toric surface is concave or convex. Let us use another approach; assume we provide a random label to the main meridians by using any pair of symbols, for example $\diamond$ and $\circ$, but not assigning a particular relation between them (as meridian $\diamond$ is steeper than $\circ$, or flatter, or whatever). The previous matrix equation would then be

$$\begin{bmatrix} P_\diamond & 0 \\ 0 & P_\circ \end{bmatrix}_\otimes = \begin{bmatrix} S & 0 \\ 0 & S+C \end{bmatrix}_\otimes - \begin{bmatrix} \frac{P_1}{1+\tau P_1} & 0 \\ 0 & \frac{P_1}{1+\tau P_1} \end{bmatrix}_\otimes,$$

and we would not have to worry about the matching of the diamond or the circle with the cylinder axis, that is, it would be irrelevant if we plug $P_\diamond$ in the first row and $P_\circ$ in the bottom row or the other way around, because the naming is arbitrary. Once the equation is solved, we could check which of $P_\diamond$ or $P_\circ$ corresponds to the flatter curve. *This* would be the base curve of the torus.

Example 19 An ophthalmic lens with power $[-3.5, 1.5 \times 125°]$ is manufactured with an inner torus whose base curve is $P_B = -6.5$ D. Compute the remaining refractive powers of the lens along with their orientations (when they apply).

As the problem specifies neither the center thickness nor the refractive index, we are forced to use the thin lens approximation in equations (4.32). Recall that they were deduced for positive cylinder and concave torus (4.33), which matches our situation in this problem. So we have

$$P_B = S + C - P_1,$$
$$P_T = S - P_1.$$

From the first one, $P_1 = S + C - P_B = 4.5$ D, and from the second, $P_T = S - P_1 = -8$ D. The orientation of the base curve is that of the axis of the negative-cylinder spherocylindrical formulation, that is, $35°$.

4.3.8 Relationship between Lens Power and Thickness

At the beginning of this chapter, Section 4.2.3, we derived an expression that gives the approximate local thickness of a lens as a function of its center thickness and the difference between the Hessian matrices of the lens surfaces,

$$t(\mathbf{r}) = t_0 - \frac{1}{2}\mathbf{r}^T (\mathbb{H}_1 - \mathbb{H}_2) \, \mathbf{r}.$$

This is a purely geometrical relationship that can be applied to any object with two surfaces whose Hessians are $\mathbb{H}_1$ and $\mathbb{H}_2$, regardless of its optical applications. However, we can readily establish a connection between the thickness of a lens and its power. Recall that the thin lens power of a lens is given by

$$\mathbb{P}_{TL} = \mathbb{P}_1 + \mathbb{P}_2$$
$$= (n - 1) (\mathbb{H}_1 - \mathbb{H}_2).$$

This is a generalization, for lenses with astigmatic surfaces, of the *lens maker equation* for thin spherical lenses:

$$P_{TL} = (n - 1) \left(\frac{1}{R_1} - \frac{1}{R_2} \right) = (n - 1) (\kappa_1 - \kappa_2).$$

It is then clear that thickness is related to thin lens power by means of

$$t(\mathbf{r}) = t_0 - \frac{1}{2(n-1)}\mathbf{r}^T \mathbb{P}_{TL}\mathbf{r}. \tag{4.33}$$

In many cases we are given the back vertex power instead of the thin lens power. We can still use the previous equations if we approximate $\mathbb{P}_{TL}$ by $\mathbb{P}_V$. In that case the quadratic form $\mathbf{r}^T\mathbb{P}_V\mathbf{r}$ reduces to $P_x x^2 + 2P_t xy + P_y y^2$, where P_x, P_y, and P_t are the components of the back vertex power matrix. The thickness is then

$$t(x,y) = t_0 - \frac{P_x x^2 + 2P_t xy + P_y y^2}{2(n-1)}. \tag{4.34}$$

If the spherocylindrical form of the back vertex power is $[S, C \times \alpha]$, and we write $\mathbb{P}_V = S\mathbb{I} + \mathbb{C}$, then

$$\mathbf{r}^T\mathbb{P}_V\mathbf{r} = S(x^2 + y^2) + \mathbf{r}^T\mathbb{C}\mathbf{r},$$

and the thickness,

$$t(x,y) = t_0 - \frac{Sr^2 + C\left(x^2 \sin^2\alpha + y^2 \cos^2\alpha - 2xy \sin\alpha \cos\alpha\right)}{2(n-1)}. \tag{4.35}$$

When the lens has revolution symmetry, the terms related to the cylinder disappear and the previous equation reduces to

$$t(r) = t_0 - \frac{Sr^2}{2(n-1)}, \tag{4.36}$$

and for a round lens with diameter D, the edge thickness and the center thickness are related by

$$t_e = t_0 - \frac{(D/2)^2}{2(n-1)}S. \tag{4.37}$$

Equation (4.36) appears in many textbooks on ophthalmic optics, especially when applied to a point of the lens contour. As it is becoming usual, the use of a matrix representation for curvature and power allows for the generalization of traditional scalar equations, only valid for lenses with revolution symmetry, to lenses with arbitrary astigmatic surfaces.

The effect of the term $\mathbf{r}^T\mathbb{P}_{TL}\mathbf{r}/[2(n-1)]$ is to reduce or increase the center thickness depending on the sign of $\mathbf{r}^T\mathbb{P}_{TL}\mathbf{r}$. For lenses with negative main meridians, this term is negative, and the thickness increases from the center toward the edge. Likewise, lenses with positive main meridians have a positive value of their quadratic form, and the thickness is reduced from its value at the center as we consider points toward the contour. The introduction of power in the thickness equation entails the introduction of the refractive index. For a given lens power, a higher refractive index will provide a smaller variation of center thickness. This is the reason the ophthalmic industry craves compatible optical materials with a high refractive index: Lenses made with these materials will be thinner.

We will finish this section with a brief analysis on the accuracy of equation (4.35), shown in Figure 4.11. We have computed and represented the thickness of a spherical negative lens at 20 and 30 mm from its center, both with the exact equation (4.1) and with the approximation (4.33) (top plot). At the bottom is a plot of the center thickness of a spherical positive lens with diameters of 40 and 60 mm, with a thickness of 0.5 mm at the

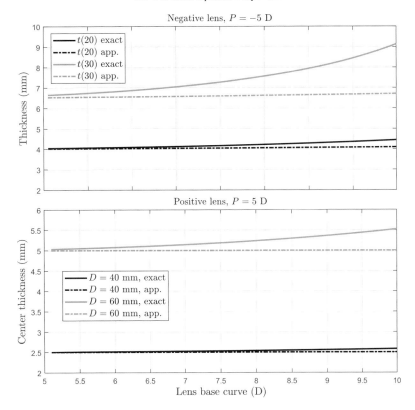

Figure 4.11 Top plot: thickness at 20 and 30 mm from the center of a negative lens with $n = 1$ and center thickness $t_0 = 2$ mm as a function of lens base curve. Solid curves have computed with exact expressions and dot-dashed lines with the approximation (4.33). Bottom plot: similar computation for a positive lens. In this case the center thickness is computed for two different lens diameters assuming an edge thickness of 0.5 mm. Like wise, we have used exact and approximate equations for the positive lens.

edge, also with the exact and approximate equations. In both cases we have computed the thickness as a function of the base curve. The approximate equation (4.33) does not depend on the base curve, or only marginally, as it depends on the thin lens power, and not on the individual lens surfaces. We see that at a distance of 20 mm the approximate expression gives good results, valid for estimation and representation purposes. For the negative lens and at 30 mm from the center, the approximate expression misses the right value by 5% or less for the base curves that are typically used, although the error grows quickly up to 25% if the lens is made with very large base curves (which is extremely uncommon). For the positive lens, the errors are somewhat smaller, about 4% for the base curves that typically would be used, between 6.5 and 8 D.

Example 20 A lens with back vertex power $[-7, -2.5 \times 45°]$ is glazed to be mounted in frame. The point on its contour further away from the center of the lens has coordinates

$(-28, -5)$ mm with respect to an XY system centered with the lens. Determine the reduction of thickness that we may expect at this point if we made the lens with a high-index material, $n_{HI} = 1.67$, instead of CR-39 $n_{CR-39} = 1.49$.

An exact calculation would require more information than is provided, so we must resort to the approximate thickness equation (4.35). The quadratic form of this lens at the specified point is

$$S(x^2 + y^2) + C(x^2 \sin^2 \alpha - 2xy \sin \alpha \cos \alpha + Cy^2 \cos^2 \alpha) = -5.66 - 0.66 = -6.32 \text{ mm.}$$

So, for the lens made of CR-39,

$$t_{CR39}(x, y) = t_0 - \frac{-6.32}{2(n_{CR39} - 1)} = t_0 + \frac{6.32}{2 \times 0.48} = t_0 + 6.85,$$

while for the high-index lens

$$t_{HI} = t_0 - \frac{-6.32}{2(n_{HI} - 1)} = t_0 + \frac{6.32}{2 \times 0.67} = t_0 + 4.71.$$

The thickness reduction is

$$\Delta t = t_{CR39} - t_{HI} = 6.85 - 4.71 = 2.13 \text{ mm.}$$

5

The Lens-Eye System

5.1 Introduction

The eyes are optical systems whose function is the formation of real images of the environment that the brain can detect, process, and interpret as visual information. Like any optical system, natural or artificial, eyes cannot produce perfect images; they have *aberration*s, the most common of which are spherical defocus and astigmatism, both known as *refractive errors*. Defocus and astigmatism are also known as *second-order aberrations*. They are nothing but errors of curvature of the wavefront refracted through the eye, and we already know that curvature is determined by the quadratic – second order – terms of the polynomial expansion used to describe the wavefronts. As eyes are biological organs, the percentage of individuals presenting significant second-order aberrations is known as the prevalence of refractive errors. Although prevalence is a clinical term typically used to measure the extent a disease affects a population, refractive errors are not diseases. An eye affected by refractive error can be perfectly healthy; it simply has a mismatch between its power and its size.

Refractive errors can be corrected by surgery, not free from risks, that changes the curvature of the cornea, henceforth changing the power of the eye. But still today a majority of refractive errors are compensated with lenses that, once annexed to the eye, create a new optical system, the lens-eye system, that should be free from second-order aberrations and should produce a sharp image on the retina. There are basically three types of these ophthalmic compensations: spectacle lenses, contact lenses, and intraocular lenses (IOL). Probably, the reader already has a general idea about the differences between the three types of lenses and may think that they mainly differ in their location with respect to the eye. Spectacle lenses are held in front of the eye, contact lenses are held in contact with the cornea, and intraocular lenses are held inside the eye, usually replacing the crystalline lens. A different position of the correcting lens will require a different lens power to achieve the desired correction. Also, the shape of the lens that will produce the best optical performance will change as a function of lens position. With respect to shape, contact lenses are special as their inner surface must fit the corneal surface. But there is a difference that goes unnoticed for the nonexpert, yet it has important consequences: While contact lenses and IOL move and turn together with the eye, spectacle lenses do not. Eyes compensated with contact lenses and IOLs can be studied using the very same techniques developed to study, design, and optimize standard optical systems, such as camera lenses or telescopes. Contact lenses may have a little play, but that is studied as a small error positioning, not as a *feature* of their design. On the contrary, if the head is still, spectacle lenses are also still when the eye scans the available field of view behind them, rotating to aim and fixate the point of interest.

We are quite used to this arrangement, as we have been using spectacle lenses for centuries, but think about it: There are very few optical systems that work like that. When we talk about field of view, magnification, and aberrations, we have to modify to a greater or lesser extent the typical definitions and methods used for optical systems whose parts "move together." The spectacle lens-eye system becomes even more different from standard optical systems when we study both eyes at the same time, as the spectacle lenses may cause different pointing errors for each eye, altering the balance of the binocular vision. In this chapter we study the spectacle lens-eye system, first with the static eye, and then with the moving eye, accurately defining and quantifying the concepts we have just outlined here.

5.2 Basic Optical Physiology

5.2.1 The Eye

The eye is an organ whose function is to provide neural information encoding planar images of the world around us. The combination of the information from the two eyes provides further information that the brain uses to construct a three-dimensional image of our environment. The image-forming system of the eye is composed of two lenses and a variable shutter, the cornea, the crystalline lens, and the iris, and the detection system is mainly formed by the retina, a tissue with photoreceptor cells and neural connections that codify the information imaged onto it. The physiology of the eye and its optical function are fascinating subjects, but we do not need a very detailed description of the eye to study most of the properties of the lens-eye system. As we will see, some fundamental principles of the theory of spectacle lenses can be established with a very simple and conceptual model of the eye, without even considering any of its actual structures. Detailed models of the eye using average parameters have been known for more than a century, but individual variability is high enough to reduce the applicability of these averaged models. Nowadays, noninvasive measuring methods such as eye aberrometry, enhanced corneal topography, or optical coherence tomography, among others, are getting closer and more easily available to the ECP, and spectacle lens design based on individual eye models is becoming a possibility. We provide next a brief summary of eye structures and optical properties: just what we need for the rest of the book. There are many excellent books in which the reader can deepen on eye physiology [50, 51] and optics of the eye [52, 53, 54].

The basic structure of the eye is shown in Figure 5.1(a). The outer layer of the eye is the *sclera*, a fibrous membrane mainly made of collagen. The sclera is continuous at the front of the eye with the cornea, and with the dural sheath of the optic nerve at the optic

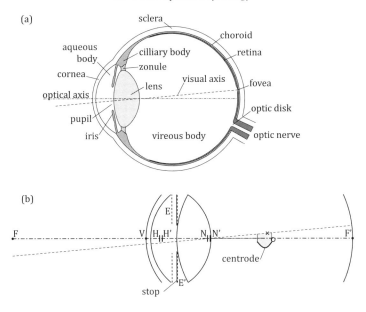

Figure 5.1 (a) Cross section of the human right eye as seen from above, showing the basic structures. (b) Plot of the exact schematic eye of Le Grand, with its cardinal points. Distances and curvature radius are all at the same scale.

disc, the exit point of ganglion cell axons. The sclera supports the external and internal forces applied to the eye globe, and provides attachments for the extra-ocular muscles that rotate the eye. The *cornea* is the transparent tissue continuous with the sclera at the front of the eye. It covers the iris and the anterior chamber, filled with the aqueous body. The cornea is the forefront lens of the eye. It is meniscus-shaped, the curvature of the inner surface being larger than that of the front surface. In air, the cornea would be a negative lens. However, the refractive index of the aqueous body is very similar to that of the cornea, so the back surface has a low refractive power, and the whole cornea is positive. The *iris* is the structure that forms the diaphragm. The *pupil* is the aperture of this diaphragm. A sphincter muscle contracts the iris and a set of dilator muscles fold it, enlarging the size of the pupil. The outer rim of the iris is attached to the sclera and the ciliary body, a structure related to the support of the *crystalline lens*, and related to the accommodation process that we will present in Chapter 8. The crystalline lens, or simply, the lens, is the second refractive component of the eye. Its form is biconvex and its refractive index is not constant but growing from the outer layers to the central ones. This variation of refractive index increases the power of the lens and seems to improve image quality by reducing spherical aberration. The lens is held in place by a fibrous annular structure called the *zonule*, which in turn is attached to the ciliary body and ciliary muscle. As we will see in Chapter 8, these holding structures determine and may change the shape of the lens, which is flexible. This accounts for the change of power of the lens, a process known as *accommodation*, which is necessary to focus near objects. In this chapter we will consider the eye in its *relaxed state*, the ciliary muscle relaxed and dilated, the lens stretched, and the eye focused to infinity.

The inner part of the sclera is covered by the *choroid*, a vascular layer continuous with the ciliary body, which provides oxygen and nutrients to the *retina*, the innermost layer of the eye on which an image is formed and detected. The retina has a complex structure with 10 sub-layers, one of which is the layer of photo-sensitive rod and cone cells (next to last closest to the choroid). The number of these cells in the retina is in the range of 90 to 150 million rods and 4.5 to 7 million cones. The distribution of cones and rods is highly nonuniform. Density of cones is around 5,000 cells/mm^2 except in a region close to the central retina and with a diameter of 5.5 mm called the *macula*. Here the density of cones abruptly rises to a sharp peak. Concentric with the macula are the fovea and the foveola with diameters of 1.5 and 0.35 mm, respectively. In the foveola there are no rods, and the cones are longer and more densely packed, with a peak density of about 150,000 cells/mm^2 (with high variability among individuals). Rod cell density, which is zero in the foveola, rises rapidly toward the edges of the macula, where it gets a similar value to maximum cone density in the foveola. From there, it steadily decays to 30,000 to 70,000 cells/mm^2 at the peripheral retina. Because the area of the fovea is much smaller than the total area of the retina, only 1% of the total number of cones is within the fovea. Typically, there are three types of cones, depending on the pigment present in their photo-sensitive region, which determine spectral responsivity. S-cones have peak responsivity at 420–440 nm (blue), M-cones at 534–545 (green), and L-cones at 564–580 (yellow). The different signals generated by each type of cone for a given stimulus allow the brain to perceive a continuous range of colors. There is only one type of rod cells, with peak sensitivity at 498 nm. As multiple rod cells converge on a single inter-neuron, their class is more sensitive to low-level light at the expense of lower spatial resolution. They also have a slower time response than cone cells. The exit and entrance point for ganglion cell axons and blood vessels is the optic disk. As there are no photoreceptor cells on top of it, a blind spot is created at its location, to the nasal side of the central retina. Simple experiments that expose the blind spot seem to locate it to the temporal side of each eye, but this is just because the retinal image is inverted.

The eye is a system with rotation symmetry, but only approximately. Neither the lens nor the cornea have perfect rotation symmetry, and there can be tilts between their surfaces and between the elements. Still, the optical axis is defined as the better linear fit passing through the (paraxial) center of curvature of the four surfaces. Besides the optical axis, a *visual axis* is defined as the lines joining the fixation point with the center of the fovea passing through the nodal points. This is not a single line, as the object and image nodal points are not coincident (though they are close). Even more important from the practical point of view is the *line of sight*, defined as the line joining the fixation point with the center of the entrance pupil. The line of sight is the central ray of the beam entering the eye, and is especially important for refraction and for measuring ray aberrations, as the pupil can be used by the observer for alignment purposes, but the nodal point cannot.

Finally, the eye can rotate by the action of six extra-ocular muscles, the *lateral*, *medium*, *superior* and *inferior rectus*, and the *inferior* and *superior oblique*. However, the rotation of the eye neither has a definite rotation center nor definite rotation axes. At each gaze direction, the eye has a local rotation center that depends on gaze direction and the instantaneous rotation axis. Spectacle lens design usually relies on the assumption that the eye has a definite rotation center [55, 56], and although the lack of it does not jeopardize lens performance in a significant way, it is convenient to have an idea of the extension of the region around which the eye rotates. Park characterized the rotation of the eye in the horizontal plane [57]. They encountered the instantaneous centers of rotation of the eye to be located to the nasal side of the visual axis, closer to the corneal vertex for large temporal gaze angles, and further away from it for large nasal gaze angles.

Figure 5.1(b) shows a schematics of the two refractive elements of the eye, the cornea and the lens, the aperture stop (iris) and the retina, along with the location of the Gaussian cardinal points: principal points H and H′, nodal points N and N′, position of the entrance and output pupils, E and E′, focal points F and F′, and the location of the average rotation center of the eye according to Fry (hollow circle) [58], Perkins (cross) [59], and the more precise centrode determined by Park, which is the thick curved line below the visual axis. All the distances are scaled according to the averaged data used in the Le Grand full theoretical eye [52]. This mention bring us to the last part of this section, devoted to a brief summary of eye models. The reader may consult the paper by Atchinson and Thibos, [60], for a larger and readable summary, or chapter 16 in [54], and the references therein.

An eye model is a physical or theoretical construct in the form of an optical system that, to a greater or lesser extent, mimics the properties and/or performance of the human eye. Physical eye models are typically used for calibrating optometric and ophthalmic instruments, also for academic use. We are more interested in theoretical eye models, which are used to understand the optical properties of the eye, predict the outcome of eye manipulation (for example, refractive surgery), and to design tests, instruments, or compensating optical components. Eye models require experimental data on the geometry of the refracting surfaces of the cornea and the lens, their thicknesses and positions, the values and distribution of refractive index of the different media, etc. The data finally fed into the model can be obtained as population averages (population eye models) or can be data from a given individual (customized or individual eye models).

From the point of view of complexity, eye models are typically separated into schematic eye models and finite models. The first are usually constructed to mimic the paraxial properties of real eyes, while the second incorporate richer information on surface geometry (asphericity that can even contain information on individual eye aberrations), the variable distribution of refracting index in the lens, and the curvature of the retina. It would be a waste of resources to analyze all this complex information within the paraxial environment, so the optical properties of finite models are always computed by means of exact ray tracing.

Eye models are a fundamental tool for the study and design of intraocular lenses and, to a lesser degree, of contact lenses as well. In particular, detailed eye models are needed to compute aberrations of the compensated eye and the tolerance of image quality to lens tilt or displacement, and in the case of contact lenses, the detailed geometry of the cornea and its flattening toward the corneoscleral rim. The use of eye models in the theory of spectacle lenses has been quite different. Some properties of the lens-eye system make use of the cardinal points of the eye and the entrance pupil, but classical lens design is based on the assumption that the whole eye can be substituted by a stop located at its center of rotation, as we will see later in detail. Advances in visual optics and vision research and the availability of advanced eye measuring systems are currently leading to different design paradigms. For example, some proposals have been made to design ophthalmic lenses according to the particular aberrations of the eye [61, 62], or taking into account the individual distance from the back vertex of the lens to the rotation center of the eye [63], or the axial length of the eye [64]. These cannot be considered applications of complete eye models to spectacle lens design, but there is one where the eye models have a definite relevance: the design of lenses intended to control the so-called *peripheral defocus* [65, 66, 67, 68], something that could be useful to control the development of *myopia* [69, 70]. We will define refractive errors in the next section, including myopia. In chapter 6 we will briefly describe the lenses proposed so far for this control of peripheral defocus.

Regardless of the direct applicability of eye models to the theory of ophthalmic lenses, it is important for the reader to be familiar with the optical system of the eye, the geometry and position of its lenses, and the Gaussian properties of the average eye. We have selected the so-called Le Grand full theoretical eye as the simplest eye that realistically reflects the overall structure of the eye. Le Grand's model neither takes into account asphericities, nor does it consider the variable refractive index of the crystalline lens. Instead, it assigns to it an effective – higher – refractive index that compensates for the power increase resulting from the gradient index of the actual lens. Despite these simplifications, Le Grand's model gives us a good grasp of the eye dimensions and basic optical properties. The model parameters are shown in Table 5.1.

5.2.2 Eye Aberrations and Refractive Errors

According to Le Grand's eye model, the distance from the back surface of the crystalline lens to the retina is 15.597 mm. When the light from an object located at infinity refracts through it, the vergence of the wavefront leaving the back surface of the lens should be the inverse of the axial length of the vitreous body, $V'_{L2} = 1/0.016597 = 60.25$ D. This is because the Le Grand full theoretical eye, like any other model of the relaxed eye, is fine tuned for the image of distant objects to form on the retina. In order to account for the actual aperture of the eye, the standard procedure is to consider the wavefront at the exit pupil. The two wavefronts are shown in Figure 5.2(a), and we name them $W_{E'}$ and W'_{L2}. The exit pupil is inside the lens, but the reader has to take into account that it is the image of the aperture in the image space, so we can get from W'_{L2} to $W_{E'}$ and the other way around by simple space propagation, that is $V'_{L2} = V_{E'}/(1 - zV_{E'})$, where z would be the distance from the exit pupil to the back surface of the lens. In subplot (a) we show a schematic of the eye

Table 5.1 *Description and Gaussian properties of the Le Grand full theoretical eye.* *The retinal radius is not used in paraxial computations, and, strictly speaking, is not part of Le Grand's model.*

Le Grand "full theoretical eye"				
System description		**Gaussian properties**		
Cornea	$R_1 = 7.8$ mm	VF	−15.089	mm
	$R_2 = 6.5$ mm	VF$'$	24.197	
	$t_0 = 0.550$ mm	VH	1.595	
	$n = 1.3771$	VH$'$	1.908	
Aqueous	$t_0 = 3.050$ mm	VN	7.200	
	$n = 1.3374$	VN$'$	7.513	
Lens	$R_1 = 10.2$ mm	H$'$F$'$	22.29	
	$R_2 = -6.0$ mm	HF	−16.68	
	$t_0 = 4.00$ mm	VE	3.038	
	$n = 1.4200$	VE$'$	3.682	
Vitreous	$t_0 = 16.597$ mm	P_E (cornea)	42.36	D
	$n = 1.3360$	P_E (lens)	21.78	
Axial length	24.197 mm	P_E (eye)	59.94	

showing the cornea and the lens (in gray) and the back of the eye. The horizontal direction is the line of sight, so it intercepts the retina at the center of the fovea. Marginal rays are traced through the system, and their virtual continuations to the entrance and exit pupils are shown with solid gray lines. The curvatures and surface positions are not uniformly scaled from a realistic eye model, in particular the vitreum has been shortened to better appreciate the variations of curvature of the wavefronts in subplots (b) to (d). Likewise, the cornea and the lens should be slightly tilted with respect to the line of sight, but this is not relevant to our discussion now, so we have not incorporated it into the plots to make them simpler. The wavefront refracted by the eye in subplot (a) is a perfect sphere and so are all the wavefronts from the exit pupil to the retina, where these wavefronts collapse into a focal point F'. An eye, or any other imaging optical system for that matter, generating such a wavefront at the exit pupil would be perfect.

Let us pay attention now to subplot (b). This "eye" is still producing a perfectly spherical wavefront $W_{E'}$, but with too large a vergence. For comparison, we are still drawing the wavefronts and rays of the previous perfect eye with dotted lines. The beam focuses into a distinct point F', but because of the larger curvature at $W_{E'}$, the focal point is located in front of the retina. After F' the beam becomes divergent, and its interception with the retina is a blur disk with size δ. When the eye's focal point does not lay on the retina, we say the eye is affected by refractive error, and the condition is known as ametropia. The particular case in which the curvature of the wavefront at the exit pupil is larger than

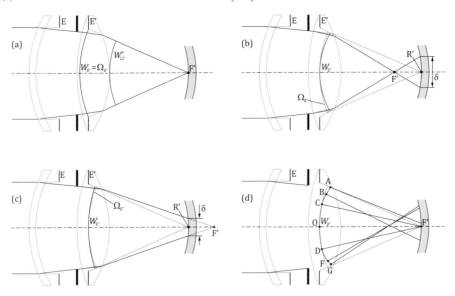

Figure 5.2 Wavefronts and marginal rays for four eyes with different refractive states. The horizontal line stands for the line of sight of the eye. The distances and curvatures are not uniformly scaled, to better appreciate the differences between the wavefronts. E and E' are the entrance and exit pupils, and $W_{E'}$ and W'_{2L} are the wavefronts at the exit pupil and the back surface of the lens, both corresponding to the beam refracted by the whole optics of the eye. (a) A perfect eye with spherical wavefronts having the right curvature to converge to the retina, that is, the image focal point of this eye is on the fovea. (b) Myopic eye. The refracted wavefront is still perfectly spherical but has excess vergence. The focal point F' is closer to the back surface of the lens, in front of the retina. (c) Hyperopic eye. Now the refracted wavefront is too flat and converges into a point F' behind the retina. (d) Emmetropic eye with higher-order aberrations. The actual wavefront at the exit pupil W'_{2L} has the right vergence between points C and D. The rays coming between these two points would focus at the F' located at the retina. However, the wavefront starts distorting above C and below D. The changes in curvature at points A, B, F, and G translate into slope changes and the corresponding rays not converging into F'.

$1/\overline{E'R'}$ is called myopia. Subplot (c) shows the opposite condition, known as hyperopia; the wavefront is still spherical but the curvature of $W_{E'}$ is smaller than $1/\overline{E'R'}$. The beam converges too slowly for the eye's axial length, and the focus would be formed behind the retina. Of course, before converging to F' the beam impinges on the retina forming a blur patch with size δ.

Cases (a) to (c) are somewhat unreal, as we are assuming the refracted wavefronts are perfectly spherical. Indeed, no biological or artificial optical system can be absolutely perfect. Even if the surfaces of the cornea and lens were absolutely spherical and perfectly centered, the refracted wavefronts would not be. Most often, real eyes even lack approximate revolution symmetry, and the refracted wavefront at the exit pupil is astigmatic. In such a case, the refracted beams do not converge to a single point, but to a Sturm's conoid, as we studied in Chapter 3.

Even more, what really happens is that the curvature of the refracted wavefront is not constant all over its surface. In the last subplot (d) we see the case of an eye whose wavefront $W_{E'}$ has constant curvature in its central region, between points C and D, but its curvature changes at its rim. The rays passing through this central region converge to a distinct focal point F', and if the curvature between C and D is equal to $1/\overline{ER'}$, the focus F' will lay at the retina. However, the depicted wavefront gets progressively more curved at F and G. This means the slope of the wavefront at F is higher than at the similar point in the perfect wavefront, and the ray passing through F is then pointing above F'. At G there is even more curvature, a bigger slope, and the corresponding ray is pointing even further up from the center of the fovea. In the upper part of the wavefront we have depicted other types of deformation. The wavefront becomes more curved at B, and the corresponding ray points below F', but then flattens at A, the slope gets more vertical, and the corresponding ray impinges on the retina above F'. In summary, the energy propagating through the center of the beam focuses on the retina, but the energy propagating through the rim surrounding the central region spreads around the focal point in a complex pattern that depends on the local distribution of curvature in this rim.

The last case illustrated in Figure 5.2(d) is indeed the actual situation in any *emmetropic* eye. The central curvature of the exit wavefront is matched with the distance from the exit pupil to the retina, and the rays refracting through this central region converge into the retina. But the rays refracted farther away from the optical axis do not point to the focus; they rather intersect the retina in a complex pattern that depends on the distribution of curvature of the exit wavefront. If the pupil has its size reduced, these peripheral rays will be blocked and the energy distribution on the retina will be sharply concentrated at F'. *Emmetropia* is then the condition in which the eye is free from refractive error, but in general it will still present some degree of third- and higher-order aberrations.

It is also possible that the curvature of the central region, though constant, does not equal $1/\overline{E'R'}$. It can even be astigmatic, in which case it requires three numbers to be fully specified and a distinct punctual focus is not formed. In any of these cases the eye is said to be ametropic. The previous discussion suggests there are three different types of ametropia:

Myopia The center of $W_{E'}$ has revolution symmetry and its curvature is bigger than $1/\overline{E'R'}$.

Hyperopia The center of $W_{E'}$ has revolution symmetry and its curvature is smaller than $1/\overline{E'R'}$.

Astigmatism The center of $W_{E'}$ does not have revolution symmetry. Its curvature can be described by a matrix $\mathbb{V}_{E'}$ with two main vergences $V_{E'+}$ and $V_{E'-}$. If both $V_{E'+}$ and $V_{E'-}$ are bigger than $1/\overline{E'R'}$, the astigmatic wavefront converges to two focal lines located in front of the retina, the condition referred as *compound myopic astigmatism*. If the two main vergences are smaller than $1/\overline{E'R'}$, the focal lines will be located behind the retina, and the condition will be *compound hyperopic astigmatism*. If $V_{E'+} > 1/\overline{E'R'}$ and $V_{E'-} < 1/\overline{E'R'}$, the retina will lay between the two focal lines and the astigmatism is said to be *mixed*. Finally, it is possible that any of the main vergences equals $1/\overline{E'R'}$, henceforth its corresponding focal line will lie on the retina. The condition is known as *simple astigmatism*. If it is the focal line corresponding to $V_{E'}$, the one lying on the retina, the eye will have *simple myopic astigmatism*. If the focal line on the retina corresponds to V_{E+}, the condition will be *simple hyperopic astigmatism*.

The deviations of the refracted wavefront from the perfect sphere that converges to the retina are known as *aberrations*. We can now proceed to a more quantitative definition of them. Our approach is a little unusual, but we think it effectively uses the techniques presented in this book and will offer an interesting perspective to the reader. A more conventional approach to aberrations based on Zernike polynomials is presented in Chapter 6 and Appendix E. Let $W_{E'}(x, y)$ be the function describing the wavefront at the exit pupil of the eye. The coordinates are measured with respect to a reference system whose plane XY contains the exit pupil and the Z axis coincides with the line of sight. Whatever the shape of the wavefront, it can be described with arbitrary accuracy by using enough terms of its Taylor expansion (See Sections 2.4.2 and C.4.1),

$$
\begin{aligned}
W_{E'}(x, y) = {} & \psi_{00} + \psi_{10}x + \psi_{01}y \\
& + \psi_{20}x^2 + \psi_{11}xy + \psi_{02}y^2 \\
& + \sum_{n\geq 3}^{n} \sum_{m=0} \psi_{n-m,m} x^{n-m} y^m.
\end{aligned}
$$

The first line contains the constant ψ_{00} and the linear terms. If the wavefront is converging to the center of the fovea, it should be perpendicular to the line of sight at the origin of coordinates. This means its first partial derivatives should be zero at the origin, and then $\psi_{10} = \psi_{01} = 0$. The constant ψ_{00} is also zero, as the wavefront is tangent to plane of the exit pupil and hence to the XY plane. The second line contains the second-order terms. We already know that the second-order derivatives at the origin are directly and uniquely related to them. The three quadratic terms can be given in terms of the Hessian matrix of the wavefront at the origin. The terms with $n \geq 3$ do not affect the curvature at the origin as their second derivatives go to zero at $x = 0$, $y = 0$. However, they do change the curvature of the wavefront away from the origin.

Let us consider now the perfect spherical wavefront that would converge to a single point at the retina, and let us name it $\Omega_{E'}(x, y)$ (in subplot (a) of Figure 5.2, we would have $\Omega_{E'} = W_{E'}$). Its polynomial expansion is

$$
\Omega_{E'}(x, y) = \frac{V_{E'}}{2}(x^2 + y^2) + \sum_{n\geq 3}^{n} \sum_{m=0} \omega_{n-m,m} x^{n-m} y^m.
$$

The quadratic term is determined by the vergence of this perfect wavefront, which we know have to be $V_{E'} = 1/\overline{E'R'}$. The higher-order terms account for the fact that the surface is spherical, not paraboloidal (recall that a paraboloid is an approximation for a real sphere, only valid in a neighborhood of the origin). Now, let us define $\Delta W_{E'} = W_{E'} - \Omega_{E'}$ as the *aberration function* of the eye. Its polynomial expansion is

$$\Delta W_{E'}(x, y) = \frac{1}{2} \begin{bmatrix} x & y \end{bmatrix} \mathbb{R}_{E'} \begin{bmatrix} x \\ y \end{bmatrix} + \sum_{n \geq 3} \sum_{m=0}^{n} w'_{n-m,m} x^{n-m} y^m,$$

where

$$\mathbb{R}_{E'} = \begin{bmatrix} 2\psi_{20} & \psi_{11} \\ \psi_{11} & 2\psi_{02} \end{bmatrix} - \frac{V_{E'}}{2} \mathbb{I},$$

and $w'_{nm} = \psi_{nm} - \omega_{nm}$. It has two differentiated terms: the quadratic term, determined by $\mathbb{R}_{E'}$, and the higher-order term, determined by the coefficients w'_{nm}. Whether the eye is emmetropic or ametropic depends on $\mathbb{R}_{E'}$. If $\mathbb{R}_{E'}$ is zero, the eye es emmetropic. If $\mathbb{R}_{E'}$ is a positive multiple of the unit matrix, the eye is myopic. If it is a negative multiple of $\mathbb{I}$, the eye will be hyperopic. Finally, if $\mathbb{R}_{E'}$ is not a multiple of the unit matrix, the eye will be astigmatic. On the other hand, the higher-order term determines the way the curvature *of the aberration function* will change as we move away from the origin. We must take into account that once the large curvature $V_{E'}$ has been removed, the remaining curvature of the aberration function is usually very small for the typical apertures in the eye, that is, the quadratic form $(1/2)\mathbf{r}^\top \mathbb{R}_{E'} \mathbf{r}$ will approximate a sphere with high accuracy, and hence any significant values of the coefficients w'_{nm} will point to variations of curvature, that is, higher-order aberrations, as we get away from the origin. We see that the role of the aperture is very important. If the pupil is small, x and y are restricted to small values, and the terms $x^n y^m$ with $n + m > 3$ become negligible with respect to the quadratic terms. If the pupil dilates and if the values of some of the coefficients w'_{nm} are not very small, the higher-order term may grow comparable to or even exceed the quadratic one. Let us note that $\mathbb{R}_{E'}$ will not change with pupil size, as it describes the curvature at the very center of $\Delta W_{E'}$, but its effect, the size of the retinal blur that it causes, does depend on pupil size. Simple inspection of Figure 5.2 (b) and (c) may help us to understand it. The position of the focus F' will not change with pupil size, as it is determined by the central curvature $\mathbb{R}_{E'}$. However, the marginal rays that determine the limits of the blur will form a smaller angle with the Z axis if the pupil is reduced, so the size of the blur, δ, will also be reduced. On the other side, the higher-order term does not affect the curvature of $\Delta W_{E'}$ at the center. It does affect the aberration function (and its curvature) at any point other than the center, the effect being larger the further the point is from the center of the pupil and the larger are the coefficients.

Now, there is an important issue we have to mention. The above discussion about refractive states and wavefront curvature has been entirely based on geometrical optics, which is fine to describe light refraction and propagation in terms of rays or wavefronts, but fails when describing the distribution of energy very close to the focal points. A more complete theory of light propagation requires taking into account its wave nature and the subsequent diffraction effects at the apertures of the optical system [71]. The effect is that even a perfect spherical wavefront does not converge to a punctual focus. Instead, energy spreads at the focal plane forming the so-called *Airy disk*. For a perfect eye with equivalent power P_E and pupil diameter D_p, the diameter of the Airy disk, for light with wavelength λ, is

$$D_{Airy} = 2.44 \frac{\lambda}{D_p P_E}.$$

The smaller the Airy disk, the better the resolution the eye can achieve. But according to the previous expression, the smaller the pupil, the larger D_{Airy} will be. We are then faced with trading both factors off. Assume an emmetropic eye with $R_{E'} = 0$. The larger the pupil, the larger the effect of the higher-order aberrations and the spreading of energy on the retina. If we reduce the pupil, the effect of the higher-order aberrations will be reduced, but then the energy spreads because of diffraction. The compromise in natural emmetropic eyes with standard levels of high-order aberrations is a pupil size between 2.5 and 3 mm. Smaller pupils will deteriorate vision because of diffraction. Larger pupils will do the same because of the growing effects of higher order aberrations. Ametropic eyes do not have such a trade-off. The presence of even small amounts of ametropia produces retinal blur patches much larger than the Airy disk at any pupil size, and so the smaller the pupil, the smaller the impairment the ametropia produces in vision. Any reader with small to medium values of ametropia will have experienced how uncompensated vision improves outdoors in sunny days, as a consequence of the reduction of pupil size.

We have defined the aberration function of the wavefront refracted by the eye and referred to the exit pupil. This approach seems natural, as we compare the eye's actual output with what should be its ideal output. It is indeed the approach typically used to introduce the concept of aberrations and different methods have been developed to measure the wavefront $W_{E'}$ and its associated aberration function $\Delta W_{E'}$. However, once the concept is established and understood, another parallel definition turns out to be more practical. Imagine we could place a punctual light source at the center of the fovea. It would shine light backward that would emerge from the entrance pupil of the eye. The situation is presented in Figure 5.3, where we show a simpler schematics of the eye just showing the first and last refracting surfaces with thick gray lines and the continuation of the rays from the object space to the entrance pupil and from the image space to the exit pupil with thin gray lines. Wavefronts are plotted now with dotted black lines. If the eye were perfect, the output wavefront would be plane, corresponding to an infinity-located image of the retinal source point. This is represented in subplot (a). In a real eye, the output wavefront W_E will not be plane. It rather may have nonzero curvature at the center and, similarly as $W_{E'}$, a term composed by third and higher combined powers of x and y that will show up as we consider points (x, y) away from the line of sight; see subplot (b) in Figure 5.3. In practice, the wavefront W_E can be directly measured [72, 73, 74, 75, 76]. A thin laser beam is focused on the fovea, the spot creating a light source tiny enough so that the wavefront at the exit pupil can be considered perfectly spherical. This spherical wavefront is refracted backward by the eye, and the emerging wavefront W_E can be measured at the plane of the entrance pupil (or any other plane, but the entrance pupil gives the right aperture). As the reference wavefront is now plane, the aberration function is W_E itself. If we use a reference system

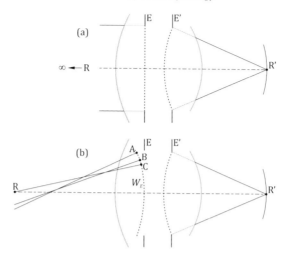

Figure 5.3 Eye aberration and refractive error referred to the entrance pupil. (a) Perfect eye. A plane wavefront at the entrance pupil will be refracted into a perfectly spherical wavefront converging at the retina. (b) Aberrated eye, with refractive error. For the refracted wavefront to be perfectly spherical and converging at the retina, the input wavefront cannot be plane. It may have some central curvature (ametropia) and further variations of curvature toward the outer rim. This wavefront would be the one refracted by the eye backward if a punctual light source were placed at the center of the fovea.

with the origin at the point where the line of sight crosses the entrance pupil and assume the aberration function is perpendicular to the line of sight at the origin, then we can write

$$W_E(x, y) = \frac{1}{2} \begin{bmatrix} x & y \end{bmatrix} \mathbb{R}_E \begin{bmatrix} x \\ y \end{bmatrix} + \sum_{n \geq 3} \sum_{m=0}^{n} w_{n-m,m} x^{n-m} y^m. \tag{5.1}$$

The quadratic term accounts for the center curvature of the exit wavefront, and it is known as the *refractive error* of the eye at the entrance pupil. If W_E were plane at its center, $\mathbb{R}_E$ would be zero and the eye would be emmetropic, otherwise it would be ametropic. Take into account that the signs of $\mathbb{R}_E$ and $\mathbb{R}_{E'}$ are opposite. A myopic eye would generate a wavefront W_E with negative curvature at the entrance pupil, that is, a wavefront that is geometrically diverging from a point R located in front of the eye. The wavefront coming backward of the hyperopic eye would have positive curvature. It then geometrically converges to a virtual point R located behind the eye. In either case, point R is known as the *remote point of the eye*. Astigmatic eyes are those for which the wavefront W_E is astigmatic, with vergence $\mathbb{R}_E$. In general it will have two main curvatures along its two main meridians. These will geometrically converge toward, or diverge from, their associated focal lines. We can say that in astigmatic eyes the remote point splits into two, one for each of the focal lines of its aberration wavefront W_E. The cases of the myopic, hyperopic, and two types of astigmatic eyes are shown in Figure 5.4.

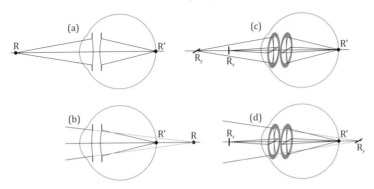

Figure 5.4 Ametropia can be defined in terms of the central curvature of the wavefront at the entrance pupil of the eye that, after refraction, converges at a distinct point on the retina. (a) For the myopic eye this wavefront is divergent. Its center of curvature is the so-called *remote point* R, located in front of the eye. (b) For the hyperopic eye the wavefront at the entrance pupil is convergent, and its center of curvature, or remote point, is in front of the eye. (c) and (d) show two types of astigmatic eyes. Entrance and exit pupils have been drawn with perspective to differentiate the two main meridians, each having its own center of curvature at the location of its corresponding focal line. In (c) we see an eye with compound myopic astigmatism, where in (d) the astigmatism is mixed. Remote points are split in R_x and R_y, as the main meridians are depicted horizontal and vertical. This is just to simplify the drawing, and any other orientation is possible.

5.2.3 Eye Aberration, Refractive Error and Zernike Polynomials

In the previous section we used the polynomial expansion of the wavefront (either $W_{E'}$ or W_E) to distinguish between the quadratic term, which describes the curvature at the center of the wavefront (which, for W_E, we identified with the refractive error), and the higher-order powers of x and y, along with the respective coefficients, which we have grouped under the general name *higher-order term*. This approach is the one leading to the definition of the Seidel aberrations [77, 78]. The accuracy of a polynomial expansion around the origin of coordinates depends on the number of terms used in the expansion (the order of the expansion) and the size of the region we want to compute. If the region is very small, the quadratic terms will be an excellent approximation to our functions. As we increase the size of the region (the pupil) we will need to incorporate higher-order terms to retain accuracy. Another approach to describe a function with polynomials is the use of *best fitting polynomials*. Assume we want to describe W_E by a fitting polynomial. First we decide the order; let us say we choose second order and let us also assume the wavefront is not tilted so we do not need linear terms. Then $W_E(x, y) \simeq e_{20}x^2 + e_{11}xy + e_{02}y^2$. The coefficients e_{nm} are no longer the second derivatives of W_E at the origin. Instead, they are computed so the paraboloid provides the best fit for the wavefront *all over the chosen aperture*. The difference between the two approaches is shown in Figure 5.5. The Taylor expansion will provide a perfect match for small pupils, providing the exact curvature of the wavefront. The fitting polynomial will give a much better overall approximation for large pupils, but its curvature at the center will not usually match the curvature of the actual wavefront.

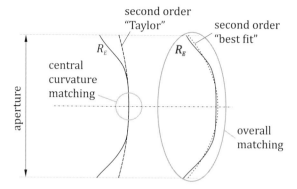

Figure 5.5 Different methods to approximate a complex wavefront. To the left, the Taylor expansion around the origin used to define refractive error in Section 5.2.2. The polynomial expansion has the same curvature as the original wavefront at the axis, but if the latter has strong asphericities at its rim, the former does not reflect them. To the right, a best fitting approximation. The low-order polynomial is more curved to better fit the whole wavefront, but the central curvature is not exact.

Seidel aberrations (briefly described in Appendix E) are polynomials similar to the monomials we have used in the previous Taylor expansions. They behave similarly in the sense they accurately describe the curvature at the center of the wavefront, but if its deformations at the edge are large, we need to add a large number of higher-order terms to describe the whole waveform accurately. Besides the very brief description of Seidel aberrations in Appendix E, we will not describe its use here, as they are seldom used in visual optics.

The fitting functions primarily used in visual optics are Zernike polynomials, also defined and briefly described in Appendix E. A more detailed overview with a thorough description of their use in describing eye wavefronts is given in the book by Dai [34]. Zernike polynomials are orthogonal in the unit circle, and are listed by two integers, $Z_n^m(x, y)$. The first, n, determines the *radial degree* or "order," which is the maximum combined power of x and y appearing in the polynomial and it goes from 0 to infinity. The second one, m, is known as *azimuthal frequency* and for a given n, it takes the values $m = -n, -n + 2, \ldots, n - 2, n$. If we convert the points in the plane (x, y) to polar coordinates (r, θ), the azimuthal frequency determines the number of oscillations of the polynomial as θ goes from 0 to 2π. Each polynomial is paired with a given type of aberration and up to fourth order they inherit the names of the closest Seidel aberrations. A table of Zernike polynomials up to fourth order is given in Appendix E in polar form, but given their importance, we will briefly describe them here giving their Cartesian form (a longer list of Zernike polynomials in both polar and Cartesian forms is given in [79]).

Zeroth order. $Z_0^0 = 1$ is called piston. It is needed for balancing aberrations. If we used just this Zernike polynomial to describe a wavefront, the coefficient would be its mean value. Unlike the Taylor expansion, even if $W_E(0, 0) = 0$, the piston term in a Zernike expansion will not be zero, as other polynomials also have constant terms that will compensate.

First order. There are two first-order polynomials: $Z_1^{-1} = 2y$ and $Z_1^1 = 2x$. They are called y-tilt and x-tilt. Their coefficients will be zero unless the wavefront presents some average tilting with respect to the XY plane.

Second order. There are three of these:

$$Z_2^{-2} = 2\sqrt{6}xy,$$
$$Z_2^0 = \sqrt{3}\left[2(x^2 + y^2) - 1\right],$$
$$Z_2^2 = \sqrt{6}(x^2 - y^2).$$

The first and the third represent astigmatism at $45°$ and $0°$, respectively. The second is called *defocus* and is a parabolic term with revolution symmetry. We may expect that the combination of the three Z_2^m somehow describes the curvature of the wavefront.

Third order. Now there are four,

$$Z_3^{-3} = \sqrt{8}(3x^2y - y^3) \text{ and}$$
$$Z_3^3 = \sqrt{8}(x^3 - 3xy^2),$$

which describe the two components of *trefoil*, and

$$Z_3^{-1} = \sqrt{8}\left[3(x^2y + y^3) - 2y\right] \text{ and}$$
$$Z_3^1 = \sqrt{8}\left[3(x^3 + xy^2) - 2x\right],$$

which describe the two components of *coma*.

Fourth order. The last order we will describe here comprises five polynomials,

$$Z_4^{-4} = 4\sqrt{10}(x^3y - xy^3) \text{ and}$$
$$Z_4^4 = \sqrt{10}(x^4 - 6x^2y^2 + y^4),$$

which describe *quadrafoil*,

$$Z_4^{-2} = \sqrt{10}\left[8(x^3y + xy^3) - 6xy\right] \text{ and}$$
$$Z_4^2 = \sqrt{10}\left[4(x^4 - y^4) - 3(x^2 - y^2)\right],$$

which describe *secondary astigmatism*, and finally

$$Z_4^0 = \sqrt{5}\left[6(x^2 + y^2)^2 - 6(x^2 + y^2) + 1\right],$$

which describes *spherical aberration*.

Although it is recommended that the Zernike polynomials are individually addressed using the two-index scheme [80], they can be rearranged for computational purposes using a single index, the one recommended by the OSA (Optical Society of America) running with increasing values of (n, m), that is, $(0, 0) \rightarrow 1$, $(1, -1) \rightarrow 2$, $(1, 1) \rightarrow 3$, $(2, -2) \rightarrow 4$,

$(2,0) \rightarrow 5$, $(2,2) \rightarrow 6$ and so on. Using this rearrangement a wavefront limited by a pupil with radius r can be fitted with the expression

$$W_E(x, y) = \sum_{n=0}^{N} c_n Z_n \left(\frac{x}{r}, \frac{y}{r}\right),\tag{5.2}$$

with all the monomials in the polynomials being dimensionless and the coefficients of the expansion having the same units as W_E, that is, length units. If we limit the number of polynomials to six and the tilts turn out to be negligible, the three coefficients c_4 to c_6 determine the average curvature of the wavefront, as we explained in Figure 5.5 and the discussion before it. This will not be the actual curvature of W_E at its center, but the average curvature all over the pupil. If more polynomials are added to the expansion, the accuracy of the local description of the wavefront will improve, but the coefficients c_4 to c_6 will still be the same. With a sufficiently high number of *modes* (a common name for the individual Zernike polynomials) the sum at the right side of (5.2) may describe W_E with high accuracy, but to compute its central curvature we either do it numerically or we unroll the Zernike expansion and reunite all the quadratic terms to get the components of the Hessian matrix of W_E. Finally, an important global figure is the wavefront *RMS* error, defined as $RMS = \sqrt{c_0^2 + c_1^2 + \cdots + c_N^2}$. RMS error can be computed for a subset of the Zernike expansion, for example the third order, and it would quantify the contribution of this particular subset to the total error. The RMS of a single mode is the absolute value of its own coefficient.

Let us summarize the material presented up to now. The wavefront at the entrance pupil of the eye (when a light source is located at the retina) determines the refractive state of the eye and the quality of the optical image at the retina. The refractive error is related to the curvature of this wavefront. But curvature changes from point to point in a surface other than a sphere, and, in general, W_E is not an sphere. We can then define the refractive error as the curvature at the center of W_E that will not depend on pupil diameter, or we can define it as the average curvature given by the second-order coefficients of the Zernike expansion, which averages the entire pupil, but depends on pupil diameter.

In either case, *the refractive error is curvature*, and as such can be described by a Hessian matrix $\mathbb{R}_E$, the coefficients c_2^{-2}, c_2^0, and c_2^2, or the spherocylindrical form $[S, C \times \alpha]$. The reader can write down the second-order term of a Zernike expansion, $c_2^{-2} Z_2^{-2}\left(\frac{x}{r}, \frac{y}{r}\right) + c_2^0 Z_2^0\left(\frac{x}{r}, \frac{y}{r}\right) + c_2^2 Z_2^2\left(\frac{x}{r}, \frac{y}{r}\right)$, use the expressions for the polynomials given above, and then rearrange the terms to get the coefficients of x^2, xy, and y^2. From there, a relation between the Hessian and the Zernike coefficients can easily be established,

$$\mathbb{R}_E = \frac{2\sqrt{6}}{r^2} \begin{bmatrix} \sqrt{2}c_2^0 + c_2^2 & c_2^{-2} \\ c_2^{-2} & \sqrt{2}c_2^0 - c_2^2 \end{bmatrix},\tag{5.3}$$

and if we use the notation

$$\mathbb{R}_E = \begin{bmatrix} R_x & R_t \\ R_t & R_y \end{bmatrix},$$

then the inverse relation reads

$$c_2^{-2} = \frac{r^2 R_t}{2\sqrt{6}},$$

$$c_2^0 = \frac{r^2 (R_x + R_y)}{8\sqrt{3}},$$

$$c_2^2 = \frac{r^2 (R_x - R_y)}{4\sqrt{6}}. \tag{5.4}$$

We already know how to transform the Cartesian form of curvature into the spherocylindrical form. To find a similar relation between Zernike coefficients and spherocylindrical components, we may compute first the main and Gaussian curvatures,

$$H = \frac{4\sqrt{3}}{r^2} c_2^0, \quad K = \frac{24}{r^4} \left[2(c_2^0)^2 - (c_2^2)^2 - (c_2^{-2})^2 \right],$$

then, using $C = 2\sqrt{H^2 - K}$ and $S = H - C/2$ from Table 4.2 and the relation $\tan 2\alpha = 2R_t/(R_x - R_y)$, which is a direct consequence of equations (C.24) in Appendix C, we get

$$C = \frac{4\sqrt{6}}{r^2} \sqrt{(c_2^2)^2 + (c_2^{-2})^2},$$

$$S = \frac{4\sqrt{3}}{r^2} c_2^0 - \frac{2\sqrt{6}}{r^2} \sqrt{(c_2^2)^2 + (c_2^{-2})^2},$$

$$\tan 2\alpha = \frac{c_2^{-2}}{c_2^2}. \tag{5.5}$$

Finally, by using the relations between the components of the curvature matrix and the spherocylindrical form that we have seen elsewhere in this book (see, for example, Table 4.2), we can compute the Zernike coefficients from the sphere, cylinder, and cylinder axis,

$$c_2^{-2} = -\frac{r^2}{4\sqrt{6}} C \sin 2\alpha,$$

$$c_2^0 = \frac{r^2}{4\sqrt{3}} \left(S + \frac{C}{2} \right),$$

$$c_2^2 = -\frac{r^2}{4\sqrt{6}} C \cos 2\alpha. \tag{5.6}$$

Example 21 When measuring a patient with an ocular aberrometer, the following values for the second-order Zernike coefficients were obtained: $c_2^{-2} = 0.443\,\mu m$, $c_2^0 = -1.461\,\mu m$, and $c_2^2 = -0.528\,\mu m$. If the pupil diameter was $D = 6$ mm, determine the refractive error of the patient.

We will first compute the cylinder. Taking $r = D/2 = 3\,mm \equiv 3 \times 10^{-3}\,m$, and expressing the coefficients in meters, we have

$$C = \frac{4\sqrt{6}}{9 \times 10^{-6}} \sqrt{\left(4.43 \times 10^{-7}\right)^2 + \left(-5.28 \times 10^{-7}\right)^2} = 1.088 \times 10^6 \times 6.89 \times 10^{-7} = 0.75\,D.$$

The sphere can be computed from equation (5.5) as

$$S = \frac{4\sqrt{3}}{r^2}c_2^0 - \frac{C}{2} = \frac{4\sqrt{3}}{9 \cdot 10^{-6}} \times -1.461 \times 10^{-6} - \frac{0.75}{2} = -1.5 \, \mathrm{D}.$$

Finally, the axis angle is

$$\alpha = \frac{1}{2}\tan^{-1}\left(\frac{0.443}{-0.528}\right) = -20° \Rightarrow \alpha = 160°.$$

Therefore, the patient's prescription is $[-1.5, 0.75 \times 160°] \, \mathrm{D}$.

5.2.4 The Nature of Curvature: Matrix and Vector Representation

For the sake of completeness, we may add another formulation of curvature (or power) known as *power vectors*. The idea of representing spherocylindrical power as vectors is from Gartner and dates back to 1965 [48]. It was further extended by Saunders and Keating [81, 82], but the current definition and interpretation of power vectors was introduced by Deal and Toop in 1993 [49]. We will use the notation introduced by Thibos et al. [83], which is the more widespread nowadays. For a given spherocylindrical form $[S, C \times \alpha]$, the components of the power vector, (M, J_0, J_{45}), are defined by

$$M = S + C/2,$$
$$J_0 = \frac{C}{2}\cos 2\alpha,$$
$$J_{45} = \frac{C}{2}\sin 2\alpha. \tag{5.7}$$

They can be interpreted as a split of the whole curvature into three components, the first being the spherical equivalent, the second a Jackson cross-cylinder with power $-C/2$ and axis $180°$ (a component $[-C/2, C \times 180°]$), and the third another Jackson cross-cylinder with power $-C/2$ and axis $45°$ ($[-C/2, C \times 45°]$). The power vector representation of curvature shares with the matrix representation the additive property: If we add surfaces (or thin lenses) perpendicular to the Z axis at their vertexes, the power vector of all the surfaces is the sum of the power vectors of each of the surfaces. Indeed, the components of the curvature matrix and those of the power vector are related by very simple linear expressions

$$M = \frac{1}{2}(R_x + R_y),$$
$$J_0 = \frac{1}{2}(R_x - R_y),$$
$$J_{45} = R_t, \tag{5.8}$$

and even easier is the relation with the second-order Zernike coefficients, as they are merely proportional

$$c_2^{-2} = -\frac{r^2}{2\sqrt{6}} J_{45},$$

$$c_2^0 = \frac{r^2}{4\sqrt{6}} M,$$

$$c_2^2 = -\frac{r^2}{2\sqrt{6}} J_0. \tag{5.9}$$

A beautiful interpretation of the power vector is the next: Let us consider the function $\kappa(\theta)$ giving the curvature along an oblique meridian forming an angle θ with the X axis (see formula C.22). This function is periodic with period π. The power vector is just the spectrum of this sine wave (see [83] for a simple proof). We have seen that the curvature matrix and the coefficients c_2^m are the quadratic terms of the Taylor and Zernike expansion of the wavefront, respectively. Another possible choice to represent the waveform is to expand it into a Fourier series [34]. It is no surprise that the power vector components are directly connected to the lower coefficients of this expansion.

Power vectors, curvature matrices, and second-order Zernike coefficients are basically the same thing [84]. Their components are homogeneous (all have the same units, diopters for power vectors and matrices, and length for Zernike coefficients) and they are all additive (as opposed to the spherocylindrical or cross-cylinder forms). These properties make them suitable for a geometrical interpretation of curvature (or power), as their components can be understood as those of vectors in 3D space. The additive property allows for simple and consistent manipulation of power and curvature. For example, the basic extension of the Abbe's invariant to astigmatic surfaces that we presented in Chapter 3, $n'\left(\mathbb{V}' - \mathbb{H}\right) = n\left(\mathbb{V} - \mathbb{H}\right)$, is a direct consequence of the additive property, and we have used this invariant to demonstrate how a lens behaves and what its power is.

Another fundamental application of the additive property is the possibility of conducting rigorous statistics of refractive errors or powers, something that would be quite complex with the spherocylindrical or cross-cylinder forms; see for example [85]. Let us consider a very simple example: Assume we have measured two refractive errors with the results $[2, -1 \times 80°]$ and $[3, 1 \times 10°]$, and we want to compute its average. The naive approach of separately averaging sphere, cylinder, and axis would lead to $[2.5, 0 \times 45°]$ and we would conclude that on average our sample has no astigmatism, which of course does not make sense. The reader may think we are performing a trick by using spherocylindrical prescriptions with different cylinder sign, and that proper averaging would require transposing all the prescription to positive (or negative) cylinder. We can do so, the sample being now $[1, 1 \times 170°]$ and $[3, 1 \times 10°]$, and the naive average $[2, 1 \times 90°]$. Now, two cylinders with almost horizontal axis average to a cylinder with vertical axis.

It is clear that because of its periodic nature, the cylinder axis cannot be used to compute averages. The reason why the naive averaging fails is simple: Scalar curvature is additive only on the same meridians. Two spherical curvatures can be added because each sphere has the same curvature along every meridian; however, the main curvatures of cylindrical curvatures cannot be added, in general, as they may lie along different meridians. When we use power vectors, we can think of the curvature (or power) as the linear combination of three base vectors of a three-dimensional vector space: If we give the name $\mathbf{R}$ to the vector $\begin{pmatrix} M & J_0 & J_{45} \end{pmatrix}$, then

$$\mathbf{R} = M \begin{pmatrix} 1 & 0 & 0 \end{pmatrix} + J_0 \begin{pmatrix} 1 & 0 & 0 \end{pmatrix} + J_{45} \begin{pmatrix} 0 & 0 & 1 \end{pmatrix}. \tag{5.10}$$

Base vector $\begin{pmatrix} 1 & 0 & 0 \end{pmatrix}$ represents a 1 D spherical curvature; vector $\begin{pmatrix} 0 & 1 & 0 \end{pmatrix}$ represents a Jackson cross-cylinder with power 1 D and axis $180°$ and vector $\begin{pmatrix} 0 & 0 & 1 \end{pmatrix}$ represents another 1 D Jackson cross-cylinder with axis at $45°$. The "coordinates" M, J_0, and J_{45} of a particular astigmatic curvature indicate the strength we have to give to each of the three base components in order to reproduce that curvature. The orientation of the base components is always the same, no matter the orientation of the astigmatic curvature they add to. This means if we add two wavefronts with curvatures $\mathbf{R}_1$ and $\mathbf{R}_2$ the resulting curvature is $\mathbf{R}_1 + \mathbf{R}_2$, the sum being understood as standard vector addition. The same can be said if instead of wavefronts we pile two thin lenses: The power vector of the stack is the sum of the power vectors of each thin lens. Of course, this can be generalized, and if we have a sample of N refractive errors, all of them described by power vectors, $\{\mathbf{R}_i\}_{i=1,\dots,N}$, we can compute their mean

$$\bar{\mathbf{R}} = \frac{1}{N} \sum_{i=1}^{N} \mathbf{R}_i. \tag{5.11}$$

Variance[1] and standard deviation can be readily computed for the components of the power vectors

$$\mathrm{Var}(X) = \frac{1}{N-1} \sum_{i=1}^{N} (X_i - \bar{X})^2, \quad \sigma_X = [\mathrm{Var}(X)]^{1/2}. \tag{5.12}$$

where X stands for any of the components M, J_0, or J_{45}.

Example 22 Compute the average of the refractive errors $[2, -1 \times 80°]$ and $[3, 1 \times 10°]$ using the matrix and vector formalisms.

Let us compute first the power matrices for the two refractive errors

$$\mathbb{P}_1 = \begin{bmatrix} 2 - \sin^2 80 & \sin 80 \cos 80 \\ \sin 80 \cos 80 & 2 - \cos^2 80 \end{bmatrix} = \begin{bmatrix} 1.0302 & 0.1710 \\ 0.1710 & 1.9698 \end{bmatrix} \mathrm{D},$$

$$\mathbb{P}_2 = \begin{bmatrix} 3 + \sin^2 10 & \sin 10 \cos 10 \\ \sin 10 \cos 10 & 3 + \cos^2 10 \end{bmatrix} = \begin{bmatrix} 3.0302 & -0.1710 \\ -0.1710 & 3.9698 \end{bmatrix} \mathrm{D}.$$

Therefore, the average power matrix is

$$\bar{\mathbb{P}} = \frac{1}{2} (\mathbb{P}_1 + \mathbb{P}_2) = \begin{bmatrix} 2.03 & 0 \\ 0 & 2.97 \end{bmatrix} \mathrm{D}.$$

Expressed in spherocylindrical form the average refractive error is $[2.03, 0.94 \times 180°]$.

---

[1] Variance is the mean of the squared deviation of each sample from the mean. When computing this mean we divide the sum of the squared deviations by $N-1$, rather than by N. This is known as *Bessel's correction* and is used to better estimate the variance of the whole population from a sample of N elements.

We will now compute the power vectors corresponding to the refractive errors whose average we want to determine. For the first refractive error

$$M_1 = 2 - \frac{1}{2} = 1.5\,\text{D},$$

$$J_{0,1} = -\frac{1}{2}\cos 160 = 0.47\,\text{D},$$

$$J_{45,1} = -\frac{1}{2}\sin 160 = -0.17\,\text{D}.$$

While for the second refractive error

$$M_2 = 3 + \frac{1}{2} = 3.5\,\text{D},$$

$$J_{0,2} = \frac{1}{2}\cos 20 = 0.47\,\text{D},$$

$$J_{45,2} = \frac{1}{2}\sin 20 = 0.17\,\text{D}.$$

Thus, the components of the average power vector are

$$\bar{M} = \frac{1}{2}(M_1 + M_2) = 2.5\,\text{D},$$

$$\bar{J}_0 = \frac{1}{2}(J_{0,1} + J_{0,2}) = 0.47\,\text{D},$$

$$\bar{J}_{45} = \frac{1}{2}(J_{45,1} + J_{45,2}) = 0\,\text{D}.$$

The refractive error, expressed in spherocylindrical form, which corresponds to this power vector is $[2.03, 0.94 \times 180°]$. So we have arrived at the same expression using matrices and power vectors.

The matrix formalism can be interpreted in the very same way. The symmetric 2×2 matrices that represent curvature form a three-dimensional vector space in which we can use the base

$$\mathbb{E}_1 = \begin{bmatrix} 1 & 0 \\ 0 & 0 \end{bmatrix}, \ \mathbb{E}_2 = \begin{bmatrix} 0 & 0 \\ 0 & 1 \end{bmatrix}, \ \mathbb{E}_3 = \begin{bmatrix} 0 & 1 \\ 1 & 0 \end{bmatrix},$$

so the curvature/power $\mathbb{R}$ can be represented as

$$\mathbb{R} = R_x\mathbb{E}_1 + R_y\mathbb{E}_2 + R_t\mathbb{E}_3.$$

In this case, $\mathbb{E}_1$ represents the cylinder $(1 \times 90°)$, $\mathbb{E}_2$ the cylinder $(1 \times 180°)$, and $\mathbb{E}_3$, which is the same as the power vector $(\ 0 \ \ 0 \ \ 1\ )$, a Jackson cross-cylinder with axis at $45°$. Of course, we can compute statistical descriptors of the components of the matrix, as we have just made with the variance and the standard deviation of the components of the power vectors, but matrix multiplication can be extended to other types of matrix operations as the square root, exponentiation, and logarithm [86], and this allows for the definition of the statistical descriptors on the matrices themselves, rather than on the individual components.

Given a symmetric matrix representing curvature, $\mathbb{R}$, its nth power $\mathbb{R}^n$ is another symmetric matrix with the same principal directions and with its principal curvatures raised to the power n (see the paper by Harris [85] for an easy proof with plenty of examples). Square roots and transcendental functions such as exponentiation can be defined as well, either through their power expansions series (which just requires integer powers of the matrix argument), or by applying the function to the principal curvatures. Of course, the square root of a matrix curvature $\mathbb{R}^{1/2}$ only makes sense if both its principal curvatures are positive, and even then a plurality of solutions can be found. As stated by Harris, logical consistency should be used to pick the right one. According to the previous discussion, mean, variance, and standard deviation of a set of symmetric matrices $\{\mathbb{R}_i\}_{i=1,\cdots,N}$ representing refractive error, power, or any other curvature, are given by

$$\bar{\mathbb{R}} = \frac{1}{N} \sum_{i=1}^{N} \mathbb{R}_i,$$

$$\mathrm{Var}(\mathbb{R}) = \frac{1}{N-1} \sum_{i=1}^{N} \left(\mathbb{R}_i - \bar{\mathbb{R}}\right)^2,$$

$$\sigma_{\mathbb{R}} = [\mathrm{Var}(\mathbb{R})]^{1/2}. \tag{5.13}$$

The interpretation of a given curvature as a "point" in a three-dimensional space leads to nice geometrical interpretations of the previous statistical descriptors as points and volumes in space, a representation brought about by Harris et al. [87, 88, 89].

Let us summarize the characteristics of the different methods explained so far to describe curvature, that is, refractive error and power:

Spherocylindrical form. Spherocylindrical and cross-cylinder forms are special notations for identifying astigmatic curvature using the main directions and curvatures. Two out of its three components are curvatures, the other one is an angle. These forms are deeply ingrained in the community of eye care professionals and lens designers. Ocular astigmatism was first understood in terms of cylindrical lenses that corrected it, and most optometric tests are formulated to find the spherical and cylindrical components of the refractive error. From the practical point of view, combination of spherical and cylindrical lenses from a trial lens case (or in a phoropter) is the simplest way to generate a huge number of different astigmatic compensations. However, computation of the optical properties of astigmatic systems from the spherocylindrical form (or the main curvatures and directions, for that matter) is usually tricky or even impossible. A spherocylindrical prescription can only describe curvature, so it can represent the curvature of a surface, the back vertex power of a lens, a refractive error, or the power of any thin lens. As such, it can describe the central curvature of the wavefront W_E (the refractive error for small pupil) or the curvature of a fitting surface to such a wavefront.

Power vectors. Its three components represent astigmatic curvature as the strength of three basic types of curvatures with fixed orientation: one sphere and two Jackson cross-cylinders. This confers additiveness to the power vectors. The three components describe curvatures, making the vectors homogeneous and suitable for mathematical manipulations, in particular statistical computations can be made on the components of the power vectors. Clinical interpretation is easy, as their base components are well known to optometrists. Finally, as the spherocylindrical forms, they may refer to central or averaged curvature.

Second-order Zernike coefficients. The three coefficients (c_2^0 c_2^2 c_2^{-2}) are basically a power vector save for proportionality coefficients. There are, however, conceptual differences that identify them as a different approach to curvature. Instead of providing the curvature of the three base components (sphere and cross-cylinders) the coefficients provide the average sagittas of the corresponding surfaces. Henceforth, they have units of length instead of inverse length. For the very same reason, they depend on the diameter of the surface to be described (curvature and sag can only be related through the distance from the vertex at which the sag is computed). They share the same mathematical properties as power vectors, and statistics can be computed on them. As they are obtained from Zernike fittings to wavefront measurements, they represent average curvature for the given pupil diameter (although if the selected pupil diameter is very small, they will eventually provide central curvature of the wavefront).

Matrix description. The matrix describing the curvature of a surface is its Hessian matrix. We have used it profusely in Chapters 3 and 4. Curvature matrices are not only additive but they also support multiplication, squaring, square rooting, and other transcendent operations. This allows for the application of most mathematical manipulations to the matrix itself, not just to their single components (for example, the computation of statistical descriptors shown in 5.13). The Hessian matrix describes equally well local curvature or average curvature. The description is exact whenever the XY plane of our reference system is tangent to the surface at the point we want to compute the curvature. Even if this condition is not met, the Hessian is still well defined (as opposed to the power vectors), but then it is just an approximation to the actual surface curvature. Probably one of the main advantages of matrices is that a neat relation between wavefront and ray direction can be established with them (see Section 3.4.2). This relation allowed us to state the propagation of astigmatic vergences and the astigmatic version of the Abbe's invariant, which in turn leads to the refraction laws for astigmatic wavefronts through astigmatic surfaces. Finally, but not less important, the symmetric matrices describing curvature are just a subset of the (2×2) nonsymmetric matrices that do not describe curvature but do describe power, as we saw in Section 4.3.5. In turn, these (2×2) matrices are components of extended (4×4) *transfer matrices* that contain all available optical information on centered paraxial astigmatic systems, and that we will define and use later in this chapter. Finally, the (4×4) transfer matrices can be further extended to (5×5) matrices to study decentered and tilted astigmatic systems. Among the many available references on the application of matrices to optics, and in particular to ophthalmic optics and optometry, the reader may consult [46, 90, 91] and references therein.

5.3 Compensation of Refractive Errors

Once we have a clear understanding of refractive error, we need a means to compensate for it. The answer to this problem is clear if we think about the definition of W_E: This is the wavefront that would come out of the eye if a punctual light source were located at the center of the fovea. We can then conclude that if we manage to convert the wavefront from a far object point into W_E at the entrance pupil of the eye, the image will also be point-like and sharp. We illustrate this compensation in Figure 5.6. Subplots (a) and (b) show two different types of compensation. In (a), a compensating element is able to change the plane wavefront from a far distant point into a complex wavefront W_E at the entrance pupil of the eye. Because of the reversibility of light propagation, this wavefront will be refracted by the eye into a perfect spherical wave that converges to R'. In case (b) a standard lens is used to generate a quadratic wavefront R_E with curvature $\mathbb{R}_E$ at the entrance pupil of the eye. This curvature can be the central curvature of W_E or the average curvature of W_E for a given pupil diameter. For small pupils, both approaches should coincide.

The reader may think that the compensation option shown in (a) is the better choice, as element PP compensates for all the eye aberrations. Indeed, some experiments demonstrate that eye aberrations can be fully compensated (within some experimental error and limitations). Liang et al. [92] accomplished full compensation of eye aberrations for the first time in a laboratory experiment using a Hartmann-Shack sensor to measure the wavefront W_E, and a deformable mirror to compensate for it [74]. The compensation allowed an auxiliary imaging system to achieve high-resolution images of the cone cells at the fovea, also for

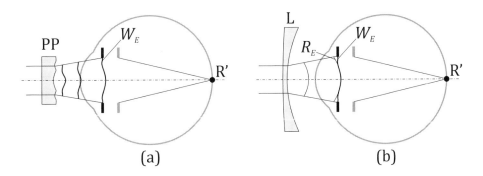

Figure 5.6 Compensation of refractive error. (a) Full compensation of the wave aberration of the eye. The compensating element PP (phase plate) is designed in such a way that it converts the plane input wavefront into W_E at the entrance pupil of the eye. Full correction of eye aberrations would be achieved. (b) Compensation of second-order aberration (either central curvature or average curvature at a given pupil size). This is accomplished by a standard lens.

the first time. Of course, these are remarkable achievements, but not practical for everyday compensation of eye aberrations, so the question is whether the compensation can be made with a simple optical element. Navarro et al. demonstrated in 2000 [93] that a compensation of up to 80% of the total aberrations of both an artificial and a natural eye were possible by means of a phase plate. This is a transparent plate with variable optical thickness, which is the product of the actual thickness times the refractive index. In some sense, a lens can be considered a phase plate, but the term is usually restricted to glass plates in which thickness variations are small. When light passes through a phase plate, the wavefront falls behind in those regions with larger optical thickness and comes ahead in those other regions with smaller optical thickness. The derivation of the output wavefront produced by a phase plate is given in Appendix D (see equation (D.12)). Other techniques have been proposed to create phase plates capable of compensating eye aberrations, for example the use of high-accuracy free-form machinery [94].

Let us take a look at the numbers involved with the wavefront sagitta and required thickness variation. Let us compare two quadratic wavefronts (or refractive errors), W_E, with revolution symmetry and curvatures of 2 and 5 D. At the edge of a 6 mm pupil, the sag of the first is $9.0\,\mu m$ while for the second the sag is $22.5\,\mu m$. The corresponding values of the Zernike coefficient c_2^0 for the two wavefronts are 2.6 and 6.5 microns, as can be deduced from (5.6). According to equation (D.12), the thickness variation required for a given wavefront is $\Delta t = W_E/(n-1)$, where n is the refractive index of the phase plate. For $n = 1.5$, the required thickness variations from the center of the plate to the edge of the region creating the 6 mm wavefronts are 18 and 45 μm. Ophthalmic lenses have much larger thickness variations because they are much larger than human pupils, and sag grows quadratically with the size of the aperture. The thickness variation from center to edge of a 60 mm diameter lens with a power of 2 D and refractive index 1.5 is 1.8 mm. A power of 5 D in the same diameter requires a thickness variation of 4.5 mm. These values were obtained for refractive errors of 2 and 5 D. What can we expect with respect to higher-order aberrations? The contribution to the wavefront sag from third- and higher-order aberrations in healthy eyes is typically below a micron, see for example [95]. The averaged RMS error over 108 eyes measured by Castejon-Mochon et al. [96] was 1.49 microns for a pupil size of 5 mm, and about 90% of it was caused by the first- and second-order modes. The inclusion of third-order terms accounted for about 97% of the total RMS error.

5.3.1 Large Field Compensation

We all know that lenses can be made with a diameter much bigger than the eye pupil. They are made that large so that the eye can rotate to aim any object in the environment and still look through the lens, and also for practical and esthetic reasons. We can see this situation in Figure 5.7. When the eye rotates, its line of sight rotates with it and crosses the lens at some point P other than its vertex V_2. Let us assume that the compensation element, either a lens or a phase plate, is generating the right wavefront W at its vertex that will become W_E at the entrance pupil of the eye in the main viewing direction. The question

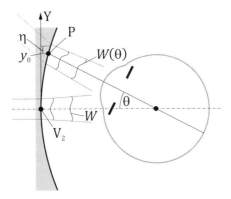

Figure 5.7 Compensation at oblique angle

now is: Would it produce also the right wavefront at P, so that when the eye rotates by angle θ the compensation is still effective, or will wavefront $W(\theta)$ be something different than W? If so, what is the difference?

A detailed answer to this question is complex and we will just give here a plausibility argument that may guide the reader. A complete demonstration would follow the same logic, but the mathematical analysis would be more elaborate. Let us assume, without losing generality, that the front surface of the lens (or the phase plate) is plane, so the thickness variations are only caused by the back surface. We will only analyze curvature along the vertical meridian, so we will reduce the problem of tackling surface curvature to an easier one-dimensional problem. Let us consider first the compensating element is a lens with a back paraboloid with vertex curvature κ. The vertical meridian is then $z = (\kappa/2)y^2$. The curvature is given by the second derivative, and of course, it is κ. Let us consider now the surface patch around point P, the one producing wavefront $W(\theta)$. Point P corresponds to the vertical coordinate y_0 and let us assign the character η to the distance from y_0 to nearby points along the Y axis. The vertical meridian can be expanded as a Taylor series around y_0,

$$z(\eta) = z(y_0) + z'(y_0)\eta + \frac{z''(y_0)}{2}\eta^2.$$

If we substitute the values of the derivatives at y_0 we get

$$z(\eta) = \frac{\kappa}{2}y_0^2 + \kappa y_0\eta + \frac{\kappa}{2}\eta^2.$$

We see that, as a function of the local coordinate around y_0, the surface is once again quadratic with the same coefficient, but it now has a constant term (the surface elevation) and a linear term (the local inclination of the surface at P). The coefficient of the linear term is κy_0, the curvature at the surface vertex times the distance to that vertex. We will see in the following sections that this is just the prismatic effect produced by the lens. Finally, the curvature at P is basically given by the quadratic term, so it must be very similar to κ^2. Now, let us introduce a cubic term in the lens surface, $z = (\kappa/2)y^2 + ay^3$, attempting to correct for some amount of coma aberration. We would need to add an extra term to the Taylor expansion at P, $(1/6) z'''(y_0)\eta^3$, which would then read

$$z(\eta) = \frac{\kappa}{2}y_0^2 + ay_0^3 + \left(\kappa y_0 + 3ay_0^2\right)\eta + \left(\frac{\kappa}{2} + 3ay_0\right)\eta^2 + a\eta^3.$$

We now see there is another linear term, so the prismatic effect will be changed with respect to the parabolic surface; we also see the same cubic term that should compensate for the third-order aberration. But now the quadratic term has increased by $3ay_0$. This means we will find at P an increase in curvature proportional to the distance y_0 and to the third-order coefficient. Correction of the third-order aberration would negatively affect correction of the second-order one! Following the same reasoning, the reader may check that a fourth order term, by^4, will further shift the curvature by $6by_0^2$ and the cubic correction by $4by_0$. In summary, the correction of a nth term at the surface vertex to compensate for a similar term of the eye aberration will introduce unwanted terms with order 1 to $n - 1$ as the eye rotates and looks through a different point on the surface. The linear terms are the only ones not producing blurring in the image, just image displacement, and this is why second-order aberrations (refractive errors) are the only errors that can be effectively compensated in large fields. Some studies have been conducted to determine the field of view within which a wavefront compensation is effective, or the tolerances to position and rotation of compensating phase plates, which is a similar problem [97, 98]. As concluded by Bara et al., a lateral displacement of 0.01 pupil diameter may reduce the effectiveness of the correction by 20%. According to Yi [94], this high sensitivity of the compensations for higher-order aberrations with spectacle lenses or plates will only pay off for eyes with large values of these aberrations and for applications where the user can hold their line of sight within a very narrow field of view.

This problem of the narrow field of view could be overcome if the phase plate and the eye could rotate together. To some approximation, this is how a contact lens behaves, so it is no surprise that considerable effort has gone into creating customized contact lenses that can compensate the complete wavefront W_E. Though some progress has been achieved in this direction, [99], the compensation of eye aberrations with contact lenses is not free from problems and limitations: lens movement, especially with RGP (rigid gas permeable) lenses, the difficulty of creating an accurate model of lens deformation for soft contact lenses, and variations of eye aberrations with accommodation and over time [100]. Other approaches for the correction of eye aberrations, including refractive errors, are refractive surgery and intraocular lenses (IOL). A basic introduction to the paraxial computation of IOLs is given in Chapter 7, but a complete review of this type of compensation is beyond the scope of this book. The interested reader may get a modern and thorough presentation of IOLs in [101].

---

[2] Let us recall that the exact computation of curvature requires the quadratic and linear terms, as shown in Appendix C, but if the vertex curvature is not too big, the deviation of the actual curvature at y_0 from κ will be small.

5.3.2 Compensation Principle for Refractive Errors

After the summary on the different alternatives to compensate the aberrations of the eye, we will now focus on the use of lenses that compensate for the quadratic components of the wavefront W_E, that is, the refractive error $\mathbb{R}_E$ (either it is central curvature or some kind of averaged curvature of W_E). Let us recall that our objective is to use a lens that will transform the plane wavefront from a distant object into a wavefront with vergence $\mathbb{R}_E$ at the entrance pupil of the eye. We already know that when a plane wavefront incides upon a lens, the vergence of the output wavefront is precisely the vertex power of the lens, $\mathbb{P}_V$. Then, if $\mathbb{R}_E$ and $\mathbb{P}_V$ are the vergences of two wavefronts of the same beam, they must have the same focus, or the same focal lines if the vergences are astigmatic. This is the so-called *compensation principle*, stating that a lens will compensate a refractive error when its focus (focal lines) coincides with the remote point (points) of the eye. Figure 5.8 illustrates the relation between the vergences involved.

Given a refractive error $\mathbb{R}_E$, the power of the lens that compensates for it is not fully determined, but depends on its position with respect to the eye, that is, depends on the distance d. By inspecting Figure 5.8, it is clear that $\mathbb{R}_E$ must equal the propagation of vergence $\mathbb{P}_V$ through a distance d, that is

$$\mathbb{R}_E = (\mathbb{I} - d\mathbb{P}_V)^{-1}\,\mathbb{P}_V, \tag{5.14}$$

or equivalently,

$$\mathbb{P}_V = (\mathbb{I} + d\mathbb{R}_E)^{-1}\,\mathbb{R}_E. \tag{5.15}$$

We have defined the refractive error as the quadratic components of the aberrated wavefront of the eye to offer a unified presentation of refractive errors and general eye aberrations; nowadays, eye aberrometers are becoming more common among ECPs (eye care professionals), and when available, the practitioner can directly obtain from them both W_E and $\mathbb{R}_E$. However, the ECP traditionally had no access to wavefront information. Instead, the refractive error is measured by optometric procedures that lead to the lens power (spherical or astigmatic) that will compensate the refractive error at a given distance. That is, the magnitude usually measured in the prescription room is $\mathbb{P}_V$ rather than $\mathbb{R}_E$. It is for this reason that ECPs usually consider refractive error *at some specified plane*. For example, the refractive error *at the spectacle plane* will exactly be the vertex power of the spectacle lens

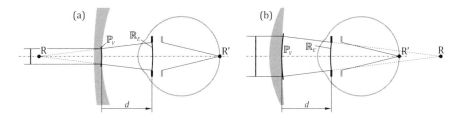

Figure 5.8 Compensation principle. (a) Myopic eye. (b) Hyperopic eye

compensating it. The refractive error at the corneal vertex would then be the power of the corresponding contact lens. These two planes are the ones we will mainly use in this book. In general, $\mathbb{R}_\#$ will mean the refractive error referred to plane #. In those cases in which the plane can be derived from the context or is not important to the discussion, we will drop the label. For a given refractive error at the entrance pupil, $\mathbb{R}_E$, the variation of the back vertex power of the compensating lens with distance d is also referred to as *lens effectiveness*.

It is worth examining the numerical difference between the refractive error at the entrance pupil and the vertex power of the spectacle lens meant to correct it. In terms of spherocylindrical power, both the sphere and the cylinder of $\mathbb{R}_E$ and $\mathbb{P}_V$ will differ, but astigmatism has no special significance here, so we may restrict the analysis to spherical power, or any of the main meridians of $\mathbb{R}_E$. The scalar relation is then $R_E = P_V/(1 - dP_V)$. The typical value of d for spectacle lenses is $d \sim 17$ mm (14 mm from the back vertex of the lens to the corneal vertex and about 3 mm more to the entrance pupil). The product dP_V will remain below 0.1 for powers up to 6 D, so we can approximate $1/(1 - dP_V) \simeq 1 + dP_V$. We then have

$$R_E - P_V \simeq dP_V^2.$$

If we assume that the sensitivity of the human eye to changes in power is 0.25 D, then the compensating power from which the variation between refractive error at the entrance pupil and lens power is bigger than 0.25 is $|P_V| \simeq \sqrt{0.25/d} \simeq 4$ D. In other words, for refractive errors smaller than 4 D in absolute value, the difference between the refractive error at the entrance pupil and the power of the lens compensating it (either contact lens or spectacle lens) will remain below 0.25 D. The exact difference between R_E and P_V computed exactly over a wide range of refractive errors is shown in Figure 5.9.

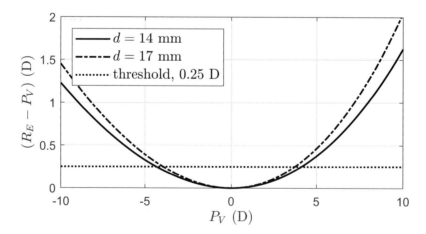

Figure 5.9 Difference between refractive error at the entrance pupil and vertex power of the compensating lens. $d = 14$ mm is the typical distance between the lens and the corneal vertex, so the corresponding curve would give the difference between spectacle lens power and contact lens power. We see that in the range $[-4, 4]$ D the difference is smaller than 0.25 D.

When refractive error at the spectacle plane is obtained with a trial frame, distance d during refraction will be similar to that of the final frame. But nowadays subjective refraction is mostly performed with phoropters. Some of these instruments may have their trial lenses farther apart from the eye than the final frame. Large refractive errors may then require correction of the vertex distance, which can be readily obtained from equation (5.14). If we call d_{ph} the distance parameter for the phoropter, d_g the distance parameter for the glasses, and $\mathbb{P}_{ph}$ and $\mathbb{P}_g$ the compensating power obtained with the phoropter and that corresponding for the glasses, then we can obtain $\mathbb{P}_g$ by simply propagating $\mathbb{P}_{ph}$ from the phoropter plane to the spectacle plane:

$$\mathbb{P}_g = \mathbb{P}_{ph} \left[\mathbb{I} - \left(d_{ph} - d_g \right) \mathbb{P}_{ph} \right]^{-1}.$$

This correction will be negligible except for large refractive errors.

5.3.3 The Size of the Eye, Emmetropization and Myopia

Let us consider now the refractive error at the corneal vertex, which we will refer to in this subsection without sub-index. We will also neglect third- and higher-order aberrations. A lens with back vertex power $\mathbb{R}$ should be located in contact with the corneal vertex to compensate for the refractive error. In this case, the image from a distant object on the line of sight should be sharply formed on the fovea. Another way of saying the eye needs this compensating lens is to say it *lacks* the power $\mathbb{R}$ to bring the image of distance objects to the retina. For example, if the eye is purely myopic, $\mathbb{R}$ can be reduced to a negative scalar R, and the eye lacks this negative power, that is, it has R diopters of power in excess. As we saw in Figure 5.2(b), this means the focus of the uncompensated eye will lie before the retina. Why is this mismatch occurring? Is it because the eye is too big or its power too strong? Indeed there is no answer to the last question. There is a large variability of eye parameters, even for emmetropic eyes [53], and so there is no "correct" size for the eye, and neither is there a "correct" eye power. Instead, the optical power of the eye and its size should properly be balanced in order for the eye to be emmetropic. Larger eyes should have weaker optics, and the other way around; smaller eyes should have stronger optics. What causes ametropia is the unbalance between the two components (and lack of revolution symmetry of the optical components in the case of astigmatism). In any case, we may compare the size of a particular ametropic eye with the average size of the human eye, for example, the size typically used in a population-based model such as the Le Grand full theoretical eye. If the size of the ametropic eye is equal or very close to the average, the ametropia is said to be *refractive*. On the other hand, if the ametropic eye is unusually large or small with respect to the average eye, we would say the ametropia is *axial*. Let us insist on the arbitrariness of these definitions, which must be understood as guides, as there is no such thing as a "correct size" for the eye.

The factors determining the characteristics of the eye, how they depend on genetics and environment, and how they evolve with growth and age are not well understood yet, despite important advances during recent decades. The distribution of the spherical component of the refractive error should be normal if the values of the parameters determining eye size and power are uncorrelated. Instead, distribution is leptokurtic: Distribution is far more concentrated than normal distribution near its maximum, and it also has higher tails [102]. This means there must be an emmetropization mechanism that guides eye growth toward emmetropization during infancy, but in some cases this emmetropization mechanism fails, leading to ametropia. Emmetropization must be optically driven: It may use the focus condition at the retina [69, 70], but it can also depend on astigmatism [103], and even on image color and luminance [104]. In any case it seems clear the evolution of refractive status with age has two main stages [105]. The first extends from birth to six years of age, approximately. Newborns present a wide Gaussian distribution of refractive error with mean of 2.5 D (hyperopic) and standard deviation of 2.2 D. The emmetropization process then takes place and the distribution of refractive errors quickly narrows with a light skew toward low hyperopia [106]. According to Flitcroft, the emmetropization process is mainly complete by this age for most populations. The second stage largely depends on the population. In some environments the refractive error distribution becomes even more leptokurtic, as if the emmetropization process were still at work. But in most populations of Western countries and specially in Asiatic countries, the opposite pattern is observed. The variance of the distribution of refractive errors grows, and both the mean and the mode get displaced toward myopic values. The typical setting of myopia starts with a rapid growth of this refractive error during childhood or adolescence, then it enters a several-year linear growth period, and finally it stabilizes during adulthood. Prevalence of myopia is reaching alarming levels in many modern societies, especially in Asia, with an average of about 47% for adults between 20 to 29 years of age and great variability among countries [107]. Despite the immense amount of research on the epidemiology and etiology about the onset, development, and arresting of myopia, and the uncovering of many optical, biological, biochemical, genetic, and environmental factors related to it, we do not fully understand how they interact, and a useful, predictive model of myopia development is lacking.

For the last 20 years, a great deal of animal experimentation has gathered evidence that hyperopic defocus on the retina triggers the axial growth of the eye [69]. On the other hand, the study of the relationship between central defocus and myopia in humans has not yielded conclusive results and a great deal of attention has gone on the peripheral defocus (see the review by Day and Duffy and references therein [108]). It seems that negative lenses typically used to compensate myopia at the fovea may induce hyperopic defocus on the peripheral retina [109], and this has triggered the development of spectacle and contact lenses that could avoid this effect [67, 110]. However, despite strong expectations on the effect of hyperopic peripheral defocus, the clinical benefits of such lenses as a means to stop myopia progression have turned out to be meager [111, 112], so clearly more research is needed.

5.3.4 Visual Acuity

Visual acuity, V, is the magnitude traditionally used for assessment of visual quality. From the optical point of view, the relevant magnitude is the *minimum angle of resolution* (MAR), which is the angle subtended by two points that can be barely seen as distinct. When stated in decimal scale, visual acuity is defined as the inverse of the MAR measured in arc-minutes. In this way we have a figure that grows as visual sharpness increases (in contrast to MAR, for which smaller values mean better vision). Visual acuity is measured by testing the ability of a person to distinguish small letters (or symbols) on a chart located at a distance of 4 to 6 meters. The letters are presented in high-contrast charts (black letters on white background) and their size has to be adjusted to the testing distance so the correct angles are subtended for each level of visual acuity. For example, the letters corresponding to $V = 1$ should have individual features subtending an angle of $1'$. The corresponding size at 6 meters is 1.75 mm, which amounts to 8.73 mm for the complete letter.

In 1976, Bailey and Lovey [113] introduced the LogMAR chart, incorporating some improvements on letter organization, font type, and font size, with logarithmic variation of the latter from line to line. For a given minimum letter size resolvable by the person, the LogMAR visual acuity is defined as the decimal logarithm of the corresponding MAR. Another advantage of the LogMAR charts is that the letter size increments from line to line correspond to linear increments of log (MAR), which makes the averaging and establishing of sub-line increments of visual acuity more precise[3].

Despite its enormous value in clinical practice, visual acuity does not uniquely determine visual quality. Vision is a complex process, and its outcome is determined by three separate stages: First, the eye forms an optical image of the external world on the retina; second, that image is sampled at the retina and encoded into nerve pulses; third, the information from the two eyes is filtered, combined, and interpreted in the visual cortex. The three stages are critical for achieving good vision. The different filtering processes, optical, sampling, and neural, do not have the same effect on high- and low-contrast images, and so a more detailed assessment of visual quality is given by the *contrast sensitivity function* (CSF), the minimal contrast of a sinusoidal grating that can be perceived at a given frequency [114]. Contrast sensitivity function and visual acuity are psycho-physical evaluations, that is, quantitatively analyzed perceptions caused by physical stimuli, chart letters, or gray-scale gratings. Of course, both visual acuity and contrast sensitivity encompasses the three stages of the visual process. Individual stages of the visual system, mainly optical image formation, can also be objectively measured with no intervening perceptions. For example, eye wavefront aberration can be measured, and from it the quality of the retinal image can be inferred using different *metrics*, some of them incorporating models of neural filtering [115, 116]. Most of these metrics correlate reasonably well with the psycho-physical parameters [117], though the latter do not predict the former with accuracy, as they do not incorporate the second and third stages of the vision process; at most they use partial models of them. In any case, there has been a lot of research on these lines during

---

[3] There is a tendency to mistakenly consider LogMAR a *linear scale* for visual acuity. The logarithm of the MAR is not more or less linear than its inverse. What is linear is the selected increments of LogMAR visual acuity from line to line in a LogMAR chart. From a scientific point of view, MAR or any function of it can be used to characterize "visual sharpness." The decimal scale will be more sensitive for small values of MAR, while LogMAR will be more sensitive for large values of the same (low vision). For clinical practice, LogMAR charts definitely have an advantage over Snellen charts.

recent years, as objective quality metrics may help to better design compensation elements (lenses) or compensation procedures (such as refractive surgery) [118, 119].

Despite these developments in the evaluation of retinal image quality, visual acuity remains the most widespread assessment of vision quality. Visual acuity is very sensitive to defocus, so it is the tool directly or indirectly used to measure refractive error and look for the best possible compensation with standard lenses. It is routinely measured in the prescription room, and its measurement is relatively simple, requiring mainstream equipment. As we will see in Chapter 6, spectacle lenses are affected by oblique aberrations. Even if the lens has the right power to compensate for the refractive error at its optical center, it is possible that this power changes considerably when the eye rotates and aims through peripheral points of the lens. In this case, the lenses oblique power does not match the refractive error, and visual acuity is reduced. Progressive lenses are also affected by unavoidable power errors in some regions of the lens. When aiming through these regions, visual acuity will, once again, be reduced. Many patents on lens design claim the use of visual acuity, or an approximation to it, as a target for lens optimization [120, 121, 122, 123], so no wonder the modelization of visual acuity is of great interest to the ophthalmic lens industry.

There are different approaches to model visual acuity. Probably the most complete model to date is that of Nestares [124, 125], which is customized and considers the three stages of the visual process, including a Bayesian discriminator to simulate optotype recognition. A completely different approach is the phenomenological model proposed by Raasch [126], set to relate visual acuity and refractive error. Raasch assigned to any refractive error a *blur* given by the length of its corresponding power vector, which in terms of the spherocylindrical prescription, $[S, C \times \theta]$, is given by

$$B_r = \sqrt{S^2 + SC + \frac{C^2}{2}}.$$

Using this single parameter as an independent variable, Raasch fit the experimental measurements of visual acuity vs. astigmatic refractive error given by Pincus [127]. Fitting models are attractive because they are simple and require very little computer power, which is a desirable property for optimization of ophthalmic compensations. However, they predict population averages; henceforth, they can neither predict the maximum visual acuity of an individual nor the reduction of visual acuity for a single individual when affected by refractive error.

An intermediate approach would use a phenomenological model that could mix experimental data (even custom data) with some knowledge about how visual acuity *should behave* in the presence of refractive error, or ill-compensated refractive error. For example, Blendowske proposed a simple model that takes into account the maximum visual acuity of an individual [128]. Using the well-known fact that for a not too small refractive error, the MAR should be proportional to it [129], Gomez-Pedrero and Alonso [130] have proposed a model that can be used to either fit population or individual visual acuity data, and can predict the small variations of visual acuity occurring when the eye is exposed to a small amount of defocus and/or astigmatism (for example, when looking through lenses with power errors). The model is embodied by the equation

$$\text{MAR} = \left\{ [\text{MAR}_0(D)]^q + (KDB)^q \right\}^{1/q}, \tag{5.16}$$

where D is the pupil diameter and $\text{MAR}_0(D)$ is the minimum MAR attainable by the individual with the best possible compensation for its refractive error (which is pupil-dependent). K is a dimensionless fitting parameter with a value close to 0.5 (see [129]) and that is also dependent on the individual. B quantifies the amount of blur, and relates to the one used by Raasch through $B = \sqrt{2}B_r$. Finally, q is another individual-dependent parameter closely related with the sensitivity to defocus. When q is small, visual acuity rapidly decays with small amounts of refractive error. A large value of q indicates the individual is tolerant to small power error over their prescription.

5.4 Prismatic Effects

Prism or prismatic effects are usually considered a side effect of the compensation of ametropia with spectacle lenses, and are linked to the fact that lenses deviate rays. Of course they do; the fact that a lens changes the vergence of a beam of light refracting through it means the rays perpendicular to the wavefronts of such a beam change direction – deviate – upon refraction. However, the term and concept of prismatic effect is missing in the study and design of most imaging optical systems, such as camera lenses, telescopes, microscopes, etc. These instruments are analyzed in terms of focal length, field of view, magnification, etc., and although there may be tilts of the wavefronts caused by misalignment or manufacturing errors, these tilts are quite different than the prismatic effects we study in ophthalmic optics. The reason for this was already mentioned in the introduction to this chapter: The spectacle lens-eye system is quite special because lens and eye do not necessarily move together. It is also special for the way human vision works, the eyes providing a very small field with high visual acuity surrounded by a large low-resolution field, and creating a brain picture of the world around us by synchronously scanning it with both eyes. In particular, the fact our vision is binocular is probably the main reason the consideration of prismatic effects is so important.

5.4.1 Propagation and Refraction of Rays through Astigmatic Systems

The approach we used in Chapter 3 to propagate light through space and refracting surfaces is wavefront-based instead of ray-based. The former is more appropriate for power definition and computation. However, for computations of prismatic effects, a ray-based approach would turn out to be far more powerful. We already had a glimpse of how ray tracing is related with wavefront propagation and refraction when defining the equivalent power $\mathbb{P}_E$ in Section 4.3.5. Now we will extend these ideas we previously used to equip ourselves with a much more powerful tool for ray propagation through astigmatic systems: 4×4 transfer matrices. This tool will be useful not just to compute prismatic effects but also magnification and field of view. These transfer matrices have been known for a long time as the natural linear approximation to the mathematical formulation of optics known as *Hamiltonian Optics* [36], and they were later proposed as efficient tools for the diffractional theory of beam propagation [38]. Transfer matrices were introduced in the field of visual optics by Keating [29, 131]. Harris has devoted an extensive work to the thorough study of their properties and applications to optometry and visual optics (see for example [46, 90, 91]).

Let us consider linear ray-tracing in three dimensions. By linear we mean that paraxial approximation is being used, so even though the systems we trace through may be astigmatic, propagation and refraction act linearly on the ray parameters. A ray can be described with four parameters. Two spatial coordinates x and y in the position vector $\mathbf{r}$, and the two components of the direction vector $\mathbf{k}$ (see Appendix B to see the relation between the two-component direction vector and its three-dimensional unitary companion).

The two basic operations for ray-tracing are shown in Figure 5.10 (a) and (b). They are propagation through space and refraction through a surface separating two media with different refractive indexes. For the space propagation, let $\mathbf{r}_1$ and $\mathbf{k}_1$ be the transverse coordinates and direction vector of the ray at a plane transversal to the Z axis. The ray propagates to a second plane, at a distance z from the first one, where it has coordinates and direction vector $\mathbf{r}_2$ and $\mathbf{k}_2$. Then it is quite clear that

$$\mathbf{k}_2 = \mathbf{k}_1,$$

$$\mathbf{r}_2 = \mathbf{r}_1 + z\mathbf{k}_1. \tag{5.17}$$

To determine the equivalent equations for the refraction operation, let us recall now the change of vergence upon refraction we studied in Chapter 3, $n'\mathbb{V}' - n\mathbb{V} = \mathbb{P}$. Let us also remember that the normal to the wavefront at a point with coordinates $\mathbf{r}$, that is, the

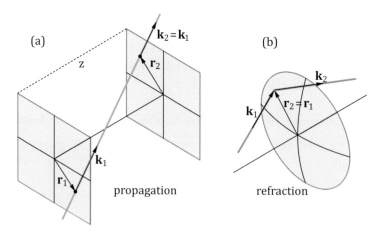

Figure 5.10 (b) refraction of a ray through a surface separating two media with different refractive index. (a) propagation of a ray through space.

direction vector of the ray that is perpendicular to the wavefront at $\mathbf{r}$, is given by $\mathbf{k} = -\nabla r$. We just have to apply all the matrix terms in the equation that give us the vergence change to the position vector $\mathbf{r}$,

$$n'\mathbb{V}'\mathbf{r} - n\mathbb{V}\mathbf{r} = \mathbb{P}\mathbf{r},$$

that is, $n'\mathbf{k}' - n\mathbf{k} = -\mathbb{P}\mathbf{r}$. On the other hand, refraction takes place on a single plane (within the paraxial approximation), so the position vectors of the incident and refracted rays are the same. Combining both results we have

$$\mathbf{k}_2 = -\frac{1}{n'}\mathbb{P}\mathbf{r}_1 + \frac{n}{n'}\mathbf{k}_1,$$

$$\mathbf{r}_2 = \mathbf{r}_1. \tag{5.18}$$

Equation pairs (5.17) and (5.18) can be rewritten as single equations if we couple the two vectors describing the ray into a single four-component vector ρ,

$$\rho = \begin{bmatrix} \mathbf{r} \\ \mathbf{k} \end{bmatrix} = \begin{bmatrix} x \\ y \\ u \\ v \end{bmatrix}.$$

Equations (5.17) and (5.18) being linear, the transformation of one of these ray vectors under propagation or refraction can be written as the multiplication of a 4×4 matrix by the initial ray vector. For example, the transformation due to spatial propagation is

$$\rho_2 = \begin{bmatrix} x_2 \\ y_2 \\ u_2 \\ v_2 \end{bmatrix} = \begin{pmatrix} 1 & 0 & z & 0 \\ 0 & 1 & 0 & z \\ 0 & 0 & 1 & 0 \\ 0 & 0 & 0 & 1 \end{pmatrix} \begin{bmatrix} x_1 \\ y_1 \\ u_1 \\ v_1 \end{bmatrix} = \mathbf{T}(z)\rho_1,$$

where $\mathbf{T}(z)$ is known as the *propagation operator*. Splitting this matrix equation into components we get

$$x_2 = x_1 + zu_1,$$

$$y_2 = y_1 + zv_1,$$

that is, $\mathbf{r}_2 = \mathbf{r}_1 + z\mathbf{k}_1$, and

$$u_2 = u_1,$$

$$v_2 = v_1,$$

or $\mathbf{k}_2 = \mathbf{k}_1$. A convenient way to write the propagation operator is as a 2×2 block matrix in which each block is in turn a 2×2 standard matrix,

$$\rho_2 = \begin{bmatrix} \mathbf{r}_2 \\ \mathbf{k}_2 \end{bmatrix} = \begin{bmatrix} \mathbb{I} & z\mathbb{I} \\ \mathbb{O} & \mathbb{I} \end{bmatrix} \begin{bmatrix} \mathbf{r}_1 \\ \mathbf{k}_1 \end{bmatrix},$$

where, as usual, $\mathbb{I}$ and $\mathbb{O}$ are the identity and the null matrices, respectively. Block matrices can be multiplied using the same procedures for standard matrix multiplication, except that order must be observed when the 2×2 blocks do not commute. Using the same idea, the refraction operation can be written as

$$\rho_2 = \begin{bmatrix} \mathbf{r}_2 \\ \mathbf{k}_2 \end{bmatrix} = \begin{bmatrix} \mathbb{I} & \mathbb{O} \\ -\frac{1}{n'}\mathbb{P} & \frac{n}{n'}\mathbb{I} \end{bmatrix} \begin{bmatrix} \mathbf{r}_1 \\ \mathbf{k}_1 \end{bmatrix} = \mathbf{Q}(\mathbb{P})\rho_1,$$

where $\mathbf{Q}(\mathbb{P})$ is known as the *refraction operator*.

The great advantage of block matrices is that any optical system made with centered lenses can be decomposed as a series of spatial propagations and refractions. For example, let us consider the refraction of a ray through a thick lens whose surfaces have refractive powers $\mathbb{P}_1$ and $\mathbb{P}_2$, with center thickness t_0 and refractive index n. The ray incides on the lens front surface with position vector $\mathbf{r}_1$ and direction vector $\mathbf{k}_1$. First operation is refraction at the front surface, $\mathbf{Q}(\mathbb{P}_1)$; second operation is propagation up to the second surface, $\mathbf{T}(t_0)$. The last operation is refraction at the back surface, $\mathbf{Q}(\mathbb{P}_2)$. The propagation of the ray through the lens requires the application of $\mathbf{Q}(\mathbb{P}_1)$, then $\mathbf{T}(t_0)$, and finally $\mathbf{Q}(\mathbb{P}_2)$, that is $\rho_2 = \mathbf{Q}(\mathbb{P}_2)\mathbf{T}(t_0)\mathbf{Q}(\mathbb{P}_1)\rho_1$. Let us call the total operator for the lens $\mathbf{L}$. The matrix product yields

$$\mathbf{L} = \begin{bmatrix} \mathbb{I} & \mathbb{O} \\ -\mathbb{P}_2 & n\mathbb{I} \end{bmatrix} \begin{bmatrix} \mathbb{I} & t_0\mathbb{I} \\ \mathbb{O} & \mathbb{I} \end{bmatrix} \begin{bmatrix} \mathbb{I} & \mathbb{O} \\ -\frac{1}{n}\mathbb{P}_1 & \frac{1}{n}\mathbb{I} \end{bmatrix}$$

$$= \begin{bmatrix} \mathbb{I} & \mathbb{O} \\ -\mathbb{P}_2 & n\mathbb{I} \end{bmatrix} \begin{bmatrix} \mathbb{I} - \tau\mathbb{P}_1 & \tau\mathbb{I} \\ -\frac{1}{n}\mathbb{P}_1 & \frac{1}{n}\mathbb{I} \end{bmatrix}$$

$$= \begin{bmatrix} \mathbb{I} - \tau\mathbb{P}_1 & \tau\mathbb{I} \\ -\mathbb{P}_1 - \mathbb{P}_2 + \tau\mathbb{P}_2\mathbb{P}_1 & \mathbb{I} - \tau\mathbb{P}_2 \end{bmatrix}. \tag{5.19}$$

According to this result, the output ray vector is related to the input ray vector with the equation

$$\begin{bmatrix} \mathbf{r}_2 \\ \mathbf{k}_2 \end{bmatrix} = \begin{bmatrix} \mathbb{I} - \tau\mathbb{P}_1 & \tau\mathbb{I} \\ -\mathbb{P}_1 - \mathbb{P}_2 + \tau\mathbb{P}_2\mathbb{P}_1 & \mathbb{I} - \tau\mathbb{P}_2 \end{bmatrix} \begin{bmatrix} \mathbf{r}_1 \\ \mathbf{k}_1 \end{bmatrix}, \tag{5.20}$$

that is,

$$\mathbf{r}_2 = (\mathbb{I} - \tau\mathbb{P}_1)\mathbf{r}_1 + \tau\mathbf{k}_1,$$
$$\mathbf{k}_2 = -\mathbb{P}_E\mathbf{r}_1 + (\mathbb{I} - \tau\mathbb{P}_2)\mathbf{k}_1. \tag{5.21}$$

We recover here the partial result we demonstrated in Section 4.3.5 related to the deviation of a ray that propagates through a thick astigmatic lens with surfaces that can both be astigmatic, and which is strongly related with the matrix $\mathbb{P}_E$, formally identical to the equivalent power of a lens with revolution symmetry.

In Section 5.4.3 we will need a relation similar to (5.20), but we will need to compute ρ_1 in terms of ρ_2. As the relation between them is linear, we can invert the lens transfer matrix to solve for ρ_1. Inverting the 4×4 operators using standard algebraic rules could seem challenging, but these matrices has a property known as *simplecticity* that makes inversion easy. We will, however, use another approach that does not rely on abstract properties and may be more instructive for the reader: We will construct operators for backward translation and refraction. It is straightforward to derive these backward operators from equations (5.17) and (5.18),

$$\rho_1 = \begin{bmatrix} \mathbb{I} & -z\mathbb{I} \\ \mathbb{O} & \mathbb{I} \end{bmatrix} \rho_2 \text{ and } \rho_1 = \begin{bmatrix} \mathbb{I} & \mathbb{O} \\ \frac{1}{n}\mathbb{P} & \frac{n'}{n}\mathbb{I} \end{bmatrix} \rho_2,$$

which we will call $\underleftarrow{\mathbf{T}}(z)$ and $\underleftarrow{\mathbf{Q}}(\mathbb{P})$, respectively. We must point out that we are not *propagating backward* actually; we are just getting the operators that give us the input ray vector given the output ray vector of a standard forward propagation, so distances and refractive powers keep their signs. We can now construct the backward operator of a lens by multiplying the three basic transfer operators, now in reverse order to before, $\underleftarrow{\mathbf{L}} = \underleftarrow{\mathbf{Q}}(\mathbb{P}_1)\underleftarrow{\mathbf{T}}(t_0)\underleftarrow{\mathbf{Q}}(\mathbb{P}_2)$. The result is

$$\underleftarrow{\mathbf{L}} = \begin{bmatrix} \mathbb{I} - \tau\mathbb{P}_2 & -\tau\mathbb{I} \\ \mathbb{P}_1 + \mathbb{P}_2 - \tau\mathbb{P}_1\mathbb{P}_2 & \mathbb{I} - \tau\mathbb{P}_1 \end{bmatrix}. \tag{5.22}$$

The reader may check that $\mathbf{L}$ and $\underleftarrow{\mathbf{L}}$ are indeed inverse operators; their multiplication in any order is the 4×4 identity matrix. The same can be said of the pairs $\mathbf{Q}, \underleftarrow{\mathbf{Q}}$ and $\mathbf{T}, \underleftarrow{\mathbf{T}}$. However, we will keep the notation $\underleftarrow{\mathbf{X}}$ instead of $\mathbf{X}^{-1}$ as we think it will be more intuitive in the next sections. It is also worth mentioning the differences between $\mathbf{L}$ and $\underleftarrow{\mathbf{L}}$: The diagonal components are interchanged and the leftmost component of the lower row is not quite $-\mathbb{P}_E$ but $\mathbb{P}_{EB} = \mathbb{P}_1 + \mathbb{P}_2 - \tau\mathbb{P}_1\mathbb{P}_2$. We see that this power differs from $\mathbb{P}_E$ in the order of the matrix product of the refractive powers. For lenses in which $\mathbb{P}_1$ and $\mathbb{P}_2$ commute (those with at least one surface with revolution symmetry or those with the two surfaces having their main meridians with the same orientation) $\mathbb{P}_E$ and $\mathbb{P}_{EB}$ would be identical.

5.4.2 Thin Plane Prisms

Before studying the prismatic effects produced by lenses, we will first study the deviation of rays by thin prisms and their effects on eye orientation. A plane prism is a slab of an optical material with plane surfaces, as shown in Figure 5.11. The angle between surfaces, α, is known as the *prism refraction angle*. The intersection between the two planes is the apex line. The base of the prism could refer to the actual base, when the prism is shaped as in Figure 5.11, but it would be a better definition to consider the base the direction perpendicular to the apex along the bisector of angle α.

We can apply techniques similar to those used in previous chapters to the study of the geometry and refractive properties of plane prisms. A plane surface containing the origin of coordinates is given by the equation $ux + vy + wz = 0$, where $\mathbf{N} = (u, v, w)$ is the vector perpendicular to the plane. If the orientation of the normal vector is close to the Z axis, then

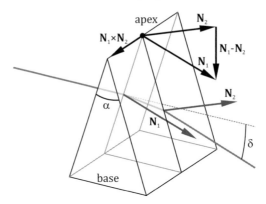

Figure 5.11 Prism geometry. A ray has been traced (thick gray line) through the center of the prism; the surface normals are depicted at the incidence points. Normals have also been translated to the apex for better visualization of their difference and cross product.

$w \simeq 1$ and $u, v \ll 1$, and the plane equation can be written $z = -ux - vy = -\mathbf{n} \cdot \mathbf{r}$, where $\mathbf{n} = (u, v)$ and $\mathbf{r} = (x, y)$ (see Section B.6).

The prism will then have a front surface $z = -\mathbf{n}_1 \cdot \mathbf{r}$ and a back surface $z = -\mathbf{n}_2 \cdot \mathbf{r} + e_0$, which has been displaced a distance e_0 along the Z axis to account for the thickness of the prism at the origin. We can study the refraction of a ray through the prism by using the three-dimensional Snell law, (3.18), but in this case the normals to the surfaces are constant everywhere. Let us assign the names $\mathbf{K}_i$ and $\mathbf{K}'_i$ to the ray direction vectors before and after refraction through surface i, $i = 1, 2$, and let n be the refractive index of the prism. The application of the Snell law to the first and second surfaces leads to

$$n\mathbf{K}'_1 - \mathbf{K}_1 = (n - 1)\mathbf{N}_1,$$
$$\mathbf{K}'_2 - n\mathbf{K}_2 = (1 - n)\mathbf{N}_2.$$

Now, combining the two equations and taking into account that $\mathbf{K}'_1 = \mathbf{K}_2$, the total deviation of the ray is given by

$$\mathbf{K}'_2 - \mathbf{K}_1 = (n - 1)\,(\mathbf{N}_1 - \mathbf{N}_2)\,, \tag{5.23}$$

which is also perfectly valid for the first two components of every vector in the equation, that is

$$\mathbf{k}'_2 - \mathbf{k}_1 = (n - 1)\,(\mathbf{n}_1 - \mathbf{n}_2)\,. \tag{5.24}$$

The approximation used for any of the two previous equations to be valid is that the angles formed by all the rays with the two prism normals are small. This requires the prism to be thin, that is, its refraction angle α should be small. Similarly, if the prism normals point close to the Z axis, we should consider rays that also form small angles with the Z axis.

The apex of the prism is the straight line resulting from the intersection of the two planes forming the prism. This line is perpendicular to the two plane normals, so its direction is given by the cross product of those two normals. If we apply the approximation $\mathbf{N}_i \simeq (u_i, v_i, 1)$, then

$$\mathbf{N}_1 \times \mathbf{N}_2 = [v_1 - v_2, -(u_1 - u_2), 0].$$

This vector is perpendicular to the difference

$$\mathbf{N}_1 - \mathbf{N}_2 = [u_1 - u_2, v_1 - v_2, 0],$$

as the reader can easily check by computing the dot product of both vectors. Similarly, it is easy to check that both vectors have the same length, as their nonzero components are the same except for they are interchanged and one of them has the opposite sign. Now, as the 3D normals are unitary vectors, the length of the cross product equals the sine of the angle formed by the two normals

$$\|\mathbf{N}_1 \times \mathbf{N}_2\| = \|\mathbf{N}_1\| \, \|\mathbf{N}_2\| \sin\left(\widehat{\mathbf{N}_1, \mathbf{N}_2}\right)$$
$$= \sin\alpha \simeq \alpha.$$

We finally conclude that the vector $\mathbf{N}_1 - \mathbf{N}_2$ (or its 2D counterpart, $\mathbf{n}_1 - \mathbf{n}_2$) is perpendicular to the apex of the prism (it points toward its base) and its length is the prism refraction angle. Following equation (5.24), we infer that the quantity

$$\delta = (n-1)(\mathbf{n}_1 - \mathbf{n}_2) \tag{5.25}$$

is the prism deviation in vector form, and it generalizes the basic equation for the deviation of a thin prism $\delta = (n-1)\alpha$, only valid for rays contained in the plane perpendicular to the apex (also called the *main section* of the prism) [132]. The vector δ wraps in the same mathematical object the total deviation and the orientation of the same. Probably, we can better understand it if we use its polar form, $\delta = (\delta, \theta)$, where the modulus δ is $(n-1)\alpha$ and θ determines the orientation of the base. The Cartesian coordinates of δ are the horizontal and vertical deviations produced by the prism, and are given by

$$\delta_x = \delta \cos\theta, \tag{5.26}$$
$$\delta_y = \delta \sin\theta. \tag{5.27}$$

Radians are a convenient angular unit to describe the deviation of rays, mainly from the mathematical point of view, as most deviation formulas will obtain their simplest expression in this way. But the deviations involved in optometry and ophthalmic lenses are small angles, typically a few cents of a radian. Back in 1890, Charles Prentice introduced the concept of prism diopter as the angle needed to displace by one centimeter the impact point of the deviated ray on a screen located 1 meter away from the prism. The symbol used to denote a prism diopter is Δ. If we call the number of prism diopters and radians contained in a given deviation angle $\delta(\Delta)$ and $\delta(\text{rad})$, the relation between them is

$$\delta(\Delta) = 100 \tan \delta(\text{rad}).$$

The use of a nonlinear function (the tangent) to define a unit may seem weird, but we have to take into account that, at least up to 20 prism diopters, psycho-physical perception of

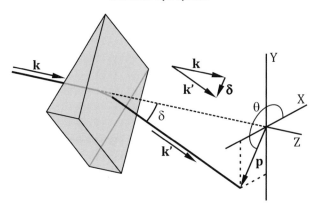

Figure 5.12 Rotated prism showing the deviation it produces in 3D, and the vector deviation, δ, that represents it.

deviation seems to be proportional to the deviation angle measured in prism diopters [133]. Also, for academic and most practical purposes, the small angle approximation is usually made, so within this environment $\tan \delta (\mathrm{rad}) \simeq \delta (\mathrm{rad})$, and a prism diopter just becomes one cent of a radian. We do not want to constantly use the labels (Δ) and (rad), so we will use a different name for prism deviation when given in terms of prism diopters, $p \equiv \delta\,(\Delta)$, and we will keep the name δ when the deviation is given in radians. Let us observe Figure 5.12 and assume the XY plane is located 1 meter behind the prism; the length of vector $\mathbf{p}$ measured in centimeters would be the number of prism diopters of angle δ. Of course, the Cartesian coordinates of $\mathbf{p}$ would also be the coordinates of the impact point of the deviated ray on the XY plane, measured in centimeters.

While the pair (δ, θ) stands as the most common way among lens designers and manufacturers to specify prisms, clinicians usually specify prisms and their effects through the horizontal and vertical components. But instead of using the arithmetic sign of Cartesian coordinates to specify direction, the clinician typically specifies the direction of the prism components with the words *up*, *down*, *nasal*, and *temporal*, the first two for the vertical component and the last two for the horizontal one. For example, a prism for the right eye "2Δ base nasal, 1.5Δ base down" means $\mathbf{p} = (2, -1.5)\Delta$. Of course, the designation "nasal" and "temporal" requires specification of the eye, otherwise the horizontal direction is left undetermined for computations. Another usual nomenclature for "base nasal" and "base temporal" is *base in* and *base out*, respectively.

Once we have modeled the prism action by the simple equation $\mathbf{k'} = \mathbf{k} + \boldsymbol{\delta}$, the combined effect of consecutive prisms with deviations $\boldsymbol{\delta}_i$, $i = 1, \ldots, N$ is given by the vector addition

$$\boldsymbol{\delta}_T = \boldsymbol{\delta}_1 + \cdots + \boldsymbol{\delta}_N.$$

In the same way, we can define the average of a population sample of vector prismatic deviations and perform any required statistical analysis over their components. One practical

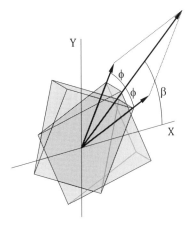

Figure 5.13 Schematics of a Risley rotary prism.

example of the additivity of prism vector deviation is the *Risley rotary prism*. This is an optical mount incorporating two identical prisms with two adjustment mechanisms. The first controls the angle between the two prisms; when operated, one prism rotates clockwise and the other counterclockwise but both by the same amount. The second mechanism rotates the two prisms together. The schematics of a Risley prism are shown in Figure 5.13. At any given position, the orientation of the whole set is given by angle β, while the bases of the individual prisms lie at $\beta - \phi$ and $\beta + \phi$. The deviation vectors of the prisms are then $\delta \left[\cos (\beta - \phi), \sin (\beta - \phi) \right]$ and $\delta \left[\cos (\beta + \phi), \sin (\beta + \phi) \right]$, where δ is the total deviation angle of the individual prisms. The total deviation produced by the Risley prism is the vector sum,

$$\delta_T = \delta \left[\cos (\beta - \phi) + \cos (\beta + \phi), \sin (\beta - \phi) + \sin (\beta + \phi) \right],$$

which simplifies to

$$\delta_T = 2\delta \cos \phi \, (\cos \beta, \sin \beta) .$$

According to this, by adjustment of the first mechanism (angle ϕ) the total deviation may change between zero ($\phi = \pi/2$) and 2δ ($\phi = 0$). The second mechanism allows us to change β continuously between zero and 2π, that is, getting an arbitrary prism orientation.

Prism thickness can be readily obtained from the equations of its surfaces and the general equation (4.1),

$$
\begin{aligned}
t(x, y) &= t_0 - \mathbf{n}_2 \cdot \mathbf{r} + \mathbf{n}_1 \cdot \mathbf{r} \\
&= t_0 + (\mathbf{n}_1 - \mathbf{n}_2) \cdot \mathbf{r} \\
&= t_0 + \frac{\delta \cdot \mathbf{r}}{n - 1} .
\end{aligned}
\tag{5.28}
$$

When $\mathbf{r}$ points in the same direction as δ the thickness grows at a rate $\delta/(n - 1)$, and if $\mathbf{r}$ points toward the apex, the thickness decreases at the same rate. At any other direction

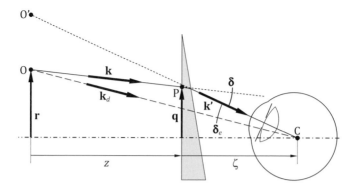

Figure 5.14 The effect of a prism in the rotation of the eye.

the rate of growth (positive or negative) is smaller, and in particular, if **r** is perpendicular to δ the thickness does not grow at all. This property leads to another definition for base orientation: the base is the direction along which the thickness of the prism grows faster, and the rays are always deviated in that direction. We will see later that this is a general property not only of prisms, but also of arbitrary lenses.

So far we have seen how prisms deviate rays and what their geometrical properties are, but we still have not studied our main concern: the effect that prisms have on eye orientation. The tools we have presented will allow us to set out the problem and solve it with ease. Let us consider the situation depicted in Figure 5.14. The eye is aiming at object point O, and we will compare the eye orientation with and without a prism interposed between the eye and the object. The distances from the object to the prism and from the prism to the rotation center of the eye are z and ζ, respectively. The line of sight to directly aim the object without the prism would be CO, where C is the approximate rotation center of the eye. Let us call the corresponding direction vector $\mathbf{k}_d$. The presence of the prism causes the aiming to change, going from C to O through point P. The ray from C to P and then to C is called the *chief* or *principal ray*. The direction vectors of the chief ray before and after refraction through the prism are **k** and **k**′, respectively. Finally, the transversal coordinates of points O and P will be **r** and **q**.

First, we will write the direction vectors in terms of the transversal coordinates and the axial distances. Taking into account that we are within the validity range of paraxial approximation, we have the relations

$$\mathbf{k}_d = \frac{-1}{z+\zeta}\mathbf{r}, \ \ \mathbf{k} = -\frac{1}{z}(\mathbf{r}-\mathbf{q}) \text{ and } \mathbf{k}' = -\frac{1}{\zeta}\mathbf{q}.$$

Now, we have two deviation equations, one for the prism, and the other accounting for the change of orientation of the eye, δ_e, when aiming at O with and without the prism,

$$\mathbf{k}' = \mathbf{k} + \delta,$$
$$\mathbf{k}' = \mathbf{k}_d + \delta_e.$$

If we substitute the equations for the direction vectors in the two previous equations, solve for $\mathbf{q}$ in each of them, and equate the results, we get

$$\delta_e = \frac{z}{z + \zeta} \delta, \tag{5.29}$$

which is known as the *effectiveness* equation. According to our definition, δ_e and δ have equal signs, but the eye rotation is toward the apex of the prism, while the ray deviation is toward the base of the prism. As $z + \zeta$ is bigger than z, the deviation of the line of sight caused by a prism is smaller than its ray deviation. When the object is at infinity, the two deviations become identical. We should notice that from the perceptual point of view, the image of the object, O', moves toward the apex of the prism, the same direction as the eye rotation.

Example 23 A plane prism with refraction angle $\alpha = 4°$ and base oriented at $135°$ has refractive index $n = 1.5$. Determine the vertical and horizontal eye rotations that it produces (in prism diopters) when the prism is located at 30 mm from the center of rotation of the eye and this eye is aiming at a point located 350 mm away from the prism.

   We will first compute the deviation angle from the basic equation $\delta = (n - 1)\alpha$, so that $\delta = 2°$. Converting to prism diopters, we get $p = 100 \tan (\pi/90) = 3.49 \,\Delta$. Given the base angle $\theta = 135°$, we can determine the Cartesian components of the ray deviation produced by the prism,

$$p_x = 3.5 \cos 135 \approx -2.47 \,\Delta,$$

$$p_y = 3.5 \sin 135 \approx 2.47 \,\Delta.$$

The eye deviation depends on the location of the object and the prism relative to the eye. We have $z = 350$ mm and $\zeta = 30$ mm, so the eye deviation is given by

$$\mathbf{p}_e = \frac{350}{350 + 30} \begin{bmatrix} -2.47 \\ 2.47 \end{bmatrix} \Delta = \begin{bmatrix} -2.28 \\ 2.28 \end{bmatrix} \Delta.$$

5.4.3 Deviation of Small Beams by Ophthalmic Lenses: Eye Deviation

We are ready now to study the deviation of rays caused by spectacle lenses, or more interestingly, the deviation that a lens causes on eye direction. This analysis, with or without matrix transfer formalisms, has been presented elsewhere [134, 135, 136, 137] with different degrees of approximation. The approach we will present here is based on backward transfer operators (a similar idea is used by Becken for computing the magnification matrix we will see in the next section [138]), which we think has an educational advantage over matrix inversion.[4] The general situation is shown in Figure 5.15, and there are many similarities with the situation shown in Figure 5.14 to illustrate prism effectiveness. There is a point O

---

[4] The same result can be obtained by using the *simplectic properties* of the transfer matrices to invert the transfer matrices, well documented from the very introduction of the latter [36] and profusely used in visual optics by Harris and others, see for example [29, 46, 134].

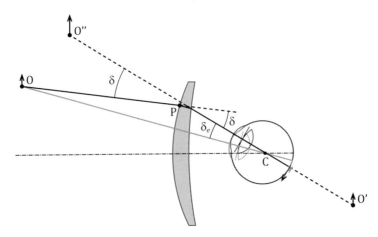

Figure 5.15 Eye rotation caused when fixating an off-axis point O through a compensating spectacle lens.

that the eye fixates with and without the lens. In the first case, the line of sight goes from O to C; in the second case it goes from O to P and then to C, requiring a rotation of the eye δ_e. The chief ray, along this line of sight, deviates the angle δ when refracting through the lens. The lens will produce an image of point O at O' (in the example of the figure, corresponding to a positive lens, the image is located behind the retina, and if the object is at infinity, the image would be located at the remote point of the eye). The eye forms an image of O' on its retina, but the object is perceived as if it were located at point O" along direction PC.

Our objective now will be to compute the eye deviation δ_e with generality, that is, for any type of lens and arbitrary position of the object and the lens with respect to the eye (though we will not consider lens tilting here). For this calculation we will use the transfer matrices described in Section 5.4.1. The points, vectors and distances involved are depicted in Figure 5.16. We will follow the same notation used in the previous section for computing prism effectiveness, the main difference here is that we are not neglecting lens thickness, so we account for the impact points of the chief ray on both surfaces and the direction vectors in each section. Object point O has transversal coordinates r_0, r_1 and r_2 are the transversal coordinates of the impact points, k_i and k'_i are the direction vectors of the incident and refracted rays at surface i, and k_0 is the direction vector of the chief ray emerging from O. As indicated in the figure, $k_0 = k_1$ and $k'_1 = k_2$. We are assuming the lens is aligned so that the rotation center of the eye, C, lies on its optical axis. The ray emerging from the lens intersects the optical axis at C, and the distance from the back vertex of the lens to C will be called l'_2. This name may seem to be rather arbitrary, but is the one typically used by lens designers; we will see the reason in the next chapter. The vector associated with direction OC will be k_d, as in the previous section. Finally, we will keep δ and δ_e for the ray and eye deviations.

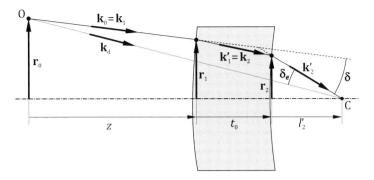

Figure 5.16 General layout for the computation of eye deviation caused by spectacle lenses.

We may construct a transfer matrix for the system extending from point O to point C. This would require multiplication of the lens transfer matrix by the translation matrices $\mathbf{T}(z)$ and $\mathbf{T}(l'_2)$. The ray at the output of this extended system would be the one passing by the rotation center of the eye (null transverse coordinates) with vector $\mathbf{k}'_2$, that is

$$\begin{bmatrix} \mathbf{0} \\ \mathbf{k}'_2 \end{bmatrix} = \mathbf{T}(l'_2)\mathbf{L}\mathbf{T}(z)\begin{bmatrix} \mathbf{r}_0 \\ \mathbf{k}_0 \end{bmatrix}.$$

On the other hand, the inverse equation can be easily written by using the backward operators we described in Section 5.4.1,

$$\begin{bmatrix} \mathbf{r}_0 \\ \mathbf{k}_0 \end{bmatrix} = \underleftarrow{\mathbf{T}}(z)\,\underleftarrow{\mathbf{L}}\,\underleftarrow{\mathbf{T}}(l'_2)\begin{bmatrix} \mathbf{0} \\ \mathbf{k}'_2 \end{bmatrix}. \tag{5.30}$$

Now, we just have to substitute the matrices given in Section 5.4.1 and carry out the multiplications, with the result

$$\begin{bmatrix} \mathbf{r}_0 \\ \mathbf{k}_0 \end{bmatrix} = \begin{bmatrix} \mathbb{L}_2 - z\mathbb{P}_{EB} & -(\tau\mathbb{I} + l'_2\mathbb{L}_2) - z\left(\mathbb{L}_1 - l'_2\mathbb{P}_{EB}\right) \\ \mathbb{P}_{EB} & \mathbb{L}_1 - l'_2\mathbb{P}_{EB} \end{bmatrix}\begin{bmatrix} \mathbf{0} \\ \mathbf{k}'_2 \end{bmatrix} \tag{5.31}$$

where $\mathbb{L}_i = \mathbb{I} - \tau\mathbb{P}_i$, $i = 1, 2$. From this equation we can obtain the position vector $\mathbf{r}_0$ in terms of the output direction vector $\mathbf{k}'_2$, the lens parameters, and the distances z and l'_2:

$$\mathbf{r}_0 = \left[-(\tau\mathbb{I} + l'_2\mathbb{L}_2) - z\left(\mathbb{L}_1 - l'_2\mathbb{P}_{EB}\right)\right]\mathbf{k}'_2. \tag{5.32}$$

Now, the eye deviation is defined by the difference

$$\delta_e = \mathbf{k}'_2 - \mathbf{k}_e, \tag{5.33}$$

and we can easily relate the direction vectors with $\mathbf{r}_2$ and $\mathbf{r}_0$,

$$\mathbf{k}_e = \frac{-1}{z + t_0 + l'_2}\mathbf{r}_0,$$

$$\mathbf{k}'_2 = -\mathbb{L}'_2\mathbf{r}_2. \tag{5.34}$$

If we plug (5.34) into (5.33) and (5.32), we get to the expression

$$\delta_e = \left\{ -L_2'\mathbb{I} + \frac{L_2'}{z + t_0 + l_2'} \left[(\tau\mathbb{I} + l_2'\mathbb{L}_2) + z \left(\mathbb{L}_1 - l_2'\mathbb{P}_{EB} \right) \right] \right\} \mathbf{r}_2, \qquad (5.35)$$

which gives us eye deviation in terms of lens parameters, object and lens positions, and the coordinates of the back surface point the eye is looking through. This equation seems rather bulky if we compare it with the typical expressions used for computing prism effects due to lenses. This is because we are not neglecting lens thickness and we are allowing the object to be at a finite distance. However, we can easily recover those simpler results by taking the appropriate limits. For example, if we neglect lens thickness, then $t_0, \tau \simeq 0$, $\mathbb{L}_i \simeq \mathbb{I}$ and $\mathbb{P}_{EB} \simeq \mathbb{P}_E \simeq \mathbb{P}_{TL}$. The previous expression is reduced to

$$\delta_e = \left\{ -L_2'\mathbb{I} + \frac{L_2'}{z + l_2'} \left[l_2'\mathbb{I} + z \left(\mathbb{I} - l_2'\mathbb{P}_{TL} \right) \right] \right\} \mathbf{r}_2$$

$$= -\frac{z}{z + l_2'} \mathbb{P}_{TL}\mathbf{r}_2, \qquad (5.36)$$

which resembles the effectiveness equation for the prism, (5.29), but now the deviation depends on the point $\mathbf{r}_2$, instead of being constant. For thin lenses, vectors $\mathbf{r}_2$ and $\mathbf{r}_1$ would be identical, and so we would not wish to tell them apart and simply write $\mathbf{r}$.

If the object is located at infinity, the first term in the square bracket of (5.35) vanishes, but not the second, which contains a z factor. The quotient $z/(z + t_0 + l_2') \to 1$ as $z \to \infty$, so we get

$$\delta_e = \left[-L_2'\mathbb{I} + L_2' \left(\mathbb{L}_1 - l_2'\mathbb{P}_{EB} \right) \right] \mathbf{r}_2$$

$$= \left(-\tau L_2'\mathbb{P}_1 - \mathbb{P}_{EB} \right) \mathbf{r}_2$$

$$= \tau\mathbb{P}_1\mathbf{k}_2' - \mathbb{P}_{EB}\mathbf{r}_2. \qquad (5.37)$$

Now we recover a result parallel to the one we obtained in Section 4.3.5, $\delta = -\tau\mathbb{P}_2\mathbf{k}_1 - \mathbb{P}_E\mathbf{r}_1$. Indeed both are the same; we just need some algebraic manipulations to relate $\mathbf{k}_1$ and $\mathbf{r}_1$ to $\mathbf{r}_2$ and $\mathbf{k}_2'$ by means of the forward and backward operators (5.19) and (5.22) to demonstrate the equivalence between both results.

Finally, if we neglect lens thickness and consider the object at infinity at the same time, we get the result

$$\delta_e = -\mathbb{P}_{TL}\mathbf{r}, \qquad (5.38)$$

which is the well-known matrix form of Prentice's rule, first brought about for the optometric community by Fick [43] and Long [44].

So far we have been using the units meters and radians for position vectors and deviations, respectively. Expression (5.35) and those derived from it by approximation can be written in terms of prism diopters. We just need to keep diopters and meters as the units for curvature and distance for all the parameters inside the braces in (5.35) and write the position vector $\mathbf{r}_2$ in centimeters. For example, the classical Prentice equation would read

$$\mathbf{p}_e = -\mathbb{P}_{TL}\mathbf{r}(\text{cm}). \qquad (5.39)$$

Example 24 Let us determine the prismatic effect (eye deviation) produced by a lens with spherocylindrical power $[-5, -2 \times 40°]$ when the eye looks through a point on the back surface with coordinates $(-5, 15)$ mm, assuming the object is located 0.9 meters away from the eye and the distance from the back vertex of the lens to the center of rotation of the eye is 27 mm. Assume the lens to be thin and give the results in prism diopters (polar and Cartesian forms).

First, the matrix form of the astigmatic back vertex power is $\mathbb{P}_V = \begin{bmatrix} -5.83 & 0.984 \\ 0.984 & -6.17 \end{bmatrix}$ D

(see, for example, the transposition Table 4.2). As the lens is assumed to be thin, we will calculate the eye deviation vector using equation (5.36). Taking $\mathbf{r} = \begin{bmatrix} -0.5 & 1.5 \end{bmatrix}^T$ cm, $z = 90$ cm, and $l_2' = 2.7$ cm, we have

$$\mathbf{p}_e = -\frac{90}{90 + 2.7} \begin{bmatrix} -5.83 & 0.984 \\ 0.984 & -6.17 \end{bmatrix} \begin{bmatrix} -0.5 \\ 1.5 \end{bmatrix} = \begin{bmatrix} -4.3 \\ 9.5 \end{bmatrix} \Delta.$$

Converting to the polar form

$$p_e = \|\mathbf{p}_e\| = \sqrt{(-4.3)^2 + 9.5^2} = 10.4 \, \Delta,$$

$$\theta = \tan^{-1}\left(\frac{9.5}{-4.3}\right) = -65.6 + 180 = 114°.$$

Notice that according to the sign of its Cartesian components, the deviation vector belongs to the second quadrant, and this is why we have added $180°$ to get the definitive base angle.

Example 25 A spherotoric lens with internal torus has refractive index $n = 1.6$, base curve $P_1 = 8$ D, center thickness $t_0 = 7$ mm, and back vertex power $[4, 2.5 \times 30°]$. The eye is looking through a point on the back surface with coordinates $(20, 20)$ mm. The object-lens distance is 55 cm, and $l_2' = 27$ mm. Let us compare the eye deviations produced by the lens using equation (5.35) and the approximated Prentice rule (5.38).

The matrix form of the back vertex power is $\mathbb{P}_V = \begin{bmatrix} 4.625 & -1.082 \\ -1.082 & 5.875 \end{bmatrix}$ D. Given the

shape factor

$$\mathbb{G} = (\mathbb{I} - \tau \mathbb{P}_1)^{-1} = \left(\begin{bmatrix} 1 & 0 \\ 0 & 1 \end{bmatrix} - \frac{0.007}{1.6} \begin{bmatrix} 8 & 0 \\ 0 & 8 \end{bmatrix}\right)^{-1} = \begin{bmatrix} 1.036 & 0 \\ 0 & 1.036 \end{bmatrix}$$

We have to also compute the backward equivalent power $\mathbb{P}_{EB}$, which is given by $\mathbb{P}_{EB} = \mathbb{G}^{-1} \mathbb{P}_V$,

$$\mathbb{P}_{EB} = \begin{bmatrix} 0.965 & 0 \\ 0 & 0.965 \end{bmatrix} \begin{bmatrix} 4.625 & -1.082 \\ -1.082 & 5.875 \end{bmatrix} = \begin{bmatrix} 4.463 & -1.045 \\ -1.045 & 5.670 \end{bmatrix}$$ D.

The matrix $\mathbb{L}_1$ is the inverse of the shape factor, so

$$\mathbb{L}_1 = \mathbb{G}^{-1} = \begin{bmatrix} 0.965 & 0 \\ 0 & 0.965 \end{bmatrix},$$

while for $\mathbb{L}_2$ we need the back surface, $\mathbb{P}_2 = \mathbb{P}_V - \mathbb{GP}_1 = \begin{bmatrix} -3.665 & -1.083 \\ -1.083 & -2.415 \end{bmatrix} \mathrm{D},$

$$\mathbb{L}_2 = \mathbb{I} - \tau\mathbb{P}_2 = \begin{bmatrix} 1.016 & 0.0047 \\ 0.0047 & 1.011 \end{bmatrix}.$$

Now, we have all the matrices needed to evaluate (5.35). Let us recall that, according to the problem statement, $\mathbf{r}_2 = \begin{bmatrix} 2 & 2 \end{bmatrix}^T$ cm, $z = 0.55$ m, $t_0 = 0.007$ m and $l'_2 = 0.027$ m, so $\tau = 0.0044$ m and $L'_2 = 37.04$ D. The matrix inside the braces in equation (5.35) is

$$-L'_2 + \frac{L'_2}{z + t_0 + l'_2} \left[(\tau\mathbb{I} + l'_2\mathbb{L}_2) + z \left(\mathbb{L}_1 - l'_2\mathbb{P}_{EB} \right) \right] = \begin{bmatrix} -5.563 & 0.992 \\ 0.992 & -6.709 \end{bmatrix} \mathrm{D}$$

and the eye deviation

$$\mathbf{p}_e = \begin{bmatrix} -5.563 & 0.992 \\ 0.992 & -6.709 \end{bmatrix} \begin{bmatrix} 2 \\ 2 \end{bmatrix} = \begin{bmatrix} -9.14 \\ -11.4 \end{bmatrix} \Delta.$$

To compute eye deviation using the simpler Prentice's rule augmented with the proximity factor,

$$\mathbf{p}_e = -\frac{z}{z + l'_2} (\mathbb{P}_1 + \mathbb{P}_2) \begin{bmatrix} 2 \\ 2 \end{bmatrix} = \begin{bmatrix} -6.2 \\ -8.6 \end{bmatrix} \Delta.$$

Finally, if instead of computing the thin-lens power of the actual lens, we assume our lens is thin but its power is the initial back vertex power, then

$$\mathbf{p}_e = -\frac{z}{z + l'_2} \mathbb{P}_V \begin{bmatrix} 2 \\ 2 \end{bmatrix} = \begin{bmatrix} -6.8 \\ -9.1 \end{bmatrix} \Delta.$$

The previous example shows that the error made when computing the eye deviation by means of Prentice's rule can be pretty high, more than 30%. We also see that it seems a better approach to compute Prentice's law with back vertex power than with the thin-lens power of the actual lens. This result is not a coincidence of our previous example, but it can be shown to hold in general for positive lenses. Prentice's rule is so extended that it is worth studying its validity in greater detail. For this we have generated an array of lenses with astigmatic power $[S, 2 \times 45°]$, where the spherical power will range from -10 D to 10 D. For each lens power we have used a base curve quite similar to those actually employed for real lenses, and the center thickness has been set to 1.8 mm for lenses with both main meridians being negative and computed with the algorithm shown in Figure 4.10 for lenses with at least one positive main meridian, assuming a lens diameter of 60 mm and minimum edge thickness of 1 mm. Refractive index has been set to 1.5 and eye deviation has been computed at the point $\mathbf{r}_2 = (1, 0)^T$ cm. Results are presented in Figure 5.17. The baselines of the two plots are the magnitude and the base orientation of the eye deviation computed with Prentice's rule using back vertex power $[S, 2 \times 45°]$, that is, $-\mathbb{P}_V\mathbf{r}_2$. The curves then represent the differences Δp (left plot) and $\Delta\theta$ (right plot) of total eye deviation and base orientation corresponding to other computation approaches with respect to the

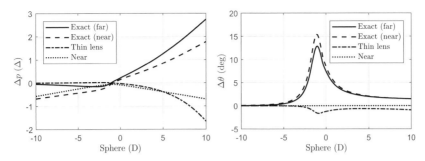

Figure 5.17 Differences in magnitude (left) and base orientation (right) between the eye deviations obtained with different approximations and as a function of spherical power. The baseline is the eye deviation obtained by using Prentice's rule assuming the lens is thin with power $\mathbb{P}_V$. All the curves represent the difference between other approximations and the base line. The solid curves correspond to the exact expression 5.35 for an object at infinity. The dashed line uses the same expression for a near object, $z = 400$ mm. The dash-dotted line corresponds to Prentice's law where the actual thin-lens power approximation of the lens is used. Finally, the dotted line stands for Prentice's law using back vertex power and corrected by the proximity factor for the same near distance. See text for more details.

baseline. The solid line represents the difference between the eye deviation given by (5.35) for far objects and the baseline. For negative lenses the difference Δp is small, but for positive lenses it is quite significant, over 2.5Δ for a 10 D lens! (or approximately the same difference for a 5 D lens when the prism is computed at a point 2 cm away from the optical center). The dashed curves represent Δp and $\Delta\theta$ when equation (5.35) is used for a near object, $z = 40$ cm. Now the difference Δp gets larger for negative lenses, and smaller for positive ones, still with quite a large difference. The dash-dot curve stands for Prentice's rule using thin-lens power, $-\mathbb{P}_{TL}\mathbf{r}_2$. Here we see the maximum difference Δp with respect to the baseline; however, base orientation is closer between $-\mathbb{P}_{TL}\mathbf{r}_2$ and $-\mathbb{P}_V\mathbf{r}_2$ than between $-\mathbb{P}_V\mathbf{r}_2$ and the exact results both for far and near objects. Finally, the dotted lines correspond to $-\mathbb{P}_V\mathbf{r}_2$ corrected by the proximity factor $-z/(z+l_2')$. In this case base orientation does not change (as the difference is a caused by a scalar multiplicative factor). Note that the errors made in base direction by using Prentice's rule may surpass $10°$. This only happens for lens powers with spherical equivalent close to zero, where the eye deviation is small.

Let us finally recall that even the "exact" result (5.35) is a paraxial one. If we are serious about accurate computing of eye deviation, we should use exact ray tracing. This, however, requires a complete description of the surfaces and their asphericities (if any) and the appropriate ray tracing software, as there are no analytical expressions for nonparaxial ray tracing. We cannot even anticipate rules about how good (5.35) is with respect to exact ray tracing, as the latter is strongly dependent on the lens base curve and surface asphericities. We can say that for spherotoric lenses or lenses with not too large asphericities, expression (5.35) is fairly accurate up to angles of $25°$(the data presented in Figure 5.17 correspond

Table 5.2 *Summary of the different expressions for the matrix operator that determines the eye deviation produced by a lens.*

Situation	Matrix used for computing eye deviation, $\mathbb{F}$
Thin lens, far vision	$\mathbb{P}_V$
Thin lens, near vision	$\frac{z}{z+l_2'}\mathbb{P}_V$
Thick lens, far vision	$\tau L_2'\mathbb{P}_1 + \mathbb{P}_{EB}$
General	$L_2' - \frac{L_2'}{z+t_0+l_2'}\left[(\tau\mathbb{I} + l_2'\mathbb{L}_2) + z\left(\mathbb{L}_1 - l_2'\mathbb{P}_{EB}\right)\right]$

to an angle of $\sim 20°$). For large asphericities or larger obliquity angles, exact ray tracing should be used to reliably compute any property of an ophthalmic lens.

To summarize, the eye deviation produced by a lens when the eye fixates an object located at some distance z from the lens, and the line of sight intercepts the back surface of the lens at a point with coordinates $\mathbf{r}_2$, is given, within the paraxial approximation, by the linear relation

$$\delta_e = -\mathbb{F}\mathbf{r}_2. \tag{5.40}$$

Matrix $\mathbb{F}$ depends on lens geometry, its refractive index, its position with respect to the eye, and the viewing distance z. Different expressions for matrix $\mathbb{F}$ are summarized in Table 5.2. Prentice's rule is presented using the back vertex power of the lens, as it gives a more accurate result than the thin-lens power $\mathbb{P}_1 + \mathbb{P}_2$.

5.4.4 Prismatic Effect and Lens Thickness

Consider the refraction of a chief ray through a lens, and let us give the notation P_1 and P_2 to the points at which the ray intersects the front and back surfaces, respectively (see Figure 5.18). The trajectory for this ray would be exactly the same if we substitute the lens by the prism formed by the planes tangent to the lens surfaces at P_1 and P_2. Of course, if we consider the rays in a bundle around the chief ray, their refraction through the lens and the prism would be different: The normals to the lens surfaces are continuously changing due to the presence of curvature, while the normals to the prism surfaces do not change. This means the lens will change the vergence of the ray bundle around the chief ray while the prism will not. But eye orientation is just determined by the chief ray, and, for it, prism and lens are equivalent. Of course, if we consider other chief rays (other eye orientations) the points P_1 and P_2 will change, and with them, the equivalent prism. The angle made by the normals at P_1 and P_2 is continuously changing in a lens. At the optical center, both normals are parallel, the planes tangent to the lens surfaces would also be parallel, and there is no ray deviation. As we move away from the center, the angle made by the normals at P_1 and P_2 continuously grows, and so does the prismatic effect. Indeed, this growth is linear, as the ray (and eye) deviation is proportional to $\mathbf{r}_2$, the x and y coordinates of P_2.

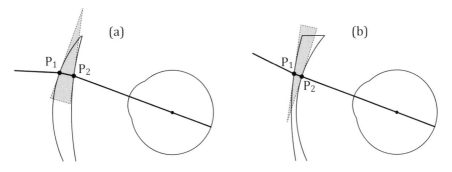

Figure 5.18 The refraction of a chief ray passing through points P_1 and P_2 of an actual lens would be exactly the same if the lens surfaces were substituted by the planes tangent to them at points P_1 and P_2. These planes form a prism with the base pointing inward in positive lenses, (a), and outward in negative lenses, (b).

Another important fact that the reader must be aware of is that the base of the prismatic effect produced by the lens always points along the direction in which lens thickness is increasing more rapidly. For plane prisms, this is a direct consequence of equation 5.28. The direction $\mathbf{r}$ along which the thickness of a plane prism changes more rapidly is that maximizing the scalar product $\delta \cdot \mathbf{r}$, that is, when $\mathbf{r}$ and δ point along the same direction. So, for the ray refracting through the lens at points P_1 and P_2, the ray (and the eye) deviates along the direction of the base of the equivalent prism, which is the direction of maximum thickness growth. As the planes forming the equivalent prism are tangent to the lens surfaces at P_1 and P_2, the ray (and eye) deviation points along the direction the lens thickness is increasing more rapidly from the segment $\overline{P_1 P_2}$. This result can be easily demonstrated for thin lenses [139]. We just need the approximate expression for the lens thickness 4.33,

$$t(x, y) = t_0 - \frac{\mathbf{r}^T \mathbb{P}_{TL} \mathbf{r}}{2(n - 1)}.$$

The gradient of the thickness function, $\nabla t(x, y) = \left(\frac{\partial t}{\partial x}, \frac{\partial t}{\partial x}\right)$, is a vector field perpendicular to the thickness iso-lines; this means the gradient vector points in the direction of maximum change of thickness. On the other hand, by taking the derivatives of the right side in the previous equation it can be shown that

$$(n - 1)\nabla t(x, y) = -\mathbb{P}_{TL} \mathbf{r},$$

and applying Prentice's rule for far objects we get to $(n - 1)\nabla t(x, y) = \delta$, so the gradient of the thickness function is proportional to the ray deviation, and henceforth the prismatic effect points in the direction of maximum thickness growth. For thick lenses the result is still valid, even if we compute the exact ray deviation, without using the paraxial approximation. We do not need mathematics for this, just the argument given above on the tangent planes at points P_1 and P_2. Figure 5.19 shows three examples of astigmatic lenses, with

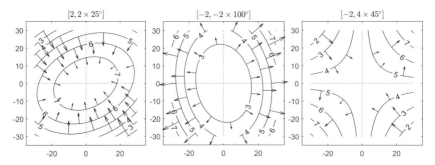

Figure 5.19 Contour plots of iso-thickness and deviation maps caused by three different astigmatic lenses.

their thickness contour plots and the field of prismatic effects they create. The two main meridians of the lens to the left are positive. Its thickness is reduced from the center faster along the steepest meridian at 115°. The thickness iso-lines are approximate ellipses with the long axis oriented along the direction of the (positive) cylinder axis, and the prismatic effect points inward, along directions of greater thickness. If the lens had been spherical, the prismatic effect would have pointed toward the very center of the lens, and the field of prismatic deviations would have a radial structure. The arrows representing the deviation are larger as we consider points further away from the optical center. This can also be seen as the relation between deviation and rate of change of thickness. At the optical center the thickness is at its maximum, and the deviation is zero. As we move away from the center, the rate of change of thickness increases and so does the deviation. The example at the center is a negative lens. The thickness grows outwards, hence the prismatic effect points outward too. As we are using a negative cylinder, the meridian with maximum thickness growth is the cylinder cross-curve, that is, at 20°. The thickness iso-lines are also approximate ellipses and the local prism is perpendicular to them, pointing outward. Finally, the example to the right is a lens whose main meridians have opposite signs. The thickness iso-lines have a saddle mount shape, with thickness growing from the center at 45° and decreasing from the center at 135°. As the thickness grows at 45°, the ray deviation points outward. At 135° the ray deviation point inward.

5.4.5 Prismatic Imbalance

Prismatic effects are unavoidable whenever spectacle lenses are used and the eye looks through points other than the optical center of the lens, as they are a consequence of the compensation of second-order aberrations. Their effects on the visual process are mainly related with *binocularity*, which, loosely speaking, is the coordinate use of two eyes to generate a single stereoscopic image [140]. When we fixate an object point, the eyes rotate so that their lines of sight intersect the fixation point. As the spatial positions of the two eyes are different, their retinal images are slightly displaced. We call this difference *retinal disparity*. In normal conditions the brain will fuse the two retinal images and use the disparity and other image data to generate the perception of depth, a process known as *stereopsis*. Various mechanisms are needed to achieve stereopsis: The individual needs simultaneous macular perception, binocular fixation, and image fusion capability. Binocular fixation involves vergence movements that are also related to accommodation, which is needed to focus near objects. Put simply: To binocularly fixate an object the brain has to get information from the individual retinas and, in a coordinate manner, it has to command the extraocular muscles and the accommodation to properly orient and focus the eyes. The prismatic effect may alter binocular fixation and with it fusion and stereopsis, endangering full binocular vision. This mainly happens when the prismatic effects are different for each eye. In that case they alter the relationship between eye orientation and disparity, hindering or even preventing binocular fixation, and henceforth, fusion and stereopsis. The prismatic imbalance, $\Delta\delta$, is defined as the difference

$$\Delta\delta = \delta_{eR} - \delta_{eL}, \tag{5.41}$$

where the sub-indexes R and L label the magnitudes for the right and left eye, respectively. The tolerance for optical imbalance depends largely on the individual, but also on the visual task and the time the individual is engaged in that task. Also, the tolerance is quite different for the horizontal and the vertical components. The human eyes constantly change horizontal direction with opposite signs (converge) to fixate near objects; however, their vertical direction has to be tightly matched. Unsurprisingly, the tolerance to horizontal prismatic imbalance is bigger than the tolerance to vertical imbalance. According to the standard ANSI Z80.1-2015, the former is 0.67Δ for $P \leq 2.75$ D and $0.25P\Delta$ for $P > 2.75$ D, while the latter is 0.33Δ for $P \leq 3.375$ D and $0.1P\Delta$ for $P > 3.375$ D.

Unexpected prismatic imbalances may occur when either one of the two lenses are not well-centered with respect to the eyes. Expression 5.35 was derived assuming the lens optical axis coincides with the line of sight at the main viewing position. When the lens is not mounted that way, the optical axis of the lens do not pass through the rotation center of the eye and prismatic effect will be produced at the main viewing position. Let $\mathbf{d}$ be the position vector from the optical center of the lens to the center of the eye pupil at the main viewing position (projected in a transverse plane). The position vectors $\mathbf{r}_i$ and the directions vectors $\mathbf{k}_i$ and $\mathbf{k}_d$ are still the same, as shown in Figure 5.20.

In this situation, $\mathbf{k}_d = -(1/z_T)(\mathbf{r}_0 - \mathbf{d})$ and $\mathbf{k}'_2 = -L'_2(\mathbf{r}_2 - \mathbf{d})$. The rotation center of the eye does not lie on the optical axis, so the relation between the input and output ray vector is

$$\begin{bmatrix} \mathbf{r}_0 \\ \mathbf{k}_0 \end{bmatrix} = \begin{bmatrix} \mathbb{L}_2 - z\mathbb{P}_{EB} & -(\tau\mathbb{I} + l'_2\mathbb{L}_2) - z(\mathbb{L}_1 - l'_2\mathbb{P}_{EB}) \\ \mathbb{P}_{EB} & \mathbb{L}_1 - l'_2\mathbb{P}_{EB} \end{bmatrix} \begin{bmatrix} \mathbf{d} \\ -L'_2(\mathbf{r}_2 - \mathbf{d}) \end{bmatrix}, \tag{5.42}$$

so now

$$\mathbf{r}_0 = (\mathbb{L}_2 - z\mathbb{P}_{EB})\mathbf{d} + L'_2\left[(\tau\mathbb{I} + l'_2\mathbb{L}_2) + z(\mathbb{L}_1 - l'_2\mathbb{P}_{EB})\right](\mathbf{r}_2 - \mathbf{d}). \tag{5.43}$$

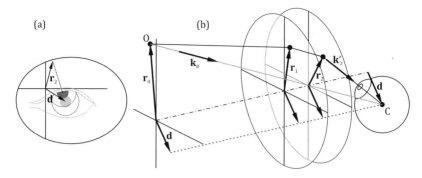

Figure 5.20 Prismatic deviation produced by a decentered lens. (a) Front view, showing the decentration vector **d**, and vector $\mathbf{r}_2$. The eye and the pupil are shown at the main viewing position in lighter gray rotated to look through $\mathbf{r}_2$ in darker gray. (b) 3D representation of the general layout showing decentration **d**.

We proceed as we did before, substituting the expressions for $\mathbf{k}_d$, $\mathbf{k}'_2$, and $\mathbf{r}_0$ into the definition for eye deviation, and after some simplification we get

$$\boldsymbol{\delta}_{eD} = \left[\left(L'_2 - \frac{1}{z_T}\right)\mathbb{I} - \frac{L'_2}{z_T}\left(\tau\mathbb{I} + z\mathbb{L}_1\right)\right]\mathbf{d} + \boldsymbol{\delta}_e, \tag{5.44}$$

where $z_T = z + t_0 + l'_2$, $\boldsymbol{\delta}_e$ is the eye deviation for the centered case (expression (5.35)) and $\boldsymbol{\delta}_{eD}$ is the eye deviation for the currently analyzed decentered case. In the case of a thin lens, the square bracket vanishes and $\boldsymbol{\delta}_{eD} = \boldsymbol{\delta}_e$. Finally, at the main viewing position, $\mathbf{r}_2 = \mathbf{d}$, so $\mathbf{k}'_2 = \mathbf{0}$, $\mathbf{r}_0 = (\mathbb{L}_2 - z\mathbb{P}_{EB})\mathbf{d}$, and

$$\boldsymbol{\delta}_{eD} = -\frac{1}{z_T}\left(\tau\mathbb{P}_2 + z\mathbb{P}_{EB}\right)\mathbf{d}. \tag{5.45}$$

Once again, for thin lenses $\boldsymbol{\delta}_{eD} = -(z/z_T)\,\mathbb{P}_{TL}\mathbf{d}$, and for far vision $\boldsymbol{\delta}_{eD} = -\mathbb{P}_{EB}\mathbf{d}$, that is, eye deviation would be the ray deviation produced by the lens at **d**. Given any pair of astigmatic refractive errors for the right and left eyes, and their respective possible decentering, we can compute with equations (5.44) and (5.45) the ray deviations for each eye, and then use (5.41) to compute the prismatic imbalance.

Another condition that will most probably cause prismatic imbalance is *anisometropia*, in which the two eyes have different refractive error. The easiest way to estimate it is by direct use of the Prentice rule to determine the prismatic effect caused to each eye. If we call $\mathbb{P}_R$ and $\mathbb{P}_L$ to the thin-lens power corresponding to the right and left eyes, then $\Delta\boldsymbol{\delta} = -(\mathbb{P}_R - \mathbb{P}_L)\,\mathbf{r}$. However, this is a simplistic approach. As the right and left powers are different, it is quite unlikely that both eyes will look through the same point **r** relative to the optical centers of their lenses. It seems more sensible to compute the prismatic effect as a function of the position of the object, that is, as a function of vector $\mathbf{r}_0$, leaving **r** (or $\mathbf{r}_2$) undetermined. Once again, the tools developed so far will allow us to make this computation readily. To abbreviate the formulas, let us assign the character $\mathbb{B}$ to the upper-right block of the operator $\underleftarrow{\mathbf{T}}(z) \underleftarrow{\mathbf{L}} \underleftarrow{\mathbf{T}}(l_2')$. The formulas we need are (5.34), (5.33), and (5.32), but now we will put $\mathbf{r}_2$ as a function of $\mathbf{r}_0$ instead of the other way around, $\mathbf{r}_2 = -l_2' \mathbb{B}^{-1} \mathbf{r}_0$. Eye deviation is now given by

$$\delta_e = \left(\mathbb{B}^{-1} + \frac{1}{z_T} \right) \mathbf{r}_0, \tag{5.46}$$

in which there are no approximations other than the paraxial environment used to model the lenses. Now, we have to apply the previous equation twice, once for each eye, so we need to take into account those magnitudes that are eye-dependent, and that we can clearly see in Figure 5.21. Eye deviation, operator $\mathbb{B}$, and the position vectors for the object point will depend on the eye and we will label them with sub-indexes R and L. We will use $\mathbf{r}_0$ to locate the fixation point with respect to the midpoint between the two eyes. Also, although z and t_0 may be different in either eye, the total distance z_T will be the same in both eyes. Finally, $\mathbf{r}_R$ and $\mathbf{r}_L$ will be the position vectors of the rotation center of either eye with respect to the midpoint between them, their lengths being the corresponding naso-pupillary distances. Applying all this nomenclature, we have

$$\delta_{eR} = \left(\mathbb{B}_R^{-1} + \frac{1}{z_T} \right) (\mathbf{r}_0 - \mathbf{r}_R),$$

$$\delta_{eL} = \left(\mathbb{B}_L^{-1} + \frac{1}{z_T} \right) (\mathbf{r}_0 - \mathbf{r}_L), \tag{5.47}$$

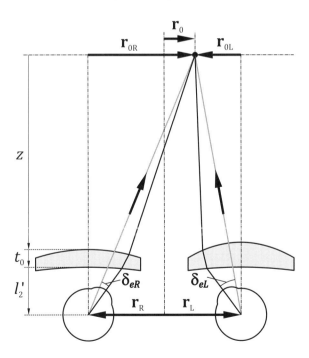

Figure 5.21 Computation of prismatic imbalance as a function of object position.

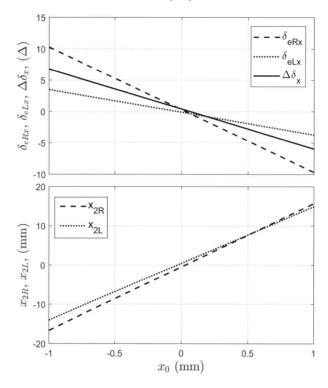

Figure 5.22 Eye deviation for the right and left eyes and prismatic imbalance (above) and horizontal coordinate of $\mathbf{r}_2$ also for both eyes (below) as a function of horizontal coordinate of the fixation point. See details in the text.

and from both,

$$\Delta\boldsymbol{\delta} = \left(\mathbb{B}_R^{-1} - \mathbb{B}_L^{-1}\right)\mathbf{r}_0 + \left(\mathbb{B}_R^{-1} + \mathbb{B}_L^{-1} - \frac{2}{z_T}\right)\mathbf{r}_R. \tag{5.48}$$

An example of the application of the previous expression is shown in Figure 5.22. We have selected a refractive error $[4, 2 \times 45°]$ for the right eye and 2 D for the left one. The lenses are assumed to be spherotoric with internal tori with base curves $P_{1R} = 8$ D and $P_{1L} = 5$ D, center thicknesses $t_{0R} = 8$ mm and $t_{0L} = 3$ mm, and refractive index $n = 1.5$. Total object distance z_T is set to 2 meters, and the horizontal position of the fixation point is scanned from -1 to 1 meters, keeping the y coordinate fixed at $y_0 = 0$. As we move the fixation object we compute individual eye deviations and prismatic imbalance, all shown in the upper plot of Figure 5.22, and compute the position vector $\mathbf{r}_2$ for the two eyes, from which we show the horizontal coordinate in the lower plot. We see that at $x_0 = 0$ there is already eye deviation and a small prismatic imbalance. This is due to the convergence necessary to binocularly fixate the object point. At reading distance the required convergence and the prismatic imbalances would have been bigger. Anyway, each eye requires a different degree of convergence due to anisometropia. As the object is moved along the horizontal direction, eye deviations and prismatic imbalance grow linearly with the object position. This is a consequence of paraxial approximation. Real ray tracing would have yield slightly curved graphs, the curvature depending on the base curve and asphericity (if any) on the lens surfaces. Prismatic imbalances reach values over 5 Δ when the object is 1 meter to either side (an obliquity angle of about 27°), a value that would probably cause perception difficulties to the person. Finally, we can see that the points through which the person is looking through either lens (vectors "$\mathbf{r_2}$") are different except when the object is about 0.5 m to the right. The difference between x_{2R} and x_{2L} can be as high as 5 mm when the object is 1 meter to the left, and this is why applying Prentice rule to this situation would have yielded inaccurate results.

5.4.6 Decentered Lenses

Eye deviation caused by spectacle lenses is a side effect that in most occasions has to be avoided, specially when large prismatic imbalances are produced as a consequence of lens decentration and/or anisometropia. However, in some cases eye deviations are sought and produced by intended lens decentration or the use of decentered lenses that we will define and describe later. Scheiman and Wick [141] identify five situations in which forcing eye deviation, which is known as the use of prisms or prismatic effects, may be helpful to treat binocular anomalies. They are: relieving prisms, horizontal and vertical, prisms as an aid to begin vision therapy, the use of prisms when vision therapy is inappropriate or impractical, and prisms used at the end of vision therapy. Our purpose here is not when or why prescribing eye deviation is beneficial, but how to achieve it with lenses.

Let us recall that eye deviation produced by a lens is given by the linear expression $\delta_e = -\mathbb{F}\mathbf{r_2}$, where $\mathbb{F}$ is any of the matrices presented in Table 5.2, depending on the circumstances and degree of approximation used. Let us assume now that, given all the parameters defining $\mathbb{F}$, we want to find the point $\mathbf{r_2}$ through which the eye should aim in order to be affected by eye deviation δ_e. We just have to solve for $\mathbf{r_2}$ in equation (5.40),

$$\mathbf{r_2} = -\mathbb{F}^{-1}\delta_e. \tag{5.49}$$

But before discussing this solution, let us remember that the expression $\delta_e = -\mathbb{F}\mathbf{r_2}$ relies on the assumption that the rotation center of the eye is on the optical axis of the lens. If we want to enforce deviation δ_e we would need the eye to rotate so that its line of sight passes through $\mathbf{r_2}$. Of course, this is not practical and we would rather decenter the lens and produce eye deviation when the eye holds the main viewing direction or is close to it. But we have already studied this situation in the previous section. When the lens is decentered a vector amount $\mathbf{d}$, the eye deviation enforced is given by expression (5.45). So, let us give the expression $\mathbb{F}_d = \frac{1}{z_T}(\tau\mathbb{P}_2 + z\mathbb{P}_{EB})$ to the operator we have to use to compute eye deviation for decentered lenses. The decentration required for producing a given eye deviation will be given by

$$\mathbf{d} = -\mathbb{F}_d^{-1}\delta_e. \tag{5.50}$$

Here we have to be cautious about the existence of the inverse $\mathbb{F}_d^{-1}$. This is a well-known problem of linear algebra, and generalized solutions can be found for the previous equation even when the standard inverse matrix $\mathbb{F}_d^{-1}$ does not exist. In this case a *pseudoinverse* matrix can be defined that will generate the whole range of solutions to equation (5.49) [142, 143]. Here, we will avoid this mathematical tools and provide a more intuitive approach to the complete set of solutions. The reader interested in the application of general algebraic techniques to the inversion of Prentice's matrix equation may consult the paper by Harris [144]. Let us assume first the lens can be considered thin with power $\mathbb{P}_V$, so $\mathbb{F}_d = (z/z_T)\mathbb{P}_V$. If the inverse $\mathbb{F}_d^{-1}$ exists (that is, $\det(\mathbb{F}_d) \neq 0$), there is one and only one solution to 5.49 and $\mathbf{d}$ is uniquely determined. Let $[S, C \times \alpha]$ be the spherocylindrical prescription of the lens. Then we have

$$
\mathbb{F}_d = \frac{z}{z_T}\begin{bmatrix} S + C\sin^2\alpha & -C\sin\alpha\cos\alpha \\ -C\sin\alpha\cos\alpha & S + C\cos^2\alpha \end{bmatrix},
$$

and its determinant, $\det(\mathbb{F}_d) = (z/z_T)^2 S(S+C)$, can only be zero if either S, $S+C$, or both are zero. In the latter case, the lens has no power and cannot produce eye deviation. There is no solution to (5.50) unless $\delta_e = 0$, which is a trivial case. If either S or $S + C$ is zero but not both, the lens is a pure cylinder. As the prismatic effect points along the direction of maximum thickness variation, the deviation caused by a pure cylinder always points along the cylinder cross-axis. We can also check this by computing the deviation produced by a cylinder for any decentration $\mathbf{d} = (d_x, d_y)^\top$,

$$
\begin{aligned}
\delta_e &= -\frac{z}{z_T}\begin{bmatrix} C\sin^2\alpha & -C\sin\alpha\cos\alpha \\ -C\sin\alpha\cos\alpha & C\cos^2\alpha \end{bmatrix}\begin{bmatrix} d_x \\ d_y \end{bmatrix} \\
&= -\frac{z}{z_T}\begin{bmatrix} d_x C\sin^2\alpha - d_y C\sin\alpha\cos\alpha \\ -d_x C\sin\alpha\cos\alpha + d_y C\cos^2\alpha \end{bmatrix}.
\end{aligned}
$$

The vector along the direction of the cylinder axis is $\mathbf{u} = (\cos\alpha, \sin\alpha)$, and we readily see that the scalar product $\mathbf{u}\cdot\delta_e$ is zero, so the deviation produced by a cylinder is perpendicular to the cylinder axis and parallel to the cross-axis. Then, coming back to our problem of finding the required decentration of a pure cylinder to produce some specified eye deviation, we see that unless this deviation points along the cylinder cross-axis, there will not be a solution for (5.50). If the deviation points along the cross-axis, it has the general form $\delta_e\mathbf{v}$, δ_e being a positive or negative scalar and $\mathbf{v} = (-\sin\alpha, \cos\alpha)$, the unitary vector along the cross-axis direction. The decentration vector must then satisfy $-\mathbb{F}_d\mathbf{d} = \delta_e\mathbf{v}$, that is,

$$
-\frac{z}{z_T}\left(d_x C\sin^2\alpha - d_y C\sin\alpha\cos\alpha\right) = -\delta_e\sin\alpha,
$$

$$
-\frac{z}{z_T}\left(-d_x C\sin\alpha\cos\alpha + d_y C\cos^2\alpha\right) = \delta_e\cos\alpha.
$$

As the determinant of the cylinder matrix is zero, the two equations are the same except for a constant (the reader can divide the first one by $\sin\alpha$ and the second one by $\cos\alpha$ to check it). Hence, we can use only one, and there are infinitely many solutions. For example, from the first equation,

$$-d_x \sin \alpha + d_y \cos \alpha = -\frac{z_T}{z}\frac{\delta_e}{C},$$

so any vector $\mathbf{d}$ whose scalar product with $\mathbf{v}$ is $-(z_T/z)(\delta_e/C)$ will produce the required deviation. Those points lie on a straight line parallel to the cylinder axis at a distance $(z_T/z)(\delta_e/C)$. Displacing the cylindrical lens along its axis does not change eye deviation, as we should expect from the symmetry properties of the cylindrical lenses.

We have kept the proximity factor z/z_T in this discussion as it does not affect the result we wanted to prove; of course, if eye deviation is intended for far vision, we can just drop it as it rapidly tends to one when the distance z gets large. In fact, even for reading distances the proximity factor is close to one and in most cases can be ignored.

For positive lenses the thin-lens approximation may not be adequate, and we have to use the complete expression $\mathbb{F}_d = \frac{1}{z_T}(\tau \mathbb{P}_2 + z\mathbb{P}_{EB})$. Now, when $\mathbb{P}_1$ and $\mathbb{P}_2$ do not commute, $\mathbb{P}_{EB}$ will not be symmetric and neither will $\mathbb{F}_d$. This matrix no longer represents the curvature of a wavefront; hence it does not correspond to a back vertex power. However, the same previous argument to analyze the existence and number of solutions to (5.50) still holds. If the determinant of $\mathbb{F}_d$ is not zero, its inverse exists and there is one and only one solution. If the determinant is zero and $\mathbb{F}_d$ is not the null matrix, there will be no solution unless the required eye deviation points along some specific direction $(\cos \beta, \sin \beta)$. If the elements of $\mathbb{F}_d$ are called f_{xx}, f_{xy}, f_{yx} and f_{yy}, then it can be shown that direction β satisfies

$$\cos \beta = \frac{f_{xy}}{\sqrt{f_{xy}^2 + f_{yy}^2}}, \quad \sin \beta = \frac{f_{yy}}{\sqrt{f_{xy}^2 + f_{yy}^2}},$$

and only when $\delta_e = \delta_e(\cos \beta, \sin \beta)$, where δ_e is a positive or negative scalar, will we find solutions to $\delta_e = -\mathbb{F}_d\mathbf{d}$. To get them, we would randomly choose one of the two scalar equations contained in the matrix equation and solve for one component of $\mathbf{d}$ as a function of the other, with the solutions being along a straight line perpendicular to $(\cos \beta, \sin \beta)$. A complete demonstration of the previous statements would require a detour on linear algebra tools that we have managed to avoid through the book, and since the detailed inversion of $\delta_e = -\mathbb{F}_d\mathbf{d}$ when the determinant of $\mathbb{F}_d$ is zero or close to zero has more of an academic interest than a practical one, we will just stop the analysis here.

From the practical point of view, producing eye deviation by means of lens decentration is only applicable when the required decentration is not too large, and that depends on the amount of required eye deviation and the lens power. For example, to get some idea of the numbers involved, let us assume the lens is thin and spherical with power P. Then $\mathbb{F}_d = (z/z_T)P\mathbb{I}$. The proximity factor cannot be smaller than 0.9, even for a viewing distance of 300 mm, so we can ignore it for this back-of-an-envelope analysis. The decentration required for a given deviation is then $\mathbf{d} = -(1/P)\delta_e$, or, if we use prism diopters, $\mathbf{d}(\mathrm{cm}) = -(1/P)\mathbf{p}_e$. If the lens has a power of 10 D, each prismatic diopter of eye deviation will require a decentration of 1 mm, which seems quite manageable. But with $P = 1$ D, each prismatic diopter of eye deviation would demand a decentering of 1 cm, and of course things get much worse if $P < 1$ D. Let us look at the problem in the next example with an actual lens with astigmatic prescription.

Example 26 A patient has a right eye prescription $[2, -1.5 \times 35°]$, and requires a horizontal eye deviation of 2Δ base for a reading distance of 350 mm. Determine the required decentration for producing the prism as well as the minimum lens diameter required for achieving decentration. Problem data: lens refractive index, $n = 1.5$; lens base curve, $B = 4$ D; center thickness, $t_0 = 3.5$ mm; distance from the back vertex to the rotation center of the eye, $l'_2 = 27$ mm. Frame data: horizontal box size, $a = 50$ mm; vertical box size, $b = 34$ mm; distance between lenses, $d = 18$ mm; User data: naso-pupillary distance, $NPD = 30$ mm; pupil height, $h_p = 20$ mm.

The matrix form of the back vertex power is

$$\mathbb{P}_V = \begin{bmatrix} 1.507 & 0.705 \\ 0.705 & 0.994 \end{bmatrix} \text{ D.}$$

The first surface is spherical, $\mathbb{P}_1 = 4\mathbb{I}$ D, so the shape factor can be computed as scalar, $G = (1-\tau B)^{-1} = 1.0093$, and the refractive power of the back surface is $\mathbb{P}_2 = \mathbb{P}_V - G\mathbb{P}_1 = \begin{bmatrix} -2.531 & 0.705 \\ 0.705 & -3.044 \end{bmatrix}$ D. We can now compute the power $\mathbb{P}_{EB} = \begin{bmatrix} 1.493 & 0.698 \\ 0.698 & 0.984 \end{bmatrix}$ D and the operator $\mathbb{F}_d = \frac{1}{z_T}(\tau \mathbb{P}_2 + z \mathbb{P}_{EB})$,

$$\mathbb{F}_d = \begin{bmatrix} 1.358 & 0.647 \\ 0.647 & 0.887 \end{bmatrix} \text{ D.}$$

In this case, the main difference between $\mathbb{F}_d$ and either $\mathbb{P}_V$ or $\mathbb{P}_{TL}$ is mainly due to the proximity factor $z/z_T = 0.92$, as the lens thickness is small and the term $(\tau/z_T)\mathbb{P}_2$ can be neglected. The determinant of $\mathbb{F}_d$ is nonzero, so its inverse exists, leading to decentration

$$\mathbf{d} = -\mathbb{F}_d^{-1} \begin{bmatrix} 2 \\ 0 \end{bmatrix} = \begin{bmatrix} -2.26 \\ 1.64 \end{bmatrix} \text{ cm.}$$

To compute the minimum lens diameter required for achieving this decentration, we need to compute the coordinates of the pupil and of the lens optical center to the frame center (*boxing center*, see Appendix A for details). Pupil coordinates are $x_P = (a+d)/2 - NPD = 4$ mm, $y_P = h_p - b/2 = 3$ mm, so the coordinates of the optical center relative to the frame, (x_O, y_O), are $x_O = x_P - d_x = 28.6$ mm and $y_O = y_P - d_y = -13.4$ mm. Finally, the minimum lens diameter needed for achieving this large decentration depends on the exact shape of the frame, as shown in Figure 5.23. The exact value for the minimum lens diameter is twice the distance OE_2, where O is the optical center of the lens and E_2 is the point in the inner contour of the frame furthest from O. As the frame contour does not have an analytical description, this distance cannot be computed with a simple expression. Direct measurement on the scaled drawing yields $OE_2 = 50.7$ mm, so the minimum diameter is $\phi_{min} \simeq 102$ mm. If the graphical means are lacking, we can also find a superior bound for

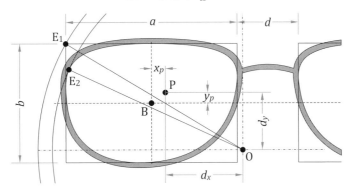

Figure 5.23 Lens decentration to produce a prescribed prismatic effect (eye deviation). The drawing is scaled according to the data in Example 26.

the minimum diameter, $\hat{\phi}_{min}$, by computing the distance OE_1 from the lens optical center to the furthest corner of the box circumscribing the frame contour. In our case, the coordinates of E_1 are $(-a/2, b/2)$, so $\hat{\phi}_{min} = 2\sqrt{(-a/2 - x_O)^2 + (b/2 - y_O)^2} \simeq 120$ mm.

The previous example illustrates a case in which a prescribed prism with medium strength is achieved by decentering a low-powered lens. The result is that the required decentration is in the range of centimeters, which in turn implies the lens diameter required for proper glazing and mounting is too large for the standard lens sizes that can be handled in the industry. For example, lens diameters larger than 80 mm are not standard in ophthalmic optics and are hardly available, yet the previous example illustrated a case in which a diameter of about 120 mm was needed. Similarly, for those cases in which one main meridian of the lens has zero power (pure cylinder) there will not be a means to generate a prism by decentering except in those rare cases in which the orientation of the cross-meridian of the lens matches the base of the prescribed prism.

It is then clear we need a way other than decentering a large lens to generate prismatic effect and the resulting eye deviation. In a centered lens, the optical axis passes through the geometrical center of the lens contour; this is why we need a large centered lens if we want to locate its optical center far from the frame center. However, it is possible to directly manufacture a decentered lens, or a lens with *ground prism*, in which the geometrical center of its contour, G, and its optical center, O, do not coincide. There are two ways of doing this: tilting one of the lens surfaces or transversally displacing one surface with respect to the other. The effect is shown, for lenses with spherical surfaces, in Figure 5.24. We already explained in Chapter 4 that the optical axis of a lens with spherical surfaces is the line connecting the curvature centers of both surfaces. If we tilt or displace one of the surface with respect to the other, the corresponding curvature center is also tilted and rotated, and that changes the orientation of the optical axis. The point at which the optical axis crosses the back surface, O, is displaced as a result of the tilt of the optical axis. This tilting/decentering of one surface with respect to the other causes the geometric center of

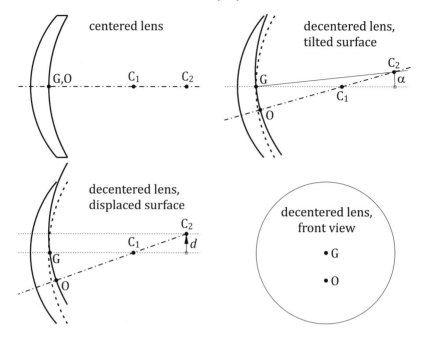

Figure 5.24 Generation of decentered lenses by means of tilt and/or displacement of one surface with respect to the other.

the lens contour; G, to separate from O. The ground prism is then defined as the deviation of a ray refracted by the lens and passing through point G.

Decentered lenses (and decentered astigmatic systems in general) can be studied by a generalization of the transfer matrix known as *augmented transfer matrices*. These are 5×5 matrices in which one column and row contains information on the tilt/decentration of the component described by the remaining columns and rows. The formalism of augmented transfer matrices have been known for a long time [145] and has been applied to the study of light beams and laser resonators [39, 146]. As pointed out by Harris [147], the formalism based in augmented transfer matrices stands out by its compactness and mathematical simplicity, and it would be the tool of choice for studying systems like the eye (in which all the components are somewhat angular or transversally misaligned, but especially with misaligned intraocular lenses). However, we would use this formalism only for the study of decentered lenses, and this need probably does not pay off the effort of introducing a new matrix formalism. We will instead characterize angular and transverse misalignment by using an extra vector along with the standard 4×4 lens transfer matrix, following Harris [148].

Let us start with tilt. If we consider the refraction of a ray by an astigmatic surface separating media with refractive index n and n', three-dimensional paraxial Snell law dictates

$$n'\mathbf{k}_2 - n\mathbf{k}_1 = (n' - n)\mathbf{n},$$

where $\mathbf{k}_1$ and $\mathbf{k}_2$ are the ray direction vectors before and after refraction and $\mathbf{n}$ is the surface normal at the incidence point. If the surface gets tilted so the normal at the Z axis points in direction $\mathbf{t}$, then the normal vector at each location is changed from $\mathbf{n}$ to $\mathbf{n} + \mathbf{t}$. Refraction at the same point of the tilted surface then produces a new $\mathbf{k}_2$, say $\hat{\mathbf{k}}_2$, so that

$$n'\hat{\mathbf{k}}_2 - n\mathbf{k}_1 = (n' - n)(\mathbf{n} + \mathbf{t}).$$

Combining the last two equations we get $\hat{\mathbf{k}}_2 = \mathbf{k}_2 + \left[(n' - n)/n'\right]\mathbf{t}$, that is, the effect of the tilt is to shift the deviation of all rays refracting through the surface by an amount $\delta = \left[(n' - n)/n'\right]\mathbf{t}$.

The effect of transversal displacement of the surface can be treated in a similar way. Let us call P the point at which the ray incides on the surface and let $\mathbf{r}$ be the position vector of P in a reference system in which the surface is centered. The normal to the surface at P follows $\mathbf{n} = -\mathbb{H}\mathbf{r}$, and the Snell law can be written

$$n'\mathbf{k}_2 - n\mathbf{k}_1 = -(n' - n)\mathbb{H}\mathbf{r} = -\mathbb{P}\mathbf{r},$$

where $\mathbb{P}$ is the refractive power of the surface. Now, if the surface is displaced a vector quantity $\mathbf{d}$ along a direction perpendicular to the Z axis, the position vector of P in the system locked to the surface vertex will be $\mathbf{r} - \mathbf{d}$, so we can write

$$n'\hat{\mathbf{k}}_2 - n\mathbf{k}_1 = -\mathbb{P}(\mathbf{r} - \mathbf{d}).$$

Once again, combining the last two equations we get to $\hat{\mathbf{k}}_2 = \mathbf{k}_2 + (1/n')\mathbb{P}\mathbf{d}$, and we conclude that the effect of a transversal misalignment of the surface is a constant angular shift amounting to $\delta = (1/n')\mathbb{P}\mathbf{d}$.

Now, let $\mathbf{Q}$ be the 4×4 transfer matrix corresponding to a centered surface separating two media with different refractive indexes. The transfer equation for this surface is written $\rho_2 = \mathbf{Q}\rho_1$, where ρ_1 and ρ_2 are the complete ray vectors before and after refraction. If the surface becomes misaligned either by rotation or displacement (or both, in which case we will add the corresponding expressions), the new transfer equation will read

$$\rho_2 = \mathbf{Q}\rho_1 + \begin{bmatrix} 0 \\ \delta \end{bmatrix}. \tag{5.51}$$

Transfer by a lens with misaligned surfaces is easily obtained by compounding the previous equation. Let us assume the deviations caused by misalignment by either surface are δ_1 and δ_2, both possibly containing angular and transversal misalignment. Let us also give the names ρ_i and ρ_i' to the ray vectors before and after refraction at surface i. Then

$$\rho_1' = \mathbf{Q}_1\rho_1 + \begin{bmatrix} 0 \\ \delta_1 \end{bmatrix}, \quad \rho_2 = \mathbf{T}(t_0)\rho_1', \quad \rho_2' = \mathbf{Q}_2\rho_2 + \begin{bmatrix} 0 \\ \delta_2 \end{bmatrix},$$

and combining the three transfers,

$$\rho_2' = \mathbf{L}\rho_1 + \mathbf{Q}_2\mathbf{T}(t_0)\begin{bmatrix} 0 \\ \delta_1 \end{bmatrix} + \begin{bmatrix} 0 \\ \delta_2 \end{bmatrix}.$$

The deviation caused by misalignment of the back surface directly adds to the output ray direction, while the deviation caused by the front surface is first transformed by propagation inside the lens and refraction through the back surface, before adding its effect to ρ_2'. To compute eye deviation caused by a lens with misaligned surfaces, we just have to use the backward operators as we did in Section 5.4.3, but introducing the constant shifts δ_i we have just used in the forward transfer. Using the same nomenclature presented in Section 5.4.3 and Figure 5.16, and applying equation (5.51) to each lens surface, we get the complete backward transfer from the rotation center of the eye to the fixation object,

$$\rho_0 = -\underleftarrow{\mathbf{T}}(z)\mathbf{Q}_1 \begin{bmatrix} \mathbf{0} \\ \boldsymbol{\delta}_1 \end{bmatrix} - \underleftarrow{\mathbf{T}}(z)\,\underleftarrow{\mathbf{L}} \begin{bmatrix} \mathbf{0} \\ \boldsymbol{\delta}_2 \end{bmatrix} + \underleftarrow{\mathbf{T}}(z)\,\underleftarrow{\mathbf{L}}\,\underleftarrow{\mathbf{T}}(l_2') \begin{bmatrix} \mathbf{0} \\ \mathbf{k}_2' \end{bmatrix}.$$

As in Section 5.4.3, we are interested in the relation between $\mathbf{r}_0$ and $\mathbf{k}_2'$. After working out the matrix multiplications we get to

$$\mathbf{r}_0 = nz\boldsymbol{\delta}_1 + (\tau + z\mathbb{L}_1)\,\boldsymbol{\delta}_2 + \mathbb{B}\mathbf{k}_2', \tag{5.52}$$

where $\mathbb{B}$ is the top-right block of $\underleftarrow{\mathbf{T}}(z)\,\underleftarrow{\mathbf{L}}\,\underleftarrow{\mathbf{T}}(l_2')$, the operator within braces in equation (5.35) that we already renamed in Section 5.4.5. Finally, using relations (5.33) and (5.34), we obtain the eye deviation produced by a lens with misaligned surfaces,

$$\boldsymbol{\delta}_e = n\frac{z}{z_T}\boldsymbol{\delta}_1 + \left(\frac{\tau}{z_T} + \frac{z}{z_T}\mathbb{L}_1\right)\boldsymbol{\delta}_2 - L_2'\left(\mathbb{I} + \frac{1}{z_T}\mathbb{B}\right)\mathbf{r}_2. \tag{5.53}$$

The relation between $\boldsymbol{\delta}_e$ and $\mathbf{r}_2$ is identical to that in equation (5.35), except for the terms depending on the angular shifts produced at each surface. Within the thin-lens approximation,

$$\boldsymbol{\delta}_e = \frac{z}{z_T}\left[(n\boldsymbol{\delta}_1 + \boldsymbol{\delta}_2) - \mathbb{P}_{TL}\mathbf{r}_2\right],$$

and if we assume both surfaces are tilted with normals $\mathbf{t}_1$ and $\mathbf{t}_2$ at $\mathbf{r} = \mathbf{0}$, then $(n\boldsymbol{\delta}_1 + \boldsymbol{\delta}_2) = (n-1)(\mathbf{t}_1 - \mathbf{t}_2)$, that is, eye deviation is obtained by adding the result from Prentice's matrix equation and the total ground prism on the lens and applying to that sum the proximity factor. Of course, equation (5.53) is general, and the angular shifts $\boldsymbol{\delta}_i$ may incorporate tilt and/or transversal displacement. If the lens is mounted so that its Z axis is aligned with the pupil, we would expect the lens to force an eye deviation equal to the ground prism, but not quite. When the line of sight is aligned with the lens Z axis, then $\mathbf{r}_2 = \mathbf{0}$, and

$$\boldsymbol{\delta}_e = n\frac{z}{z_T}\boldsymbol{\delta}_1 + \left(\frac{\tau}{z_T} + \frac{z}{z_T}\mathbb{L}_1\right)\boldsymbol{\delta}_2,$$

so that only for thin lenses and objects at infinity would eye deviation exactly match ground prism. For near objects and positive lenses with a large base curve and thickness there is a significant difference between eye deviation and ground prism. Finally, it is worth mentioning that if we set $\mathbf{r}_2 = \mathbf{k}_2'$ to zero, by virtue of equation (5.52) we are also locking the position of the fixating object, $\mathbf{r}_0$. In practice, the user with prescribed prism will fixate different objects in the visual field. By repeating the approach used in Section 5.4.5, we

can show that eye deviation produced by the misaligned lens when the eye fixates an object
with position $\mathbf{r}_0$ is given by

$$\delta_e = \left(\mathbb{B}^{-1} + \frac{1}{z_T}\right)\mathbf{r}_0 - \mathbb{B}^{-1}\left[n\frac{z}{z_T}\delta_1 + \left(\frac{\tau}{z_T} + \frac{z}{z_T}\mathbb{L}_1\right)\delta_2\right], \tag{5.54}$$

which, within the paraxial approximation is completely general.

5.5 Magnification and Field of View

5.5.1 Field of View

The field of view of the eye can be defined in two ways depending on whether we are
considering whole retinal vision or just foveal vision. In the first case, we suppose the eye
is fixed at the main sight direction. The resulting visual field, which we will call the *static* or
peripheral visual field, is given by three factors: location of the eye pupil, the full extension
of the photosensitive layer of the retina, and some facial anatomical structures. The static
field of view is nonuniform, with a progressive reduction of image quality as we move away
from the fovea. This is due to the inhomogeneous distribution of the photoreceptors along
the retina, both in type and density, and to the effect of the ocular aberrations. When we use
spectacles, the lens contour (determined by the inner edge of the frame rim) is perceived
defocused, but it limits the static field of view and behaves as a field diaphragm. The static
field of view is quantified by the angle φ as shown in Figure 5.25(a).

Rather than the static field of view, we are usually more interested in the *dynamic field
of view*, defined as the volume formed by the conjugate points of the fovea for all possible
gaze directions. For the naked eye, the dynamic field of view is limited by facial structures
such as the nose and the supra-orbital ridge, as well as by the maximum ocular rotation
allowed by the extraocular muscles. If the user wears spectacles, the dynamic visual field
with compensated vision encompasses those directions for which the principal ray passes
through the lens. The dynamic field is therefore limited by the shape and dimensions of the
inner rim as shown in Figures 5.25(b) and 5.26.

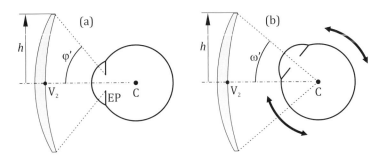

Figure 5.25 Definition of (a) the static and (b) dynamic field of view for the lens-eye system.

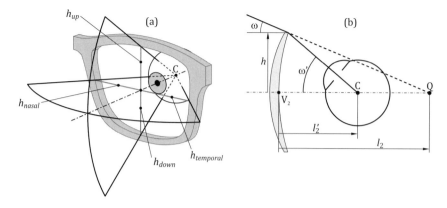

Figure 5.26 Real (ω) and apparent (ω') fields of view.

We can also distinguish between image and object dynamic fields, also known as *apparent* and *real*, respectively. The limits of the apparent field would be determined by the angles ω' shown in Figures 5.25(b) and 5.26. The real dynamic field, 2ω, is the actual field available to the user and depends on the power of the compensating lens, as shown in Figure 5.26(b). Either type of dynamic field can be easily computed by observing the geometry in Figure 5.26. Although we could compute the actual 3D volume of the dynamic fields using transfer matrices and any mathematical description of the frame contour, knowing the (maximum) fields with such a detail is rarely necessary. The apparent semi-field is independent of the compensating power and is given by

$$\tan \omega' = \frac{h}{l'_2} = hL'_2,$$

where h is the height corresponding to the border of the rim, l'_2 is the distance from the back vertex of the lens to the rotation center of the eye, and $L'_2 = 1/l'_2$. Similarly, ω satisfies the relation

$$\tan \omega = \frac{h}{l_2} = hL_2,$$

where l_2 is defined in Figure 5.26(b) and $L_2 = 1/l_2$. Now, to compute L_2 we need the lens power. For astigmatic lenses the incident and refracted rays are not necessarily coplanar. In this case the virtual prolongation of the incident ray will not intersect the line passing through V_2 and C (which we will assume is the line of sight at the main viewing position). We will ignore this possible skewness and, in case of astigmatism, will assume Q is the point at which the virtual ray prolongation and the line of sight get closer. Under this assumption, Q is object conjugate of the rotation center of the eye if we use the scalar lens power along the meridian contained in the drawing plane, P, and in that case

$$L'_2 = P + L_2,$$

and the object semi-field of view is

$$\tan \omega = h \left(L_2' - P \right). \tag{5.55}$$

Notice that, according to equation (5.55), the real field of view is smaller than the apparent one for positive lenses, and the other way around for the negative ones. The eye enters the uncompensated dynamic field of view when it rotates by angles larger than ω'. For myopes, the apparent and uncompensated fields of view overlaps. When the myopic eye aims toward any point of its frame, it can observe a blurred image of the object behind just outside the frame, and a sharp image of the same object just inside the frame. For hyperopes, the apparent and uncompensated fields of view not only do not overlap, but there is a gap between them (the region between angles ω and ω') that cannot be accessed without moving the head. For moderate levels of hyperopia this region lies in the geometrical shadow of the frame, and goes unnoticed. But for high levels of hyperopia the gap can be noticeably wide, and is known as a *ring scotoma*.

Example 27 Compute the real (object) and apparent (image) semi-fields of view for the right eye of a user whose refractive error is compensated with a lens of power $P_{TL} = -6$ D, assuming his naso-pupillary distance is 30 mm, the pupil height is 21 mm, and he wears spectacles with rectangular rim, size 50□20, and rim height of 30 mm. Suppose that $l_2' = 27$ mm and that the lens optical center is located in front of the eye pupil for the primary sight direction.

The coordinates of the pupil are given by (see Appendix A)

$$x_p = \frac{a+d}{2} - DNP_D = 5 \,\text{mm}, \; y_p = h_p - \frac{b}{2} = 6 \,\text{mm}.$$

Now, let us call the distances from the pupil to the respective edges of the frame h_{up}, h_{down}, h_{nasal}, and $h_{temporal}$, as shown in Figure 5.26(a). As the frame in this example is rectangular, the distances can be directly computed from the pupil coordinates and the frame box sizes. For the horizontal distances,

$$h_{temporal} = x_p + \frac{a}{2} = 30 \,\text{mm}, \; h_{nasal} = \frac{a}{2} - x_p = 20 \,\text{mm},$$

while for the vertical ones

$$h_{up} = \frac{b}{2} - y_p = 9 \,\text{mm}, \; h_{down} = y_p + b/2 = 21 \,\text{mm}.$$

Once we have the distances, we can directly apply (5.55) to compute the semi-field along the temporal direction,

$$\omega_{temporal} = \arctan \left[h_{temporal} \left(L_2' - P_T \right) \right] = 53°.$$

Operating in the same way we get the values of the remaining semi-fields: $\omega_{nasal} = 41°$, $\omega_{up} = 21°$, and $\omega_{down} = 42°$.

5.5.2 *Magnification*

Another side effect of the compensation of refractive errors with spectacle lenses is magnification. The retinal image of a compensated eye will be sharp, but with a different size than the blurred image of the same uncompensated eye. Compensated myopia will produce smaller images, while compensated hyperopia will yield larger ones. Astigmatism will produce images with different magnification along the main meridians, so the image of a circle in a compensated astigmatic eye will be an ellipse. This magnification effect may cause surprise and some adaptation difficulty but most of the time the ametropic person compensated with spectacles will adapt and get used to the new size image to the point it will go totally unnoticed, even when putting glasses on and off. This does not mean magnification does not have any effect. For example, it reduces visual acuity in myopes and increases it in hyperopes: A myope with refractive error of -10 D and visual acuity 0 LogMAR with contact lenses may have its visual acuity reduced to 0.176 LogMAR (0.67 decimal) when using spectacles (see for example [149]). Also, the anisotropic (or anamorphic) magnification caused by astigmatic lenses may cause more adaptation problems than the isotropic magnification of spherical lenses. Finally, the presence of anisometropia will cause different magnifications to be present in each eye. This condition of different image size projected on either retina is known as *aniseikonia*. Large enough values of aniseikonia may prevent fusion and, henceforth, binocular vision.

There is no unique way to define magnification, and different proposals have been presented, especially for providing a unified approach to the anisotropic magnification created by astigmatic lenses. For spherical power, or for providing magnification along the main meridians of astigmatic lenses, the most accepted definition is *spectacle magnification*. It is defined as the quotient between the sizes of the images in the compensated and uncompensated eyes. To determine the size of a blurred image formed by an uncompensated eye, distances are measured between the centers of the confusion blurs. The classical result, a demonstration of which can be found elsewhere (see for example [56] or [132]), is that spectacle magnification, M, is given by the expression

$$M = \frac{1}{1 - \tau P_1} \frac{1}{1 - dP_V},\qquad(5.56)$$

where, as usual, P_1 is the refractive power of the lens front surface, P_V is its vertex power, $\tau = t_0/n$, and d is the distance from the vertex of the back surface to the entrance pupil of the eye. When $M > 1$, the image is bigger with compensation, whereas $M < 1$ indicates the opposite. We already know the first factor; it is the *shape factor* of the lens, $\mathbb{G}$, though in this expression it is scalar, as we are assuming the lens surfaces to have revolution symmetry. The second factor is known as the *power factor*, and depends on the lens power and its position with respect to the eye. The shape factor is the magnification caused by a plano lens ($P_V = 0$) with a curved front surface. We already know that ophthalmic lenses invariably have convex front surfaces, so the shape factor is always bigger than 1, and it grows with τ and P_1. The power factor can be smaller or bigger than 1, depending on the sign of P_V. Positive lenses provide a power factor bigger than one; negative lenses work the other way around.

Another approach to define the change of size of retinal images is *relative magnification*. It is defined as the quotient between the size of the image in the compensated eye, and the size of the image in the emmetropic average eye. This definition leads to different magnification behavior for eyes with axial or refractive ametropia. A rule has even been established according to which refractive ametropia should be corrected with spectacle lenses to reduce the impact of magnification, a result known as *Knapp's law*. Indeed, in the comparison of the magnification of a given system (the uncompensated eye) with another system (the emmetropic average eye) it makes little sense to study the effect of compensation on the first system, and as stated by many authors, Knapp's law has no clinical relevance, so we will not bother here with it. See, for example, the discussion in [132].

Yet another definition is *dynamic magnification*, introduced by Remole [135, 136, 150]. Remole links magnification with eye deviation caused by the prismatic effect, and defines the former as the quotient between the eye orientation angles with and without the compensating lens. Barbero and Portilla deeply generalize on this idea, establishing the relation between local magnification and local blur [151, 152]. Their approach has a great interest for lens design, especially for multifocal lenses that we will study in Chapter 8, but it is beyond the paraxial approximation that we are using now to model the lens-eye system.

The main problem with the extension of (5.56) to astigmatic compensation is that the entrance pupil, which the distance d is referred to, is not well defined in astigmatic eyes. Keating was the first author to apply transfer matrix analysis to the magnification problem in visual optics [153, 154]. His approach consisted of the determination of the impact points on the retina of two rays passing through the center of the actual pupil, one for the uncompensated eye, the other for the compensated eye. The linear map transforming the former into the latter would be the magnification matrix. His approach, though limited to distant objects, was fully general, in the sense that no approximations other than the paraxial one were assumed. The price to pay for using the actual aperture stop of the eye (instead of a nonexistent entrance pupil) is that Keating's magnification matrix depended on the characteristics of the eye's anterior chamber. Nevertheless, Keating demonstrated that under certain approximations involving the matrices of the eye components, his magnification matrix could be made independent of the eye characteristics. It would only depend on the compensating lens and its position with respect to a sort of "averaged" entrance pupil of the eye and it would take the form $\mathbb{M} = (\mathbb{I} - d\mathbb{P}_V)^{-1} (\mathbb{I} - \tau\mathbb{P}_1)^{-1}$, that is, a matrix generalization of equation (5.56).

In 2001, Harris generalized Keating's approach allowing for the lens-eye system to be not fully compensated, for the usage of rays not passing through the center of the pupil, and for misalignment and tilts of the lens and the eye components [155]. Harris did not bother to obtain a generalization of equation (5.56) but provided a full analysis of the relation between the bundle of rays reaching the retina with and without the compensating element, leading to four different matrices describing four types of magnification: *blur, size, spread,* and *directional magnification*, in Harris' terminology. These matrices are also dependent on eye characteristics, but provide the most complete picture of ray transformation (and hence lateral and angular magnification) that can be given for astigmatic systems within the paraxial approximation. These results were generalized for near objects in another paper [156]. Becken et al. [138] faced the magnification problem in 2008 with a hybrid approach partly based on the so-called *step-along method for vergence propagation* from Acosta and Blendowske [157]. This is an alternative method to compute the paraxial properties of optical systems. Becken recovered the generalization of equation (5.56) to astigmatic systems with the add of proximity factors that allowed for the computation of the magnification matrix for near objects. Also, Becken's method was intended to be generalized outside the paraxial approximation [158].

Our approach to spectacle lens magnification will look for a balance between generality and helpfulness and we expect to clarify for the reader the meaning of the direct matrix generalization of (5.56). In Figure 5.27 we see the typical arrangements used to define visual magnification (a) and dynamic magnification (b). In the first case, the magnification would be quotient y'_L/y_L, while in the second case, the magnification would be the quotient between the eye orientation angles needed to fixate an object with and without the lens. Within the paraxial approximation, these angles are proportional to the object heights, so

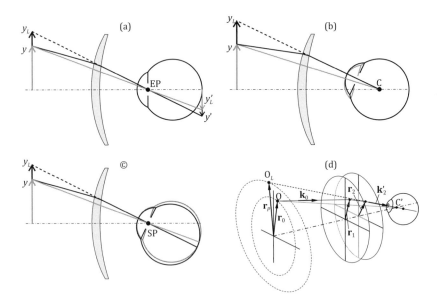

Figure 5.27 Different approaches to define magnification. (a) Visual magnification. Static eye, we compare the the size of the retinal images with and without the lens, y'_L and y', for rays passing through the center of the entrance pupil. (b) Dynamic magnification. Magnification is defined as the quotient between the rotation angles needed to fixate the object with and without lens. (c) Dynamic magnification with a modified rotation center (for example, rotating the eye around the stop pupil). (d) Layout to compute dynamic magnification.

the dynamic magnification would be the quotient y_L/y. Subplots (a) and (b) are restricted to a single section of the lens-eye systems in terms of clarity, but in the general case the eye and the lens can be astigmatic, and a ray tracing will require the layout shown in subplot (d). Let us also recall that the entrance pupil is the image of the actual pupil as formed by the cornea. A ray directed to the center of the entrance pupil will be refracted by the cornea and will hit the center of the actual pupil. Now, an astigmatic cornea cannot form a sharp image of the pupil. This is why Keating and Harris had to incorporate information on eye optics to control the way rays cross the stop pupil. Dynamic magnification is free from this problem, as we are always using chief rays that propagate along the line of sight of the eye (instead of tilting the rays, we tilt the eye).

Let us then compute the size of the perceived vector $\mathbf{r}_p$ as a function of the object vector $\mathbf{r}_0$, for an arbitrary astigmatic lens (we will assume the lens is centered and its optical axis matches the line of sight of the eye at the main viewing position). However, instead of tilting the eye around its actual rotation center, we will rotate it around another point C' on its line of sight and on the optical axis of the lens. By now, we do not need to relate the exact location of this modified rotation center to any particular eye element; we will just give the notation d to the distance from the vertex of the back surface to the new rotation center C'. When the ray from O and position vector $\mathbf{r}_0$ reaches C', its transversal coordinates will be zero, and its direction vector $\mathbf{k}_2' = -(1/d)\mathbf{r}_2$. The backward transfer between the transversal planes containing C' and O is given by the operator $\underleftarrow{\mathbf{T}}(z)\,\underleftarrow{\mathbf{L}}\,\underleftarrow{\mathbf{T}}(d)$, identical to (5.30) except for the first distance being d instead of l_2'. In terms of 2×2 blocks, the transfer reads

$$\begin{bmatrix} \mathbf{r}_0 \\ \mathbf{k}_0 \end{bmatrix} = \begin{bmatrix} \mathbb{L}_2 - z\mathbb{P}_{EB} & -(\tau\mathbb{I} + d\mathbb{L}_2) - z(\mathbb{L}_1 - d\mathbb{P}_{EB}) \\ \mathbb{P}_{EB} & \mathbb{L}_1 - d\mathbb{P}_{EB} \end{bmatrix} \begin{bmatrix} \mathbf{0} \\ -(1/d)\mathbf{r}_2 \end{bmatrix}, \tag{5.57}$$

so

$$\mathbf{r}_0 = \frac{1}{d}\left[(\tau\mathbb{I} + d\mathbb{L}_2) + z(\mathbb{L}_1 - d\mathbb{P}_{EB})\right]\mathbf{r}_2. \tag{5.58}$$

Now, the perceived position of the fixation object, $\mathbf{r}_p$, is related to $\mathbf{r}_2$ and $\mathbf{k}_2'$ through

$$\mathbf{k}_2' = \frac{1}{z + t_0}\left(\mathbf{r}_2 - \mathbf{r}_p\right),$$

from which

$$\mathbf{r}_2 = \frac{d}{z + t_0 + d}\mathbf{r}_p. \tag{5.59}$$

Substituting this result into (5.58), we get the relation

$$\mathbf{r}_0 = \frac{1}{z + t_0 + d}\left[(\tau\mathbb{I} + d\mathbb{L}_2) + z(\mathbb{L}_1 - d\mathbb{P}_{EB})\right]\mathbf{r}_p, \tag{5.60}$$

and so the magnification matrix can be defined as

$$\mathbb{M} = (z + t_0 + d)\left\{[\tau\mathbb{I} + d(\mathbb{I} - \tau\mathbb{P}_2)] + z(\mathbb{I} - \tau\mathbb{P}_1 - d\mathbb{P}_{EB})\right\}^{-1}. \tag{5.61}$$

The magnification matrix relates the size of the perceived vector $\mathbf{r}_p$ with the object vector, $\mathbf{r}_0$, and we have obtained it for arbitrary values of z and d. In the case of a distant object, $z \to \infty$, and from (5.60) we find that $\mathbb{M} = (\mathbb{L}_1 - d\mathbb{P}_{EB})^{-1}$. Now, matrix $\mathbb{L}_1$ is just the inverse of shape factor $\mathbb{G}$, so we can operate

$$
\begin{aligned}
\mathbb{L}_1 - d\mathbb{P}_{EB} &= \mathbb{G}^{-1} - d\,(\mathbb{P}_1 + \mathbb{P}_2 - \tau\mathbb{P}_1\mathbb{P}_2) \\
&= \mathbb{G}^{-1} - d\big(\mathbb{P}_1 + \mathbb{G}^{-1}\mathbb{P}_2\big) \\
&= \mathbb{G}^{-1}\,[\mathbb{I} - d\,(\mathbb{G}\mathbb{P}_1 + \mathbb{P}_2)] \\
&= \mathbb{G}^{-1}\,(\mathbb{I} - d\mathbb{P}_V)\,.
\end{aligned}
$$

Henceforth, for a distant object, the magnification matrix can be written as the (matrix) product of a power factor and the shape factor,

$$
\mathbb{M} = (\mathbb{I} - d\mathbb{P}_V)^{-1}\,\mathbb{G}, \tag{5.62}
$$

which is the result obtained by Keating [153]. We have thus demonstrated that the magnification matrix given by (5.62) for far objects, or by (5.61) for near objects, describes the dynamic magnification for an eye that rotates around a point located at distance d from the vertex of the back surface of the lens. In our derivation we have not made use of any further approximation, and the result does not depend on the optics of the eye, because we have defined the magnification as the quotient between line of sight angles with and without the lens. Of course, if we make d equal to the distance from the vertex of the back surface to the center of the pupil, an actual eye cannot rotate around this point. But if it could,[5] the chief ray we use to compare sizes would enter the eye along the line of sight, henceforth passing through the pupil center, as shown in Figure 5.27(c). The visual magnification defined by Keating would ensure the rays refracted through the nonrotating eye would also pass through the center of the pupil but that would require knowledge of the anterior optics of the eye. In any case, Keating demonstrated with an example that for a typical astigmatic eye the differences between his "exact" method and (5.62) are negligible. Further, from the practical point of view we should not even bother about these differences, as a complete characterization of the optics of the eye would be extremely difficult: The pupil is not totally round and is not centered, the eye lacks an optical axis and the line of sight does not exactly pass through the center of the pupil, the compensating lens will usually be tilted, etc. Besides, the standard error in the measurement of d will produce larger deviations in the magnification matrix than all the other factors together. In any case, and assuming all these minute factors could be measured, the treatment that could account for all them, within the paraxial approximation, would be Harris'.

The meaning of the magnification matrix is illustrated in Figure 5.28. When we apply it to the base vectors $(1,0)$ and $(0,1)$, they are transformed to the vectors formed by its columns, (m_{11}, m_{21}) and (m_{12}, m_{22}). The unit square is transformed into the parallelogram

---

[5] To rotate the eye with respect to the lens around a point other than the actual rotation center of the eye, we would have to rotate the head around this point, the eye rotating together with the head and keeping the compensating lens still.

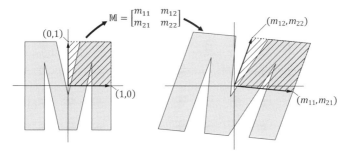

Figure 5.28 Meaning of the magnification matrix and its components.

with vertexes at $(0,0)$, (m_{11}, m_{21}), (m_{12}, m_{22}) and $(m_{11} + m_{12}, m_{21} + m_{22})$, and the same geometrical transformation is applied to any object perceived through the lens, such as the letter "M" in Figure 5.28. When $\mathbb{P}_1$ and $\mathbb{P}_2$ commute, all the matrices $\mathbb{P}_E$, $\mathbb{P}_{EB}$, $\mathbb{P}_V$, and $\mathbb{M}$ also commute between them and with $\mathbb{P}_1$ and $\mathbb{P}_2$. All these matrices are then symmetric, and the effect of $\mathbb{M}$ is just stretching or shrinking by a different factor along each main meridian. Let us recall that the Hessian matrices representing curvature have associated with them two main meridians along which the curvatures are extrema. In the same way, any symmetric matrix representing magnification has two main meridians along which the stretching/shrinking is maximum. If our object is a circle, it would transform into an ellipse under the action of the symmetric magnification matrix, with the major and minor axis of the ellipse oriented along the main meridians of the magnification matrix. It is also clear that if $\mathbb{M}$ is symmetric, its main meridians are the same as those of $\mathbb{P}_V$.

If $\mathbb{P}_1$ and $\mathbb{P}_2$ do not commute, $\mathbb{P}_V$ will still be symmetric, but it will no longer commute with $\mathbb{P}_1$, and $\mathbb{P}_E$, $\mathbb{P}_{EB}$, and $\mathbb{M}$ will not be symmetric. It can be demonstrated that this asymmetry translates into a rotation of the magnified object. The components in the diagonal of $\mathbb{M}$, m_{11}, and m_{22} represent magnification along the horizontal and vertical directions, as we would expect. In general, these components will be numbers close to 1. If they are larger than 1, it means there is stretching along the corresponding direction. If they are smaller, they will represent shrinkage. The nondiagonal components m_{12} and m_{21} will usually be close to zero and can either be positive or negative. If the magnification is symmetric, they will be identical. This means the vectors (m_{11}, m_{21}) and (m_{12}, m_{22}) will both be in the first quadrant if $m_{12} = m_{21} > 0$ or in the fourth and second quadrants, respectively, if $m_{12} = m_{21} < 0$. According to this, the situation depicted in Figure 5.28 cannot correspond to a symmetric matrix, as it requires $m_{21} < 0$ and $m_{12} > 0$. The magnification matrix of the example is then compounding a rotation and a stretching operation. It can be shown that for a nonsymmetric magnification matrix $\mathbb{M}$, there exist a rotation $\mathbb{R}(\epsilon)$ and a symmetric magnification matrix $\hat{\mathbb{M}}$, such that $\mathbb{M} = \mathbb{R}(\theta)\hat{\mathbb{M}}$, [159]. The latter will stretch or shrink so that maximum and minimum magnification is achieved along some orthogonal main meridians, and the former will rotate the magnified image by angle ϵ. The maximum and minimum magnification of $\hat{\mathbb{M}}$ are obtained with the same techniques used for curvature: If we name its components $\hat{m}_{ij}$, first we compute $M = (1/2)(\hat{m}_{11} + \hat{m}_{12})$ and

$K = \hat{m}_{11}\hat{m}_{12} - \hat{m}_{12}^2$, and then obtain the minimum and maximum magnifications as usual, $m_1 = M - \sqrt{M^2 - K}$ and $m_2 = M + \sqrt{M^2 - K}$. Similarly, the orientation at which magnification m_1 applies is $\theta_1 = \arctan[(m_1 - \hat{m}_{11})/\hat{m}_{12}]$. Adopting the nomenclature used by [138], we may simply give the name *magnification* to the average M, and *anamorphic distortion* to the difference $N = m_2 - m_1$. If we want to determine the four values (rotation ϵ, magnification M, anamorphic distortion N, and main orientation θ_1) from the components of the original nonsymmetrical matrix $\mathbb{M}$, we may use [138, 159]

$$\tan\epsilon = \frac{m_{12} - m_{21}}{m_{11} + m_{22}}$$

$$M = \frac{1}{2}\sqrt{(m_{11} + m_{22})^2 + (m_{12} - m_{21})^2}$$

$$N = \sqrt{(m_{11} - m_{22})^2 + (m_{12} + m_{21})^2}$$

$$\tan\theta_1 = -\frac{m_{11}^2 - m_{22}^2 + m_{21}^2 - m_{12}^2 + 2MN}{2(m_{11}m_{12} + m_{21}m_{22})}. \tag{5.63}$$

Example 28 An ophthalmic lens has a back vertex power $[4, 2 \times 60°]$, center thickness $t_0 = 6.5$ mm, and refractive index $n = 1.5$. Its base curve is $P_1 = 9$ D and the approximate distance from the vertex of its back surface to the stop pupil of the eye is 16 mm. Determine the magnification matrix and the magnification components given in equation (5.63).

As usual, we first compute the matrix form of the back vertex power, $\mathbb{P}_V = \begin{bmatrix} 5.5 & -0.866 \\ -0.866 & 4.5 \end{bmatrix}$ D. The power factor is then

$$(\mathbb{I} - d\mathbb{P}_V)^{-1} = \begin{bmatrix} 1.0967 & -0.0164 \\ -0.0164 & 1.0778 \end{bmatrix}.$$

Regarding the shape factor, as the front surface of the lens is spherical we can compute it as a scalar,

$$g = (1 - \tau P_1)^{-1} = 1.0406.$$

The magnification matrix is

$$\mathbb{M} = (\mathbb{I} - d\mathbb{P}_v)^{-1} g = \begin{bmatrix} 1.141 & -0.017 \\ -0.017 & 1.121 \end{bmatrix}.$$

Now, direct application of equations (5.63) gives us

$$\tan\epsilon = 0, \ M = 1.131, \ N = 0.0394 \text{ and } \tan\theta_1 = 1.745,$$

so there is no rotation ($\mathbb{M}$ is symmetrical), and the orientation of the meridian with the smallest magnification is $\theta_1 = \arctan(1.745) \simeq 60°$.

Example 29 Repeat the previous example assuming the front surface is now a ring toric surface with horizontal base curve $B = 8$ D and vertical cross curve $T = 12$ D, and the thickness of the lens is $t_0 = 10$ mm.

The refractive power of the front surface is now astigmatic, $\mathbb{P}_1 = \begin{bmatrix} 8 & 0 \\ 0 & 12 \end{bmatrix}$ D, so we have to compute the shape factor as a matrix,

$$\mathbb{G} = (\mathbb{I} - \tau \mathbb{P}_1)^{-1} = \begin{bmatrix} 1.056 & 0 \\ 0 & 1.087 \end{bmatrix}.$$

The power factor does not change, so the magnification matrix is now

$$\mathbb{M} = (\mathbb{I} - d\mathbb{P}_v)^{-1} \mathbb{G} = \begin{bmatrix} 1.1585 & -0.0178 \\ -0.0173 & 1.1716 \end{bmatrix}.$$

Notice that the magnification matrix is not symmetrical. The magnification components are now

$$\tan \epsilon = 2.57 \times 10^{-4}, \ M = 1.165, \ N = 0.03754 \text{ and } \tan \theta_1 = 0.6964.$$

The rotation angle is now $\epsilon = 0.01°$, and the orientation of the meridian with the smallest magnification, $\theta_1 = \arctan(0.6964) \simeq 35°$. We see that the asymmetry of the magnification matrix has more of an academic interest than a practical one. First, bitoric lenses are very rare, only provided by a very small fraction of workshops making specialty lenses. Second, the rotation caused by the asymmetry is very small. We could have chosen a more "extreme" lens with larger refractive powers and greater thickness, yet the rotation associated to the magnification matrix would still be quite small.

We have seen that M and the diagonal components of the magnification matrix are close to 1. It is more practical to specify the magnification as a percentage, $m_{ii}(\%) = (m_{ii} - 1) \times 100$. The same applies to the average magnification, $M(\%) = (M - 1) \times 100$. The anamorphic distortion and the nondiagonal elements of the magnification matrix, m_{12} and m_{21}, representing deformations not present in the original object, can be expressed as a percentage by simply multiplying their value by 100. In Figure 5.29 we show average magnification and anamorphic distortion as a function of the object distance according to expression (5.61) and for two spherotoric lenses: To the left a positive one with vertex power $[8, -2 \times 180°]$ base curve 10 D and center thickness 9 mm, and to the right a negative lens with vertex power $[-6, -2 \times 180°]$, base curve 2 D and center thickness of 2 mm. The refractive index for both lenses, $n = 1.5$. The plots in the upper row represent averaged magnification and those in the lower row anamorphic distortion. We see that both magnification and anamorphic distortion are smaller (in absolute value) for near objects, but the effect is small, slightly less than 1% in magnification change for the positive lens and slightly above 1% for the negative, the range for z extending from 100 mm to 3 m. The absolute values of the main meridians of both lenses are identical, but the positive lens produces substantially more magnification than the negative. Of course, this is due to the shape factor, which as long as the front side is convex is always bigger than one, and especially bigger for the positive lens, which has a larger base curve and a larger center thickness.

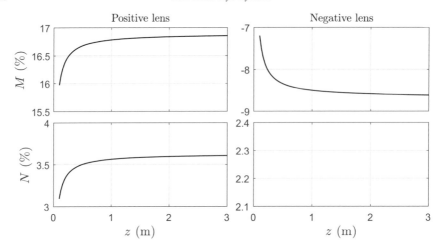

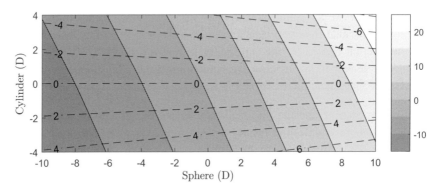

Figure 5.29 Effect of object distance in magnification and anamorphic distortion. To the left, positive lens $[8, -2 \times 180°]$. To the right, negative lens $[-6, -2 \times 180°]$.

Figure 5.30 Magnification and anamorphic distortion (percentage) as a function of the spherical and cylindrical components of lens power.

Figure 5.30 represents averaged magnification and anamorphic distortion at far distance for spherotoric lenses with spherical power ranging from -10 D to $+10$ D and cylindrical power ranging from -4 D to $+4$ D. The base curve was provided to each lens in a continuous way, interpolating the base curve typically used nowadays by the ophthalmic industry but allowing it to change continuously. Center thickness was frozen to 2 mm for lenses with negative larger meridian, and computed using the algorithm shown in Section 4.3.6 for the remaining lenses, assuming a lens diameter of 60 mm. The average magnification is shown in the gray-colored contour map with solid lines, while the anamorphic distortion is shown in a superimposed contour map with dashed lines. Average magnification goes from -15% to over 20% and basically depends on the spherical equivalent. Anamorphic distortion basically depends on the cylinder, although it also shows a slow dependence on the spherical power.

5.5.3 Aniseikonia

Aniseikonia is the condition by which the images from the two eyes have different shape and/or size. These images are not the actual retinal images but their representations in the visual cortex. A small amount of aniseikonia is always present as part of the binocular disparity, and the brain uses it to generate depth perception. Its origin is of geometrical nature and depends on the relative position of the eyes and the fixated object. Any other source of size or shape differences between the images of the two eyes is known as *anomalous aniseikonia* [132]. Aniseikonia can be *anatomical* if the retinal receptors present different spatial density in each eye or somehow the visual cortex assigns different sizes to the images coming from each eye, or *optical*, when the actual sizes of the optical images on each retina are different. Finally, the optical aniseikonia can be *inherent* or *induced*. In the first case the different image size will come from different eye size or other characteristics of the eyes. In the second case, aniseikonia is caused by the use of compensating lenses with different power, as a consequence of the user having anisometropia. We will explore here the second of these, first studying the way we can quantify the induced aniseikonia. We will also briefly present the way aniseikonia can be compensated or reduced by means of afocal magnifiers (basically by changing the base curve and center thickness of the lenses). This process can be used also for those cases in which inherent or anatomical aniseikonia is causing binocular problems. We will mainly follow the nomenclature and definitions given by Keating in his paper about aniseikonia [154].

We will define the aniseikonic matrix as the linear map that transforms the perceived object position for the right eye, $\mathbf{r}_{pR}$ to the perceived object position for the left eye, $\mathbf{r}_{pL}$. From the definition of the magnification matrix,

$$\mathbf{r}_{pR} = \mathbb{M}_R \mathbf{r}_0,$$
$$\mathbf{r}_{pL} = \mathbb{M}_L \mathbf{r}_0. \tag{5.64}$$

Solving for $\mathbf{r}_0$ in the second equation and substituting in the first one, we get $\mathbf{r}_{pR} = \mathbb{M}_R \mathbb{M}_L^{-1} \mathbf{r}_{pL}$, so the aniseikonia matrix is defined as $\mathbb{M}_A = \mathbb{M}_R \mathbb{M}_L^{-1}$. It is clear that if were able to induce an extra magnification $\mathbb{M}_A$ to the left eye, or $\mathbb{M}_A^{-1}$ to the right one, the aniseikonia would be compensated.

Another way to approach the problem of compensating aniseikonia is by forcing the magnification for the two eyes to be equal, $\mathbb{M}_R = \mathbb{M}_L$, that is

$$(\mathbb{I} - d\mathbb{P}_{VR})^{-1} \mathbb{G}_R = (\mathbb{I} - d\mathbb{P}_{VL})^{-1} \mathbb{G}_L.$$

The back vertex powers of either eye cannot be modified (or only very slightly) if we want the user to be properly compensated, but we have some freedom to change the shape factors. As they only provide positive magnification (greater than one), we may try to increase the shape factor of the lens producing a smaller average magnification. For example, if $M_L > M_R$, we may try to find $\mathbb{G}_R$ so that

$$\mathbb{G}_R = (\mathbb{I} - d\mathbb{P}_{VR}) (\mathbb{I} - d\mathbb{P}_{VL})^{-1} \mathbb{G}_L.$$

Solving for the refractive power in $\mathbb{G}_R$, we get

$$\frac{t_{0R}}{n_R}\mathbb{P}_{1R} = \mathbb{I} - (\mathbb{I} - \tau_L\mathbb{P}_{1L})(\mathbb{I} - d\mathbb{P}_{VL})(\mathbb{I} - d\mathbb{P}_{VR})^{-1}, \qquad (5.65)$$

where we are assuming distance d to be equal for the two eyes, but the remaining parameters may be different, the sub-indexes R and L indicating right or left eye. In general, the previous equation does not have an exact solution, as $\mathbb{P}_{1R}$ has to be symmetric but its right side does not. Even for spherotoric lenses for which $\mathbb{P}_1$ and $\mathbb{P}_V$ commute, the powers for the right and left eye may have different orientation of their main meridians and they may not commute. Moreover, if the asymmetry of the right side of equation (5.65) is zero or negligible, the complete compensation of aniseikonia usually demands unpractical values of thickness t_{0R}, curvatures in $\mathbb{P}_{1R}$, or both. Let us see this with a simple example.

Example 30 A person has spherical refractive errors $P_{VR} = 2\,\mathrm{D}$ and $P_{VL} = 4\,\mathrm{D}$. The lenses are made with refractive index 1.5, and the distance from the back vertex to the entrance pupil of the eye is 15 mm. Compute the induced aniseikonia in the case that two lenses have a base curve of 6 D, with central thicknesses $t_{0R} = 3.5$ mm and $t_{0L} = 4.6$ mm. Then, assuming the left lens is made with base curve of 5 D, compute the relationship between thickness and base curve of the right eye lens required to avoid aniseikonia.

As the lenses are spherical we will operate with scalar magnitudes. The shape factors are

$$g_R = (1 - \tau_R P_{1R})^{-1} = 1.0142, \quad g_L = (1 - \tau_L P_{1L})^{-1} = 1.0187.$$

Similarly, the power factors are given by

$$(1 - dP_{VR})^{-1} = 1.031, \quad (1 - dP_{VL})^{-1} = 1.064,$$

and the corresponding magnifications are $M_R = 1.047$ and $M_L = 1.083$. The induced aniseikonia is then

$$M_A = M_R M_L^{-1} = \frac{1.047}{1.083} = 0.967.$$

The relationship center thickness and base curve for the right lens to be iseikonic is given by equation (5.65). In our case, we have the following scalar equation

$$\tau_{0R}P_{1R} = 1 - (1 - \tau_L P_{1L})\frac{1 - dP_{VL}}{1 - dP_{VR}}.$$

Plugging the problem data into the previous equation (recall that the left eye base curve is now $P_{1L} = 5$ D), we get $t_{0R}P_{1R} = 0.06876$. We can get a better grasp of this condition by expressing the thickness in mm, in which case, t_{0R} (mm) P_{1R} (D) $= 68.6$. If we make the right lens with a center thickness of 10 mm, its base curve should be 6.86 D to compensate for aniseikonia. Instead, we can use a base curve of 10 D with a center thickness of 6.86 mm. We can see that in either case the lens is going to be unsightly and heavy and not every patient will accept the iseikonic compensation. From the clinical point of view, the ECP may look for a balance between partial compensation of the aniseikonia and acceptable lens thickness and base curve dissimilarity.

The previous example illustrates that achieving perfect compensation for aniseikonia is quite unlikely. Most of the time, the optometrist may expect to achieve an approximate compensation, and, for this, some approximations to the previous expressions will turn out to be quite useful. First, the power and shape factors can be approximated, $\mathbb{M}_{shape} \simeq \mathbb{I} + \tau \mathbb{P}_1$ and $\mathbb{M}_{power} \simeq \mathbb{I} + d\mathbb{P}_V$. The corresponding percentage magnifications are $\mathbb{M}_{shape}(\%) = 100\tau\mathbb{P}_1$ and $\mathbb{M}_{power}(\%) = 100d\mathbb{P}_V$. The total magnification is

$$\begin{aligned} \mathbb{M} &\simeq (\mathbb{I} + \tau\mathbb{P}_1)(\mathbb{I} + d\mathbb{P}_V) \\ &= \mathbb{I} + \tau\mathbb{P}_1 + d\mathbb{P}_V + \tau d\mathbb{P}_1\mathbb{P}_V \\ &\simeq \mathbb{I} + \tau\mathbb{P}_1 + d\mathbb{P}_V, \end{aligned}$$

as the numerical values of τ and d are in the order of 10^{-2} or less and those of $\mathbb{P}_1$ and $\mathbb{P}_V$ in the order or 10 or less. We conclude that, in percentage, $\mathbb{M}(\%) \simeq \mathbb{M}_{shape}(\%) + \mathbb{M}_{power}(\%)$.

An afocal magnifier is an optical element that does not change the vergence of light refracting through it, but does change the magnification. A lens with $\mathbb{P}_1 \neq \mathbb{O}$ and $\mathbb{P}_V = \mathbb{O}$ is an afocal magnifier. Any ophthalmic lens can be considered the superposition of a thick afocal magnifier with front surface $\mathbb{P}_1$ and magnification $\mathbb{M}_{shape}$ followed by a thin-lens with power $\mathbb{P}_V$ and magnification $\mathbb{M}_{power}$. The combined magnification is the product of the two matrices in reverse order, but if we express them as percentages, it will approximately be the sum of the two. Take into account that within the last approximation, the percentage total magnification matrix will always be symmetric, as it is the sum of two symmetric matrices $\mathbb{M}_{shape}$ and $\mathbb{M}_{power}$. This is not a big deal: The asymmetry of spectacle lenses, if present, will usually be small, as even thicker lenses are relatively thin from the optical point of view. According to this, interposing additional afocal magnifiers in front of the lens would have the effect of additively combining their percentage magnifications to that of the lens. Another way to see it is as modifications of the intrinsic lens afocal magnifier. For example, doubling the value of $\mathbb{P}_1$ will multiply the percent magnification of $\mathbb{M}_{shape}$ by two, and this is equivalent to adding a second afocal magnifier with the original values for τ and $\mathbb{P}_1$. In the same way, when we use percentages, the aniseikonic matrix can be expressed as the difference between the magnifications affecting both eyes. To show this, take into account that for any magnification matrix, $\mathbb{M} = \mathbb{I} + 0.01\mathbb{M}(\%)$, so the product defining the magnification matrix can be written as

$$\begin{aligned} \mathbb{M}_R\mathbb{M}_L^{-1} &= [\mathbb{I} + 0.01\mathbb{M}_R(\%)][\mathbb{I} + 0.01\mathbb{M}_L(\%)]^{-1} \\ &\simeq [\mathbb{I} + 0.01\mathbb{M}_R(\%)][\mathbb{I} - 0.01\mathbb{M}_L(\%)] \\ &\simeq \mathbb{I} + 0.01\mathbb{M}_R(\%) - 0.01\mathbb{M}_L(\%), \end{aligned}$$

so $\mathbb{M}_A(\%) = \mathbb{M}_R(\%) - \mathbb{M}_L(\%)$. Magnification and aniseikonia can be written in "sphero-cylindrical" notation, and clinicians dealing with aniseikonia usually do so. For example, the power factor of a lens with back vertex power $[2, 2 \times 45°]$ worn at a distance from the pupil $d = 15$ mm, can be written as [3%, 3% × 45°]. But take into account that this is only correct under the previous approximation; for this example, exact computation of maximum and minimum magnifications yields 3.1% and 6.4%. Under the conditions explained in Chapter 2, curvature is additive, so we can split it into sphere and cylinder, but magnification is only additive under the low magnification approximation. The advantage is that it allows much faster computation of an estimate of the required modifications to either $\mathbb{M}_R$ or $\mathbb{M}_L$ (or both) to reduce aniseikonia to tolerable levels. For example, let us consider an anisometric person with refractive errors 0 D and -2 D in the right and left eyes, respectively. The lenses are made in base 3 D, with refractive index 1.5 and central thickness 2.4 mm. Distance d is assumed to be 15 mm. As the refractive errors are spherical, the magnifications turn out to be scalar, and we have $M_R = 0.48\%$ and $M_L = -2.44\%$. The aniseikonia is then 2.9%. To compensate for it, we should either add a -2.9% magnifier to the right eye or 2.9% to the left. We could also split the correcting magnification between the two eyes, adding -1.45% to the right eye and 1.45% to the left. To produce $\pm 1.45\%$ magnification we require an afocal magnifier having a product $\tau P_1 = \pm 0.0145$. The positive one can be achieved with a center thickness of 5 mm and a base curve of 4.35 D. The negative one would require a *negative base curve*, which are seldom used in spectacle lenses to avoid optical aberrations. In a practical case we cannot add more lenses to the frame, and we would have to shape the lenses to be *iseikonic*, that is, to compensate or mitigate the effect of aniseikonia. Let us assume our anisometropic person can tolerate 1% of aniseikonia, so we have to remove two-thirds of the aniseikonia produced by the original lenses. To avoid positive magnification over the right eye, we will use $P_1 = 0$ for the right lens, a truly parallel plate. Then, maximum negative magnification for the left eye will be -1%. The power factor amounts to $M_{Lpower} = -2.9\%$, so we need a shape factor with magnification 1.9%, that is, $\tau P_1 = 0.019$. Assuming a center thickness of 5 mm, the lens would need $P_1 = n \times 0.019/0.005 = 5.7$ D. We see that even for this mild case of anisometropia, and even assuming the person can handle a level of aniseikonia of 1%, the required lenses would be quite unalike: totally flat (with arbitrary thickness) for the right eye, and quite curved, 5.7 D, for the left eye, with a central thickness of 5 mm that would grow to about 6 mm at 20 mm from the optical center of the lens.

6

Aberrations and Lens Design

6.1 Introduction

In previous chapters, we studied and analyzed the optical properties of ophthalmic lenses under the paraxial approximation of geometrical optics. This theory constitutes a useful and powerful way to analytically describe the image-forming properties of ophthalmic lenses, and it allows the computation of important magnitudes such as lens power, ray or eye deviation (prismatic effect), or magnification. Within the framework of paraxial optics, any system behaves as an ideal imaging system, fulfilling the so-called Maxwell's conditions for a perfect imaging system. However, paraxial approximation is not accurate enough to completely describe the behavior of most real systems, including ophthalmic lenses. In general, an optical system will present some degree of image degradation due to the effect of aberrations. Aberrations appear because the trajectories actually followed by rays, determined by applying Snell's law at the points where rays intersect with the surfaces of the optical system, differ from those predicted within the paraxial approximation. Indeed, a fundamental part of geometric optics is the so-called *nonparaxial optics*, also known as the *theory of aberrations*, which studies the imaging properties of optical systems outside the scope of the paraxial approximation.

It is important to understand the difference between paraxial and nonparaxial optics. As part of geometric optics, both theories share a common methodology: the study of optical systems through the analysis of the trajectories followed by the light rays propagating within them. In paraxial optics, a first-order approximation to the Snell law is used and the surface sags are neglected when computing the intersection of rays with surfaces. In nonparaxial optics, the degree of approximation is reduced and, consequently, the light trajectories are more realistic. Most of the important properties of the optical system can be computed analytically within the framework of paraxial optics. This means we can find manageable formulas that allow us to compute and predict these properties: If we want a lens with a given power, paraxial formulas will allow us to easily determine the required curvatures its surfaces should have. There are two main consequences to the partial or total removal of paraxial approximations: First, the theory becomes mathematically intricate, and it is not always possible to obtain analytical expressions that link image properties and parameters of optical systems. Second, some paraxial definitions lose meaning, or at least have to be redefined. Despite these drawbacks, the nonparaxial study of optical systems is rewarding, as it shows the relationship between image quality and the actual configuration of an optical system, thus constituting the basis of the optical design process [160].

These general principles also apply to ophthalmic lenses. In fact, a simple experiment dramatically shows how image quality is related to lens design: It consists of flipping the eyeglasses so that the temples point outward the face. The lenses are just the same, but the order in which the light passes through their surfaces is reversed. Anyone trying this experiment will experience a similar image quality at the very center of the lenses, but quality rapidly degrades toward the periphery. Images will be greatly distorted, becoming barrel-shaped for myopes and the other way around for hyperopes. Besides, strong image blurring will be observed as the eye rotates a few degrees away from the optical center. The standard shape and orientation of ophthalmic lenses is then the result of a design (or optimization) process intended to provide the best possible image quality. Obviously, this design process can only be tackled within the framework of nonparaxial optics, as the theory we studied in previous chapters would not predict any change in image quality when flipping the lenses, just a change in magnification. Therefore, the goal of this chapter is to describe the design of ophthalmic lenses within this framework of nonparaxial optics.

The classic theory of ophthalmic lens design was based on the so-called third-order approximation to the Snell law, and therefore was focused on the study of third-order or Seidel's aberrations. The main achievement of this model was the justification of the meniscus (or convex-concave) shape as the optimum form for a spherical ophthalmic lens. To some extent, this model was also applied to lenses presenting nonspherical surfaces, such as astigmatic or aspheric ones, and constituted a basic reference for lens manufacturing throughout the first half of the twentieth century.

With the advent of computers, optical design experienced a complete revolution. The use of more and more powerful processors allowed the numerical computation of the exact trajectories for millions of rays through complex optical systems in very short times. This fact, combined with the application of numerical optimization algorithms, enabled the design of optical systems with greater image quality than before. These new design techniques were also adopted by the ophthalmic lens manufacturing industry, adapting them to the existing manufacturing process, which limited somewhat the ultimate quality of the lenses produced, as it was not possible to optimize each of the possible prescriptions while, at the same time, maintaining acceptable production costs.

The lens manufacturing industry has experienced a major change in the past fifteen years with the introduction of the so-called free-form technology (see Chapter 10). This new technology overcomes the limitations presented by the traditional manufacturing process: Current free-form generators and polishers can produce surfaces with arbitrary curvature distribution (within some range), with a similar throughput as traditional surfacing technology. This allows the manufacturing of lenses with unique aspherical surfaces, individually optimized for each prescription and each user.

We will begin this chapter with a brief description of the particularities of the lens-eye system that will lead to the classical procedure of substituting the rotating eye for a stop

placed at the rotation center of the eye. Afterward, we will describe the classic design
of ophthalmic lenses. Finally, we will qualitatively describe the computer-assisted design
process of ophthalmic lenses using exact ray-tracing software with free-form surfaces,
including some application examples. If the reader is not yet familiar with the basic theory
of aberrations, we recommend, before tackling this chapter, the reading of Section 5.2 and
Appendix E.

6.2 Aberrations of the Lens-Eye System

Before we enter into ophthalmic lens design and its historical development, we must first
understand the particularity of the lens-eye system. We already know the eye is an optical
system that is not free from aberrations, so we first should understand, at least qualitatively,
how the compensating lens could modify, for better or worse, the final quality of the retinal
image. There are two different "modes" or "regimes" relative to the performance of the
lens-eye system, and these are depicted in Figure 6.1. In the static-eye mode, we consider
the whole retinal image as the eye keeps still at the main viewing position behind the
ophthalmic lens. The static field of view of the eye is so large that normally most of the
ophthalmic lens, with the only probable exception of the nasal side, will be inside this field
(this can be observed by anyone wearing glasses, paying attention to the peripheral image
and observing the – defocused – rim of the spectacle within it). In this operating mode,
we expect the light pencils refracting through the lens and the eye at any angle within the
static visual field will focus on the retina. The first thing we observe is that the image
should not lie on the paraxial plane, represented with a dotted line tangent to the retina in
Figure 6.1. This means the lens-eye system should have plenty of field curvature aberration (see Appendix E) for the peripheral retinal image to be correctly focused on the
curved retina.

In the second mode, the eye rotates behind the lens about its rotation center. We already
explained in Section 5.2 that there is no real rotation center, but a region known as the
centrode containing the rotation centers of each infinitesimal eye rotation. However, we will
ignore this fact and assume the size of the centrode is negligible with respect to the lens-
centrode distance. Eye rotation brings into the fovea the image of the object point being

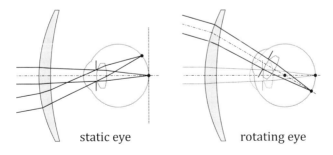

static eye   rotating eye

Figure 6.1 Lens-eye system. Static vs rotating eye.

fixated. Under these circumstances, we expect the pencil refracted through the lens and then through the eye will focus on the fovea. We should be tempted to hypothesize that the two working modes are incompatible. If a lens could be designed to provide optimum performance in the static mode, it would probably fail to provide optimum performance for the rotating eye, and the other way around. We will revise the state of the art on this issue later, but the hypothesis seems to be correct, and both modes cannot be made compatible [161].

The traditional choice for lens design has been to ignore the static mode, and the reason is the way the visual system works. Even for subjects with emmetropic eyes and high visual acuity, the peripheral retinal image is known to be of poor quality. And if it were not, the low resolution of the peripheral retina would severely limit the detail of the perceived peripheral image. It would make no sense for the eye to evolve to produce a sharp peripheral image that would be wasted because of a low-resolution peripheral retina, and the other way around. The sharp and wide field of view perceived by an emmetropic person with high visual acuity is not backed by a wide-field sharp image; it is rather the result of neural processing of many small-field foveal images generated as the eyes rotate and scan the dynamic field of view. Anyone can notice this because even if we perceive the whole field of view as sharp, when we need to get actual information from any noncentral point within the field of view, we automatically rotate the eyes (or the head) and bring this point to the center of the field of view, so that its image will lie on the fovea. According to this, it seems quite reasonable to abandon the static mode as a paradigm for lens design and concentrate on the rotating-eye or dynamic mode.

We remind the reader that, even if we restrict the analysis to foveal vision, the eye is not free from aberrations. The most extended way to define and manage eye aberrations is through the wavefront at the entrance pupil of the eye that would produce a sharp image point at the fovea, W_E. In a perfect eye, this wavefront would be plane, so the wavefront error is the wavefront itself. We saw in Section 5.2 that we can describe the wavefront aberration with an expansion using Zernike polynomials,

$$W_E\,(d\,x, d\,y) \approx \sum_{n=1}^{N} \sum_{m=1}^{M} a_{nm} Z_{n,m}\,(x,y)\,, \tag{6.1}$$

where d is the pupil radius and the coordinates x and y are limited to the unit circle, $x^2 + y^2 \le 1$. In general, we may ignore piston and tilt, the coefficients for the polynomials with $n = 0$ and $n = 1$; the coefficients of the three $n = 2$ Zernike polynomials describe the average curvature of the wavefront, which for small pupils matches the clinical refractive error (see Section 5.2). Finally, all the terms with $n \ge 3$ are named *higher-order aberrations* (HOA).

The measurement of ocular aberrations and their effect on vision have been important research topics in recent decades. The combination of compact light sources and image sensors based on solid-state electronics, along with the high capabilities of modern computers, has allowed the development of a number of experimental techniques for measuring the aberrations of a human eye in a highly controlled and repeatable manner. See, for example, particular developments [72, 73, 74, 75, 76, 162] or a general review [95]. Numerous clinical studies have been carried out in order to determine the average values of the aberrations in a selected human population [96, 163, 164] and to extract statistical relations between eye aberrations and different factors such as eye aging [165] or accommodation [166].

Clinical studies on ocular aberrations conducted on a general population [96, 163] show some interesting aspects of human eye aberrations. First, the predominant values of the Zernike expansion are those corresponding to second-order terms ($n = 2$), particularly spherical defocus, given by the value of the coefficient $a_{2,0}$, and "with the rule" astigmatism, given by the coefficient $a_{2,1}$. According to Porter et al. and Castejón-Mochón et al. [96, 163], second-order terms explain up to 91% of the RMS of the wave aberrations for a pupil of around 5 mm. Second, there is almost no correlation in the values of the Zernike coefficients among the population, which means that HOA are different for each individual. However, there is some similarity (correlation) between the coefficients corresponding to the left and right eye of a given individual, which indicates certain symmetry. Third, there is little or no correlation between the refractive error and the HOA [167]. Finally, the only HOA whose mean value within the studied population departs significantly from zero is the spherical aberration [95].

All the facts described in the previous paragraph suggest a difficult prospect for correction of full wavefront eye aberration with a single lens. As most of the error comes from second-order terms, these should be the ones we have to focus on first. The lack of correlation between HOA and refractive error means that we cannot expect to find surface families that would automatically compensate both at the same time, and the compensation should be uniquely computed and manufactured for each individual. Last, but not least, even if we could compute and manufacture these individual surfaces, they would only work at a very small field of view, as we explained in Section 5.3.1. Large-field compensation of eye aberration in the rotating-eye mode can only be achieved for the second-order terms of the Zernike expansion, which, when computed with small pupils (2 to 3 mm), should coincide with the refractive error of the eye.[1]

In summary, the objective of the spectacle lens designer will be to provide a lens that compensates the refractive error of the eye according to the rotating eye scheme. We can define this scheme more precisely by using the remote point of the eye, as shown in Figure 6.2. In the case of an eye with a spherical refractive error, its remote point is the paraxial conjugate of the retina through the optics of the eye. A lens compensates this refraction error when its focus is located at the remote point. In this way, a far object is imaged by the lens at the remote point, and the eye, in turn, images this image on its retina. The eye, its fovea, and its remote point rotate together, so the remote point moves on a spherical surface whose center is located at the rotation center of the eye. For any oblique

---

[1] For large pupils the second-order terms provide a large-aperture averaged curvature that may not match the clinically measured refractive error.

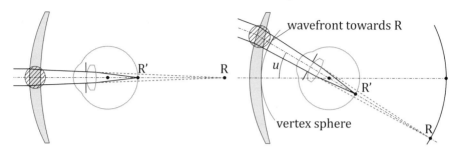

Figure 6.2 Compensation principle for the rotating eye in terms of the remote point and the remote point sphere.

direction, we expect the lens will keep focusing at the remote point, so those light pencils whose central ray passes through the rotation center of the eye after refraction through the lens should focus on this sphere of remote points. For any given gaze angle, the region of the lens through which the pencil of rays forming the foveal image refracts is relatively small, and is indicated with hatched circles in Figure 6.2. Not only is this region small, but for each of these regions, we are only concerned with an incoming direction: that of the ray which after refraction passes through the rotation center of the eye. We will see that under these working parameters, it is possible to achieve lens shapes and surfaces that comply quite well with the compensation principle, not only at the main viewing direction, but for a large range of oblique gaze directions.

In terms of wavefronts, let us give the notation $W_{LE}(u)$ to the wavefront, at the entrance pupil of the eye, of the light pencil refracted by the lens at a given oblique direction characterized by the eye rotation angle u. According to the previous discussion, it will be possible to achieve lens designs that will make the small-aperture second-order terms of $W_{LE}(u)$ identical to the small-aperture second-order terms of W_E (compensation of refractive error). We have also stated that in general it will not be possible that the higher-order terms in $W_{LE}(u)$ match the higher-order terms in W_E. We may still wonder how the HOA in $W_{LE}(u)$ will affect vision quality. In general, the curvature radii of ophthalmic lenses are much larger than the pupil and the curvature radii of the cornea and the crystalline lens. This difference in scale suggests that the HOA in $W_{LE}(u)$ will be negligible with respect to its second-order terms. In any case, although we cannot hope for the compensation of the HOA in W_E, we do have some control on the HOA in $W_{LE}(u)$. For example, the coma and spherical aberration of the wavefronts $W_{LE}(u)$, being small, will depend on the lens bending, but the general trend is that the type of lenses and surfaces that provide the smallest HOA are not those providing the best second-order compensation at different gaze angles. There is some modern research on the effect of the combinations of the higher-order aberrations from the eye with other higher aberrations from a compensating element [168, 169], but, to the knowledge of the authors, there are no clear consequences for lens design yet.

The classic theory of ophthalmic lens design was established by the work of several authors, with progress extending through the nineteenth century and the beginning of the twentieth. William Wollaston was the first to propose a lens shape that would have better imaging performance [170]. He worked with Thomas Young, who had made some important contributions to the mathematical description of the astigmatism generated by the oblique refraction of rays through spherical surfaces [171, 172], but Wollaston's work was mainly experimental, and he proposed strongly curved meniscus-shaped lenses that would provide improved peripheral vision and that he called *periscopic*. Wollaston thought that it was necessary that the light should cross the two surfaces of the lens at almost perpendicular angles, and he was fully aware that the region of the lens producing the foveal image for any given viewing direction was scarcely bigger than the pupil of the eye, so looking for a lens shape that would provide good image quality when used at full aperture was unnecessary. Almost 30 years later, Airy provided a mathematical theory that partially explained the performance of Wollaston's periscopic lenses in "camera obscuras" and in spectacles as well. Johannes Muller, in 1826, and especially August Muller, in 1889, realized the role of the rotation center of the eye and its importance to define lens performance. The lack of advance during the nineteenth century was probably due to a poor understanding of astigmatism, which is the main aberration occurring at oblique sight directions. Almost 30 years after Muller's contributions, the French ophthalmologist Ostwalt realized that good image quality when looking at oblique angles could be obtained with two different lens bendings: highly curved periscopic lenses from Wollaston but also another shallower shape [173].

The first to provide a complete solution to the problem was Marius Tscherning in 1904 [174]. In this and subsequent works, he applied Coddington's equations and third-order theory, along with thin-lens approximation, to compute oblique astigmatism, field curvature, and distortion. He realized that the first two had a quadratic dependence with the refractive power of the back surface of the lens. According to this, there could be up to two refractive powers (and hence two lens bendings), for which at least one of the aberrations could be zeroed, and it showed graphical means from which the two bendings for either aberration could be easily read off, later known as *Tscherning ellipses*. Tscherning proposed base curves that would correct curvature and distortion, both for far and near distances, and called his lenses *orthoscopic*. From 1908, when Tscherning's work was published in the German journal *Archiv für Ophtalmologie*, to halfway through the twentieth century, many authors proposed lens bendings intended for different compromises regarding the compensation of astigmatism, field curvature, and distortion, with Percival and Whitwell being among the first [175]. In 1908, Moritz von Rohr undertook the first trigonometric computation of lenses free from astigmatism [176]. He was working with Carl Zeiss, and his work brought about the collection of *Punktal* lenses, optimized to remove oblique astigmatism at 35° for plus lenses and 30° for minus lenses. The lenses were both spherical and spherotoric, the latter with external toric surface. They were computed taking into account center thickness, and each lens had a different base curve, optimal for the cancellation of astigmatic error at the target obliquity angle. As a result, the manufacturing process was slow and expensive, due to the huge amount of tools needed for the surfacing process of the two sides of the lens. Von Rohr and Carl Zeiss also contributed to the standardization of back vertex power as the magnitude to quantify power in ophthalmic lenses, an idea apparently brought about by Badal in 1883 [177]. Carl Zeiss also commercialized the first collection of lenses with an aspherical surface, according to designs from Gullstrand. In 1919 Edgar Tillyer, working for American Optical, patented a particular lens design in which both oblique astigmatism and field curvature were considered, allowing for larger tolerances of both aberrations [178]. This small relaxation led Tillyer to the idea of grouping some lens powers to be produced with the same base curve, laying the foundations for the concept of clustered base curve series the industry has used since then.

Tscherning, von Rohr, Gullstrand, Tillyer, and others set the basis for ophthalmic lens design. Since their pioneering work, there have been no substantial breakthroughs in the design of single vision lenses. Of course, the field has been continuously improving quantitatively, as a consequence of the advance of manufacturing technologies and computing capabilities, but the limited number of degrees of freedom in the ophthalmic lens system[2] has prevented the development of qualitatively new design paradigms. The reader may find interesting the complete review by Davis et al. [179] or the more recent and brief review by Atchison [55]. Another comprehensive resource for classic lens design is Jalie [56].

6.3 Classical Theory of Ophthalmic Lens Design

6.3.1 Lenses with Revolution Symmetry

As described in the previous section, the classical theory of ophthalmic lens design is based on assumptions we describe next. The lens will be corrected for the rotating eye, so any narrow beam whose principal ray passes through the rotation center of the eye after refraction through the lens should focus on the spherical surface centered at the rotation center of the eye and formed by the rotating remote point. The diameter of the beam will be limited by the pupil of the rotating eye. These assumptions can be formulated in an equivalent and more precise way. If an ophthalmic lens is to be optimized for the rotating eye, the eye can be substituted by a fixed stop located at its rotation center, see Figure 6.3(a). This stop limits the diameter of the beams whose principal ray, after refraction through the lens, passes through this rotation center. Any beam whose principal ray does not pass through the rotation center of the eye will not be considered. If the beam inciding upon the lens is made from parallel rays (from an object located at infinity), the refracted beam should focus on the *remote-point sphere* of the eye (also known as the *far-point sphere*). In this section, we will assume the lens has revolution symmetry (so its two surfaces have the

---

[2] Free-form technology has increased the number of degrees of freedom, as it allows for real-time production of arbitrary surfaces. However, the ophthalmic lens is still just a single lens. Other instruments such as microscopes and telescopes can benefit from the use of many lenses, some of them with aspherical surfaces, that allow the reduction of aberrations to extremely low levels. However, the ophthalmic lens is limited by its weight, size, and position relative to the eye, and ergonomic and esthetic constraints limit the number of lenses to just one.

same symmetry) and its optical axis matches the line joining the rotation center of the eye and the back vertex V_2, that is, the lens is not tilted.

According to the compensation principle enunciated in Section 5.3.2, the back vertex power of the lens should be equal to the inverse of the distance from the back vertex of the lens, V, to the remote point of the eye, R. In a similar way, if we define the *vertex sphere* as the sphere centered at the rotation center of the eye passing through the back vertex of the lens, the distance from point V_s, determined by the *obliquity angle* u'_2, to the rotated remote point $R(u'_2)$ should be the same as the distance from V to R, as the former pair is obtained from rotation of the latter about the rotation center of the eye. In other words, the lens power for the oblique direction determined by angle u'_2 will be measured from point V_s on the vertex sphere instead of point B on the back surface of the lens; in this way we can make a direct comparison with the paraxial back vertex power. We will give the notation l'_2 to the distance from the back vertex of the lens, V_2, to the rotation center of the lens, C, and the angles formed by the incident and refracted ray and the normal to the surface, i_k and i'_k, where the subscript will be 1 for the front surface and 2 for the back surface, as shown in Figure 6.3(b). We will give the notation y to the height of the principal ray at the exit point on the back surface.

Given the geometry depicted in Figures 6.2 and 6.3, and the small size of the eye pupil compared to the lens diameter and lens curvature radii, we can assume that the effect of those aberrations that depend on the pupil size, such as spherical aberration and coma, is negligible. This fact was numerically confirmed by Bourdoncle et al. [180], in which spherical and other higher-order aberrations are computed for progressive lenses, which have higher values for these aberrations than single vision lenses, and yet they are very small compared to astigmatism and field curvature. A similar result was experimentally obtained by Villegas and Artal [181]. According to this, the Seidel's aberrations that have a significant effect on the image quality of an ophthalmic lens are oblique astigmatism, field curvature (which we will redefine to account for the curvature of the remote-point sphere), and distortion. The presence of the first two will blur the image at oblique gaze directions.

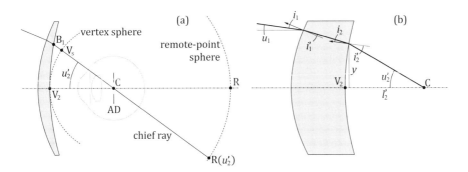

Figure 6.3 (a) geometry of the lens-eye system used in lens design. (b) nomenclature for angles and some distances used in the text.

Distortion can also be defined for the static and the rotating eye, but in any case it does not introduce image blurring, but geometrical deformation of the image, or nonlinear eye deviation.

Oblique Astigmatism and Oblique Power Error

Next, we will provide an outline of the computation of oblique astigmatism and field curvature in lenses with revolution symmetry. We will mainly follow Atchison's approach [182], which extended a classical analysis by Whitwell [183] to include aspheric surfaces. However, Atchison defines oblique powers with respect to point B on the back surface of the lens, while we will use Jalie's choosing of point V_s on the vertex sphere as the reference point. The complete derivation of the final expressions does not involve complex mathematics, but the calculations are lengthy and somewhat tedious, so we will just focus on the main steps and results. The reader interested in step by step derivation may consult papers by Smith and Atchison [184, 185], and books by Hopkins [186] and Jalie [56].

In Appendix E we give expressions for the wavefront errors in the case of astigmatism and field curvature. By combining both, we can compute the vergence (curvature of the wavefront) at V_s, but still we would need to compute the coefficients b_a and b_f. For this, we will use Coddington equations, from which we will directly obtain the main curvatures of the wavefront at V_s, that is, the oblique powers of the lens, as a function of the lens characteristics, the position of the object, the obliquity angle, and the distance l'_2. Coddington equations are a generalization of the refraction equation we studied in Chapter 3, $n'\mathbb{V}' - n\mathbb{V} = (n' - n)\mathbb{H}$, which in turn is a generalization for astigmatic vergences and surfaces of the basic refraction equation through a spherical surface with radius R separating media with refractive indexes n and n', $n'/s' - n/s = (n' - n)/R$. These previous equations are paraxial, which means that the principal ray incides perpendicular to the surface, and the beam aperture is small with respect to R and the object and image distances.

Now assume the beam is still narrow (small aperture) but it incides obliquely to the surface normal. Let us also assume the incident ray is coplanar with the revolution axis of the surface, in which case the ray is called *meridional*. The shared plane between a meridional ray and the revolution axis is known as the *tangential plane*. Because of the revolution symmetry of the surface, an incident meridional ray stays in the same tangential plane after refraction. The planes containing either the incident or refracted ray and perpendicular to the tangential plane are called *sagittal planes*. When the beam incides upon the surface at some oblique direction, the broken symmetry will produce astigmatism, even if the surface is spherical. If the incident beam is already astigmatic, its astigmatism will change after refraction. In the standard use of an ophthalmic lens, the beam inciding on the front surface is not astigmatic, as it comes from an object point. After refraction, it becomes astigmatic, and as such it incides upon the second surface, which will change the astigmatism again.

Consider the situation depicted in Figure 6.4, either the side view shown in (a) or the perspective view in (b). The surface has revolution symmetry and a meridional ray is shown, with the tangential plane being that of the paper in subplot (a). An astigmatic beam with

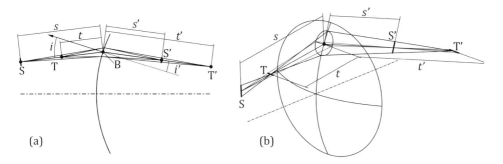

(a)   (b)

Figure 6.4 Astigmatic pencil refracting through a surface with revolution symmetry.

focal lines S and T incides on the surface, with its principal ray forming angle i with the surface normal at the incidence point. The angle for the refracted beam is i', and its focal lines are S' and T'. As we would expect, S stands for sagittal focal lines while T stands for tangential focal lines. The first are formed by rays in the sagittal plane and contained in the tangential plane. Tangential focal lines are formed by tangential rays and are contained in the sagittal planes. The distances s, t, s', and t' are measured from the incidence point to the respective focal lines. The classical form of Coddington equations is

$$\frac{n'}{s'} - \frac{n}{s} = \frac{n'\cos i' - n\cos i}{R_S},$$

$$\frac{n'\cos^2 i'}{t'} - \frac{n\cos^2 i}{t} = \frac{n'\cos i' - n\cos i}{R_T}, \tag{6.2}$$

where n and n' are the refractive indexes at either side of the surface and R_S and R_T are the curvature radii of the surface along the sagittal and tangential sections at the point of incidence (the surface has revolution symmetry, but it is not necessarily spherical, so both radii can be different). Note the similarity between both Coddington equations and the purely paraxial refraction equation $n'/s' - n/s = (n' - n)/R$. For sagittal focal lines, the left side is identical, while the right side resembles a refractive power depending on the incidence and refraction angles and the local radii of curvature. The research that led to Coddington's equations trace back to Barrow, Newton, and Young [187], but it was Coddington who compiled them in their modern form in 1829.

Now, let us apply Coddington equations repeatedly to the two refractions at the two lens surfaces, see Figure 6.5. We first start with a single object distance $s_1 = t_1$, and the associated vergence, $L_1 = 1/s_1 = 1/t_1$. Let κ_{1S} and κ_{1T} be the sagittal and tangential curvatures of the front surface at the incidence point, B_1. The first application of Coddington equations leads to s'_1 and t'_1, which will be equal to the object distance for the second surface, s_2 and t_2, except for the distance from B_1 to B_2, which can be neglected if we consider the lens to be thin. Then, a second application of Coddington equations with the curvatures κ_{2S} and κ_{2T} at B_2 yields the output vergences

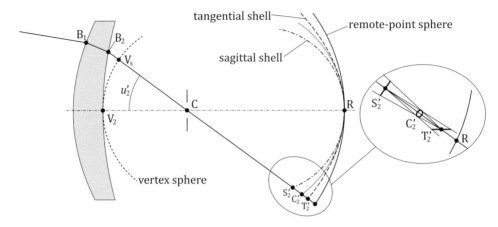

Figure 6.5 Refraction of a narrow oblique beam. An astigmatic pencil (see detail in the inset) is formed around the principal ray as a result of the combination of oblique astigmatism and field curvature. As a result, two image lines, T'_2 and S'_2, appear. Note that the tangential line is perpendicular to the tangential plane that is the drawing plane.

$$L'_{2S} = L_1 + (n\cos i'_1 - \cos i_1)\kappa_{1S} + (n\cos i'_2 - \cos i_2)\kappa_{2S},$$

$$L'_{2T} = \frac{\cos^2 i_2}{\cos^2 i'_2}\left[\frac{\cos^2 i_1}{\cos^2 i'_1}L_1 + \left(\frac{n\cos i'_1 - \cos i_1}{\cos^2 i'_1}\right)\kappa_{1T} + \left(\frac{n\cos i'_2 - \cos i_2}{\cos^2 i'_2}\right)\kappa_{2T}\right].$$

Now, the incidence and refraction angles are determined by the condition for the output principal ray to pass through the rotation center of the eye but also depend on the characteristics of the surfaces, the lens thickness, and Snell's law. We will assume the surfaces are conicoids, with their equation being (2.4) in Chapter 2. As stated previously, we will neglect the lens thickness, so $B_1 = B_2 \equiv B$, and the same for their heights, $y_1 = y_2 \equiv y$. We will also limit the oblique gaze angle u'_2 so that y is small enough for the Taylor expansion

$$z_{1,2} = \frac{\kappa_{1,2}}{2}y^2 + \frac{p_{1,2}\kappa_{1,2}^3}{8}y^{4'}$$

to be valid, where κ_j and p_j are the center curvature and asphericity parameter of surface j (see equation (2.5)). Similarly, we recall Example 4 in Chapter 2 for the approximate expressions giving the tangential and sagittal curvatures of the conicoid surface,

$$\kappa_{(1,2)S} = \kappa_{1,2}\left[1 + \frac{1}{2}(p_{1,2} - 1)\kappa_{1,2}^2 y^2\right],$$

$$\kappa_{(1,2)T} = 3\kappa_{(1,2)S} - 2\kappa_{1,2}.$$

Finally, all the angles involved in the problem, i.e. the inclinations of the normals at the incidence points, the obliquity angle u'_2, and the incidence and refraction angles i_j and i'_j, can be related with the incidence height y through geometry and Snell's law. However, we want to simplify this dependence so we can get manageable expressions but still keep the essential optics stemming from Coddington equations. This can be achieved by expanding the cosine and sine functions of all the angles as Taylor series on variable y, up to second order. The result is summarized in the four relations

$$\cos i_1 = 1 - \frac{y^2}{2}J^2, \ \cos i_1' \simeq 1 - \frac{y^2}{2n^2}J^2,$$

$$\cos i_2 \simeq 1 - \frac{y^2}{2n^2}F^2, \ \cos i_2' \simeq 1 - \frac{y^2}{2}F^2,$$

where $J = L_2' - P - \kappa_1$, P being the thin-lens power of the lens, and $F = L_2' - \kappa_2$.

Now we can substitute the expressions for the cosine functions and the tangential and sagittal curvatures into the equations for L_{2S}' and L_{2T}'. The multiplications of the cosine terms and the curvatures will produce fourth-order terms on the variable y. Also, the expression for the tangential vergence will have quotients involving different powers of y. We then repeat the same procedure of expanding each term as a Taylor series on y and retaining only the terms up to second order. The final result is

$$L_{2S}' = P + L_1 + \frac{(n-1)y^2}{2n}\left[J^2\kappa_1 - F^2\kappa_2 + n\kappa_1^3(p_1 - 1) - n\kappa_2^3(p_2 - 1)\right], \quad (6.3)$$

for the sagittal vergence, and a little longer

$$L_{2T}' = P + L_1 + \frac{(n-1)y^2}{2n^2}\left\{J^2\left[(n+2)\kappa_1 - 2L_1(n+1)\right]\right.$$

$$+ F^2\left[2L_1(n+1) - n\kappa_2(2n+1) + 2\kappa_1(n^2 - 1)\right]$$

$$\left. + 3n^2\left[\kappa_1^3(p_1 - 1) - \kappa_2^3(p_2 - 1)\right]\right\}, \quad (6.4)$$

for the tangential vergence. Although the expressions are long, they are pretty simple. The refracted sagittal and tangential vergences are given by the paraxial term $P + L_1$ plus a term accounting for the oblique astigmatism and field curvature that is proportional to y^2. This term is different for the tangential and sagittal vergences and depends on the lens refractive index, the lens power, the curvatures of the two surfaces, their asphericities, and the distance l_2' (embedded in parameters J and F).

L_{2S}' and L_{2T}' are the inverse of the distances from B_2 to the image lines S_2' and T_2', respectively. The last step is to propagate these vergences to the vertex sphere. If we give the notation δ to the distance from B_2 to V_s, it can be shown that, up to second order on y,

$$\delta = \left(\frac{L_2'}{2} - \kappa_2\right)y^2. \quad (6.5)$$

Let us give the notation L_S' and L_T' to the vergences at the vertex sphere. The propagation is performed with the standard techniques we saw in Chapter 3,

$$L' = \frac{L_2'}{1 - \delta L_2'} \simeq L_2'(1 + \delta L_2').$$

In general, δ will be very small with respect to the distances from B_2 to the tangential and sagittal images, so the rightmost expression can be used here. By inserting (6.3) and (6.4), along with (6.5), into the previous equation, and once again retaining only the terms up to second order on y, we get to the final result,

$$
L'_S = P + L_1 + y^2 \left\{ \left(\frac{L'_2}{2} - \kappa_2 \right) (L_1 + P)^2 \right.
$$
$$
+ \frac{(n-1)}{2n} \left(J^2 \kappa_1 - F^2 \kappa_2 \right)
$$
$$
\left. + \frac{(n-1)}{2} \left[\kappa_1^3 (p_1 - 1) - \kappa_2^3 (p_2 - 1) \right] \right\}, \tag{6.6}
$$

and

$$
L'_T = P + L_1 + y^2 \left\{ \left(\frac{L'_2}{2} - \kappa_2 \right) (L_1 + P)^2 \right.
$$
$$
+ \frac{(n-1)}{2n^2} J^2 \left[(n+2)\kappa_1 - 2L(n+1) \right]
$$
$$
+ \frac{(n-1)}{2n^2} F^2 \left[2L_1(n+1) - n(2n+1)\kappa_2 + 2(n^2-1)\kappa_1 \right]
$$
$$
\left. + \frac{3(n-1)}{2} \left[\kappa_1^3 (p_1 - 1) - \kappa_2^3 (p_2 - 1) \right] \right\}. \tag{6.7}
$$

There are some important remarks about the result we have just presented:

1. Although the expressions were obtained neglecting third and higher powers of the incidence height y, the approach is known as third-order theory of oblique aberrations. This is because the wavefront having main vergences L'_S and L'_T is affected by third-order Seidel astigmatism and field curvature.
2. The error that both vergences present with respect to the paraxial vergence (which is just one, as the lens has rotation symmetry) is proportional to the incidence height squared. As $\tan u'_2 \simeq yL'_2$, we can conclude that the errors are proportional to $\tan^2 u'_2$.
3. According to the definitions of parameters J and F and equations (6.6) and (6.7), the oblique vergences depend on the two curvatures and the lens power. By using the definition of the refractive powers of the surfaces, $P_1 = \kappa_1(n-1)$ and $P_2 = (1-n)\kappa_2$, renaming $B \equiv P_1$ as the base curve of the lens, and using the thin-lens power $P = B + P_2$, we can set $\kappa_2 = (P-B)/(1-n)$ and write the oblique vergences as a function of P, B, L'_2 and n. The last terms in both equations (6.6) and (6.7) depend on the asphericity parameters of the surfaces and on the third power of the base curve. The remaining terms in both equations only contain first- and second-order powers of B (we could think that the term $J^2 \kappa_1 - F^2 \kappa_2$ in L'_S and $J^2(n+2)\kappa_1 - F^2 \left[(2n^2+n)\kappa_2 + 2(n^2-1)\kappa_1 \right]$ in L'_T may have cubic terms in B, but it can be easily shown that they cancel out, and the dependence is just quadratic). In summary, if the lens has spherical surfaces with no asphericity, $p_1 = p_2 = 1$, both L'_S and L'_T are given by a quadratic polynomial on the

base curve. If at least one of the surfaces is aspherical, with either p_1 or p_2 different than 1, then both L'_S and L'_T are given by cubic polynomials on the base curve B.

4. Regardless of the asphericity of the surfaces, the two oblique vergences L'_S and L'_T depend cubically on the lens power.

5. Setting the object vergence to zero, $L_1 = 0$, is the same as locating the object at infinity. In this case, the oblique vergences turn into oblique powers P_S and P_T. Similarly, the images S' and T' turn into the sagittal and tangential focal lines.

6. *Oblique astigmatism* is defined as the difference between the oblique vergences or powers,

$$OA = L'_T - L'_S,$$

and for distance vision,

$$OA = P_T - P_S.$$

7. The vergence associated to the circle of least confusion, represented by C'_2 in Figure 6.5, is given by $L'_C = (L'_S + L'_T)/2$. The *mean power error* is defined as

$$MPE = \frac{L'_S + L'_T}{2} - (P + L_1),$$

and for distance vision,

$$MPE = \frac{P_S + P_T}{2} - P.$$

When the mean power error is zero and for distance vision, the circle of least confusion is formed on the remote-point sphere. When $L_1 \neq 0$, the eye must accommodate to bring the image of the object into focus (see Chapter 8 for further details). Under accommodation, the conjugate of the retina moves to the right of the remote point. At maximum accommodation, the conjugate of the retina is known as near point. At any level of accommodation, the conjugate of the retina is located at a distance $1/(P+L_1)^3$. The mean power error for near objects quantifies the distance between the circle of least confusion and this point.

Once we have expressions for computing the oblique astigmatism and mean power error of an ophthalmic lens with a given power, refractive index, and base curve, the next step is to ask for the conditions under which these oblique aberrations are zeroed or minimized. The refractive index and the distance l'_2 cannot be considered design parameters, as they cannot be changed arbitrarily. The refractive index depends on the material chosen for the lens, and just a handful of them are available. The distance l'_2 is basically determined by the shape of the frames we use nowadays. It may change a few millimeters from one person to another, depending on the detailed shape of the nose and the frame bridge, but once again,

---

[3] This distance is based on the thin-lens approximation. If the lens thickness is taken into account, a small error defined by Jalie as *near vision effectivity error*, or NVEE, will appear [56], but it is only significant for very large lens powers.

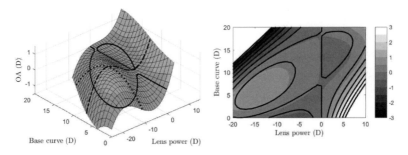

Figure 6.6 Oblique astigmatism as a function of lens power and base curve, for lenses with $n = 1.5$, $l'_2 = 27$ mm, and obliquity angle $u'_2 = 30°$. The black curve is given by $OA = 0$, and is a Tscherning ellipse.

it cannot be arbitrarily varied. Power is given by the prescription, and once the power is set, we can choose the base curve, which in turn determines the center curvature of the back surface. Finally, asphericity parameters p_1 and p_2 are arbitrary. So, for spherical lenses, the only design parameter is the base curve. Any asphericity parameter on either surface will be added as a new design parameter.

According to point 3 in the previous list, oblique astigmatism (OA) and mean power error are quadratic functions of the base curve for spherical lenses, that is,

$$OA = y^2(a_2B^2 + a_1B + a_0),$$
$$MPE = y^2(e_2B^2 + e_1B + e_0),$$

where the coefficients a_j and e_j are functions of P, n, and l'_2. Setting any one of them to zero yields a quadratic equation that may have two, one, or no solutions. If we focus on the oblique astigmatism, the behavior is shown in Figure 6.6. In the left plot we can see a surface representing oblique astigmatism as a function of lens power and base curve, for distant objects, with refractive index $n = 1.5$ and $l'_2 = 27$ mm. The vertical sections of the surface are the parabolas given by the previous equation (dotted curve). The horizontal sections are cubic curves (dash-dotted curve). The solid elliptical curve corresponds to the setting $OA = 0$, which establishes a relationship between base curve and lens power known as a *Tscherning ellipse*. The vertical black line corresponds to plano lenses, which within this third-order theory are free from astigmatism for any base curve.

A surface plot like the one in Figure 6.6 vividly shows the variation of oblique astigmatism (and mean power error) as a function of base curve and lens power, but the use of contour plots is more practical, such as the one shown on the right. Here, the color map represents oblique astigmatism computed by exact ray tracing, while the black isolines represent oblique astigmatism computed with equations (6.6) and (6.7). The frontiers between regions with different gray tone indicate integer or half-integer values of oblique astigmatism. We can see that the third-order theory is accurate to within 0.25 D for most values of power and base curve, with errors getting bigger for positive lenses with high power values or high base curves, i.e. for lenses with greater thickness.

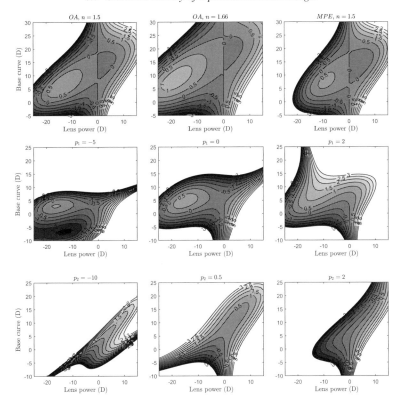

Figure 6.7 Contour maps of oblique astigmatism and mean power error. First row: spherical lenses. Second row, lenses with aspherical front. Third row, lenses with aspherical back.

The contour plots shown in Figure 6.7 illustrate other aspects of the theory. The first two plots in the first row show the contour plots of oblique astigmatism for spherical lenses with refractive indexes $n = 1.5$ (leftmost plot) and $n = 1.66$ (center plot). As before, the elliptical curves corresponding to the contour line $OA = 0$ are Tscherning ellipses. We see that the ellipse is larger for the bigger refractive index. For $n = 1.5$, the ellipse approximately extends from -22 D to 7 D. Outside this range, the parabolas $a_2 B^2 + a_1 B + a_0$ do not cross the axis $OA = 0$, and there is no lens shape with spherical surfaces for which oblique astigmatism is zeroed. We can still look for the base curve that minimizes it (see [56]). With higher refractive indexes, the range for which there are lens shapes with zero astigmatism is expanded, but the required base curves are larger. The plot on the right represents the mean power error in spherical lenses. Once again, the curve $MPE = 0$ is elliptical, but with a size and orientation slightly different than that for oblique astigmatism.

For a given power inside the range determined by the ellipses, there are two solutions (two base curves) with either $OA = 0$ or $MPE = 0$. The shallower ones are known as Ostwalt solutions, while the steeper ones are named Wollaston solutions. In general, Wollaston solutions are too steep for practical purposes, so actual lenses use base curves closer to the Ostwalt form. When the lens shape cancels oblique astigmatism the lens is said to be *punctual*, meaning it has a single point-like focus. On the other hand, if the lens shape cancels mean power error, it is said to be a *Percival lens*. The two aberrations cannot be canceled at the same time with spherical surfaces. We must choose one or the other. Fortunately, if the oblique astigmatism is zero, the mean power error is small, and the other way around. Percival lenses require base curves slightly shallower than punctual ones.

Similarly, the optimal base curve depends on the object vergence. If $L_1 < 0$, which means the object is not at infinity, the required base curves for zeroing either *OA* or *MPE* are somewhat shallower than the corresponding base curves for distant objects. For example, a punctual lens for a distant object becomes approximately a Percival lens for near objects. For the first half of the twentieth century some lens manufacturers launched what was called *corrected lens forms*, i.e. sets of base curves intended to eliminate oblique astigmatism, mean power error, or something in between. The theory just revealed demands a different base curve for each lens power, as in the first generation of Punktal lenses by Zeiss. However, this manufacturing approach required a very large amount of tools to cut and polish different lens surfaces for each possible prescription. Edgard Tillyer realized that some oblique astigmatism and power error could be tolerated and divided the whole range of powers into nonoverlapping intervals. Lenses with powers within each interval would be manufactured with the same base curve, and this would bring the cost of manufacturing down. Still, the set of base curves would adhere to the Ostwalt region of the Tscherning ellipses in most cases. However, large frames came into fashion during the 1980s. These frames required lenses with larger diameters, and making flatter lenses became a practical and cosmetic issue. The idea that flatter lenses were better from the esthetic point of view permeated deeply in the industry and in the optometric and optical communities, and this led to a shift toward flatter base curves, even for small frames or for prescriptions for which the esthetic gain would be none or negligible and even when the off-axis performance of the lenses could be significantly degraded. Nowadays, the trend is still strongly implanted.

A solution to produce flatter punctual or Percival lenses is the use of aspheric surfaces. As mentioned previously, if at least one of the surfaces is aspheric, then *OA* and *MPE* are cubic functions of the base curve, and hence there is always at least one solution to the equations $OA = 0$ or $MPE = 0$. In the second row of plots in Figure 6.7, the oblique astigmatism is represented for lenses with aspherical front surface, with asphericities $p_1 = -5$ (hyperbolic surface), $p_1 = 0$ (parabolic), and $p_1 = 2$ (ellipsoidal). The third row contains plots of oblique astigmatism for lenses with aspherical back surface. In the second and third rows, we have assumed the values $n = 1.5$ and $l'_2 = 27$ mm. We can see how different the behavior of the oblique astigmatism is when either surface is made aspheric, and we also see that for a given asphericity, the base curve required for eliminating oblique astigmatism (or similarly mean power error) is completely different than the base curve required with spherical surfaces. However, asphericity is not always a panacea for zeroing oblique astigmatism and mean power error. Although using one conicoid surface increments by one the degrees of freedom of the design problem, in general it is not possible to eliminate at the same time the two oblique aberrations, or it requires an inconvenient or unpractical base curve. However, in general it is possible to achieve punctual or Percival lenses with flatter surfaces by appropriate selection of the asphericity parameter of one of the surfaces, see for example [32, 56].

The first aspheric surfaces were computed by Gullstrand and used by Carl Zeiss in their *Katral* lenses. More recent studies on the possibilities offered by aspheric lenses are those from Katz [188], Atchison [182], Smith [184], Malacara [189, 190], and Miks [191], aside from tens of patents in which exotic surfaces are used to optimize ophthalmic lenses. The aspheric surfaces that are employed the most are the conicoids we have used in the previous development and the generalized conicoids that we studied in Chapter 2.

Distortion

Distortion can be computed with a stationary eye, in which case it is called *peripheral distortion,* or with a rotating eye, in which case it is called *dynamic (or rotary) distortion.* The first consists of a variation of magnification with the transversal location of the object (or with object size). When magnification grows with object size, we say the distortion is pincushion-type. When magnification is reduced with object size, we say the distortion is barrel-type. Rotary distortion is related to the prismatic effect suffered by the rotating eye. According to Prentice's rule, the prismatic effect should grow linearly with the distance from the optical axis to the point through which the gaze direction crosses the lens, which for small distances is also proportional to the obliquity angle. If the change of the prismatic effect with the obliquity angle is nonlinear, we say the lens has rotary distortion. Both types of distortion are computed in a similar way, comparing the size of the actual image with the size of the paraxial image. The only difference between them is the location of the aperture. For peripheral distortion, the stop is located at the entrance pupil of the eye. For rotary distortion, the stop is placed at the rotation center of the eye.

Atchison and Smith proved that under third-order approximation, distortion also depends quadratically on the base curve [192]; the difference with respect to the oblique errors is that distortion does not become zero for any value of the base curve, and the only option left for the lens designer is to look for the base curve that produces minimum distortion. When the surfaces are spherical, the base curves that minimize distortion are too steep and do not match the solutions for zeroing astigmatic error or mean power error. However, if one of the surfaces is a conicoid, Atchison and Smith proved that it was possible to eliminate one of the oblique errors and one of the distortions, although the required base curves were too steep to be a practical solution.

Jalie has provided approximated expressions for the value of the base curves that minimize rotary or peripheral distortion for a given lens power [56]. These expressions are

$$B = L_2'(n-1) - \frac{P}{2}(n-2), \tag{6.8}$$

for rotary distortion and

$$B = \frac{1}{d}(n-1) - \frac{P}{2}(n-2), \tag{6.9}$$

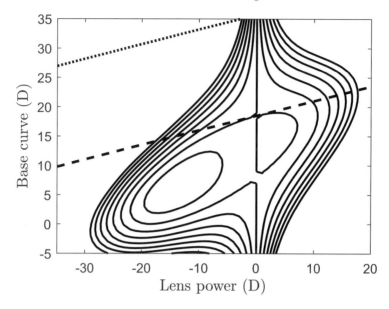

Figure 6.8 The solid lines are contour curves of oblique astigmatism. The dashed line indicates the approximate relation between power and base curve that minimizes rotary distortion. The dotted line does the same for peripheral distortion.

for peripheral distortion, where d is the distance from the vertex of the back surface of the lens to the entrance pupil of the eye. In both cases, there is a linear dependence of the optimal base curve to the lens power, but as $d < l'_2$, the optimum bases for peripheral distortion are more curved than the corresponding ones for rotary distortion. In fact, both solutions are too curved, as we mentioned previously, and as shown in Figure 6.8. The line corresponding to minimum rotary distortion is approximately tangent to the Tscherning ellipse for punctual lenses and index 1.5, but at the Wollaston part of the ellipse, and only for low-powered lenses.

In summary, we cannot hope to significantly reduce either rotary or peripheral distortion with spherical lenses, or even with aspherical ones, as the required base curves are impractical. Nevertheless, it is important to consider that neither type of distortion impairs the image sharpness as oblique aberrations do. Distortion changes the perception of the environment, but the brain can usually adapt to it, to the point that the user stops noticing distortion at all. Le Texier et al. [193] propose a merit function that will take into account distortion, but with a smaller weight than those used for oblique errors.

It is worth mentioning the work of Katz [188]. He used exact ray tracing to design lenses with complex aspheric surfaces at the two sides, which would be significantly corrected from both rotary distortion and the two oblique aberrations. However, Katz's lenses were extremely curved, which, once again, renders the solution impractical for actual use.

Finally, the reader should be aware that distortion in any optical system is strongly dependent on the position of the stops. The same lens will have a totally different distortion depending on whether the stop is located in front or behind. It can even change from pincushion to barrel distortion just by changing the position of the stop. In some cases, ophthalmic lenses are shown in showcases with a rectangular grid a few centimeters behind, for customers to assess how little distortion the lens has. Potential users are invited to look at the grid through the lens, but of course the encased lens is about a meter in front of the eye. With this arrangement, the distorted grid observed through the lens has little to do with the actual distortion, either rotary or peripheral, perceived when the lens is used "as worn." Hence, these setups cannot demonstrate the actual lens performance with regard to distortion.

6.3.2 Design of Astigmatic Lenses and Exact Ray Tracing

The design of astigmatic lenses has some differences compared to the design of spherical ones. The first, and most obvious, is that power is now defined by three parameters. Second, astigmatic lenses can be manufactured in many different ways, even if we discard the use of aspherical or atoric surfaces. Astigmatic lenses can be spherotoric, bitoric, or sphero-cylindrical, and even though the standardization of the manufacturing process makes the spherotoric format the more widespread, the torus can be barrel type or ring type, and it can be internal or external. Although this flexibility can be used to make astigmatic lenses with smaller values of oblique aberrations, by the 1980s, the internal torus format had already become the standard in astigmatic lens manufacturing. The advantages derived from this simplification were much smaller numbers of tools for grinding and polishing lenses, and the esthetic advantage of the internal torus: The edge thickness variation is concealed behind the frame. Another difference between the design of spherical and astigmatic lenses is the way in which the aberrations are computed. Coddington's equations can still be used to compute oblique errors, but now we also have surface astigmatic power (due to the toric surface), which is combined with the oblique astigmatic error. As a result, the classical Coddington equations can only be used when the tangential and sagittal planes are aligned with the main meridians. In this case, the whole procedure shown for rotationally symmetric lenses can be repeated for each main meridian. In fact, the exact same third-order theory we have used for aspherical lenses with revolution symmetry can be used for the main meridians of the astigmatic lens, changing the sagittal and tangential curvatures at the point of incidence by the main curvatures of the astigmatic surface.

Assume the astigmatic prescription of the lens is $[S_0, C_0 \times \alpha]$. This is the paraxial back vertex power of the lens when the user looks through its center and the lens is not tilted. Assume now that the eye rotates, with its visual axis coplanar with the optical axis and the lens section at angle α. If the angle made by the line of sight with the optical axis is u'_2, the user will perceive powers $S(u'_2)$ and $S(u'_2) + C(u'_2)$, where we may name these functions *oblique sphere* and *oblique cylinder*, and they meet $S(0) = S_0$ and $C(0) = C_0$. The obliquity is introducing errors in both meridians, the error along the cylinder axis, $\Delta S = S(u'_2) - S_0$, and the error along the crossed direction, $S(u'_2) + C(u'_2) - (S_0 + C_0)$, which can be written as $\Delta S + C(u'_2) - C_0 = \Delta S + \Delta C$. We see then that obliquity introduces errors in sphere and cylinder. Note that this error in cylinder is different than the oblique astigmatism in lenses with revolution symmetry. For the latter, there is no paraxial cylinder, and the obliquity introduces cylinder. Astigmatic lenses do have paraxial cylinder, and obliquity modifies it.

Let us assume we have selected the spherocylindrical prescription with positive cylinder, and let us call the powers along the main meridians $P_+ = S_0 + C_0$ and $P_- = S_0$. Now, let us assume we may find a base curve for which the tangential power is P_- and the sagittal power P_+. In this case, the lens would be free from sphere and cylinder error along the axis meridian. If we move now to the crossed curve meridian, the same lens with the same base curve should have tangential power P_+ and sagittal power P_-, which seems quite unlikely. We can see a clear example of this problem in Figure 6.9. We have computed the off-axis performance, i.e. the oblique powers $S(u_2', v_2')$ and $C(u_2', v_2')$, for a lens with back vertex power [3, 2 × 180°], refractive index 1.5, and $l_2' = 27$ mm. We use two angles u_2' and v_2' because we are not restricting to the main meridians, and we need to specify the horizontal and vertical angles formed by the sight direction with ZY and ZX planes, respectively. In the plots, the angles have been translated to the Cartesian coordinates of the impact point of the principal ray on the back surface. Each row contains the sphere and cylinder maps, computed with a different base curve. With $B = 6$ D, the sphere behaves quite well, becoming too large at the edges of the crossed meridian, while the cylinder has very little error along the axis and a strong oblique error along the crossed meridian. With $B = 10$ D (third row), the cylinder behaves the other way around, nicely along the cross axis, and with an oblique error of 1 D at the limits of the plotted region along the axis. The sphere is good inside a circle with a radius of 10 mm and then quickly drops about 1 D with radial symmetry. With the intermediate base curve, $B = 8$ D, the behavior is also in between the other two, both for the sphere and for the cylinder.

The example presented in Figure 6.9 reveals that spherotoric astigmatic lenses have more conditions to meet, but still a single free parameter: the base curve of the lens. This fact leads us to the concept of full field optimization, which consists of evaluating the oblique performance of the lens for many different sight directions, as in the maps in Figure 6.9. The performance of the lens will be summarized by a single number that balances all the errors in the maps and is known as the *merit function*, and the free parameters (the base curve for a standard internal torus astigmatic lens and the asphericity parameters if there are any) are scanned looking for the smaller value of the merit.

Full field optimization requires exact ray tracing, [189, 190], or so-called generalized ray tracing [194, 195, 196, 197]. The procedure for exact ray tracing can be outlined as follows:

1. We set the problem parameters: refractive index, distance from the lens to the rotation center of the eye, lens geometry (type and location of the surfaces and lens thickness).
2. We specify a grid of gaze directions. As the lens no longer has revolution symmetry, two angles are needed to fully specify an oblique gaze direction.

   2.1 For each gaze direction in the grid, we trace a single ray backward, starting from the rotation center of the eye and using the obliquity angles to define the ray direction.

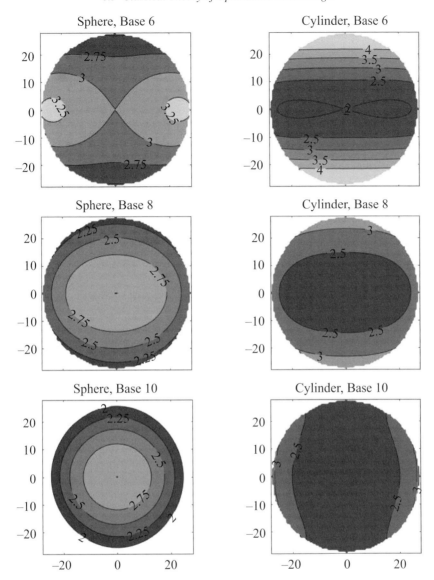

Figure 6.9 Oblique performance of a spherotoric lens with prescription $[3, 2 \times 180°]$ manufactured with base curves 6, 8, and 10 D.

The intersection of the ray with the second surface is computed, then the vector Snell law is used to refract the ray. Next, the ray is propagated to the first surface and finally refracted again. This step gives us the direction of the incoming ray that, after refraction, will pass through the rotation center of the eye with the selected obliquity angles.

2.2 Now we proceed with a forward ray trace. With the standard ray-tracing technique, some rays are traced forward around the principal ray we traced backward in the previous step. The new rays will be parallel if we simulate far vision, or will diverge from a point at a finite distance, for near or intermediate vision. The rays are traced through the two lens surfaces and up to the vertex sphere.

2.3 Using the direction information and impact points on the vertex sphere of the rays traced forward, the astigmatic power of the lens is computed at the vertex sphere. Steps (a) to (c) are repeated until the oblique power is known for the whole grid.

3. To define and compute the merit function, we use the matrix form of the astigmatic power. Let $\mathbb{P}_0$ be the paraxial power of the astigmatic lens and $\mathbb{P}(u, v)$ the oblique power of the lens computed at the vertex sphere for the gaze angles u and v (we will drop the subindexes and the prime symbols for convenience). In other words, $\mathbb{P}(u, v)$ is the vergence of the beam refracted by the lens and propagated up to the vertex sphere (point A in Figure 6.3). The power error can be defined as

$$\Delta \mathbb{P}(u, v) = \mathbb{P}_0 - \mathbb{P}(u, v).$$

$\Delta \mathbb{P}(u, v)$ can be understood as an induced ametropia: if the prescription is correct, the eye will be well compensated at the lens center. At any other sight direction the lens provides power $\mathbb{P}(u, v)$, creating an error $\Delta \mathbb{P}(u, v)$ that produces over- or under-refraction. The reader can readily verify that the blur $\sqrt{2S^2 + 2SC + C^2}$ that we used to construct a model of visual acuity in Section 5.3.4 is nothing but the Frobenius norm of $\Delta \mathbb{P}(u, v)$ (see Section B.4). The larger the norm of $\Delta \mathbb{P}(u, v)$, the worse the oblique performance of the lens. The advantages of using matrices for describing astigmatic powers is that we can subtract paraxial and oblique power, and the operation is perfectly well defined. Now we may construct a merit function as follows

$$U(\mathbb{P}, B, \alpha) = \sum_{j=1}^{N} w_j^2 \left\| \Delta \mathbb{P}(u_j, v_j) \right\|^2 + w_\Phi^2 \, \Phi(\mathbb{P}, B, \alpha)^2, \tag{6.10}$$

where the index j runs over the N gaze directions in the grid, and where w_j and w_Φ are numbers called *weights* such that $\sum_j w_j^2 + w_\Phi^2 = 1$, and Φ is a function that quantifies any nonoptical effect we want to reduce, for example thickness or weight. Finally, α represents all the parameters that can be used to optimize the lens. For example, if the lens is spherotoric, α would just be the base curve. If the lens has a front spherical surface and a back atoric surface, α will stand for the set of parameters (B, p_B, p_T), with the last two being the asphericity parameters of the atoric surface. Having a different weight for each sight direction may help us to improve some regions of the lens at the expense of reducing the quality at other regions. Finally, we may add as many functions "Φ" as needed.

4. Once we have computed the off-axis power of the lens $\mathbb{P}(u, v)$, and from it the merit function, we change the free parameters in α and repeat the process, steps 2 and 3.

Of course, there are optimization algorithms that use well-established mathematical rules to modify the free parameters according to the values the function $U(\mathbb{P}, B, \alpha)$ has in each step. The objective is to find the set of parameters α that render a minimum value for $U(\mathbb{P}, B, \alpha)$.

Function (6.10) is just an example of a possible merit function, but other approaches are possible. For example, instead of computing the norm of $\Delta\mathbb{P}(u_j, v_j)$, we may compute from it the corresponding errors in sphere and cylinder, and use them directly in the merit function with independent weights. Also, we may compute the blur, but taking into account accommodation in the model of visual acuity described in Section 5.3.4, and then introduce visual acuity in the merit function. Similarly, other optical aberrations can be considered, such as distortion or transverse chromatic aberration. See, for example, the patent [198].

Sub-steps (b) and (c) within step 2 would be slightly modified if we use generalized ray tracing. This method employs a generalization of the Coddington equation established by Stavroudis [199]. The method consists of the exact refraction of vergences through arbitrary astigmatic surfaces. Only one ray is needed to complete the generalized trace, but the curvatures of the surface must be computed. The obtained result is identical to that of a standard ray trace [197, 200].

Generalized ray tracing especially benefits from the matrix description of curvature. Although it can be considered an advanced optical tool for optical design and assessment, the background that the reader should have gained by now studying the matrix methods presented in the book should be enough to grasp the generalized Coddington equations. Let us consider a beam refracting through an arbitrary (but smooth) surface, and let us focus on its principal ray. As we already saw in Section 3.4.1, the direction vectors of the incident and refracted rays and the surface normal at the incidence point are coplanar (the three vectors are contained in the so-called incidence plane). Let us then choose three reference systems, one for the incoming beam, the second for the surface, and the third for the refracted beam, such that their Z axes are contained in the incidence plane and their X axes are perpendicular to this plane. Next, let us give the notation $\mathbb{V}$, $\mathbb{V}'$, and $\mathbb{H}_S$ to the Hessian matrices of the wavefront incident at the surface, the refracted wavefront, and the surface, respectively. These Hessian matrices are given in the corresponding reference systems just described. Then, if the angles of incidence and refraction are i and i', the refracted vergence can be computed from the surface curvature and the input vergence by means of the matrix equation

$$n'\begin{bmatrix} 1 & 0 \\ 0 & \cos i' \end{bmatrix}\mathbb{V}'\begin{bmatrix} 1 & 0 \\ 0 & \cos i' \end{bmatrix} - n\begin{bmatrix} 1 & 0 \\ 0 & \cos i \end{bmatrix}\mathbb{V}\begin{bmatrix} 1 & 0 \\ 0 & \cos i \end{bmatrix} = \tag{6.11}$$

$$= \left(n'\cos i' - n\cos i\right)\mathbb{H}_S.$$

The first thing we see is that, if the incidence is normal then $\cos i = \cos i' = 1$, and it reduces to the well-known paraxial equation $n'\mathbb{V}' - n\mathbb{V} = (n' - n)\mathbb{H}_S$. Second, the structure of the previous equation clearly resembles the structure of (6.2). Indeed, if the system had revolution symmetry and we choose the tangential plane to be vertical, the vergence matrices would be

$$\mathbb{V}' = \begin{bmatrix} 1/s' & 0 \\ 0 & 1/t' \end{bmatrix}, \ \mathbb{V} = \begin{bmatrix} 1/s & 0 \\ 0 & 1/t \end{bmatrix}, \ \mathbb{H}_S = \begin{bmatrix} \kappa_S & 0 \\ 0 & \kappa_T \end{bmatrix}$$

and then (6.2) would be recovered.

The great advantage of equation (6.11) is that it can be applied even when the tangential and sagittal planes are not defined. The main curvatures of the surface and the wavefronts can be outside the plane of incidence, and still equation (6.11) can be perfectly applied.

6.3.3 Lens Optimization for the Static Eye: Control of Peripheral Defocus

As explained at the beginning of this chapter, classic lens design was always based on the assumption of the rotating eye. The image at the peripheral retina was largely ignored, due to both the low quality of this peripheral image and also the low resolution of the retina. However, a relatively new hypothesis relating to the peripheral image and the onset of myopia has sparked interest in the peripheral image. Myopia is currently considered an imminent epidemic. Its prevalence is extremely high in all countries with a high degree of literacy, especially in some Asian countries [201]. Many different optical methods have been proposed and tried to slow down the axial growth of the eye and the subsequent onset of medium to high myopia: over-correction, under-correction, prescription of bifocal or progressive lenses, etc., but none of them has turned out to be really successful [202]. However, basic research conducted during the 1990s established a clear link between the focus status on the retina and biochemical mechanisms that trigger axial eye growth [69]. Many of these experiments with chickens and monkeys demonstrate that retinal hyperopic defocus created by external lenses or any other means causes the eye to grow. It was suggested that in myopic human eyes compensated with negative lenses, there could be large regions of the peripheral retina affected by hyperopic defocus, even if the foveal image was in focus or slightly defocused toward the myopic side. The much smaller density of ganglion cells responsible for this homeostatic mechanism could be compensated by the much larger area of the peripheral retina suffering from hyperopic defocus [69].

Some patents have been issued on lenses that would correct the hyperopic peripheral defocus while keeping a focused foveal image. They would do so by somehow increasing its power along the radial direction [67, 110]. These first patents did not establish a precise relationship between the peripheral defocus and the power distribution of the lens; they rather claimed the general idea of using lenses with power distributions that grew outward radially or along some specific directions, mainly horizontally and downward. Since 2007, many clinical trials have been conducted with results that are not fully conclusive [70], and some authors have undertaken research on the optical properties of the lenses intended to compensate peripheral error.

Atchison proposed an analysis similar to the classical theory described in this chapter, but moving the stop from the center of rotation to the entrance pupil of the eye [204].

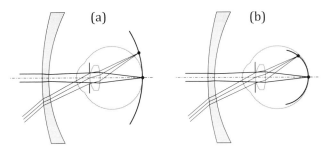

Figure 6.10 (a) Myopia compensation with a standard lens. The peripheral image presents hyperopic peripheral defocus. (b) Compensating lens that produces myopic peripheral defocus.

Rojo et al. [68, 200] proposed the use of a surface conjugate to the retina that would play the role of the remote-point sphere in classical lens design. This conjugate of the retina should have to be measured by peripheral refractometry. Finally, Barbero and Faria-Ribeiro used the retinal conjugate plus a particular optimization method to design lenses that would produce no peripheral defocus (see Figure 6.10). Barbero also finds that lens optimizations for both the rotary eye and the static eye are incompatible.

As stated by Charman in his review [70], the hypothesis that peripheral hyperopia is responsible for axial eye growth, and hence for the development of myopia, is not fully proven. The characteristics of peripheral defocus are very complex and difficult to measure with accuracy. However, this is an active field of research that may prove to be fruitful in the near future.

6.4 Modern ("Free-Form") Lens Design

Equation (6.10) describes a merit function that allows for general lens optimization. However, the full potential of this approach could not be used until the end of the twentieth century and the arrival of free-form technology. Prior to this technological revolution, lens generators and polishers used in RX-labs were only able to shape spherical and/or toric surfaces. Lens manufacturers could use aspherical semifinished blanks, but their flexibility was highly reduced by the small number of different base curves and fixed values of asphericity. Equation (6.10) or others similar to it could be used by the manufacturer of blanks, but the final performance of these blanks would be limited by the fact that each blank had to be shared by a large number of different prescriptions as well as different frame and user characteristics.

The first equipment for fast free-form surfacing of ophthalmic lenses was launched in 1998 by the German company Schneider. Free-form surfacing basically consists of applying CNC (computer numerically controlled) techniques that can guide the cutting tools and polishing pads to shape – almost – arbitrary surfaces. The technology is limited by the maximum local curvature it can create (around 15 D, although this number depends on the cutting and polishing speed). Free-form technology was available to the precision optics industry many years before 1998, but it was a lengthy process. The real breakthrough for the ophthalmic industry was the ability to cut and polish arbitrary surfaces in less than 2 minutes. The surfaces this technology can produce may need a large set of parameters to be fully specified. A free-form lab can shape a spherical surface, a conicoid, or a complex surface defined by 150 Zernike polynomials or an even larger number of spline coefficients, as described in Chapter 2.

The main characteristic of the free-form manufacturing process is the capacity for creating customized lenses. Customization starts with the actual reduction of oblique errors. While the standard manufacturing process can only hope to provide a lens with a bending factor that will avoid very strong oblique errors, and that bending factor, or base curve, is shared with many other prescriptions with different values of sphere and cylinder, free-form technology allows the generation of the specific surface for which oblique errors are reduced the most.

Other factors that can be accounted for when computing a free-form surface for a customized lens are:

Pantoscopic tilt. In the theory presented in this chapter, we have always assumed the line of sight, at the main viewing position, is perpendicular to the lens surfaces. In fact, most frames are designed in a way that the lenses are tilted with respect to line of sight at the main viewing position. When the lens is rotated around a horizontal axis parallel to the body sagittal plane, we say it has pantoscopic tilt. Most frames hold lenses with tilts between 0 and $10°$. Pantoscopic tilt introduces oblique astigmatism at the center of the lens and typically increases the oblique aberrations in the upper half of the lens.

Wrapping angle. Also known as facial angle. It is a rotation of the lens around a vertical axis, the temporal side of the lenses getting closer to the face. The wrapping angle is quite common in sports frames and sunglasses. It has similar effects to the pantoscopic tilt, but along the horizontal direction. Vision quality typically worsens toward the nasal side.

Distance to the rotation center. We are always assuming that l'_2 is approximately 27 mm. This is an average value, but in fact this distance can change to between 24 and 30 mm. The optimal base curves depend on this value. If a base curve is optimal for, say 25 mm, but the lens is worn at 30 mm, the off-axis performance will be changed, normally for the worse.

Prisms. If the lenses incorporate prescription prisms, oblique aberrations will be different.

Naso-pupillary distances. These user parameters have little effect in single vision lenses, but they have nonnegligible effects in progressive lenses. Ideally, they should be taken into account during the computation of the free-form surface.

Refractive index. Of course, refractive index is taken into account during calculation of the optimal free-form surfaces, but it is not usually considered a custom parameter.

Frame shape. High-quality lens design usually takes into account the shape and size of the frame, to optimize the free-form surface only on these regions inside the rim. Reducing the size of the optimization region usually improves the quality.

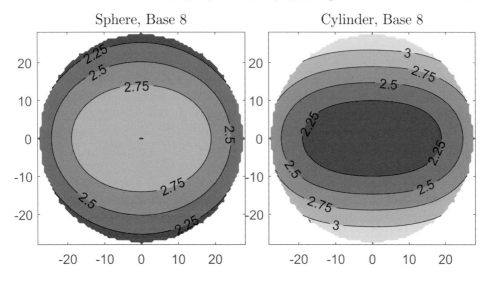

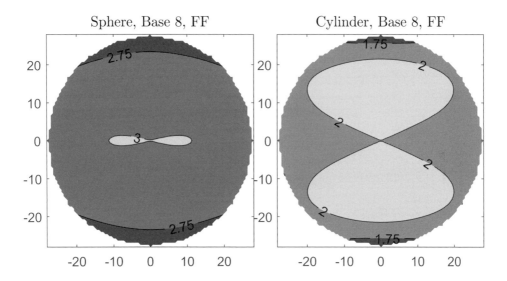

Figure 6.11 Comparison between a standard lens (upper row) with power $[3, 2 \times 180°]$ made with base curve $B = 8$ D and the corresponding free-form design (bottom row).

We will finish this chapter with a pair of examples demonstrating the possibilities of free-form technology. In Figure 6.11 we present maps of the same astigmatic lens used in the previous section, with power $[3, 2 \times 180°]$, refractive index $n = 1.5$, and $l_2' = 27$ mm. The first row shows the sphere and cylinder maps for the standard base made in base 8. We see that the cylinder grows faster vertically, from the nominal 2 D to above 3 D at 22 mm above the optical center of the lens. The sphere error is about 0.75 D. Now, we optimize the same prescription using a hybrid free-form surface composed of a biconic plus a polynomial term formed by 32 Zernike polynomials. The free parameters are then $\alpha = (p_B, p_T, c_1, \cdots c_{32})$ (see Chapter 2 to refresh the meaning of the asphericity parameters of the biconic surface). The center curvature radius of the biconic are not free parameters, as they are needed to provide the correct paraxial power. Also, the base curve has been

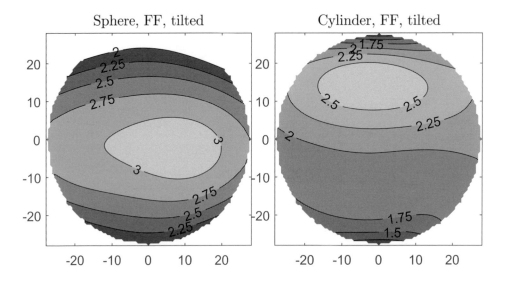

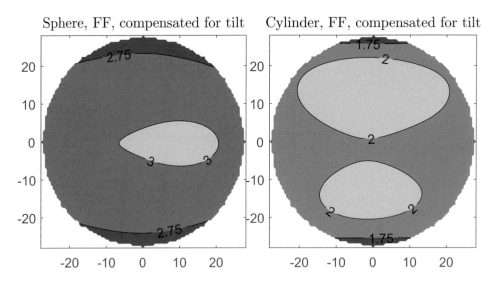

Figure 6.12 Comparison between the performance of the free-form lens computed for Figure 6.11, but with pantoscopic tilt of 10° and wrapping angle of 5° and the same lens optimized to compensate for the same tilts.

fixed to 8 D, for a fair comparison with the standard lens. The free-form lens has cylinder error smaller than 0.25 D in practically all the optimization region (in this case a circle with a diameter of 55 mm). Similarly, the sphere error is smaller than 0.25 inside 95% of the optimized region. This is the performance a user of high-quality free-form lenses should currently expect.

In Figure 6.12 we show, in the first row, the free-form lens just computed, but tilted with pantoscopic and wrapping angles of 10° and 5°, respectively. The performance has degraded as a consequence of the tilts, with the cylinder error reaching 0.5 D in the upper part of the lens and the sphere error reaching 1 D. Now, if we optimize the lens again, this time considering the tilts, we basically recover the same performance as before, that is, oblique errors smaller than 0.25 in most of the lens surface.

7

Optics of Contact and Intraocular Lenses

7.1 Introduction

In this chapter, we will provide a brief overview of the optical properties of contact and intraocular lenses. These devices are used as alternatives to ophthalmic lenses for the compensation of ametropia, and some conditions of the human eye. Therefore, for the sake of completeness, it is necessary to study the optical properties and design of contact and intraocular lenses. In this study, we will see that there are similarities but also remarkable differences between spectacle lenses and contact and intraocular lenses. For example, although one may think that a contact lens is similar to an ophthalmic lens in all but its location closer to the eye, this is not the whole story. There are many factors to be considered such as whether the contact lens is a rigid or a soft one, whether the user presents corneal astigmatism, the quality, and thickness of the user's tear film, the tighter or looser fit of the contact lens to the wearer's cornea, etc. Although some of these considerations are fitting-related issues outside the scope of this book (see, for example, Mandell's classic work [205]), there are a number of important differences between the optical properties of contact (or intraocular) lenses and those of spectacle lenses that should be known by the practitioner.

Thus, we will start our study with the description of the optical properties of contact lenses. We will focus on the effect of the tear film located between the lens and cornea that gives rise to the so-called *tear lens*. As with ophthalmic lenses, we will make use of the matrix formulation of power and we will operate within the framework of paraxial optics. Once we have described the paraxial properties, we will focus on the design of contact lenses, with the aim of providing a brief description of multifocal designs, stressing, particularly, the role played by pupil size, depth of focus, and lens aberrations. Afterward, we will center our attention on the basic optical properties of intraocular lenses, and we will derive an original expression based on the power matrix formulation to find the power of an intraocular lens from the data collected before surgery. Finally, we will end the chapter with a brief description of intraocular lens design, with a special mention for modern multifocal and extended focus intraocular lenses.

7.2 Optical Properties of Contact Lenses

A *contact lens* is a lens worn in front of the eye in close contact with the cornea for compensating ocular ametropia. The contact lens covers a great percentage of the corneal external layer, the corneal epithelium. This fact is of great importance because the corneal epithelium cells take their oxygen supply directly from the surrounding air. Thus, when the user wears contact lenses, this supply is interrupted unless the contact lens is made of an oxygen-permeable material. However, the first widely employed contact lenses were made of PMMA polymethyl methacrylate and similar materials [79], which are not oxygen-permeable. So, in order to avoid damage to the corneal tissue due to oxygen deprivation, those lenses could be worn only for a limited amount of time. Nowadays, contact lenses are manufactured with oxygen-permeable materials in both rigid and flexible configurations, which are also denominated hard and soft, respectively. Hard contact lenses do not change their form when fitted to the eye, and they are usually made of rigid gas-permeable material (RGP). On the other hand, soft contact lenses are made of flexible materials (typically hydrogels), which adopt the form of the cornea [79]. The optical properties of a contact lens, particularly when it is fitted to astigmatic corneas, depend on its flexibility. Therefore, it is customary to consider hard contact lenses separately from soft ones. Moreover, we will also distinguish between spherical and toric (RGP) contact lenses, and we will begin our study with the former.

7.2.1 Power of Rigid (RGP) Contact Lens with Spherical Surfaces

In Figure 7.1, we represent the situation that occurs when a hard contact lens is fitted on the eye. Although we may think that the lens back surface and the front surface of the cornea would be in contact, this is not the case, as there is always a space between the back surface of the contact lens and the anterior surface of the cornea. This void is usually filled

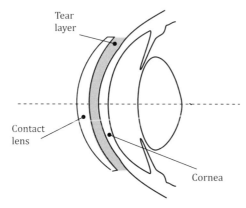

Figure 7.1 Representation of a contact lens-eye fitting. As can be seen, a volume of tear fills the space between the lens and the cornea forming the so-called tear layer. In this figure, and subsequent ones, the width of the tear layer has been greatly exaggerated for illustration purposes.

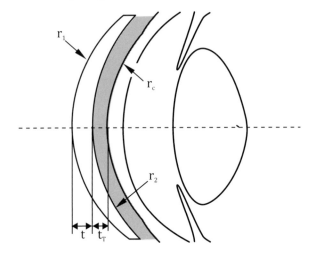

Figure 7.2 Geometrical parameters (curvature radius and thickness) of the contact lens, tear layer, and cornea system. r_1 and r_2 are the front and back curvature radii of the lens, r_c is the curvature radius of the cornea, t is the lens thickness, and t_T is the thickness of the tear layer.

with tear, giving rise to the so-called *tear layer* (or tear meniscus). The tear layer appears because the cornea is not perfectly spherical; it is flatter at the border, so it is not possible for all the points of the back surface of the lens to be in touch with the cornea. Besides, as the lens is rigid, it cannot accommodate its shape to that of the cornea as soft lenses do.

In order to obtain the optical properties of the lens-eye system it is necessary to set the way in which the power of the lens plus the tear layer is computed. In the literature, two main techniques have been described for calculating the power required in order to compensate user ametropy. The first one is the so-called *in-situ* technique [132], and it corresponds with the actual case in which a tear layer is located between the contact lens and the cornea. In Figure 7.2, we show the optical system formed by those elements with its main dimensional parameters (curvature radius and thickness). In these conditions we can apply Abbe's invariant and vergence propagation to compute the exit vergence at the cornea, which will allow us to establish the compensation principle for contact lenses.

Let us consider an incident parallel beam of light with vergence $V_1 = 0$ on the first surface of the lens. This beam is refracted by the first surface of the lens so that the exit vergence would be

$$V_1' = \frac{P_1}{n} \equiv \frac{(n-1)}{nr_1}, \qquad (7.1)$$

where n is the refractive index of the contact lens, P_1 the refractive power of the first lens surface and r_1 the curvature radius of the first surface of the lens[1] (see Figure 7.2). After propagation, the incident vergence at the second surface of the lens would be

---

[1] Throughout this chapter we will denote the curvature radius with lower case "r" to avoid confusion.

$$V_2 = \frac{P_1}{n\left(1 - \frac{t}{n}P_1\right)} = \frac{g_1}{n}P_1. \tag{7.2}$$

Now we have a refraction at the back surface of the lens, but instead of air we have a tear layer with refractive index n_T, so the exit vergence at the back surface of the lens would be

$$V_2' = \frac{n}{n_T}V_2 + \frac{P_{2T}}{n_T} = \frac{g_1}{n_T}P_1 + \frac{P_{2T}}{n_T}, \tag{7.3}$$

where P_{2T} is the refractive power of the back surface of the lens when it is surrounded by the tear film. As V_2' is the vergence after refraction, it should be propagated up to the anterior surface of the cornea, so the incident vergence at this surface would be

$$V_3 = \frac{V_2'}{\left(1 - t_T V_2'\right)} = \frac{g_1 P_1 + P_{2T}}{n_T\left\{1 - \frac{t_T}{n_T}\left[g_1 P_1 + P_{2T}\right]\right\}}. \tag{7.4}$$

Finally, after refraction at the anterior surface of the cornea we have

$$V_3' = \frac{n_T}{n_c}V_3 + \frac{K_T}{n_c} \equiv \frac{g_1 P_1 + P_{2T}}{n_c\left\{1 - \frac{t_T}{n_T}\left[g_1 P_1 + P_{2T}\right]\right\}} + \frac{K_T}{n_c}, \tag{7.5}$$

with n_c and K_T being the refractive index and the refractive power of the cornea when the incident medium is the tear layer. Let us consider now the following relationships

$$P_{2T} = \frac{n_T - n}{r_2} \equiv \frac{n_T - n}{1 - n}P_2,$$

$$K_T = \frac{n_c - n_T}{r_c} \equiv \frac{n_c - n_T}{n_c - 1}K,$$

relating the refractive powers of the back surface of the lens and front surface of the cornea referred to the tear layer as incident medium to those referred to air where r_c is the front surface curvature radius of the cornea. Substituting these expressions in equation (7.5) we find, after some algebra, the following expression:

$$V_3' = \frac{P - \frac{n_T - 1}{n - 1}P_2}{n_c\left(1 - \frac{t_T}{n_T}\left[P - \frac{n_T - 1}{n - 1}P_2\right]\right)} + \frac{n_c - n_T}{n_c(n_c - 1)}K, \tag{7.6}$$

where $P = g_1 P_1 + P_2$ is the back vertex power of the contact lens in air. We will now apply the compensation principle; to do so, let us suppose an incident beam coming from the remote point of the subject. In these conditions, the incident vergence at the cornea would be the refractive error at the corneal vertex, $\mathcal{R}$, the exit vergence at the front surface of the cornea would be $V_c' = \mathcal{R}/n_c + K/n_c$, and a sharp image would be formed at the retina. Therefore, by equating the right-hand side of equation (7.6) with vergence V_c', we can state the compensation principle for a rigid contact lens as

$$\frac{P - \frac{n_T - 1}{n - 1}P_2}{1 - \frac{t_T}{n_T}\left[P - \frac{n_T - 1}{n - 1}P_2\right]} + \frac{n_c - n_T}{n_c - 1}K = \mathcal{R} + K. \tag{7.7}$$

Although it is a somewhat complicated expression, equation (7.7) yields the exact value of a rigid contact lens power necessary for the compensation of a given ametropy, as we will see in the next example.

Example 31 Compute the back vertex power (in air) of a rigid contact lens necessary for compensating a user with $\mathcal{R} = -3.5$ D. The corneal radius is $r_c = 7.81$ mm, the tear layer thickness is $t_T = 100\ \mu$m, and the values of the refractive indexes of the cornea, tear film, and contact lens are $n_c = 1.376$, $n_T = 1.3375$, and $n = 1.490$, respectively. Suppose that the second surface of the lens has the same curvature as the cornea.

From the problem data we have

$$K = \frac{n_c - 1}{r_c} = \frac{(1.376 - 1)}{7.81} \times 1000 = 48.14\,\text{D}.$$

On the other hand, if $r_2 = r_c = 7.81$ mm, then

$$P_2 = \frac{1 - n}{r_2} = \frac{(1 - 1.490)}{7.81} \times 1000 = -62.74\,\text{D}.$$

Substituting these data into equation (7.7), we have

$$\frac{P + 43.21}{1 - 7.48 \times 10^{-5}\,(P + 43.21)} + 4.93 = -3.5 + 48.14,$$

after rearranging and simplifying we have

$$P + 43.21 = 39.71 - 0.003\,(P + 43.21).$$

Finally, after some further algebra, we obtain the resulting power, $P = -3.62$ D, of the contact lens.

In many practical cases, the thickness of the tear layer can be neglected in a first approximation, so we will have $t_T \approx 0$. In addition, in order that the lens can fit properly, it is necessary that $r_2 \approx r_c$. Therefore, we will write the back radius of the lens as $r_2 = r_c + \Delta r$, with $\Delta r < r_c$ so that

$$P_2 = \frac{(1 - n)}{r_c + \Delta r} \approx \frac{(1 - n)}{(n_c - 1)}K - \frac{(1 - n)}{(n_c - 1)^2}K^2\Delta r. \tag{7.8}$$

Therefore, after substituting equation (7.8) in (7.7) and taking $t_L = 0$, we arrive at the following result

$$P \approx \mathcal{R} + \frac{n_T - 1}{n_c - 1}K^2\Delta r. \tag{7.9}$$

This simpler (and more useful) equation allows for the computation of the contact lens power as the refractive error plus a correction that depends on the flatness (or steepness) of the back surface of the lens compared to the cornea. Notice that, as we have neglected the tear film thickness, if the lens is perfectly fitted to the cornea, the tear lens has no effect, so the power of the contact lens necessary to compensate the patient should be just the refractive error. According to equation (7.9), if the back surface of the lens is steeper than the cornea, the lens should be more positive than the refractive error and, conversely, if the back surface is flatter than the cornea, the lens must be more negative, as we will see in the following example.

Example 32 Compute the power of a rigid contact lens for a subject whose refractive error is $\mathcal{R} = 3.75$ D if the back surface of the lens is 0.1 mm steeper than the cornea. Suppose that the thin tear layer approximation holds and $n_c = 1.376$, $n_T = 1.3375$, and $K = 44$ D.

As the back surface is steeper, then δr is negative, so, according to the statement of the problem, $\Delta r = -0.1 \times 10^{-3}$ m. Therefore, we have

$$P = 3.75 + \frac{0.3375}{0.376^2} \times 44^2 \times \left(-0.1 \times 10^{-3}\right) = 3.75 - 0.46 = 3.29 \text{ D}.$$

Thus, if the lens is not properly fitted to the cornea we must correct the power of the contact lens with a factor that depends on the index of the tear film and the cornea, and the value of K. This correction can be estimated as 0.5 D for each 0.1 mm of ΔR. The power should vary in such a way that, the steeper (flatter) the back surface the lower (higher) the contact lens power should be.

An alternative way for computing contact lens power is the so-called exploded technique. This method is based on the exploded view of the contact lens, tear film, and cornea depicted in Figure 7.3. In this method, both the contact lens and the tear film are considered lenses separated by a zero-thickness air layer. The tear film, known as the *tear lens* within this technique, is also separated from the cornea by a zero-thickness air layer. In these conditions, the power of the system formed by the contact lens and tear lens is the sum of their powers computed in air. Moreover, as the contact and tear lens system is separated from the cornea by an extremely thin layer of air, the application of the compensation principle results in the following condition

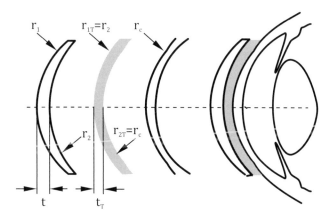

Figure 7.3 Exploded view of the contact lens, tear layer, and cornea layers. This is the basis of the "exploded" technique for computing contact lens power.

$$P + F = \mathcal{R}, \tag{7.10}$$

where $\mathcal{R}$ is the refractive error at the corneal vertex, P is the contact lens power (in air), and F is the power of the tear film lens (in air). The contact lens power can be written as

$$P = \mathcal{R} - F = \mathcal{R} - \frac{P_{1T}}{1 - \frac{t_T}{n_T}P_{1T}} - P_{2T}, \tag{7.11}$$

with P_{1T} and P_{2T} being the front and back refractive powers of the tear lens, respectively. Taking into account that, by definition of the tear lens geometry, $r_{1T} = r_2$ and $r_{2T} = r_c$, and supposing the thickness of the tear lens is zero, we have

$$P = \mathcal{R} - F = \mathcal{R} - \frac{n_T - 1}{n - 1}P_2 - \frac{n_T - 1}{n_c - 1}K. \tag{7.12}$$

Example 33 Calculate the power of the contact lens necessary for compensating a patient whose refractive error is $\mathcal{R} = -3.25$ D when the tear layer is thin and the lens is perfectly fitted to the cornea of the subject in such a way that $r_2 = r_c$. Consider that the queratometry yields a value of $K = 48$ D, and the refractive indexes of the tear film, cornea, and contact lens are $n_T = 1.3375$, $n_c = 1.376$, and $n = 1.490$, respectively.

We will compute first the corneal radius $r_c = 376/48 = 7.833$ mm. Afterward, we obtain $P_2 = (1 - n)/r_2 = -490/7.833 = -62.55$ D and we can substitute the data into equation (7.12) for computing the power of the contact lens as

$$P = -3.25 - \frac{0.3375}{0.490} \times -62.55 - \frac{0.3375}{0.376} \times 48 = -3.25 + 43.08 - 43.08 = -3.25 \text{ D}.$$

Therefore, if the back surface of the lens is perfectly fitted to the cornea, then

$$P_2 \equiv \frac{1 - n}{n_c - 1}K,$$

and equation (7.12) becomes

$$P = \mathcal{R}. \tag{7.13}$$

Before proceeding further, we must take into consideration some facts. The first is that, although we have followed a different path to derive equations (7.9) and (7.12), they are completely equivalent, as can be easily proved by substituting equation (7.8) into equation (7.12) taking into account that $r_2 = r_c + \Delta r$. On the other hand, it is important to notice that we are assuming that K is the actual refractive power of the cornea. However, the value of K depends on the measurement technique employed, and it may differ from the actual value due to a number of approximations. Further details on corneal measurement can be found in reference [206].

7.2.2 Power of RGP Contact Lenses with Toric Surfaces

Corneal astigmatism is quite a common finding in eye examination. Due to a number of factors, the corneal surface becomes an astigmatic one presenting two principal meridians with different refractive power: $K_- @\theta_-$ and $K_+ @\theta_+$ with $K_+ - K_- > 0$ and $\theta_+ = \theta_- \pm \frac{\pi}{2}$. As a consequence, an eye with corneal astigmatism will have an astigmatic refractive error. In many cases, the total amount of astigmatic refractive error is due to corneal astigmatism; however, we can also find individuals who, in addition to corneal astigmatism, also present *internal astigmatism*. Therefore, in the general case, *ocular astigmatism* is the result of the combination of corneal plus internal astigmatism. When dealing with contact lens fitting, it is quite important to make this differentiation between corneal astigmatism and internal astigmatism, because if the patient only presents corneal astigmatism, she can be compensated with a spherical rigid contact lens fitted with the back surface of the contact lens parallel to the flatter meridian of the eye [132, 206]. This compensation is possible due to the effect of the tear lens, which also becomes astigmatic as is shown in the following example.

Example 34 Compute the power of the tear lens (in air) when a subject whose queratometry is 44 D @ 35°, 45.25 D @ 125° is compensated with a rigid contact lens whose back radius is parallel to the flattest meridian of the cornea. Suppose the refractive index of the tear $n_T = 1.3375$ and the calibration index of the queratometer $n_c = 1.376$.

The flattest meridian (base curve) of the cornea is the one corresponding to K_-. According to the problem data, $K_- = 44$ D, so $r_{cB} = 376/44 = 8.54$ mm, while the other principal radius of the cornea is $r_{cT} = 376/45.25 = 8.31$ mm. Thus, the front surface radius of the tear lens is $r_1 = 8.54$ mm with a refractive power $F_1 = 337.5/8.54 = 39.51$ D. Regarding the back surface of the tear film, it is a toric surface with the same shape as the cornea, so its main radii are $r_{2T} \equiv r_T = 8.31$ mm for the cross meridian and $r_{2B} = 8.54$ mm for the base, which are equivalent to the back refractive powers of $F_{2B} = (1 - n_T)/r_{2B} \equiv -337.5/8.54 = -39.51$ D and $F_{2T} = -337.5/8.31 = -40.61$ D. Thus the tear lens is a sphero-toric one, and its main powers are

$$F_- = P_1 + P_{2T} = 39.51 - 40.61 = -1.1 \text{ D @ } 125°$$
$$F_+ = P_1 + P_{2B} = 39.51 - 39.51 = 0 \text{ D @ } 35.$$

According to these data, the power of the tear lens, when expressed in spherocylindrical form, is $[0, -1.1 \times 35°]$ D. Note that, as given by the queratometry, the corneal power would be $[44, 1.25 \times 35°]$, so the cylindrical power of the tear lens almost compensates that of the cornea.

When internal astigmatism cannot be compensated with a spherical rigid contact lens, we must use toric contact lenses. In this case, we can use both spherotoric or bitoric lenses. For spherotoric lenses, the toric surface is the front one, while the back surface is spherical in order to compensate the corneal astigmatism with the tear lens. On the other hand, bitoric lenses present two toric surfaces, with the back surface having its main curvature radius parallel to that of the cornea for compensating corneal astigmatism, while the remaining ocular astigmatism is compensated by the front surface of the lens.

In any case, to study astigmatic lenses it is more practical to make use of the power matrix formalism. We will focus on the study of bitoric lenses, and we consider an astigmatic subject whose refractive error is given by matrix $\mathbb{R}$, while the refractive power of the cornea (in air) is $\mathbb{K}$ with a corneal index n_c. In these conditions, we will use a bitoric contact lens with a perfect fit between the back surface of the lens and the cornea. Therefore, the principal meridian of the contact lens and that of the cornea are parallel so that $\mathbb{P}_2 = \frac{1-n}{n_c-1}\mathbb{K}$ where $\mathbb{P}_2$ is the back surface power matrix. Moreover, as the back surface of the lens fits perfectly with the cornea, equation (7.13) holds, so $\mathbb{P} = \mathbb{R}$. Thus, it is possible to compute the refractive power of the front surface of the contact lens as

$$\mathbb{P}_1 = \left[\mathbb{I} + \frac{t}{n}\left\{\mathbb{R} + \frac{n-1}{n_c-1}\mathbb{K}\right\}\right]^{-1}\left[\mathbb{R} + \frac{n-1}{n_c-1}\mathbb{K}\right], \tag{7.14}$$

where $\mathbb{I}$ is the identity matrix and t is the central thickness of the contact lens.

Example 35 Determine the refractive power of the front surface of a bitoric contact lens calculated for compensating a subject whose refractive error is $\mathcal{R} \equiv [2.25, 1.75 \times 55°]$ if the back surface of the lens fits the patient's cornea perfectly. The patient queratometry is 46 D @ 25, 46.75 D @ 115°. Suppose that the lens thickness is $t_L = 0.55$ mm, and the lens and corneal refractive indexes are $n = 1.46$ and $n_c = 1.376$, respectively.

The first task will be the computation of the corneal refractive power from queratometric data

$$\mathbb{K} = \left[\begin{array}{cc} K_- + (K_+ - K_-)\sin^2\theta_- & -(K_+ - K_-)\sin\theta_-\cos\theta_- \\ -(K_+ - K_-)\sin\theta_-\cos\theta_- & K_- + (K_+ - K_-)\cos^2\theta_- \end{array}\right]$$

$$= \left[\begin{array}{cc} 46 + 0.75 \times \sin^2 25 & -0.75 \times \sin 25 \times \cos 25 \\ -0.75 \times \sin 25 \times \cos 25 & 46 + 0.75 \times \cos^2 25 \end{array}\right] = \left[\begin{array}{cc} 46.134 & -0.287 \\ -0.287 & 46.616 \end{array}\right] \text{D},$$

next, we compute the refractive error matrix:

$$\mathbb{R} = \left[\begin{array}{cc} 2.25 + 1.75 \times \sin^2 55 & -1.75 \times \sin 55 \times \cos 55 \\ -1.75 \times \sin 55 \times \cos 55 & 2.25 + 1.75 \times \cos^2 55 \end{array}\right] = \left[\begin{array}{cc} 3.42 & -0.822 \\ -0.822 & 2.826 \end{array}\right] \text{D}.$$

Thus, we have

$$\mathbb{R} + \frac{n-1}{n_c-1}\mathbb{K} = \left[\begin{array}{cc} 59.860 & -1.173 \\ -1.173 & 59.856 \end{array}\right] \text{D},$$

so the refractive power of the front surface can be calculated as

$$\mathbb{P}_1 = \left(\left[\begin{array}{cc} 1 & 0 \\ 0 & 1 \end{array}\right] + \frac{5.5 \times 10^{-4}}{1.46}\left[\begin{array}{cc} 59.860 & -1.173 \\ -1.173 & 59.856 \end{array}\right]\right)^{-1} \times \left[\begin{array}{cc} 59.860 & -1.173 \\ -1.173 & 59.856 \end{array}\right]$$

$$= \left[\begin{array}{cc} 1.022 & -4.4 \times 10^{-4} \\ -4.4 \times 10^{-4} & 1.022 \end{array}\right]^{-1} \times \left[\begin{array}{cc} 59.860 & -1.173 \\ -1.173 & 59.856 \end{array}\right] = \left[\begin{array}{cc} 58.544 & -1.122 \\ -1.122 & 58.536 \end{array}\right] \text{D},$$

note that we have to be careful with the units in order to avoid mistakes. From the result obtained, we see that the first surface of the lens is toric with principal refractive powers $P_{1-} = 57.42$ D @ 45.1° and $P_{1+} = 59.66$ D @ 135.1°. Those principal powers are computed, as customary, through the determinant and trace of $\mathbb{P}_1$ as we explained in Chapter 4.

7.2.3 Power of Soft Contact Lenses

Soft contact lenses are, nowadays, made with oxygen-permeable materials such as hydrogels that present different mechanical properties than rigid contact lenses. In particular,
soft contact lenses are flexible and adaptable in such a way that we can assume a perfect fit
between the back surface of the lens and the front surface of the cornea, provided there is
not much difference between the back curvature radius of the lens and that of the cornea.
Therefore, when fitting a spherical soft contact lens to a spherical cornea, we can safely
suppose that the power of the tear lens is zero, $F = 0$ D. In this case, the lens power must
be just the refractive error of the user $P = \mathcal{R}$ (see equation (7.13)).

The problem becomes more difficult when the subject presents corneal astigmatism.
In this case, contrary to what happened with RGP lenses, a spherical soft lens would not
compensate ocular astigmatism. The lens, being flexible, would change its form to that of
the cornea and it would become astigmatic too. In this case, a number of solutions can be
devised depending on the value of the corneal and total astigmatism of the eye, which are
the parameters available after the optometric examination. If the whole ocular astigmatism
is corneal and its magnitude is not too high, it is possible to fit a spherical lens and leave
the astigmatism uncorrected provided that the user can deal with the loss of visual acuity.
If this is not possible, we can use a soft toric lens, but we must be aware that soft contact
lenses are more prone to move when the eye blinks than RGP lenses. To solve this problem,
most soft toric contact lenses are manufactured either truncated or with a ballast prism in
order to keep the lens in its proper position [132, 206]. As with toric rigid contact lenses,
soft ones can also be manufactured in both spherotoric and bitoric configurations [132].

7.3 Multifocal Contact Lens Designs

Contact lenses can also be used for the compensation of presbyopia. For a proper introduction to it, the reader is directed to Chapter 8. For the remainder of this chapter, a basic
knowledge of presbyopia and its compensation with spectacle lenses will be assumed.
Nowadays, we can find several solutions, ranging from bifocal to multifocal design both in
RPG and hydrogel [207]. Several geometries are also available for each kind of lens. For
bifocal lenses, we have rotationally symmetric, non rotationally symmetric, and simultaneous designs [207]. The first two classes are similar to spectacle bifocals, as they provide the
user with a distance zone located in the central position of the lens, and a lower near zone
that is reached when the pupil is lowered in order to gaze at near objects. The difference
between rotationally symmetric and nonsymmetric designs is that the former present an
annular near zone, while the shape of the near zone for the latter does not present rotational
symmetry, as it is similar to a small bifocal segment without nasal inset [207].

An alternative to classic bifocal designs are simultaneous ones. As shown in Figure 7.4,
simultaneous bifocal contact lenses do also present two zones with different powers, a
central circular zone with a diameter between 3 to 4 millimeters and a second annular
zone. In the center near configuration [207], the power at the annular zone is lower than

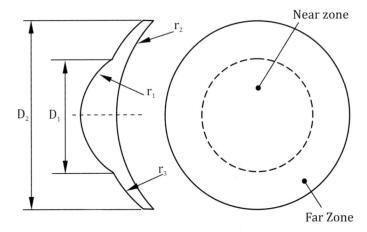

Figure 7.4 Configuration of an RPG bifocal contact lens designed for center near simultaneous vision. Notice that, for the first surface of the lens, the curvature radius R_1 of the central zone is lower than that which corresponds to the annular zone of the lens, R_3, so the lens has more power at the center. The difference between the two radii of curvature has been exaggerated in order to illustrate the geometry of the lens.

the power at the central zone that carries the addition of the lens. The rationale behind this configuration is that for a dilated pupil and either for far or near objects, part of the energy refracted by the lens will be in focus, while the other part will be out of focus. The contrast of the resulting image will be reduced, but high-contrast visual acuity will keep, at least, good enough values. It is also assumed that, for near tasks, the illumination will be high enough to keep the eye pupil small so that only the near portion of the lens is used. This solution has a number of potential drawbacks. The first is that image quality will be dependent on pupil size, and, consequently, the user must adapt to differences in image quality. For example, the user may experience some blur when reading a book outside if the day is cloudy. More important, when looking at far objects, the user will experience some loss of contrast due to the effect of the two different power zones.

In order to show these effects, we have carried out a computer simulation. Using commercial ray-tracing software (Zemax OpticStudio), we have modeled a bifocal contact lens with $A = 2.50$ D of addition fitted to a myopic eye with no accommodation. We have selected Goncharov's eye model [208] for simulating the optics of the eye. We have computed the image of a letter "F" formed by the contact lens-eye system for different object distances and pupil diameters. In Figure 7.5, we show the results obtained. The first conclusion that we can draw is the noticeable variation of the image quality with object distance and pupil size. For far objects, there is a slight defocusing of the image when the pupil size decreases, while the contrary happens for the intermediate object distance. Finally, as the eye has no accommodation, the image corresponding to the near object is slightly blurred even for the lower pupil size. Therefore, if no accommodation is present the user will experience problems in focusing near objects. On the other hand, as can be

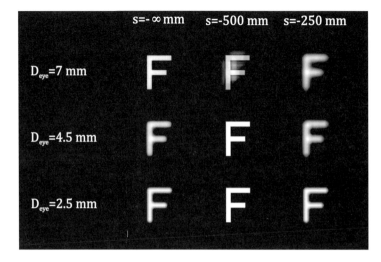

Figure 7.5 Numerical simulation of the image of a letter "F" formed by a bifocal contact lens with addition $A = 2.50$ D. Here we have used Goncharov's eye model for different distances and pupil sizes. The data have been obtained through the ray tracing software Zemax OpticStudio©.

seen in the upper row of Figure 7.5, the passing of light through two different power zones produces, on some occasions, a characteristic halo that may impair image contrast and resolution. The situation depicted in Figure 7.5 is quite common for other multifocal lens designs such as trifocals or axicons [209].

Besides bifocal and trifocal contact lenses, a number of highly aspherical multifocal designs are available [210]. All of them operate within a kind of "*negative spherical aberration principle.*" In this approach, the ultimate design goal is to have an optical system with extended depth of focus but with the central rays focusing closer to the lens than the peripheral ones. In current practice, we can find two kinds of designs: soft and hard. The soft, or progressive, design has a smooth variation of power between the center and the periphery of the lens. On the other hand, hard designs have a brusque power variation characteristic of bifocal lenses, although with a different distribution [210]. In any case, the dynamic of the user's pupil (size, light response, etc.) constitutes one of the most important parameters to be taken into account in the fitting of multifocal contact lenses.

7.4 Optical Properties of Intraocular Lenses

An *intraocular lens* (IOL) is a lens implanted in the eye after a surgical procedure [211]. The more common type of IOL is the pseudophakic IOL which is designed to replace the natural crystalline lens when it is removed as a result of cataract surgery. Theoretically, the power of an IOL can be computed before surgery for compensating the refractive error of the eye. In practice, emmetropization of a pseudophakic eye is challenging due to the changes induced by the surgery and by the intrinsic difficulty of positioning the IOL accurately within the eye, particularly for toric IOLs.

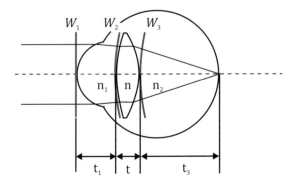

Figure 7.6 Diagram of a pseudophakic eye showing the location of the intraocular lens. Also depicted are the wavefronts W_1 and W_2 incident at the cornea and the lens, respectively, and the wavefront W_3 at the exit of the lens.

In order to compute the power of an IOL, it is necessary to know some critical parameters. In Figure 7.6, we have depicted a schematic pseudoaphakic eye formed by the cornea and the IOL. As can be seen in this figure, the distance between the cornea and the IOL is t_1, the IOL thickness is t, and the distance between the IOL and the retina is t_2. In these conditions, when a wavefront with vergence $\mathbb{V}_1$ incides over the cornea the refracted vergence would be

$$n_1 \mathbb{V}_1' = \mathbb{K}^* = \frac{n_1 - 1}{n_c - 1} \mathbb{K}, \tag{7.15}$$

where $\mathbb{K}$ is the refractive power of the cornea given by the queratometer and n_c the calibration index of the queratometer. In equation (7.15), $\mathbb{K}^*$ would be the corneal refractive power corrected to the actual index of the cornea n_1. $\mathbb{V}_1'$ is propagated in such a way that the vergence of the incident wavefront at the front surface of the IOL is

$$\mathbb{V}_2 = \left(\mathbb{I} - \frac{t_1}{n_1} \mathbb{K}^* \right)^{-1} \frac{\mathbb{K}^*}{n_1}. \tag{7.16}$$

On the other hand, the vergence of the exit wavefront $\mathbb{V}_3'$ at the back surface of the lens is given by the following equation:

$$n_2 \mathbb{V}_3' = \mathbb{P}_2 + \left[\mathbb{I} - \frac{t}{n} (\mathbb{P}_1 + n_1 \mathbb{V}_2) \right]^{-1} (\mathbb{P}_1 + n_1 \mathbb{V}_2), \tag{7.17}$$

where $\mathbb{P}_1 = (n - n_1)\mathbb{H}_1$ and $\mathbb{P}_2 = (n_2 - n)\mathbb{H}_2$ are the refractive powers of the lens submerged between two liquids (the aqueous and vitreous humours) whose respective refractive indexes are n_1 for the aqueous and n_2 for the vitreous, while the refractive index of the lens is n, and its thickness t. As can be seen in Figure 7.6, a sharp image will be formed at the fovea if the exit wavefront is spherical with a curvature radius equal to the distance t_2 between the back surface of the lens and the fovea. Therefore, the compensation principle for IOL can be stated as

$$\frac{n_2}{t_2}\mathbb{I} = \mathbb{P}_2 + \left[\mathbb{I} - \frac{t}{n}(\mathbb{P}_1 + n_1\mathbb{V}_2)\right]^{-1}(\mathbb{P}_1 + n_1\mathbb{V}_2). \tag{7.18}$$

Multiplying this equation by factor $\mathbb{I} - \frac{t}{n}(\mathbb{P}_1 + n_1\mathbb{V}_2)$, we find that

$$\frac{n_2}{t_2}\mathbb{I} - \frac{n_2}{n}\frac{t}{t_2}(\mathbb{P}_1 + n_1\mathbb{V}_2) = \mathbb{P}_2 - \frac{t}{n}\mathbb{P}_2(\mathbb{P}_1 + n_1\mathbb{V}_2) + \mathbb{P}_1 + n_1\mathbb{V}_2, \tag{7.19}$$

and, after rearranging the terms, we found that

$$\mathbb{P}_E = \frac{n_2}{t_2}\mathbb{I} - \frac{n_2}{n}\frac{t}{t_2}(\mathbb{P}_1 + n_1\mathbb{V}_2) - n_1\mathbb{V}_2\left(\mathbb{I} - \frac{t}{n}\mathbb{P}_2\right). \tag{7.20}$$

where $\mathbb{P}_E$ is equivalent power of the lens given by $\mathbb{P}_E = \mathbb{P}_1 + \mathbb{P}_2 - \frac{t}{n}\mathbb{P}_1\mathbb{P}_2$, with $\mathbb{P}_1$ and $\mathbb{P}_2$ being the refractive powers defined previously. Notice that the lens power given by equation (7.20) corresponds to the lens power when it is placed within the eye, and thus it is surrounded by two fluids (the vitreous and aqueous humours).

Due to the complexity of equation (7.20), it is necessary to pose some hypotheses. We can suppose first that the refractive index of aqueous and vitreous humour are the same, and we will refer to this index as n_a. The next hypothesis regards the shape of the IOL. A quite common shape for IOL lenses is the biconvex form for which $\mathbb{P}_1 = \mathbb{P}_2$ and $\mathbb{P}_{1,2} \cong \mathbb{P}_E/2$. Therefore, for a biconvex IOL, equation (7.20) can be written as

$$\left(\mathbb{I} + \frac{1}{2}\frac{n_a}{n}t\left(\frac{1}{t_2}\mathbb{I} - \mathbb{V}_2\right)\right)\mathbb{P}_E = \frac{n_a}{t_2}\mathbb{I} - \left(1 + \frac{n_a}{n}\frac{t}{t_2}\right)n_a\mathbb{V}_2. \tag{7.21}$$

Finally, after substituting equation (7.16) in (7.21), and performing a further rearrangement of the terms, we arrive at the final approximate equation that gives the power of a biconvex IOL lens as:

$$\mathbb{P}_E = \left[\mathbb{I} + \frac{1}{2}\frac{t}{n}\left(\frac{n_a}{t_2}\mathbb{I} - \left(\mathbb{I} - \frac{t_1}{n_a}\mathbb{K}^*\right)^{-1}\mathbb{K}^*\right)\right]^{-1}$$
$$\times \left[\frac{n_a}{t_2}\mathbb{I} - \left(1 + \frac{n_a}{n}\frac{t}{t_2}\right)\left(\mathbb{I} - \frac{t_1}{n_a}\mathbb{K}^*\right)^{-1}\mathbb{K}^*\right]. \tag{7.22}$$

It is important to note that, under the approximations made, the IOL power depends on the precise measurement of the following parameters:

- Corneal queratometric data, given by the corneal refractive power $\mathbb{K}^*$.
- Depth of anterior chamber, (post-surgical) given by the distance t_1.
- Lens data, particularly its refractive index n and central thickness t.
- Finally, axial length of the eye l_o, which allows us to obtain distance t_2 through the equation $l_o = t_1 + t + t_2$.

We will see in the next example how equation (7.22) can be used to determine the approximate power of a toric IOL necessary for compensating an individual who, before surgery, presented axial myopia and corneal astigmatism.

Example 36 Compute the power of a biconvex IOL necessary for compensating a subject whose corneal queratometry is 45.5 D @ 65°, 47.75 D @ 155°. The axial length of the eye is $l_0 = 24$ mm. Suppose that the refractive index of the aqueous and vitreous humor is the same $n_a = 1.336$, that the calibration index of the queratometer is $n_c = 1.376$, and that the depth of the anterior chamber is $t_1 = 4.5$ mm, while the lens thickness is $t = 2$ mm and its refractive index $n = 1.469$.

Let us calculate first the corneal refractive power

$$\mathbb{K} = \begin{bmatrix} 45.5 + 2.25 \times \sin^2 65 & -2.25 \times \sin 65 \times \cos 65 \\ -2.25 \times \sin 65 \times \cos 65 & 45.5 + 2.25 \times \cos^2 65 \end{bmatrix} = \begin{bmatrix} 47.348 & -0.862 \\ -0.862 & 45.902 \end{bmatrix} D.$$

After applying the refractive index correction, the corrected corneal refractive power is

$$\mathbb{K}^* = \frac{n_1 - 1}{n_c - 1} \mathbb{K} \equiv \frac{0.3376}{0.376} \begin{bmatrix} 47.348 & -0.862 \\ -0.862 & 45.911 \end{bmatrix} = \begin{bmatrix} 42.513 & -0.774 \\ -0.774 & 41.241 \end{bmatrix} D,$$

and

$$\left(\mathbb{I} - \frac{t_1}{n_a} \mathbb{K}^* \right)^{-1} \mathbb{K}^* = \begin{bmatrix} 0.857 & 0.003 \\ 0.003 & 0.861 \end{bmatrix}^{-1} \begin{bmatrix} 42.513 & -0.774 \\ -0.774 & 41.240 \end{bmatrix} = \begin{bmatrix} 49.621 & -1.049 \\ -1.049 & 47.896 \end{bmatrix} D.$$

Given the anterior chamber and IOL thickness, we can compute $t_2 = l_0 - t_1 - t = 24 - 4.5 - 2 = 17.5$ mm. Substituting $\mathbb{K}^*$ and t_2 in equation (7.22), we have

$$\mathbb{P}_E = \begin{bmatrix} 21.5118 & 1.1151 \\ 1.1151 & 23.3831 \end{bmatrix} D,$$

which corresponds to the spherotoric prescription [21.0, 2.9 × 155°] D. Using an exact ray tracing software in Goncharov's eye model [208] with the same physiological parameters, the value of the optimum biconvex lens power that emmetropizes the eye is [21.2, 3.2 × 155°] D, so we have an error of about 0.2 D in the spherical power and 0.3 D in the cylinder.

The calculation of the power of toric IOLs using equation (7.22) is based on several assumptions that must be carefully reviewed. First, we have employed a paraxial formula, so aberrations have not been taken into account. However, the only way to consider accurately the effect of aberrations is through numerical ray tracing, which can be employed to check the accuracy of the solution obtained [212]. We have also assumed a schematic eye model with a thin cornea. Although we have included a correction for the keratometric value, it is just an approximation that can be overcome by the use of a thick model of the cornea. In addition, there is another source of error due to the astigmatism induced by the surgical incision on the cornea. This change must be predicted in advance. Furthermore, we have considered a biconvex lens, which is just one of the many possible shapes available, and we have made the approximation $\mathbb{P}_{1,2} \approx \mathbb{P}_E/2$.

However, the main point of contention, and the reason why so many different formulas for IOL power have been published in the literature [211], is the determination of the exact postoperative values of the constants that define lens power, particularly anterior chamber depth t_1, eye length l_o, and corneal power K^*. The accurate prediction of postsurgical *anterior chamber depth* (ACD) has been the subject of numerous research papers [211, 212]. It is generally admitted that a correlation exists between the ACD and the axial length of the eye that can be accurately measured by means of ocular biometry. For example, using the partial coherence interferometric technique reported by Drexler et al. [213], it is possible to obtain a good estimate of the ACD. Besides the effect of the axial length of the eye on the post-surgical ACD, other magnitudes such as the preoperative ACD and lens thickness also influence the value of this magnitude. We can find some statistical models in the literature that allow for the computation of postoperative ACD from the values of several of those parameters. This has allowed for a significant reduction in IOL power errors [214]. On the other hand, both corneal power and eye length have been measured more accurately in recent years with the development of new measurement techniques. In general, the practitioner must be aware of the wide range of existing formulas for computing IOL lens power, their limitations, and the conditions in which they operate in order to avoid mistakes in the actual implantation of an IOL.

7.5 Design of Multifocal Intraocular Lenses

Once the natural crystalline lens is removed, the patient loses all ability to accommodate, regardless of age. In these conditions, if a monofocal IOL calculated for correcting distance vision is implanted in the eye, the user will not be able to focus near objects without a reading addition. Obviously, it would be a great advantage to have an IOL able to focus at different distances, and a lot of effort has been devoted in order to develop such a lens. However, the design of multifocal IOLs faces a number of challenges. First, with no remaining natural accommodation the amount of addition would be large (around 3 D or more). Another problem is the difficulty to predict the exact location and orientation of the IOL before surgery. Finally, it is necessary to balance pupil size with lighting conditions for each working distance. Therefore, the ultimate goal of design for multifocal IOLs is to achieve an increased depth-of-focus without compromising either the resolution or the contrast of images viewed through the IOL [215].

Possible multifocal IOLs include bifocal/trifocal refractive and diffractive designs. The basis of bifocal refractive IOLs is the formation of a double image due to the different refraction at the center and at the periphery. When a subject looks at a distant object, the far (central) portion of the lens forms a sharp image on the retina, while the near (peripheral) portion forms a secondary image before the retina. Therefore, the user perceives simultaneously the sharp image of the object produced by the far zone and a halo due to the peripheral zone of the IOL (see Figure 7.7). For near objects, the situation is the contrary,

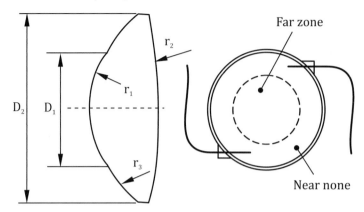

Figure 7.7 Scheme of a bifocal IOL. It is formed by a central zone for far distance and an annular zone that contains the lens' addition. Note the two hooks (known as "haptics") used to fix the IOL to the eye walls. Compare this configuration with the simultaneous contact lens design shown in Figure 7.4.

as the peripheral zone of the IOL forms a well-focused image on the retina while the central portion of the IOL forms a defocused image. The ratio between the area of the far and near zones of the lens is highly correlated to the contrast and brightness of the sharp image formed at the retina [215]. It is also possible to design the far zone as the outer zone and the near as the inner one. In this case, it is assumed that most of the near tasks are performed with high illumination so the pupil is small. Trifocal designs add a focus for intermediate distances but, usually, with a greater loss of contrast due to the formation of three images of the same object.

The philosophy of diffractive multifocal IOLs is the same as that of bifocal designs, but the physical principle under which they operate is different. Diffractive IOLs are lenses with a diffractive optical element (DOE) engraved on one or two of the lens surfaces. A DOE is any optical element that uses light diffraction as its operating principle. A well-known DOE example is the Fresnel zone plate [15, 216, 217], which is a set of concentric rings alternately opaque and transparent that focus light in several locations, producing multiple images from a single object. To achieve this result, the radius of each ring must be

$$\rho_n^2 = nf_o\lambda \quad n = 1, 2, 3, \ldots \tag{7.23}$$

where ρ_n is the radius of the nth ring, λ is the wavelength, and f_o is the main focal length of the zone plate that is the point at which the zone plate focuses the light of a collimated beam with maximum intensity and n is the index of the consecutive rings. Besides this main focal length, a Fresnel zone plate focuses the light of an incident collimated beam at a set of secondary focii, with corresponding focal lengths given by

$$f_k = \frac{f_o}{2k+1} \quad k = 1, 2, 3, \ldots, \tag{7.24}$$

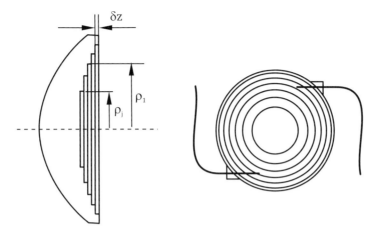

Figure 7.8 Profile (left) and top view (right) of a diffractive IOL with a Fresnel zone plate engraved on the posterior surface of the lens. The optical properties of this lens will depend on the radius of the central zone ρ_1 and the depth of each zone δz. This depth has been exaggerated in the drawing.

where k is the so-called difractive order of the Fresnel zone plate. Notice that the number of rings, and, thus, the number of focii are limited by the diameter of the zone plate. Instead of the set of alternately opaque/transparent rings, it is possible to engrave the Fresnel rings on the surface of a transparent material. In this case, all the rings are transparent, and the multiple image formation is obtained through the depth and the radius of each ring [216, 217]. A representation of an IOL with a convex anterior surface and a Fresnel zone plate engraved on the back surface is shown in Figure 7.8. Therefore, a diffractive IOL forms multiple images of a given object with different degrees of sharpness, the contrast of the final image being accordingly reduced.

Difractive IOLs are designed for a given wavelength, and they are very dispersive, presenting high longitudinal chromatic aberration. Fortunately, this aberration has the opposite sign to the aberration of the refractive elements of the eye, and it is even possible to compensate the chromatic aberration of the eye. The main drawback of difractive IOLs is the loss of contrast due to the distribution of the incident energy among multiple images. Another common problem is the presence of halos due to diffracted light coming from lateral light sources (such as streetlights or car's headlights). The design of diffractive IOLs is not restricted to zone plate designs. The high sophistication of the available numerical algorithms for designing and simulating DOE and the enhancement experienced by IOL manufacturing techniques allow for a wide range of multifocal diffractive IOLs [217]. Fresnel rings can also be engraved on a curved surface; in this case, the power of the surface is the sum of a refractive power given by the mean curvature of the lens and a diffractive power given by the Fresnel zone plate. This kind of design is commonly found in modern IOLs.

One of the most important topics in current multifocal IOL design is the extended depth of focus [218]. The depth of focus can be defined as the range of distance in the image space for which a well-focused image is formed by an optical system. For a given optical system, its depth of focus determines the focusing range, known as depth of field in the literature. The depth of field is the range of distances, measured from the optical system, at which an object can be placed so that the system produces a well-focused image of the object. Currently, there are three strategies to achieve an IOL with extended depth-of-focus [218]:

- *Reduced pupil IOLs*: It is well known that reducing the aperture of any optical system increases the depth of focus. This principle is used in IOLs with reduced pupils (or pinhole lenses), which have a small diameter aperture that only lets rays that are closer to the optical axis pass, while the remaining rays are blocked. An additional benefit of a reduced pupil is the diminishing of glare and halos.
- *Diffractive IOLs with chromatic compensation*: The compensation of spherical and chromatic aberrations of the eye results in an increment in the visual acuity of a an individual [219]. Therefore, a number of modern IOLs try to improve the depth of focus, compensating the spherical aberration and the chromatic aberration of the cornea [220, 221]. This type of IOLs present a complex design, in some cases with two different diffractive zones in order to manage the energy balance between the different focii [220, 222, 223].
- *Accommodating IOLs*: This last group of lenses tries to imitate the behavior of the natural human lens, which changes its power depending on the object distance. To do so, accommodating IOLs are made with flexible hooks (or haptics) that can push forward the optical portion of the IOL when the cilliary muscle is contracted [224]. Thus, the distance between the lens and the cornea is changed and the eye power varies. An alternative solution are IOLs made of flexible materials, which can be deformed when the cilliary muscles are contracted, behaving like the natural crystalline lens [218].

It is important to take into account that the optical quality of all multifocal IOLs is strongly affected by tilt and decentrations produced during their surgical insertion. In any case, increased research on multifocal IOLs is expected in the next few years, given the progressive aging of the world population with the consequent increase in patients with cataracts.

8

Multifocal Lenses

8.1 Introduction

What we expect from a lens intended to compensate for a refractive error at far vision, in most cases, is that its power remains constant at all gaze directions within the visual field provided by the frame. We saw in Chapter 6 that oblique aberrations may cause this power to deviate from its expected value, mainly toward the periphery of the lens, but also at its vertex or around it when tilts or ground prism are present. These power deviations are not intended; they are caused by oblique aberrations. However, there are other types of lenses for which we may want the power to purposely change from one point to another, or from one region to another. The changes can be continuous or discontinuous, but the fact is that these lenses have a plurality of powers, and so, a plurality of focii, hence the name *multifocal lenses*. Some of these lenses have a finite number of regions in which the power is intended to be constant. Among these lenses, those with just two regions are, by far, the more popular. They are called *bifocals*. Although scarcely used, there have been lenses with three regions, trifocals, and very rare designs with even more than three regions. Power can also be forced to change in a continuous way. Lenses having this characteristic are generally known as progressive power lenses (PPL). There is no universal agreement about the use of the term *multifocal*. Some authors use it only to refer to lenses with a finite number of constant-power regions, such as bifocals or trifocals. Others include progressive power lenses as a type of multifocal lens, where the prefix "multi" would mean "infinitely many." In this chapter we study both types of lenses, those with a finite number of constant-power regions and those with continuously varying power. As the name of the chapter suggests, we decided to use the term "multifocal" in the more general way. In the end, the name is not that important, and there are even lenses in which there is a region with continuously changing power and a region with constant power, a sort of mix between a progressive and a bifocal lens. With this extended acceptation, any lens with an intentional power variation, whether the variation is continuous, discontinuous, or both, can be named in a unified way.

What are these lenses for? Multifocal lenses were invented and are mainly used for the compensation of *presbyopia*, which is the loss of the ability to focus on near objects. Bifocals can also be used for some specific activities requiring special magnification at near vision, regardless of whether users are presbyopes or not, for example watchmakers or dentists. Nowadays, with the flexibility brought about by free-form technology, other applications can be thought of for multifocal lenses, for example special lenses for wearable devices, or for virtual reality systems, etc. By now, all these applications are just incidental if we compare them with the compensation of presbyopia. Indeed, presbyopia is the more prevalent refractive error as it is due to aging, a process we all are involved in, much more so if we take into account the aging tendency of modern societies. As presbyopia compensation is the main application for multifocal lenses, the first section of this chapter will be devoted to it. Then we will study bifocal lenses, and then progressive lenses. We will not dedicate separate sections to trifocals and other even rarer multifocals, as their properties can be computed using the same tools presented for bifocals, and their use is so limited. Both bifocal and progressive lenses have important advantages for the presbyope, but they also have strong limitations, if we compare the near and intermediate vision capabilities of the young, nonpresbyopic eye with those of the multifocal-compensated presbyopic one. These limitations are always driving the industry to develop new and sometimes quite disruptive lens designs for the presbyope. Although there are some innovative approaches, none of them offers performance levels, in terms of optics and comfort, beyond the well-known bifocal and progressive lenses.

8.2 Presbyopia and the Compensation of Near Vision

In any focused optical imaging system, object distance, focal length (or power), and image distance are linked by a mathematical relationship. If one of them changes and another is kept constant, the third must change. For example, if object distance changes and image distance is kept constant, the power of the imaging system must change in order to keep the image in focus. The emmetropic eye receives a focused image of distant objects. The ametropic eye may do so with the help of a compensating optical element, for example an ophthalmic lens. Both can also focus near objects by using an internal mechanism known as accommodation. Let us recall the physiology of the eye, which we presented in Chapter 5. The eye has two lenses, the cornea and the crystalline lens. In order to bring the image of a near object into focus, the eye could change in many different ways: The curvature of the cornea could increase, the axial length could also increase, the lens could axially move, and finally the lens could change its shape to increase its power. Each of these mechanisms has been considered by different authors, but from the experiments by Young back in 1801 and a lot of subsequent experimental and theoretical work, [225], it is widely accepted that any change other than the increase in power of the crystalline lens has a negligible effect on accommodation.

8.2.1 Accommodation and the Presbyopic Eye

Accommodation is amazingly complex. Although there is a proposal for an accommodation mechanism accepted by the majority of scientists, von Helmholtz's theory, still there are competing theories in debate. And even within the framework of von Helmholtz's theory, there are many details that are poorly known and open to debate. This state of the knowledge is due to the large number of structures playing a role in the accommodation process, the complexity of these structures, and the difficulty of getting "in vivo" accurate measurements of their geometrical, optical, and mechanical properties. Here we will present a rather brief and simplified description of accommodation and presbyopia, just the minimum required to get the chapter up and running. We recommend the interested reader to delve into the subject in books on physiological optics [52, 53], or in reviews of the subject, such as Charman [226].

The structures of the eye actively taking part in the accommodation process are, broadly speaking, the ciliary body including the ciliary muscle, the zonule, the capsule, and the crystalline lens; see Figure 8.1(a). According to the mechanism proposed by von Helmholtz and further developed by many others, the lens and the capsule surrounding it have elastic properties; when both elements are not subjected to external forces, the shape of the lens is imposed by the elastic properties of the capsule, and that shape corresponds to the fully accommodated state. The capsule is held in place within the eye by the fibrous zonule, which in turn is affected by the ciliary muscle. We will not describe here the detailed structure of the ciliary body; the important fact is that when the ciliary muscle is relaxed, the zonular fibers holding the capsule are under tension, and so they stretch the capsule and the lens inside it. This is the relaxed state in which the emmetropic eye is focused to infinity. When the ciliary muscle contracts, it moves forward and reduces its diameter, releasing the tension on the zonular fibers and allowing the capsule to force the lens into the accommodated state.

As the lens bends into the accommodated state, its surfaces become more curved, with smaller curvature radius, though the change is bigger at the front surface than at the back. The lens diameter is reduced, the vertex of the front surface bulges forward, and the lens thickness increases. According to the measurements reported by Rosales et al. on a group of 11 young people, [227], the averaged front radius of the lens changes from 11 mm

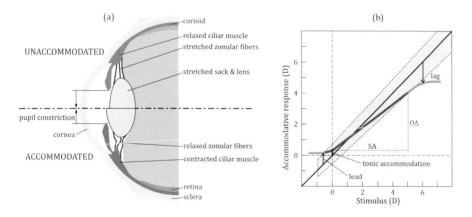

Figure 8.1 (a) Eye structures taking active part in the accommodation process. The upper half of the picture represents an unaccommodated eye, whereas the lower part represents the accommodated eye. (b) Accommodative response vs stimulus curve.

at the relaxed state to 6.75 mm under an accommodative stimulus of 8 D. Under the same circumstances, the back radius of the lens changes from 7 to 5.5 mm, with all the measurements having standard deviations of the order of 0.5 mm. Accommodation is linked to convergence and pupil constriction, the three of them known as the *near triad*. Part of the accommodation and the convergence effort are triggered by a feedback mechanism based on an error signal: accommodation is triggered by defocus caused when fixating a near object, and convergence by fixation disparity. The two feedback mechanisms are interconnected, so that an accommodative stimulus will trigger not only accommodation but also convergence, and the other way around, a convergence stimulus will trigger convergence and accommodation. Pupil constriction under accommodative stimulus is not linked to the other two responses in the same direct way and the relation is not fully understood. As pupil constriction increases depth of field, it probably helps in undertaking near work with smaller levels of accommodation.

We have just suggested that not all the accommodation may be caused by a response to defocus. There are also other mechanisms, with their own innervation ways, that contribute to the accommodation level of the eye. These mechanisms are known as the *components* of accommodation. Heath [228] identified four of them, and nowadays some authors add *voluntary accommodation* as a different one:

Reflex accommodation. Is the response of the accommodative system to a change in the vergence from the fixation object. It is controlled by the parasympathetic nervous system. The mechanisms identifying the direction that the accommodative system has to work in order to bring a defocused image into focus are not well known, but it seems longitudinal chromatic aberration of the eye could play an important role. Recently, some evidence has been found that the eye's spherical aberration could also be used.

Proximal accommodation. Also known as psychic accommodation. This component is triggered by the information the brain has about the distance to the object (whether correct or incorrect).

Convergence accommodation. This is a consequence of the convergence system trying to reduce the fixation disparity. As previously mentioned, this error signal also enters the accommodative feedback loop, producing a degree of accommodation.

Tonic accommodation. This is the accommodation level present when there are no visual stimuli, either in darkness or when the eye is presented with an empty field of view, without any structure that could trigger reflex, proximal, or convergence accommodation.

Voluntary accommodation. It is possible to train an individual to voluntarily change the amount of accommodation, either relaxing or increasing it. Heath and others considered voluntary convergence/accommodation (for example, when crossing eyes) as part of proximal accommodation, as somehow requiring "thinking on near." However, voluntary relaxation of accommodation could suggest an independent pathway.

A useful description of the accommodation mechanism is offered by the accommodative response vs stimulus curve, shown in Figure 8.1(b). The thin black line at 45° represents the "ideal" response: The accommodative system provides as many diopters of accommodation as required by the stimulus (a negative vergence from a near object or generated with negative lenses in front of the eye). In general, the response of an actual eye is not like that. For small stimuli, accommodation may be bigger than needed. This is called an accommodation *lead* and is related to tonic accommodation. For larger stimuli, the response of the accommodation system will *lag* behind the stimulus. The consequence of this is that the slope of the curve will be smaller than 1. There is a region around the ideal response in which, either with leads or lags of accommodation, the image seems to be in focus, mainly due to the depth of field (colored in light gray in the figure). When the response curve lies on this area, the image has no perceptible blur. The maximum accommodative effort that can be put in by the eye is called *amplitude of accommodation*. For stimuli larger than this amplitude, the response curve turns flat. It is important to take into account that accommodation lags may be present even within the range given by the amplitude of accommodation. The slope of the response curve will depend on a number of factors, the most important being the age of the individual, the luminance of the target, and its contrast.

The amplitude of accommodation decreases with age and this relationship has been thoroughly tested and measured by many researchers, see [226] and [53] and references therein. Cross-sectional studies in which clinical measurements of amplitude of accommodation are averaged among individuals of the same age predict an approximately linear reduction of the amplitude of accommodation with age until 40 years of age. Then the negative slope starts flattening until the age of about 60, at which accommodation becomes minimum and stable. This minimum value ranges from 1 to 2 diopters, depending on the clinical tests used by the researcher to determine both the refractive error and the amplitude of accommodation. However, by using a stigmatoscopic technique, Hamasaki and coworkers [229] demonstrated that this minimum accommodation was due to the depth of field of the eye linked to pupil size. Once the effect of depth of field was removed, Hamasaki's results, lately confirmed by Sun [230], strongly suggest that accommodation becomes negligible by the age of about 55. In 1965 Hofstetter published a longitudinal study of the amplitude of accommodation of two subjects over a period of eight years [231]. He found that amplitude decreased linearly with age, and then hypothesized that the smoothing slopes of cross-sectional studies could be caused by statistical artifacts. Charman [232] later analyzed Hamasaki's data under the same assumption, reaching the same conclusion as Hofstetter: linear reduction of the amplitude of accommodation with age and total loss of accommodation ability by the age of about 50.

Presbyopia can be defined as the condition by which the available accommodation of the aging eye is not enough to perform tasks requiring near vision. Lack of accommodation is compensated by "adding" some positive power to the presbyopic eye. In the case of emmetropia, plus lenses should be used to focus on near objects. For ametropic eyes, a determined lens power should be used to correct the refractive error at far vision, and a more positive lens should be used for near vision. The difference between the near vision prescription and the far vision prescription is called *addition*.

8.2.2 Near Vision and Lens Effectiveness

Let us consider eyes without astigmatism. The far point, $\mathcal{F}$, was defined in Chapter 5 as the conjugate with the retina when the eye is relaxed. In the same way, the *near point, $\mathcal{N}$,* is defined as the conjugate with the retina when the ciliary muscle is maximally contracted and the lens is left to its relaxed (accommodated) shape. The distance from the former to the later is called the *range of accommodation, $\mathcal{A}_r$*. The precise definition of the amplitude of accommodation, $\mathcal{A}_a$, is the difference between the vergences from the far and near points, both computed at some reference plane. If we use the spectacle plane, S, and call these vergences $V_F = 1/\overline{SF}$, and $V_N = 1/\overline{SN}$, then the amplitude of accommodation referred to the spectacle plane is $\mathcal{A}_a = V_F - V_N$. The astigmatic eye has two pairs of focal images conjugate with the retina, the first one for the unaccommodated state and the second one for the maximally accommodated state. The study of accommodation in astigmatic eyes is a complex subject. Questions about the variation of astigmatism with accommodation, as well as the effect of astigmatism in accommodation response, have been addressed for a long time and by many authors, with the results having a lot of variation, probably due to the different approaches and techniques employed and the variability of the changes across subjects, see for example the general descriptions at [226, 233] and the more detailed works of Brzezinski, Ukay, Stark, and Radhakrishnan, [234, 235, 236, 237]. On average, with-the-rule astigmatism of the uncorrected eye seems to grow with accommodation at a rate of about 0.04 D per diopter of accommodation. Besides this intrinsic variation, Brzezinski acknowledged the variation of lens effectivity at near vision as one of the possible needs for change in astigmatic prescription. Of course, this variation of effectiveness has been known for a long time, see for example Hofstetter [238]. As this is significant to the next section, we will derive a simple expression for lens effectiveness at near vision using the tools presented in Chapters 3 and 4.

Although matrix description of astigmatism allows us to propagate vergences and compute effectiveness in a compact way, we need some approximations if we want to get to manageable and useful expressions. In his analysis of lens effectiveness at near vision, Hofstetter referred the refractive error to the corneal vertex, assuming that a large percentage of astigmatism originates at the cornea and hence the astigmatism at the corneal vertex would not change under accommodation. This type of analysis just allows for the computation of the change in astigmatism at the spectacle plane, and does not relate it to accommodation. To keep track of the possible changes that could also affect the spherical component of the refractive error, and to relate them to accommodation, we will refer the refractive error to the principal plane of the eye. Of course, the astigmatic eye lacks true principal planes (see Section 4.3.5); we can rather say that there will be a pair of principal planes per main meridian. Besides, these principal planes move closer to the crystalline lens with accommodation. Nevertheless, we may neglect these shifts, which will take place well within 1 mm. In this way, if $\mathbb{P}_e$ is the equivalent power of the relaxed eye, then $\mathbb{P}_e + \mathcal{A}\mathbb{I}$

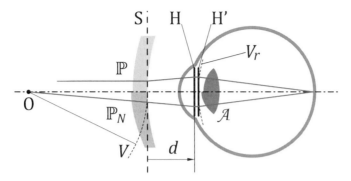

Figure 8.2 Schematic paraxial ray tracing to compute lens effectiveness at near vision. The upper half of the drawing corresponds to the unaccommodated eye using the right lens to compensate its refractive error (far vision). In the lower half, we see the eye using some accommodation $\mathcal{A}$ to bring object O into focus.

will be the equivalent power of the accommodated eye. The proposed approximation is equivalent to substituting the real eye (or any *exact* schematic eye model) by Emsley's reduced eye keeping the retinas of the two models coincident. In the reduced eye all the principal planes are actually located at the corneal vertex, approximately 22 mm in front of the retina. Accommodation of the reduced eye should be accomplished by an equal increase of curvature of the two corneal main meridians. In any exact schematic eye model, the principal planes would lie between 22 and 22.5 mm from the retina, depending on the accommodation and the equivalent power of each meridian. On the other hand, we will assume the spectacle lens to be thin. This approximation can be removed with more ease than that on the position of the principal planes, but we will still keep it to derive clearer expressions. In any case, thorough paraxial ray tracing demonstrates that the next, simplified, approach is accurate within a few cents of diopter.

Let's consider then the paraxial layout shown in Figure 8.2. As mentioned in the previous paragraph, we assume the lens to be thin and located at plane S, and we assume the principal planes of the eye to be located at H and H', for both the two main meridians and for any value of accommodation. The lens power that compensates for the refractive error of the eye is $\mathbb{P}$, and the coupling distance from the spectacle plane to the front principal plane of the eye, $d = \overline{SH}$. Our hypothesis is that for near vision we may need a different power, $\mathbb{P}_N$, and we want to relate it to lens power, accommodation, coupling distance, and vergence from the near object at the spectacle plane, V. At far vision, the output vergence from the lens is $\mathbb{P}$, which upon reaching H has changed to $\mathbb{P}(1 - d\mathbb{P})^{-1}$. Refraction through the eye gives

$$n_v V_r \mathbb{I} = \mathbb{P}_e + \mathbb{P}(1 - d\mathbb{P})^{-1},$$

where n_v is the refraction index of the vitreum and V_r is the spherical vergence at H' of the beam focusing on the retina. For near vision, the output vergence from the lens is $\mathbb{P}_N + V\mathbb{I}$, so the imaging equation is now

$$n_v V_r \mathbb{I} = \mathbb{P}_e + \mathcal{A}\mathbb{I} + (\mathbb{P}_N + V\mathbb{I})[1 - d(\mathbb{P}_N + V\mathbb{I})]^{-1}.$$

From both equations we get the result

$$\mathbb{P}(1 - d\mathbb{P})^{-1} = \mathcal{A}\mathbb{I} + (\mathbb{P}_N + V\mathbb{I})[1 - d(\mathbb{P}_N + V\mathbb{I})]^{-1}. \tag{8.1}$$

As $\mathbb{P}$ and $\mathbb{P}_N$ have the same main meridians (lens effectiveness will not change their orientation), the matrices in the previous equation and their inverses commute, and it can be solved for $\mathbb{P}_N$,

$$\mathbb{P}_N = [\mathbb{P} - V\mathbb{I} - \mathcal{A}(1 - dV)(\mathbb{I} - d\mathbb{P})][1 - d\mathcal{A}(\mathbb{I} - d\mathbb{P})]^{-1}. \tag{8.2}$$

Far more useful than the previous expression is its Taylor expansion up to first order on d,

$$\mathbb{P}_N = \mathbb{P} - (\mathcal{A} + V)\mathbb{I} - d\mathcal{A}^2 + d\mathcal{A}\mathbb{P}, \tag{8.3}$$

from which we can draw some conclusions. First, if we neglect d and the accommodation equals the stimulus, $\mathcal{A} = -V$, then $\mathbb{P}_N$ equals $\mathbb{P}$. Second, if we do not neglect the coupling distance but still $\mathcal{A} = -V$, then lens effectiveness would require a near vision power $\mathbb{P}_N = \mathbb{P}(1 - dV) - dV^2$. The change with respect to $\mathbb{P}$ is a shift dV^2 that affects the spherical component and a proportionality factor $(1 - dV)$ that affects both sphere and cylinder. However, there is no need for the accommodation to perfectly match the stimulus; we could consider equation (8.3) the other way around and obtain from it the eye accommodation that brings $\mathbb{P}_N$ as close as possible to $\mathbb{P}$. Let us split the lens power into its spherical and cylindrical components, $\mathbb{P} = S\mathbb{I} + \mathbb{C}$, $\mathbb{P}_N = S_N\mathbb{I} + \mathbb{C}_N$ (see Chapter 4, equation (4.28)) and let us substitute these into equation (8.2). The equation itself then splits into a scalar equation for the spherical components

$$S_N = -(\mathcal{A} + V) - d\mathcal{A}^2 + (1 + d\mathcal{A})S,$$

and a matrix equation for the cylindrical components

$$\mathbb{C}_N = (1 + d\mathcal{A})\mathbb{C}.$$

From the scalar equation we may solve for the accommodation value that will make $S_N = S$. That is a second-order equation on $\mathcal{A}$ with solution

$$\mathcal{A} = \frac{2dS - 1 + \sqrt{(2dS - 1)^2 - 4dV}}{2d}$$
$$\simeq -V\left[1 + d\left(2S + V^2\right)\right], \tag{8.4}$$

and we finally may substitute this accommodation value to compute the correction needed for the cylinder, which simplifies to $\mathbb{C}_N = (1 + dV)\mathbb{C}$ if we neglect second- and higher-order terms in d. These equations are illustrated in Figure 8.3. The plot to the left shows the accommodation demand for a near object vergence $V = -3$ D as a function of the spherical component of the refractive error, assuming this sphere is kept the same for far and near vision. With no spherical refractive error the accommodative demand is slightly smaller than the vergence at the spectacle plane, as a consequence of the vergence propagation $V/(1 - dV)$. But the interesting effect is the dependence with S. For a given object

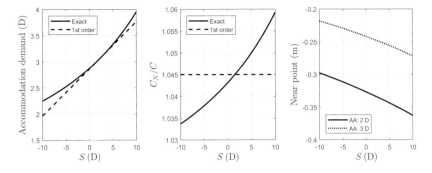

Figure 8.3 Lens effectiveness at near vision. The plot to the left shows the accommodation demand for a stimulus of −3 D as a function of the spherical component of the prescription. Very low hyperopes and myopes require less accommodative effort to focus a near object. The central plot shows the relative variation of cylinder prescription at near vision with respect to far vision. We can see that the cylinder required at near vision is always bigger than the cylinder needed at far vision, with the effect being bigger for hyperopes than for myopes (in this case for a stimulus of −3 D). In both plots, the continuous line represents the magnitudes computed from equation (8.2), whereas the dashed line represents the magnitudes computed up to first order in the coupling distance. The plot to the right shows the position of the near point as a function of the prescription for amplitudes of accommodation of 2 D (dotted) and 3 D (solid).

vergence V, myopic eyes will require a considerably smaller accommodative demand than hyperopic eyes. It is because of this effect that among early presbyopes, myopes corrected with glasses require less addition than hyperopes, or also they seem to need compensation for presbyopia at a later stage. Effectiveness of cylinder is shown in the central plot. Here we plot the factor $(1 - d\mathcal{A})$ and its first-order approximation, $(1 - dV)$. The factor is bigger than 1 for any value of the prescription. This means the ideal prescription for the cylinder at near vision should always be slightly larger than for far vision. Of course, the effect is small, between 4% and 5% for most prescriptions. A change of cylinder in 0.25 D would require a prescription cylinder above 3 D (after rounding to the closer quarter of diopter). As we can see, the correction for cylinder effectiveness is negligible for most prescriptions, and the interest is mainly academic.

8.2.3 Compensation of Presbyopia: Ranges of Clear Vision

The effect of presbyopia is the reduction of both the range and amplitude of accommodation with age. At some point along the aging process, the amplitude of accommodation will become smaller than the accommodation demand for near work. At that time, the individual will no longer be able to sharply focus near objects. The addition of some positive power to the regular prescription will reduce the accommodation demand, and will extend the range of clear vision, bringing the near point closer to the eye and back to the working region. We will use the concepts and developments of the previous section to compute this new range of accommodation, when the ametropic eye is fully corrected but some positive spherical power is added to the compensating lens, as in any multifocal lens that we will study later.

Instead of computing the accommodative demand, we will examine range of clear vision as a function of the amplitude of accommodation. We start from equation (8.2), but in this case we substitute $\mathbb{P}_N$ by $\mathbb{P} + A\mathbb{I}$, where A is the spherical power added to the prescription lens. Also, once the cylinder is compensated, the location of the point conjugate with the retina only depends on the spherical component of the prescription. Changing $\mathbb{P}$ by S, $\mathbb{P}_N$ by $S + A$, reducing the equation to scalar, and solving for the vergence from the object, we have

$$V = \frac{-A - \mathcal{A}(1 - dS)\,[1 - d(S + A)]}{1 - d\mathcal{A}(1 - dS)}. \tag{8.5}$$

If we neglect the coupling distance d, we obtain the approximate and expected result $V = -(A + \mathcal{A})$, that is, the eye will focus at a distance that is the inverse of addition plus accommodation. Also, if accommodation is zero or very small, then $V = -A$, with the coupling distance playing no role, as we are referring V to the spectacle plane. The first-order approximation to the solution for V is

$$V \simeq -(A + \mathcal{A}) - \mathcal{A}^2 d + 2\mathcal{A}Sd. \tag{8.6}$$

The main effect of the coupling distance on the range of clear vision is that it shifts the vergences of the near point by $2\mathcal{A}Sd$, which depends on the prescription. For myopes, the shifted near point would be closer to the eye, with the opposite being true for hyperopes. Of course, this is the same effect as the variation of the accommodative demand with the prescription we studied in the previous section. The effect fades away as presbyopia progresses and the amplitude of accommodation reaches zero, the range of clear vision reducing to $-1/A$ plus minus the depth of field. The position of the near point as a function of the spherical ametropia is shown at the rightmost plot of Figure 8.3, for amplitudes of accommodation of 2 and 3 D and for a coupling distance of 15 mm.

In summary, compensation of presbyopia requires the addition of some positive power to the far vision prescription that will extend the range of clear vision of the presbyopic eye. For very high astigmatic prescriptions the addition could incorporate some cylindrical power to compensate for lens effectiveness, but as we have seen the effect is rather small, and probably other effects such as oblique astigmatism or variation of eye astigmatism with addition could be higher. In general, the addition is assumed to be purely spherical and we will adhere to that assumption.

Example 37 Let us compute the range of clear vision for both myopic and hyperopic eyes of -5 and $+5$ D having amplitude of accommodation of 3 D, if they are corrected with spectacle lenses having 1 D of addition and with a vertex distance of 12 mm.

We may assume the principal object plane lies 3 mm behind the corneal vertex, so $d = 15$ mm in this example. Now, with no accommodation the far point would be at $1/A = 1$ m for both the myopic and hyperopic eye. At maximum accommodation, we can either use (8.5) or (8.6) to compute the vergence, getting $V = -4.64$ D for the myope and $V = -3.68$ D for the hyperope. That would mean a range of clear vision from 22 cm to 1 m for the myopic eye and 27 cm to 1 m for the hyperopic eye. This interval does not include depth of field, so in a practical case it would be larger. We may also consider what would happen if we remove the addition. In such a case the myopic eye would have its near point vergence at $V = -3.64$ D, whereas for the hyperopic eye $V = -2.68$ D, with corresponding near point distances of 27 and 37 cm, respectively. Without addition the myopic eye could handle near vision well in most environments (even under a low-light, large pupil situation), while the hyperopic eye would be in trouble at near distance, specially under low-light levels.

8.3 Bifocal Lenses

According to the previous discussion, the presbyopic eye needs, at least, two different prescriptions: one for far vision and the other for near vision, the difference between them being the positive spherical power added to the later. If the presbyope is emmetrope, then only eyeglasses mounting positive spherical lenses would be needed for near vision, and these are known as *readers*. Otherwise, the presbyope will require two eyeglasses, one for far vision mounting the prescriptions $\mathbb{P}_R$ and $\mathbb{P}_L$ for the right and left eye, respectively, and another pair for near vision with the corresponding prescriptions $\mathbb{P}_R + A\mathbb{I}$ and $\mathbb{P}_L + A\mathbb{I}$. In terms of sphere and cylinder, the far vision prescription for either eye would be $[S, C \times \alpha]$, whereas for the near vision prescription it would read $[S + A, C \times \alpha]$. The presbyope needs both eyeglasses on hand, and needs to alternate them whenever the visual task requires it. It is easy to imagine how uncomfortable this is, especially in activity environments in which frequent and quick changes between far vision and near vision are needed. No doubt this led to the invention and diffusion of the bifocal lens, in which there are two distinct regions with different power. One region has a prescription for far vision, and the other has a prescription for near vision. Just changing the gaze and possibly adjusting the head will allow the user of these lenses to select the right prescription.

The invention of bifocal spectacles is credited to Benjamin Franklin, by the end of the eighteenth century [239, 240]. From letters sent to and from friends, it is pretty clear that he was using bifocals by 1785, and it is possible that he had been using them many years before. It is also clear that he helped introducing bifocals in the United States by the beginning of the nineteenth century, as his name was associated with their invention as soon as 1790 [240]. However, there is no sound evidence that he, or any other famous contemporary, were the actual inventors. Split lenses were already being produced in London for artistic purposes by the fourth quarter of the eighteenth century, and it is possible that Franklin saw them and ordered and adapted his own for visual purposes. At that time, the name used for such eyeglasses was split lenses, double spectacles, or divided glasses. The term bifocal was coined by John I. Hawkins, who issued the first patent for trifocal lenses in 1827.

The Franklin bifocal was made by cutting in half the round lenses for far and near vision prescriptions, and combining each half in a single frame ring. The far vision half would sit in the upper part of the ring, and the near vision in the lower part. From the optical point of view, this is a good solution: Each lens can be manufactured with the adequate base curve and the location of the optical centers of each portion of the bifocal can be easily controlled. With appropriate technology, the upper and lower parts of the bifocal lens could be cut to any desired shape. But the combination making a Franklin bifocal had some practical drawbacks, specially for the technological level of the nineteenth century: The mounting process was laborious, the assembly was unstable, the half lenses difficult to hold in the frame, even if they were cemented to each other, and the edge along which the half lenses were joined together tended to scatter light and collect dirt. These concerns led to the development of many new types of bifocal lenses throughout the nineteenth century and the first half of the twentieth. We will not give a thorough account of the large number of bifocal and trifocal lenses developed since 1800, but a glimpse of the soundest designs that set the course of development of multifocal lenses. The interested reader may consult the excellent book by Rosenthal or other classical books on ophthalmic lenses [132, 241] for a deeper look into bifocal technology and its developments.

8.3.1 Introduction

We can divide bifocal designs into two main groups: one-piece bifocals manufactured with a single piece of a single material, and bifocals made from two or more pieces of the same or different materials. Though Franklin split lenses obviously belonged to the second type, one-piece bifocals are not a modern development; quite the contrary, designs from either group have appeared interspersed for the last 200 years. The common feature shared by one-piece bifocals is that either their front or back surface presents two regions with different curvature, while the other is continuous. Depending on the geometry of the split surface, the edge between the two regions can be more or less conspicuous. On the other hand, bifocals made with two or more pieces can be divided into split, cemented, and fused bifocals. Split bifocals (including Franklin's and some modifications of them) were in use for approximately the whole nineteenth century, then cemented bifocals and fused bifocals appeared at the end of this century and the beginning of the twentieth. Patents for the first cemented bifocal and the first fused bifocal were issued by A. Morck in 1888 and J. Borsch in 1908, respectively. Both technologies used a principal or major lens and a small one providing the addition, called the *segment*, either cemented or fused to the main one.

In Morck's cemented bifocal the segment was made with the same glass as the main lens and was cemented to the bottom of its back surface. This cemented lens is depicted in Figure 8.4. The front surface of the segment had the same curvature as the back surface of the main lens. The back surface of the segment was concave for most prescriptions and base curves, so as much as the edges of the segment were well crafted, the near region was less conspicuous than a split bifocal. The adhesive used for cementing the segment was Canada balsam, a turpentine-based natural glue used at that time in the optical industry. However, Canada balsam aged quickly, developing some yellowness, and was sensitive to heat. Because of these reasons, cemented bifocals gave way to fused bifocals.

Fusing technology employs a main lens intended for far vision, but in an unfinished form. On either of its surfaces, typically the front, a depression with a different curvature is

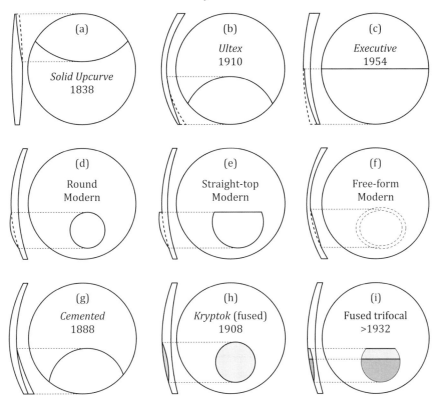

Figure 8.4 Different types of bifocal lenses. The first two rows, (a) to (f), are one-piece bifocals. In the third row, (g) is the cemented bifocal patented by August Morck, (h) is the Kryptok bifocal patented by John Borsch Jr., and (i) illustrates a modern fused trifocal.

ground and polished, and on this depression a segment made with a glass having a higher refractive index is positioned. As expected, the curvature of the side of the segment in touch with the main lens basically matches that of the depression. Once the segment is in place, the set is exposed to high temperature, 600 to 700° C depending on the glasses used for the two parts. At this temperature a fusion process happens at the interface between the two glasses, combining them as if they were a one-piece. Once the fusion is done and the lens cooled, the whole surface is reshaped, generating exactly the same curvature on the main lens and the segment. This leaves a round segment buried into the main lens with a perfect knife-edge, so perfect that it is hard to detect by the inexperienced observer. In modern fused bifocals, the back surface of the lens just produced is still to be surfaced to the right prescription, so the process just described generates a fused bifocal semifinished blank. This type of fused bifocal is depicted in Figure 8.4(h), the grayed segment indicating a different (higher) refractive index.

In 1905, Henry Courmettes patented a variation of the fusing process that allowed for noncircular segments. Courmettes constructed the segment as a split lens, the lower portion having the higher refractive index, and the upper portion with the same material as the main lens. After fusing this split segment with the main lens, the upper portion simply ceased to be something different than the main lens, as it comes to the same material with the same refractive index and an interface between the two parts that just vanish with the fusion process. That left a nonround higher refraction index segment buried in the main lens. The noncircular part of the segment contour was always at the top of the segment and could be straight or slightly curved. These types of segments were called "D" shaped segments. Variations of the Courmettes technique gave way, in the twentieth century, to many different types of fused bifocal segment shapes, some of them also made with different sizes. Similarly, another variation of the Courmettes technique gave way to the trifocal lens, with the patent issued to Virgil H. Hancock in 1934. In Hancock's patent, the split segment was also formed by two different materials, one for the upper part with a refractive index in between the index of the lower part and that of the principal lens. The resulting segment would be round, but it would have two different regions with two different materials. Full addition would be obtained at the lower region, and intermediate addition at the upper region. Figure 8.4(i) shows a further variation of Hancock's trifocal. The lens shown has a two-region segment with a straight top. To achieve this shape, the segment is made from three parts: the lower with the highest refractive index, the intermediate with an intermediate refractive index, and the upper made with the same glass as the main lens. Once again, this upper portion of the segment would melt into the main lens, its frontiers with it disappearing, and leaving as the visible top of the segment its frontier with the intermediate-index material.

Examples of different one-piece bifocals are shown in Figure 8.4 (a) to (f). As mentioned before, these lenses are made from a single piece of glass or plastic, the near region formed by providing a different curvature to some region of either the front or the back surface of the lens. The dashed line in each of the six sections indicates what the main surface would be if the bifocal modification was not there. The first lens in subplot (a) is the *Solid Upcurve* invented by Schnaitmann in 1838. The manufacturing process started with a standard lens having the prescription for near vision, typically a biconvex lens. The upper part of its back surface was reshaped to a flatter curve, in most cases a concave one, leaving a region with the prescription for distance vision. The edge of this region, the intersection between the lower spherical portion and the also spherical upper portion is circular, curving up. This lens is worth mentioning as the first one-piece bifocal, but it had serious problems. First, the geometry of the lens was strongly limited by the surfacing capabilities of the time, and this imposed a limitation on the available range of prescriptions. Second, the distance vision region was affected by a strong base-down prismatic effect. Its great advantage was that, once manufactured, it could be handled and glazed like any single vision lens, simplifying the laborious mounting process of split bifocals.

Since the introduction of the *Upcurve* bifocal, there were many patents issued for manufacturing one-piece bifocals, but none of them succeeded until Charles Conner's invention in 1909. His patent gave way one year later to the *Ultex*, represented in Figure 8.4(b). The bifocal was cut on the back (concave) side of the lens, so in order to get positive addition, the curvature had to be reduced in the near region. This was achieved by producing a large saucer-like glass disk with revolution symmetry, a circular central region with lower curvature, and an external ring-shaped region with larger curvature. From each disk two *Ultex* lenses could be glazed. The front side of the lens would then be resurfaced to the right prescription, with external torus if astigmatism was present. The optical center of the segment was typically at the bottom of the near region, so it would produce a base-down prism at the reading position. Probably one of the most successful bifocals was the *Executive*, introduced in 1954 by American Optical, and shown in Figure 8.4(c). The *Executive* is a one-piece bifocal with the near region on the front surface, meaning it requires higher curvature than the far region. The line separating the two regions is totally straight, along the horizontal diameter of the lens, meaning there must be a ledge separating the two regions, which is its main disadvantage. However, the existence of this ledge provides two important optical advantages: First, the optical centers of the far and near regions are very close to the ledge, leading to a negligible change of prismatic effect across the dividing line (called *image jump*, as we will see later); second, the field of view of the near region is as wide as in a single vision lens. The *Executive* became so popular that once the patents expired, many manufacturers started making their own, using the generic term *Franklin-style* bifocals.

As manufacturing techniques improved, and especially as plastic lenses started generalizing, different types of one-piece bifocals with round and straight-top segments appeared in the second half of the twentieth century. Examples are shown in Figure 8.4 (d) and (e). These segment configurations followed those developed some years earlier with fusing technology, and their optical properties were pretty much the same, except for the higher chromatic aberration of fused bifocals, especially for negative prescriptions. Another important difference is that the surface of a fused bifocal in which the segment is embedded is smooth and with continuous curvature, while in those one-piece bifocals we have seen so far, a sudden change of curvature, with or without a ledge, is always found at the dividing line. It is no surprise that there were many efforts to create a lens with a smooth transition between the far and near regions. A blending ring continuously joining the two regions would make the line disappear, with the resulting bifocal being more esthetic and free from the light scattering that takes place at discontinuities of sag, orientation, or curvature. Examples of these attempts were the patents by L. Bugbee and H. Beach, issued in 1923 and 1946, respectively. However, these ideas could not be applied until Norman and Irving Rips patented in 1955 a method more simple and efficient for doing so, leading to the *Seamless* bifocal, a lens still available today. These lenses are front-side bifocals, and in order to smoothly connect the larger curvature at the near region with the lower one at the far region, the connecting ring undergoes continuous but strong curvature changes, even with two changes of sign of the curvature along the direction perpendicular to the ring. This means vision through the ring is not possible, as the image would be extremely distorted. This is not a real hindrance, as the ring is just 2 or 3 mm wide.

Finally, the same idea of the *Seamless* bifocal is used today on the back surface of modern lenses surfaced with free-form technology, as shown in Figure 8.4(i). As more and more types of bifocal lenses were developed since their invention, some of them had the segment on the front surface (for example, the *Executive*, the *Kryptok*, and most fused bifocals, the *Seamless*, etc.), and others had the segment on the back (*Cemented*, *Ultex*, some fused bifocals, etc.) As the manufacturing started to heavily rely on front-side blanks, most bifocals with the segment on the back surface were retired. Since the sixties to approximately 2005, nearly all bifocals were made as front-side blanks, with the back surface resurfaced at the RX-lab as a spherical or toric surface to provide the lens with the correct power (prescription). Addition cannot be modified for these types of blanks, as it is a property of the front surface. Free-form technology is partially changing this way of manufacturing. A one-piece bifocal can be surfaced at the RX-lab starting from a spherical blank and cutting the back surface with a bifocal design. The overall geometry for these lenses is similar to that of the *Ultex*: The segment is a region on the back surface in which the curvature is smaller than in the rest of the main lens. However, there are two big differences with respect to the *Ultex*. First, if the software computing the back surface allows for it, the manufacturer of free-form bifocals may have a good degree of control of the shape of the segment. Typically, they can be made round or elliptical, and their size and exact location can be controlled as well. This means each lens can be personalized to the desire of the eye care professional (ECP), or to the needs of the final user. Second, as free-form bifocals are back-side, the region in the segment has to be flattened with respect to the rest of the surface. This means the transition between the far region and the near region is smoother than in a front-side *Seamless* bifocal, with the curvature along the radial direction varying without change of sign. The third main difference between a free-form bifocal and any other previous type is that addition and astigmatism are located at the same surface, so both the far and near regions of the back surface may be astigmatic and aspherical. The differences between the blending curvatures in front-side and back-side bifocals is shown in Figure 8.5.

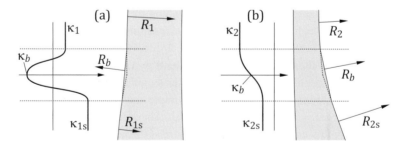

Figure 8.5 Different behavior of blending geometry in front-side, (a), and back-side, (b), blended bifocals. The plots represent vertical section of blended bifocals. The blending region is contained between the dashed horizontal lines. The approximately vertical dashed lines represent the surfaces if no blending were present. In a front-side blended bifocal with full round segment, the curvature of the blending region must become negative, forcing a great change in curvature in a narrow blending region. In a back-side blended bifocal, the curvature changes much more smoothly between the far and near region. These allow for narrower blending regions, or better optical properties of the same.

Throughout the second half of the twentieth century, many specialty bifocals were developed for special applications, either with fusing or one-piece technologies. A fine example are bifocals with two segments, one at the top and one at the bottom of the lens, for activities that require near vision with both upward and downward gazing (e.g. pilots, mechanics, electricians, etc.). Another example are *minus add lenses*. These are near vision lenses with a relatively small segment, typically at the top, in which the plus power of the near prescription is reduced to provide clear vision at intermediate or far distances. See [132] for more details.

Bifocal lenses currently in use

Bifocals have been the only multifocal solution for presbyopes from their discovery to the last quarter of the twentieth century, when progressive power lenses started to overtake them in the market. Among the available compensations for presbyopia, the percentage constituted by bifocal lenses has continuously been shrinking since then. Another factor contributing to this trend is the strong reduction of the use of glass as an optical material for ophthalmic lenses. It is difficult to say that some technology or material has been completely replaced, as there may be labs sticking with it because of special requests from particular ECPs. Also, the presence of bifocals in the market depends on the country. For example, in some Western European countries bifocals amount to a small percentage of multifocal lenses, whereas in Germany, or in the United States, they still have a significant market share, though smaller than that of progressive lenses.

In the mid-twentieth century, the RX-lens industry became more uniform in making lenses from front-side blanks (see Chapter 10) and by the last quarter of the century almost all RX-labs produced finished lenses by resurfacing the back surface of front-side semifinished blanks. Because of this, back-side bifocals such as the original *Ultex* practically disappeared from the industry, and all the bifocal geometries, whether in glass or plastic, fused or one-piece, were manufactured as front-side bifocal blanks, to be stored at RX-labs, and to be surfaced on the back with a proper spherical, toric, or, in some cases,

Table 8.1 *Summary of bifocal lenses currently in use (ST and MST stands for straight-top and modified straight-top).*

Location	Material	Shape	Technology
Front-side	Glass	Round ST/MST	Fused
		Franklin type	One-piece
	Plastic	Round Seamless ST/MST Franklin type Specialty	One-piece (cast)
Back-side	Plastic	Various	One-piece (FF)

atoric surface to create the finished lens with the user prescription. Free-form technology is changing this paradigm. It allows the RX-lab to buy a less expensive spherical blank and reshape the back surface with a progressive or bifocal geometry, with this surface incorporating the prescription data. The great advantage of this approach is that the amount of stock the RX-lab needs is greatly reduced, as the same blank can be used to produce many different segment geometries and all the necessary add powers. Besides, a free-form lab can resurface standard front-side bifocal blanks with toric or atoric back surfaces, tailored to the prescription and other custom parameters of the user.

8.3.2 Thin Lens Model of Bifocal Lenses: Power Relations

The performance of bifocal lenses is determined by their optical properties and by the shape and location of the segment. Main optical properties would include power, prismatic effect and oblique aberrations, as in any single vision lens.[1] There is an important concern though: With single vision lenses the user is free to look through any point of the lens. Thus, if the base curve is not well selected and the lens has large oblique aberrations, the user can still look through the optical center to avoid them. If the location of the target to be fixated requires an oblique gaze, the user can move the head so the visual axis aiming at the target passes through or close to the optical center of the lens. A presbyope using bifocal lenses *is forced to look through the segment* in order to achieve the correct power for near vision. As the segment is typically out of the main viewing direction (which is reserved for far vision), looking through it will require a certain amount of obliquity. Then, if the geometry of the lens produces oblique aberration at the segment, the user will simply be unable to avoid it. Something similar can be said about prism. In a single vision lens the prismatic effect grows from its optical center according to Prentice's rule. In bifocal lenses, the shape of the segment, its position, and eventually the tilt of its surface determine the prism that the user will experience at near vision.

Using a thin-lens model, the relationship between the curvatures of the surfaces of a bifocal lens and its far and near prescription is straightforward. We will assume the spherocylindrical form of the far vision prescription is $[S, C \times \alpha]$, whereas for the near vision we have $[S + A, C \times \alpha]$. In matrix form, we will use $\mathbb{P}$ for the far power and $\mathbb{P}_N = \mathbb{P} + A\mathbb{I}$ for the near power. As usual, it is more useful to split the complete power matrix into the spherical and cylindrical components, $\mathbb{P} = S\mathbb{I} + \mathbb{C}$ and $\mathbb{P}_N = (S + A)\mathbb{I} + \mathbb{C}$. Let us analyze each of the three types of bifocals represented in Figure 8.6:

Front-side, one-piece. Both the far and near regions at the front surfaces are spherical. Their refractive powers are, respectively, $P_1 = (n - 1)\kappa_1$ and $P_{1s} = (n - 1)\kappa_{1s}$. The former is generally referred to as the lens base curve, and we will also denote it with the letter B. The base curve, or simply the base, is a degree of freedom for the manufacturer when choosing the semifinished blank. Any power derived for

---

[1] Transverse chromatic aberration can also be important in bifocal lenses, especially in negative fused bifocals. See [56].

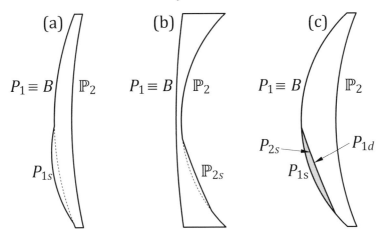

Figure 8.6 Refractive powers determining the prescription and addition of the bifocal lenses currently produced. (a) One-piece, front-side bifocal. The figure represents a round segment, but it could be Franklin-type, straight-top, modified straight-top, seamless, etc. (b) One-piece, back-side bifocal, produced by free-form technology. (c) Fused bifocal, only available as a front-side lens. The curvature of the surfaces has been selected for drawing purposes, but any power from negative to positive within the range of manufacturing capabilities can be made with any of the three technologies.

the lens will come as a function of B. Using thin-lens approximation, it is clear that $S\mathbb{I} + \mathbb{C} = B + \mathbb{P}_2$, and so we get the required surface on the back

$$\mathbb{P}_2 = (S - B)\mathbb{I} + \mathbb{C}.$$

Obviously, the back surface is totally defined by the far prescription and the base curve, so it has nothing to do with the addition. For the near region we have $(S + A)\mathbb{I} + \mathbb{C} = P_{1s}\mathbb{I} + \mathbb{P}_2$, so comparing the two equations we get $S - B = S + A - P_{1s}$, that is

$$A = P_{1s} - B$$
$$= (n - 1)(\kappa_{1s} - \kappa_1). \tag{8.7}$$

Back-side, one-piece. For this type of lenses, the front surface is spherical and so we have thin-lens equations for the far and near regions, $B + \mathbb{P}_2 = S\mathbb{I} + \mathbb{C}$, and $B + \mathbb{P}_{2s} = (S + A)\mathbb{I} + \mathbb{C}$. Subtracting both equations we get $\mathbb{P}_{2s} = \mathbb{P}_2 + A\mathbb{I}$, that is

$$\mathbb{P}_{2s} = (S - B + A)\mathbb{I} + \mathbb{C}. \tag{8.8}$$

As expected, the cylindrical component of the back surface is identical in both the near and far regions, the only difference being the spherical component. The previous matrix equation holds for any reference system, but if we use the one parallel to the main meridians (X axis parallel to the cylinder axis, as usual), then

$$\mathbb{P}_{2s}|_\otimes = \begin{bmatrix} S - B + A & 0 \\ 0 & S - B + A \end{bmatrix} + \begin{bmatrix} 0 & 0 \\ 0 & C \end{bmatrix}_\otimes.$$

Thus, along the axis meridian the back surface power is $S-B+C$ and $S-B+A+C$ along the cross-axis meridian.

Front-side, fused. The analysis is now somewhat more complex, as there are four surfaces involved. Still, within thin-lens approximation is just a matter of adding and subtracting refractive powers. At the far region we have the same equation than in previous lenses, $S\mathbb{I} + \mathbb{C} = B\mathbb{I} + \mathbb{P}_2$. At the near region we have

$$(S + A)\mathbb{I} + \mathbb{C} = (P_{1s} + P_{2s} + P_{1d})\,\mathbb{I} + \mathbb{P}_2,$$

where P_{1s} and P_{2s} are the refractive powers of the two surfaces of the segment and P_{1d} is the refractive power of the depression in the main lens. As in the one-piece bifocal at the front side, the back surface only depends on the far vision prescription, and the addition only depends on the refractive powers of the segment and the base curve of the lens. Direct substitution of $\mathbb{P}_2$ from the far region power relation leads to

$$A = P_{1s} + P_{2s} + P_{1d} - B.$$

Now, we have to be cautious with the refractive index of each region. If we give the notation n to the refractive index of the main lens and n_s to that of the segment, then

$$P_{1s} = (n_s - 1)\,\kappa_1, \quad P_{2s} = (1 - n_s)\,\kappa_d,$$
$$B = (n - 1)\kappa_1, \quad P_{1d} = (n - 1)\kappa_d,$$

where κ_d is the curvature of the depression carved to lodge the segment. Combining the five previous equations we have

$$A = (n_s - n)\,(\kappa_1 - \kappa_d)\,. \tag{8.9}$$

As the addition must be positive, the two factors in the previous equation must be positive as well. This means the refractive index of the segment must be bigger than that of the main lens, and the depression curve, if convex, must be shallower than the base curve; otherwise it must be plano or concave.

Fused trifocal lenses use the same depression curve, but the segment is made in two portions, each with a different refractive index, n_i for the intermediate addition and n_s for the full addition. The full addition then would follow the same equation (8.9), and the intermediate addition would be given by $A_i = (n_i - n)\,(\kappa_1 - \kappa_d)$. Obviously, if we want $A > A_i$ then $n_s > n_i$.

8.3.3 Thin Lens Model of Bifocal Lenses: Prismatic Effects

We will investigate now the prism properties of bifocal lenses. Before going into the mathematics, let us discuss in a qualitative way the parameters affecting the prismatic effects and the locations of the optical centers in bifocals. Let us consider first front-side one-piece bifocals with spherical surfaces. The intersection between two spherical surfaces is a circumference. In a one-piece bifocal, the spherical cap corresponding to the segment intersects with the front surface of the main lens generating the circular contour of the segment. A small relative displacement between the spheres may produce a large variation of the segment contour. We can see this in Figure 8.7(a). In the subplot we appreciate that the displacement of the segment spherical cap can be understood as a rotation of the same around point B. The consequence of this rotation is a bigger segment whose upper part can be removed to free the far region of the lens, leaving a straight-top style bifocal. Of course, the straight line does not belong to the intersection between spheres, and as a consequence a ledge is created. Also, the tilting of the segment spherical cap introduces a constant prism component all over the segment, in this example a base-up prism.

We can approach this problem in a different way. Consider the subplots (b) and (c) in Figure 8.7. Once again the lenses are front-side one-piece bifocals, the first with positive far power, and the second being negative. As we saw in Chapter 5, the optical axis of a lens is the line joining the centers of curvature of the two surfaces of the lens. The optical center is located at the optical axis and on any of the lens surfaces, as long as we consider the lens to be thin. Prismatic effect at any other point is related to lens power and the distance from the point to the optical center. To lighten the notation, we have given the letters F, N, and S to, respectively, the optical centers of the far vision region, the near vision region, and the segment, which is considered the lens formed by the front surface of the main lens and the surface of the segment. The optical center of the segment is located at the center of the circumference resulting from the intersection of its two spherical surfaces, but the location of the optical center of the near region largely depends on the design of the bifocal (in this case a round segment vs a straight-top) and the lens power at the far region. In the round bifocal with positive far vision power shown in subplot (b), N is located between F and S. However, in the negative straight-top bifocal shown in (c), the center of the near region is outside the segment, and even outside the main lens, which means that there is prismatic effect at any point in the near region.

To put numbers into the previous reasoning, we will use the simple model of bifocal shown in Figure 8.8 and the approach presented by Harris [242]. Regardless of the geometry and technology of the bifocal, it can always be modeled as the superposition of two lenses: the main one with power $\mathbb{P}$ and a spherical lens with power A providing the addition. Take into account that this addition lens is not always real, but its effects are. For example, in a one-piece round bifocal we can imagine the segment as a "separate" entity from the main lens, which only happens to be in full contact with it. In a fused bifocal, the power of the segment does not match the addition. It is the whole combination of segment and main lens with depression that makes for the near power $A\mathbb{I} + \mathbb{P}$. However, within the paraxial and thin-lens approximations, the optical effect of the superposition of the two thin lenses shown in Figure 8.8 is the same as if it were the real lens.

Let us use a coordinate system whose origin is located at the optical center of the main lens, and let us compute the prismatic effect at any point Q on the lens. If Q belongs to the near region, the prism $\mathbf{p}_Q$ is given by

$$\mathbf{p}_Q = -\,(A\mathbb{I} + \mathbb{P})\,\mathbf{r}_{NQ},$$

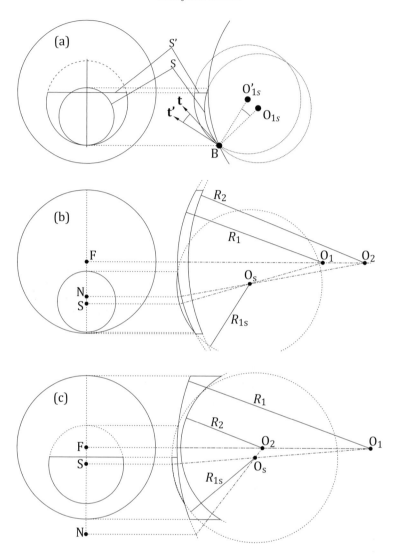

Figure 8.7 (a) Intersection of two spherical surfaces generating a circular contour. To the right we see a side view of the front curve of the lens and two possible spheres for the segment. The first has the center of curvature at O_{1s}. Its intersection with the main curve produces the small segment S. If we rotate this sphere around point B, the center moves to O'_{1s} (see the tangent vector **t** and its rotated **t'**) and the effect is a larger segment S'. We could remove the upper part of the bigger segment leaving a straight top, but then we will have created a ledge. (b) A round segment bifocal with plus distance power. In the lateral view we see the positions of the centers of curvature of the three spherical caps conforming the lens. (c) A straight-top bifocal with negative distance power. As the positions of the centers of curvature, O_1 and O_2, are inverted with respect to the plus lens, the optical center of the near region is displaced downward, even outside the main lens. This means that prismatic effect will be present everywhere in the segment.

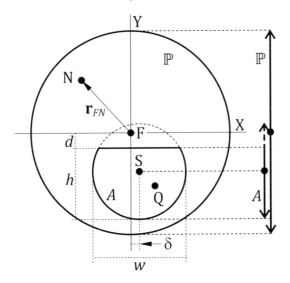

Figure 8.8 Thin bifocal model consisting of the superposition of two thin lenses: The main one with the far power $\mathbb{P}$, and the addition lens, with center at S and power A. F, S, and N are, respectively, the optical centers of the far region, the segment, and the near region.

where $\mathbf{r}_{NQ}$ is the vector from the optical center of the near region to Q. This prism can also be computed as the sum of the prisms produced by the main lens and the segment. For this we apply Prentice's rule to each lens, with the coordinate vector going from the respective optical center to Q,

$$\mathbf{p}_Q = -\mathbb{P}\mathbf{r}_{FQ} - A\mathbf{r}_{SQ}.$$

Now, given three arbitrary points in the plane, A, B, and C, the vector from A to B can be split as the sum of the vector from A to C plus the vector from C to B. Applying this property to our problem, $\mathbf{r}_{NQ} = \mathbf{r}_{NF} + \mathbf{r}_{FQ}$ and similarly, $\mathbf{r}_{SQ} = \mathbf{r}_{SF} + \mathbf{r}_{FQ}$. Equating the two expressions for the prism at Q and using the vector relationships we get,

$$-\left(A\mathbb{I} + \mathbb{P}\right)\left(\mathbf{r}_{NF} + \mathbf{r}_{FQ}\right) = -\mathbb{P}\mathbf{r}_{FQ} - A\left(\mathbf{r}_{SF} + \mathbf{r}_{FQ}\right),$$

which reduces to

$$-\left(A\mathbb{I} + \mathbb{P}\right)\mathbf{r}_{NF} = -A\mathbf{r}_{SF}. \tag{8.10}$$

Finally, let us remember that for any two points A and B, $\mathbf{r}_{AB} = -\mathbf{r}_{BA}$, so we finally have

$$\mathbf{r}_{FN} = A\left(A\mathbb{I} + \mathbb{P}\right)^{-1}\mathbf{r}_{FS}. \tag{8.11}$$

From a practical point of view, the position of the optical center of the segment is obtained from the parameters d, h, w, and δ shown in Figure 8.8. The vertical distance from the optical center of the far region to the topmost point of the segment is known as the *seg drop*, d. The total vertical size of the segment, h, is called *seg depth*; its maximum horizontal width, w, *seg width*. Finally, the horizontal displacement of S with respect to F is named the *seg inset*, δ. For the majority of segment shapes, the seg width is just the diameter of the segment. Then, also for the majority of the segments, the distance from the top of the segment to its optical center would be $h - w/2$. According to this, the coordinates of S relative to F are

$$\mathbf{r}_{FS} = \left(-d - h + \frac{w}{2}, \pm\delta\right),$$

where the plus sign should be used for right eye bifocals and the minus one for left eye bifocals. This expression is not fully general, and there can be specialty bifocals for which it does not apply (for example, those with the segment at the top of the lens, or for Ultex-type bifocals for which the segment diameter is bigger than the seg width) but it is correct for the majority of bifocals intended for standard near vision use.

Expression (8.11) gives us the location of the optical center of the near region with respect to the optical center of the main lens in terms of the lens power, the addition, and the vector from the optical center of the main lens to the optical center of the segment. For lenses without astigmatism, (8.11) reduces to

$$\mathbf{r}_{FN} = \frac{A}{A + P}\mathbf{r}_{FS},$$

which means that, for spherical prescriptions, the position of the optical center of the near region should lie on the line joining F and S. We can use this expression to figure out the approximate position of N depending on the type of ametropia and the value of the addition. To shorten the notation, let us give the notation γ to the factor $A/(A + P)$:

Hyperopia. In this case $P > 0$, so $0 < \gamma < 1$ and $\mathbf{r}_{FN}$ points in the same direction as $\mathbf{r}_{FS}$ being shorter. This means N is somewhere on the line joining F and S. If the hyperopia is large, γ will be small and N will be close to F. If the hyperopia is much smaller than the addition, N will be closer to S.

Emmetropia. In this case $P = 0$, F is not defined, and the only optical center is that of the segment, that is, $\gamma = 1$ and N and S will be the same point.

Myopia, $|P| < A$. In this case $A + P$ is still positive but smaller than A, so $\gamma > 1$. N is still on the line joining F and S, but below S, and further from the reading region of the segment (which should be just below the segment top) than in the previous cases.

Myopia, $P = -A$. Now the near region has zero power and its optical center is not defined. Still, there can be a constant prismatic effect all over the segment, which should be computed using $\mathbf{p}_Q = -\mathbb{P}\mathbf{r}_{FQ} - A\mathbf{r}_{SQ}$, for any Q inside the segment. The fact that a zero power region presents nonzero prism is due to this region being created by two lenses with exactly opposite power but with optical centers at different locations. In effect, as $\mathbb{P} = -A\mathbb{I}$, the prismatic effect is $\mathbf{p}_Q = -A\left(\mathbf{r}_{FQ} - \mathbf{r}_{SQ}\right)$, that is, $\mathbf{p}_Q = -A\mathbf{r}_{FS}$.

Myopia, $|P| > A$. In this final case the myopia is strong enough for the near region to be negative. Now $\gamma < 1$ and $\mathbf{r}_{FN}$ points opposite to $\mathbf{r}_{FS}$. The optical center of the

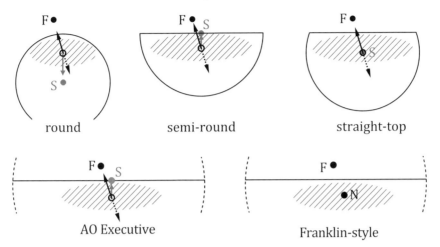

round semi-round straight-top

AO Executive Franklin-style

Figure 8.9 Prismatic effect built-up in the reading region of a bifocal, represented by the shadowed ellipses. F and S are the optical centers of the far region and the segment, respectively. Black solid arrows represent the prismatic effect caused by the main lens for hyperopes, while the shaded black arrow corresponds to myopes. In gray is the prism produced by the segment.

near region is above F. Strong myopia will bring N close to F, whereas myopia just above the addition value (in absolute value) will move N upward and away from F.

When the far vision prescription is astigmatic, the optical center of the near region, N, no longer lies on the line joining F and S. If we write equation (8.11) in the reference system formed by the principal meridians of the main lens, and give the notation α and β to the angles made by vectors $\mathbf{r}_{FS}$ and $\mathbf{r}_{FN}$ with the cylinder axis, respectively, it is not difficult to show that $\tan(\beta) = (1 + C) \tan(\alpha)^2$. In the special cases where the cylinder axis is parallel or perpendicular to $\mathbf{r}_{FS}$, then $\mathbf{r}_{FN}$ still points along the same direction. When the cylinder axis is close to perpendicular to $\mathbf{r}_{FS}$, $\tan(\alpha)$ is a large number and the effect of the cylinder in the direction of $\mathbf{r}_{FN}$ will be small: The previous classification on the position of N as a function of the spherical ametropia will still be valid. In any other case, especially when the angle made by $\mathbf{r}_{FS}$ with the cylinder cross-axis is small, the cylinder will force N to move away from the line joining F and S, the previous classification will not necessarily be valid, and the location of N should be analyzed using equation (8.11) on a case-by-case basis.

A clear idea of the location of the optical center of the near region is useful to figure out the prismatic effects at the reading region. We can graphically show how the prismatic effect builds up in the reading region of the segment, that is, the region within the segment the user is going to use with greater probability. The analysis is illustrated in Figure 8.9.

---

[2] Just write equation (8.11) in the coordinate system of the principal meridians of the main lens, as introduced in Section 4.3.7, and compute the polar angles of $\mathbf{r}_{FS}$ and $\mathbf{r}_{FN}$ in this reference system.

There we show three types of segment: round, straight-top semi round, and straight-top, as well as two Franklin-style bifocals by way of comparison. The shaded ellipses indicate the regions the user of bifocals will use with greater probability, and where we should be more concerned about prismatic effects. As the addition is always positive, the prismatic effect caused by the segment (represented by gray arrows) always points toward S. This means it points downward in the round segment, upward in the straight-top semi-round, and for the straight-top, the reading region is around S, so the prismatic effect caused by the segment there is zero or very small. The prismatic effect caused by the main lens points toward F for positive lenses, and away from it for negative. Then we can infer the next approximate rules:

Round segment. The prism from the segment is base down. For hyperopic prescriptions, the prism from the main lens is base up, so there is partial compensation and the total prism is negligible or very small. For myopic prescriptions the prism from the main lens is base down, so the addition increases this prism at the reading region.

Straight-top, semi-round segment. Now the optical center of the segment is above the reading region, so the prism from the segment is base up. The final result is opposite to that for round segments: Hyperopes get a stronger base up prism than with single vision lenses while myopes get partial cancellation of the prismatic effect they perceive with monofocals.

Straight-top segment. For this type of bifocals the optical center of the segment is located in the reading region. The add contributes then with very little prism, and the lens behaves, in this region, as a single vision lens, both for myopes and hyperopes. This type of segment produces the lesser changes in the vertical prismatic effect at the reading region.

Franklin-style bifocals. These lenses do not have the constraint of a round intersection between the surface defining the segment and the surface of the main lens that accommodates the segment. The straight ledge separating the far and near regions allows for a greater flexibility in the positioning of the optical centers. For example, the *Executive* bifocal from American Optical was manufactured in such a way that the center of the segment was located at the middle point of the dividing line in the blank. During lens processing the optical center of the far region could be displaced as needed [243]. With regard to prismatic effects, this lens would behave just the same as a semi-round segment bifocal. Other lens manufacturers were making front-side Franklin-style bifocals for the last two decades of the twentieth century, in which the location of S was base curve dependent so that the optical center of the near region could be approximately located at the reading region (for this S should be moved upward for plane base curves intended for negative prescriptions, and moved down for stepper base curves intended for positive prescriptions).

In any case, we must take into account that the presence of prismatic effect is not necessarily a problem. When a user of single vision lenses starts using bifocals, these will produce different eye deviations at the reading region than those produced by the previous single vision lenses. Likewise, benefiting from the addition in bifocals requires a downward gaze the user of single vision lenses is not used to. Both facts require a learning process to readjust the head and eye inclinations required for a typical near vision task. In any case, if there is no prismatic imbalance between the two eyes, the user should adapt to the new situation. The later the presbyopic person starts using bifocal lenses, the harder the adaptation will be. On the one hand, higher addition will produce larger changes of prismatic effect and magnification. On the other hand, the advanced presbyope will have smaller ranges of clear vision and heavier reliance on the addition. Finally, the advanced presbyope will usually be among the elderly, with reduced adaptation and learning capabilities. These facts indeed apply to any type of multifocal lenses, and the conclusion is that adaptation to the visual and ergonomic changes imposed by multifocals will always be easier the earlier the presbyope starts using these lenses.

There is, however, a situation in which prismatic effects at the reading position may prevent adaptation to bifocal lenses. When the user has a significant amount of anisometropia and rotates their gaze out from the optical center of the lenses, eye deviation grows at a different rate for each eye. To keep this imbalance within tolerable levels, the anisometropic user of single vision lenses avoids looking through points too far away from the optical center. As a rule of thumb, if we assume a tolerable vertical prism imbalance no larger than 1Δ, and the anisometropia along the vertical direction is δP_v, the vertical extension of the region around the optical center of the lens that the anisometropic person may use is about $1/\delta P_v$ cm (which directly follows from Prentice's rule). For example, with a vertical anisometropia of 2 D, the vertical region the user can "comfortably" use with good binocular vision extends just 5 mm below the optical center. These patients may find it difficult or even impossible to adapt to bifocal lenses, as the segment is necessarily located outside the region in which prismatic imbalance remains small. The addition may change the prismatic effect within the segment, but the change will be similar for both eyes, and the imbalance will still be there. Adaptation will require at least one of the next conditions or modifications:

1. The user has large tolerance to prismatic imbalance, or the amount of anisometropia is small.
2. The user does not have binocular vision at near vision, and the prismatic imbalance is not causing diplopia.
3. Using a Franklin-style bifocal such as the one shown in Figure 8.9, if still available. Locating S above the dividing line for myopes and below the reading region for hyperopes would reset the prism at the dividing line and locate the optical center of the near region at the useful spot. The anisometric patient still must learn to rotate their gaze downward to achieve the addition (they would not do this with single vision lenses to avoid prismatic imbalance).
4. Different segment shapes can be prescribed to help balance the eye deviations at the reading region. As shown in Figure 8.9, round segments will increase the base down

component of the prism, while straight-top semi-round segments will slightly increase the base up component. Both could be combined to decrease the vertical imbalance from the anisometropia. It goes without saying the patient must tolerate the unsightly effect of using different-shaped bifocals for each eye (and for this, the improvement of vision quality *must* be sound).

5. Bifocals with different vertical positions of the optical center of the segment can be manufactured. In this way, two bifocals with the same segment shape may produce different prismatic effects at the reading region. There is, however, a price to pay; the modified bifocals will have a ledge at the segment contour. For straight-top bifocals, the width of the ledge at the top of the bifocals will change if the optical center S is moved up or down. This change goes mostly unnoticed in Franklin-style bifocals, but can be quite conspicuous in smaller-segment bifocals with some portion of a round contour.

6. Some manufacturers can cement a prismatic component on top of the segment. This prismatic component would have the same shape as the segment. It would have a ledge in most of its contour, and from the optical point of view, this option is indistinguishable from the previous one.

7. The same as before but using Fresnel prisms. These are soft plastic components that can be cut with scissors to any shape, and can be adhered like a suction cup on top of the segment. The advantage is that they can be removed. The main drawbacks are low optical quality and unsightly appearance.

8. Bicentric lenses. A base down or base up prism can be cut on the lower half bifocal. This prism will be computed to cancel the prism the bifocal produces at the reading region. The two techniques for creating bicentric lenses are known as slab-off and reverse slab-off. The first produces base up prism, while the second generates base down, in both cases at the bottom half of the lens.

8.3.4 Image Jump

Image jump is a popular name given to the sudden variation of prismatic effect occurring when the line of sight crosses the contour of the segment. Image jump happens at any point at the segment contour, but it is typically noticed and evaluated at its top edge, which the line of sight is constantly crossing when changing from far to near vision and vice versa. Not only does prismatic effect change when crossing the top of the segment; magnification also changes, introducing a bigger perceptual impact on image jump. The effect is shown in Figure 8.10. Let us assume the line of sight crosses the segment contour at point K. Just above K, the prismatic effect is determined by the main lens. Below K, the prismatic effect from the segment suddenly adds to the previous one. The result is either a scotoma, or loss of visual field, when the prism from the segment at K is base down, or double image when the prism from the segment at K is base up. In Figure 8.10 the first case is represented, typical for round segment bifocals. The scotoma is a portion of the field of view that is not available for a given head inclination. The user's point of view is also shown in the figure. The contour of the segment is perceived defocused. The letters "ima" and part of the letter

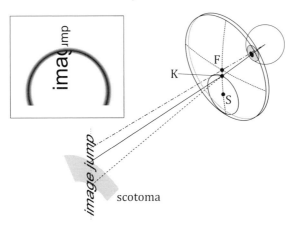

Figure 8.10 Image jump in bifocals.

"g" are seen from within the segment, while the letters "ump" (only part of the "u") can be seen through the main lens. The remaining letters or part of letters are not visible unless the user changes their head orientation.

The image jump, S, can be easily estimated with Prentice's rule,

$$S = -A\mathbf{r}_{SK}. \tag{8.12}$$

It is just the prism produced by the segment at K. If K happens to be directly below or above S, the image jump will be vertical. But in general K can be to the right or left of S, and then image jump will have some horizontal component. The vertical extension of the scotoma (or the double vision region) for a given viewing distance z is approximately given by $z\,|\mathbf{S}|$. If we measure z in meters and $|\mathbf{S}|$ in prism diopters, their product will be given in centimeters. A more precise estimation of the extension of the scotoma or double vision region can be obtained by applying equation (5.32) to the main lens and equation (5.43) to the near region, taking into account that we must use N as the optical center for the latter. As vector $\mathbf{r}_2$, we should use $\mathbf{r}_{FK}$ for the main lens and $\mathbf{r}_{NK}$ for the near region. The two outputs, $\mathbf{r}_{0F}$ and $\mathbf{r}_{0N}$, would be the fixation points when the user looks through K just above it and just below it, respectively.

Early presbyopes usually adapt easily to image jump, but the task is not so easy for the elderly who start using bifocals for the first time. Stumbling into unseen objects and tripping on ladder steps are commonly reported issues that can be traced back to image jump. Of course, the easiest way to get rid of it is by prescribing bifocals with the optical center S as close as possible to the center of the top edge of the segment. For example, straight-top semi-round and Executive bifocals would be free from image jump at the center of the straight top. As we move left or right along this line, horizontal image jump would develop, but it would have a lesser impact than the vertical image jump of round segment bifocals. A thorough analysis of the clinical implications of the image jump can be found in Fannin and Grosvenor [132].

8.4 Progressive Lenses

8.4.1 Introduction

We have just seen that bifocal lenses have some problems related to the sudden change in power happening at the contour of the segment, such as jump images, ghost reflections at the segment line, and an unsightly segment, especially when the line is clearly visible. From the optical point of view, probably the main limitation of bifocal lenses is the lack of intermediate addition values needed for advanced hyperopes (partially solved by trifocal lenses, at the cost of adding more unsightly lines and jumps). No wonder that from a long time ago, opticians, optometrists, and optical designers were looking for a multifocal lens in which the power could change smoothly from the far to the near prescription. This smooth change would encompass all of the intermediate powers, allowing the user to adjust its gaze to find the right amount of addition. A smooth transition would also remove image jump, bifocal or trifocal lines, and no segment would be visible. But, is it possible to produce a lens with such a smooth change in power? Indeed, it is not difficult to recall some surfaces in which curvature changes smoothly from some points to others, for example a revolution hyperboloid: At its vertex the two curvatures will be equal, but as we move away from it, both the tangential and the sagittal curvatures decrease, the tangential getting smaller at a faster rate (see Chapter 2). As long as the two main curvatures (tangential and sagittal) are not equal, the surface presents some cylinder. So, if we intend to make a progressive lens using a hyperboloid as a front surface, we would have some astigmatism at the flatter region (the far vision region), and no or little astigmatism at the more curved region (near vision). We could also use an oblate ellipsoid, for which the vertex is less curved than the periphery. Locating the vertex of a front-side oblate ellipsoid at the far vision region of the lens would give us no astigmatism at this point, but then we would have astigmatism at the near region. Take into account that this astigmatism we are talking about is not prescription astigmatism, neither is it caused by oblique aberration. It is simply caused by the variation of curvature of the surface. We could choose the asphericity parameters of the previous conicoids so that the variation of mean sphere across a fixed distance is the one we need to provide the right addition, but the rigidity of the geometry of the conicoids does not allow us to reduce or modify the resulting astigmatism.

A simple model of a progressive surface

We can easily understand progressively varying curvature in a cylinder, as depicted in Figure 8.11(a). In this case an ellipse is parallel-transported to generate a progressive cylinder. The upper part of the grayed surface has cylindrical power with radius R_F, whereas the lower part has radius R_N. We can easily make a tangible progressive cylindrical surface by bending a sheet of paper in a similar way. Of course, a progressive cylinder is not enough; we need to increase spherical power, not just cylindrical power. If instead of a sheet of paper we had used an elastic membrane to make the progressive cylinder, we would have also been able to bend it along the horizontal direction. In principle, we could

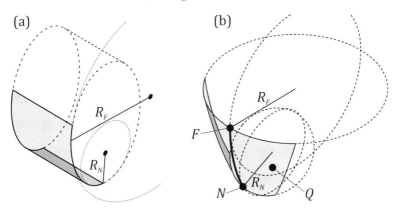

(a)   (b)

Figure 8.11 (a) Progressive cylinder generated by parallel transport of an ellipse. (b) A more general progressive surface in which the far, intermediate, and near points are free from astigmatism.

set two points, F and N, and bend the surface on them along the horizontal direction with the same amount of bending present along the vertical direction, getting to the surface in Figure 8.11(b). By matching the vertical and horizontal bending we would have a surface with at least two points, F and N, without surface astigmatism and with increased curvature from the former to the latter. If this surface is possible, what are its properties? Does it have surface astigmatism at points other than the vertical meridian, for example, at point Q?

The first documented proposal for a progressive lens was presented by the Briton Owen Aves in a UK patent filed in 1907 [244]. At that time numerical computation tools required for describing and handling surfaces such as the one shown in Figure 8.11(b) were lacking. Aves described a lens in which the front surface was a progressive cylinder curving along the vertical meridian, similar to the one shown in Figure 8.11(a), and the back surface was a conical patch with a vertical cone axis, which provided progressive curvature along the horizontal direction. A lens like that had large amounts of astigmatism, and astigmatic prescription could not be easily incorporated into either surface.

Progressive lenses have come a long way since Aves' time, but we can still use an idea based on his and presented by Volk and Weimberg [245] in 1962 to explore the nature of progressive power. Assume we have a progressive cylinder with horizontal axis such as the one shown in Figure 8.11(a). The curvature can increase anyway we want from top to bottom, but to keep it simple, let us assume the upper third of this surface has constant curvature $\kappa_F = 1/R_F$, the lower third also has constant curvature $\kappa_N = 1/R_N$, and the middle third is the progressive section in which curvature changes from the upper to the lower value. We will give the notation $\kappa_A = \kappa_N - \kappa_F$ to the total increase of curvature. Let us have another surface identical to the first one except for the curvature having opposite sign. Now, let us rotate the first one by $45°$ and the second one by $-45°$, and make a lens with them. Being these surfaces the front and back sides of our lens, both will contribute to the lens power with progressive cylindrical power with the same sign. In Figure 8.12(a)

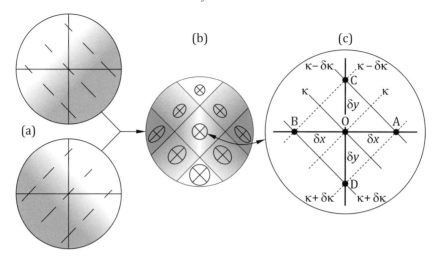

Figure 8.12 Progressive lens composed of two progressive cylinders with axis at 45° and 135°.

we can see the two rotated cylinders, with the oblique lines inside them oriented along the power meridian and having lengths proportional to the local curvature of the cylinder (the shading also represents local curvature.) The superposition of the two cylinders is shown in Figure 8.12(b). Despite the cylinders being progressive, their axes are perpendicular to each other everywhere, so the final power of the combination can be read as a cross-cylinder prescription. The ellipses shown in the figure represent the local power of the resulting lens, the axes of the ellipses being proportional to the local main powers. It is easy to check that along the Y axis the cylinders superimpose with exactly the same curvature, their combined effect being that of spherical power (see Chapter 2). As cylindrical curvature grows downward, the same happens to the resulting spherical power. At points outside the Y axis the cylinders do not combine with the same curvature. For example at the left part of the resulting lens, the first cylinder shows $(\kappa_F \times 45°)$ while the second provides $(-\kappa_N \times 135°)$. The resulting spherocylindrical power is then $(n-1)[\kappa_F, \kappa_A \times 135°]$, n being the refractive index of the lens material.

With the "simple" technique just described, we have created a progressive lens with an arbitrary continuous power increase along the vertical axis and free from astigmatism in the upper and lower regions of the lens, where the cylinders combine with stable curvature (and the same curvature for both of them). The drawback is that the curvatures of the cylinders only match on the vertical meridian or at regions where both cylinders have stable curvature. Anywhere else there is an imbalance between the cylinders that renders the local power astigmatic. With this construction, the maximum value for this unwanted astigmatism is exactly that of the addition, and is generated at the nasal and temporal sides of the lens. The orientation axis for this unwanted astigmatism is either 45° or 135°. We may also consider how astigmatism builds up as we move away from the Y axis. For this, let us check the drawing in Figure 8.12(c). The vertical line stands for the Y axis, and we consider any point O on it, in the region in which power is changing. As explained previously, there is no astigmatism at O. Now consider point A, to the right of O and at a small distance δx along the horizontal direction. The cylinder with axis at 45° presents the same curvature at A and D, whereas the other cylinder has the same curvature at A and C. From construction, it is clear that $\delta x = \delta y$. Let us call κ the curvature of both cylinders at O, and $P = (n-1)\kappa$ the lens power at the same point. The power at C is $P - \delta P = (n-1)(\kappa - \delta\kappa)$, and similarly the power at D, $P + \delta P$. The cross-cylinder prescription for the lens power at A is then $((P + \delta P) \times 45°)((P - \delta P) \times 135°)$, that is, $[(P - \delta P), 2\delta P \times 45°]$. Similarly, the power at B is found to be $[(P - \delta P), 2\delta P \times 135°]$. What we learn from this is that when power increases δP across a distance δy along the Y axis, the cylinder will grow $\delta C = 2\delta P$ when we move the same distance $\delta x = \delta y$ along the horizontal direction. If we divide the increments in spherical and cylindrical power by the distance increment, $\delta C/\delta x = 2\delta P/\delta y$, and take the limit when $\delta x \to 0$, then we have

$$\frac{dC}{dx} = 2\frac{dP}{dy}. \tag{8.13}$$

This can be read as follows: This particular lens with perfectly spherical varying power along the Y axis presents astigmatism outside this axis, and the rate of growth of this astigmatism is twice the rate of growth of power along the Y axis. We may think this is a consequence of the particular construction of the lens, but we will see in the next sections that the previous expression embodies a completely general result for progressive surfaces.

There is still another approach using this idea of 45° and 135° rotated cylinders. We may recall from Chapter 2 that the optical effect of the superposition of two crossed cylinders, where they are two plano-cylindrical lenses, or the two surfaces of a single lens, is the same as if we combine (add) the two cylindrical surfaces into a single one. This is true in paraxial terms and if the surfaces are not too curved (and, of course, providing the right sign to the curvature and the change in refractive index across the surfaces.) The low curvature assumption is necessary so that the curvature of the surface can be accurately described by the Hessian matrix. Let us then give the notation $z_1(x, y)$ and $z_2(x, y)$ to the rotated progressive cylinders $(C_p(x, y) \times 45°)$ and $(C_p(x, y) \times 135°)$, now both having positive curvature, and let us create the new surface $z(x, y) = z_1(x, y) + z_2(x, y)$. The Hessian of the resulting surface is the sum of the Hessians of each cylinder, and if both are flat enough, we may expect the resulting surface will have the properties shown in Figure 8.12(b). In effect, if we use $\kappa_F = 2$ D and $\kappa_A = 4$ D, which with $n = 1.5$ would give us $P_F = 1$ D and $P_N = 3$ D, the resulting surface is a nice progressive one, whose exact main powers (in terms of sphere and cylinder) are shown in Figure 8.13. With this surface at the front side of a lens, we could use a toric surface at the back to account for the prescription of the user, generating a fully functional progressive lens. We still have to solve some questions: The distribution of unwanted astigmatism and mean sphere is linked to the construction technique (the crossing cylinders) and the characteristics of the progressive section. If we want to change this distribution (which we will have to, depending on the lens' intended use) we have to resort to other construction techniques. Also, the results are valid only for

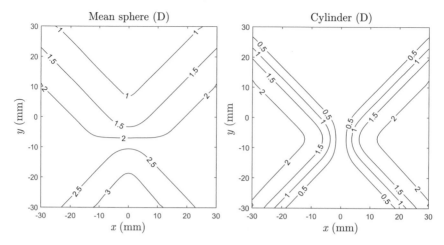

Figure 8.13 Curvature maps of the surface created as the sum of two progressive cylinders rotated by 45° and 135°. These are exact curvature maps of a well-defined mathematical function, and as such it could be used for a real lens. However, the shown distribution of sphere and cylinder will degrade as we increase the overall curvature of the forming cylinders, and then other construction methods should be used.

low curvature surfaces. If the lens prescription corresponds to a low myope or a hyperope, larger curvatures on the front surface would be required, and another surface construction method would be needed.

The whole example with the crossed cylinders is intended as a first introduction to progressive power. The reader can check how easily this power variation can be achieved and what are the expected consequences of doing so. In particular, the main lesson is that vertical progression of spherical power can be achieved without any limitation on the details for this progression. However, this progression of spherical power along the Y axis causes a large part of the remaining surface to be astigmatic. Astigmatism grows faster from those points on the vertical axis at which the rate of growth of spherical power is bigger. In those points on the Y axis at which spherical power is stable, astigmatism doesn't grow, though you may encounter astigmatism some distance from the Y axis. Progressive lens design is about deciding which is the best progression rate, where it should start and where it should finish, and also about deciding how the inevitable unwanted astigmatism may and should be distributed. This astigmatism is not related to the oblique astigmatism we studied in Chapter 6; it is just linked to the progression of spherical power. In an actual progressive lens, both types of astigmatism will be present, and they will mix together. Throughout this chapter we will try to understand the first one and how the two types interact.

8.4.2 Low-Curvature Model of a Progressive Surface

One of the major problems when studying progressive lenses is the lack of a useful analytical model from which we can derive their main properties. A single vision lens with spherical or even aspherical surfaces can always be studied in the paraxial region, in which we can approximate the surfaces by paraboloids. What we indeed do is assuming that the actual properties at the vertex of the lens holds in a small region around it. In this region, we can compute curvatures in scalar or matrix form, prismatic effect, magnification, or any other paraxial property. The deviations of performance that any lens suffers as we move away from the paraxial region are then considered aberrations. However, this approach is not valid for progressive lenses. Its very definition demands that power changes across a significant region of at least one of their surfaces, and these changes cannot be considered aberrations; they are basic properties of the lens. There is no way to create an analytical model of a progressive surface using exact curvatures. Even for the simpler progressive surfaces we can think of, the equations describing their exact curvatures grow absurdly large and unmanageable. But we can use a flat surface approximation that simplifies the curvature equations and allows us to build a meaningful model.

Let us consider a surface defined by a Monge chart, as explained in Appendix C, $z(x, y)$. In preceding chapters we have used the Hessian matrix evaluated at the surface vertex as the mathematical entity that describes the curvature of the surface. That is exactly at the vertex if the surface is tangent to the XY plane, and approximate at any other point, with the quality of the approximation at a certain distance from the vertex depending on how flat the surface is: the flatter the surface, the better the approximation at any given distance. Now we will take a slightly different approach. As we want a surface with varying power, we cannot restrict ourselves to constant vertex power. We need a curvature that changes along the x and y coordinates. So we will assume the surface is flat enough so that its Hessian matrix describes the curvature *everywhere* with enough accuracy. This means the Hessian matrix is no longer a constant matrix, but its elements are functions of the x and y coordinates

$$\mathbb{H}(x, y) = \left[\begin{array}{cc} z_{xx}(x, y) & z_{xy}(x, y) \\ z_{xy}(x, y) & z_{yy}(x, y) \end{array} \right]. \tag{8.14}$$

From this, and recovering the definitions and equations presented in Appendix C, the Gaussian and mean curvatures are given by

$$K = z_{xx}z_{yy} - z_{xy}^2, \quad H = \frac{1}{2}\left(z_{xx} + z_{yy} \right), \tag{8.15}$$

the main curvatures by

$$\kappa_{\pm} = H \pm \sqrt{H^2 - K}, \tag{8.16}$$

and the orientation axis of the local spherocylindrical prescription with positive cylinder by

$$\tan(\theta_+) = \frac{z_{xy}}{\kappa_+ - z_{yy}}, \tag{8.17}$$

the five of them also being functions of the x and y coordinates.

Just to have an idea of the goodness of this approximation, consider a round surface with a diameter of 60 mm. If the curvature radius does not get smaller than around 200 mm, the maximum error we may expect from the previous equations is around 0.1 D at any point on the surface (2%). However, with smaller radius of curvature, say, 50 mm, the error at the edge of the surface can grow up to 2 D (10%). The model we present next relies on this flat surface approximation, but extensive numerical computation conducted by the authors shows that the main properties we will derive from it are still valid with more general steep surfaces. We have already listed the tools; let us now summarize our objectives:

1. We want to construct a lens having the far vision prescription at some point and the near vision prescription at another point, with the power changing smoothly between these two points.
2. We want the far and near vision points to be on the Y axis. Modern progressive lenses have the near point shifted toward the nasal side to account for eye convergence, but this adds mathematical complication to our model and does not offer insight about the properties derived from the imposed curvature variation, so we will keep it simple and will assume a perfectly vertical power variation.
3. All the power variation will be produced at one surface; the other one will have constant power. The remaining section will deal with the properties of this progressive surface.
4. The progressive surface will not have astigmatism on the Y axis. In other words, the two main curvatures will be identical on that axis: $\kappa_+(0, y) = \kappa_-(0, y)$ for any y. If the main curvatures of a surface at some point are identical, we call this point *umbilical*. We can also say that the surface is *locally* spherical at that point. We say a curve contained in a surface is umbilical if all its points are also umbilical. In our model, the section of the surface generated by the plane YZ, that is, the curve $z(0, y)$, will be an umbilical curve. To simplify the language, sometimes we will just say that the Y axis is "the" umbilical line.

Let us describe the progressive surface as a polynomial in the x variable whose coefficients are functions of the y variable, $z(x, y) = \sum_{n=0}^{N} q_n(y)x^n$. This description is completely general and, as long as the $N + 1$ functions $\{q_n(y)\}$ are sufficiently smooth and N is large enough, the sum can describe any continuous surface that could be used for progressive lenses. By construction, we know that the two main curvatures along the Y axis are equal, and hence they also equal the mean curvature. We will give a special name to this one-dimensional function: the *main curvature profile*, $\kappa_p(y) = H(0, y) = \kappa_+(0, y)$. Also, umbilicality requires that the cylinder along the Y axis is zero, so the torsion component of the Hessian matrix has to be zero at $x = 0$. Both conditions can be written

$$z_{xx}(0, y) = z_{yy}(0, y) = \kappa_p(y),$$
$$z_{xy}(0, y) = 0. \tag{8.18}$$

Now, the second derivatives of our surface are easy to compute

$$z_{xx} = \sum_{n=2}^{N} n(n-1)q_n(y)x^{n-2},$$

$$z_{yy} = \sum_{n=0}^{N} q_n''(y)x^n,$$

$$z_{xy} = \sum_{n=1}^{N} nq_n'(y)x^{n-1}. \tag{8.19}$$

If we substitute these derivatives into the umbilicality conditions (8.18), then we get some constraints on the functions $q_n(y)$ that must hold if $z(x, y)$ has to describe a surface with a vertical umbilical line. These are

$$2q_2(y) = q_0''(y) = \kappa_p(y),$$
$$q_1'(y) = 0.$$

The second equation means q_1 must be any constant, that is, the linear term in x doesn't depend on y. As a linear term in the x variable can be understood as tilt around the Y axis, we can just set $q_1 = 0$, without losing generality. Using this information on the definition of $z(x, y)$, we get

$$z(x, y) = \iint \kappa_p(y)dy^2 + \frac{\kappa_p(y)}{2}x^2 + q_3(y)x^3 + \cdots + q_N(y)x^N, \tag{8.20}$$

which is the *general expression for a surface with a vertical umbilical line along the Y axis*[3] and such that its curvature along this umbilical line is $\kappa_p(y)$. Of course, the expression is approximate and it is valid as long as the overall surface curvature is small, but a lot of information can be obtained from this simple model. First, the $\{q_n\}_{n\geq 3}$ functions are arbitrary; we can start setting them all to zero and the surface will still have a perfect umbilical meridian

$$z(x, y) = \iint \kappa_p(y)dy^2 + \frac{\kappa_p(y)}{2}x^2. \tag{8.21}$$

This simple function stands for a progressive surface with parabolic horizontal curves, each with center curvature $\kappa_p(y)$ and with vertical profiles given by $\iint \kappa_p(y)dy^2$. Its Hessian matrix is

$$\mathbb{H}(x, y) = \begin{bmatrix} \kappa_p & \kappa_p'x \\ \kappa_p'x & \kappa_p + \frac{1}{2}\kappa_p''x^2 \end{bmatrix}, \tag{8.22}$$

---

[3] A progressive surface with an expression very similar to that in equation (8.20) is disclosed in US Patent 2005/0083482A1, 2005; we may assume that the inventors were aware of the model presented here or at least a similar one, though to our knowledge, a description of such a model has not been previously published.

and when computed at the Y axis, $x = 0$,

$$\mathbb{H}(0, y) = \begin{bmatrix} \kappa_p(y) & 0 \\ 0 & \kappa_p(y) \end{bmatrix} = \kappa_p(y)\mathbb{I},$$

which confirms the umbilical nature along the Y axis. We can also take a look outside the umbilical line. The mean curvature is just half the trace of the Hessian matrix, $H = \kappa_p + (\kappa_p''/4)x^2$. It has a parabolic dependence on the horizontal coordinate, and it will grow or decrease from its value at the umbilical line depending on the sign of κ_p''. The cylinder takes a little more work. First, let us remember that the (positive) cylinder is given by $C = \kappa_+ - \kappa_- = 2\sqrt{H^2 - K}$. In terms of the second-order derivatives this formula has $C^2 = (z_{xx} - z_{yy})^2 + 4z_{xy}^2$, which is valid for any low-curvature surface. Upon substitution of the derivatives in the last expression we reach

$$C^2 = 4\kappa_p'^2 x^2 + \frac{1}{4}\kappa_p''^2 x^4. \tag{8.23}$$

If we only consider small values of the x coordinate, corresponding to points close to the umbilical line, the fourth-order term can be neglected with respect to the second-order term. This leaves us with the expression

$$C(x, y) = 2|\kappa_p'(y)x|, \tag{8.24}$$

where we apply the absolute value to both x and κ_p' because we just chose the positive cylinder. This formula embodies the so-called Minkwitz's theorem, [246], which we already encountered in the simple model using two crossed cylinders, (equation (8.13)), and that we will formulate later more precisely. Now, we will say that for the type of surfaces described by equation (8.21), which belong to a particular type within the family described by equation (8.20), the astigmatism of the surface at a point (x, y) is proportional to the distance to the umbilical line, x, with the proportionality constant being twice the slope of the main curvature profile at point y. This property plays a fundamental role in the characteristics and design criteria of progressive lenses.

Example 38 Let us apply the previous results to two particular profiles. First, let us assume the vertical curvature profile is linear, $\kappa_p(y) = my + \kappa_0$, κ_0 being the curvature at the origin and m the rate of change of curvature along the umbilical line (the slope, which in this example will be constant). To obtain the corresponding progressive surface, we have to integrate the profile twice,

$$\iint \kappa_p(y)dy^2 = \int \left(\frac{m}{2}y^2 + \kappa_0 y\right) dy = \frac{m}{6}y^3 + \frac{\kappa_0}{2}y^2.$$

Integration constants can be set to zero, as they only add a linear and a constant term that under the flat surface approximation only accounts for displacement and tilt, without any effect in surface curvature. The surface is then

$$z(x, y) = m\left(\frac{y^3}{6} + \frac{yx^2}{2}\right) + \frac{\kappa_0}{2}(x^2 + y^2) + \sum_{n=3}^{N} q_n(y)x^n,$$

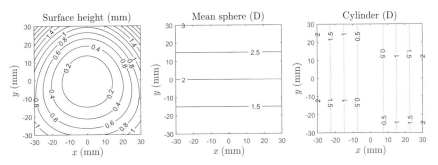

Figure 8.14 Maps of surface height, mean sphere, and cylinder of the progressive surface with a linear profile determined by $\kappa_0 = 2$ D and $m = 0.033$ D/mm, with the curvature growing two diopters from $y = -30$ mm to $y = 30$ mm. Curvatures are computed with the flat surface approximation.

where the q_n are, as we already know, arbitrary functions of y. As they multiply powers of x starting from order 3, these functions influence the surface shape and properties only at some distance from the umbilical line. The shape and properties near the umbilical line are determined by the first polynomial terms. If we set all the q functions to zero, it is easy to verify that the mean curvature is $H = \kappa_0 + my$, the Gaussian curvature $K = (my + \kappa_0)^2 - x^2m^2$, and the astigmatism, if we select positive sign, $C = 2|xm|$. The main curvatures are given by $\kappa_+ = \kappa_0 + my + m|x|$ and $\kappa_- = \kappa_0 + my - m|x|$, and finally, the axis orientation of the positive cylinder, $\tan(\theta_+) = x/|x|$. This last expression just returns the sign of x, so the orientation angle is 45° for positive x and 135° for negative x. The maps of surface height, mean curvature, and cylinder are shown in Figure 8.14. Within the flat surface approximation, the distribution of mean curvature and cylinder corresponding to a perfectly linear progression are quite simple: mean curvature only depends on the y coordinate, whereas the cylinder only depends on the x coordinate, growing at a rate $2m$. The maps in the figure have been obtained with a center curvature $\kappa_0 = 2$ D and assuming that the curvature linearly grows two diopters across 60 mm, which gives a slope $m = 0.0333$ D/mm. As the curvature grows from the bottom of the surface to its top, the surface turns out to be flatter at the bottom, which we can appreciate by the larger spacing between iso-lines in the surface height map. If we had added some q functions to the surface, we would have obtained different maps of cylinder and mean sphere, for the worse or the better, though in the region around the umbilical line, which in a practical application would go from $x \sim -10$ mm to $x \sim 10$ mm, the maps (and the surface) would have been pretty much the same as those we show in Figure 8.14.

Next example shows a more interesting, nonlinear profile.

Example 39 Let us consider the profile

$$\kappa_p(y) = \begin{cases} \kappa_0, & y \geq 0 \\ \kappa_0 - \kappa_A p(y/L), & y < 0, \end{cases}$$

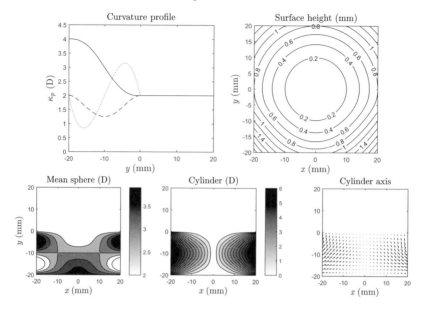

Figure 8.15 Details of the progressive surface corresponding to the piecewise profile defined in Example 2, for $\kappa_0 = 2$ D, $\kappa_A = 2$ D and $L = 20$ mm. Top: To the left, curvature profile (continuous line), its first derivative (dashed line), and second derivative (dotted line). Both derivatives are scaled and vertically shifted for the sake of visualization. To the right, surface height map. Bottom: Mean sphere, cylinder, and axis.

where p is the polynomial $p(u) = 10u^3 + 15u^4 + 6u^5$. The profile along with its derivatives (scaled and shifted) are shown in the top-left plot in Figure 8.15. It depends on three constants, κ_0, κ_A, and L. As in the previous example, κ_0 stands for the curvature at $y \geq 0$. In the upper half of the XY plane, the profile keeps constant curvature. In the lowest half, curvature increases from κ_0 at $y = 0$, to $\kappa_0 + \kappa_A$ at $y = -L$. This profile could be used in a real-life progressive lens (with progressive front surface) with appropriate selections for κ_0, A, and L. The polynomial p has been selected so its first- and second-order derivatives are zero at $y = 0$ and $y = -L$, so it smoothly joins the constant curvature section of the upper plane, and gets to the maximum curvature with horizontal slope. As the profile is a piecewise defined polynomial, the surface also has two distinct parts. In the upper half plane, the curvature is constant, and the surface is given by the paraxial (paraboloidal) approximation to a sphere, $z(x, y \geq 0) = (\kappa_0/2)(x^2 + y^2)$. For the lower part, we use equation (8.21) to readily find

$$z(x, y < 0) = \frac{\kappa_0}{2}(x^2 + y^2) - \frac{\kappa_A x^2}{2}p(u) - \kappa_A L^2 \left(\frac{u^5}{2} + \frac{u^6}{2} + \frac{u^7}{7} \right),$$

where $u = y/L$. The two parts join smoothly, and we can see the height iso-lines in the upper side have perfect circular symmetry, whereas at the bottom the surface becomes more curved and distorted. Mean sphere, cylinder, and axis are shown in the bottom row in Figure 8.15. Mean curvature is $H = \kappa_p(y) - (\kappa_A/4)x^2 p''(u)$, and cylinder $C^2 = (\kappa_A^2/4)[(4xp'(u)/L)^2 + (x^2 p''(u)/L^2)^2]$. We see the mean sphere reaches large maxima and minima at both sides of the plot. These points correspond with the extrema of the second derivative of the profile we can see in the profile plot. In a similar way, the cylinder has the shape of a "valley" whose bottom goes along the umbilical line. This bottom of the "valley" is usually called the *corridor*. The narrower point along the corridor is located at $y = -10$ mm, the same location at which the absolute value of the profile slope has its maximum. Although the profile is not far from a realistic one, the curvature maps are far from optimum. Mean sphere changes too much and too fast to the left and right sides of the surface, and cylinder keeps its linear growth toward these sides. Indeed, the cylinder keeps its linear growth all over the surface, as the term $x^2 p''(y/L)/L^2$ is much smaller than the linear one (and at $y = -10$ mm it is strictly zero). This behavior generates over 6 D of cylinder for a total increment in power along the umbilical line of 2 diopters.

The profiles we have chosen for the previous examples were based on polynomials, which we chose because they are easy to integrate so the reader can work the examples out. Most progressive lenses have a sigmoidal profile as the one shown in Example 2, but in general other functions are preferred to generate the profiles. Little can be done with respect to the slope. We will always face the fact that if we want to change the power along a short corridor, the slope will be larger and the astigmatism will grow faster. On the other hand, a long corridor with a small slope, hence a slower growth of astigmatism, may be impractical: The total power increase will be obtained in a very low position on the lens surface and the user will have to tilt his gaze downward by a large, uncomfortable angle. To improve the curvature maps right and left from the umbilical line, more degrees of freedom are required in the surface descriptions: We need to incorporate q functions with higher powers of x. Let us remark that the computation method we have presented in this section is based on the flat surface approximation. Although Minkwitz's theorem and the curvature behaviors we have presented are basically correct, the method cannot be used to design real-life lenses. The reason why the method can predict well the surface behavior in the neighborhood of the umbilical line is described next. At any point along the umbilical line, we may rotate the coordinate system so the XY plane is tangent to the surface and the Y axis parallel to the umbilical line. In this new reference system, the Hessian matrix is the correct description of the surface curvature at that point, even if the surface curvature is large at it. In its neighborhood, equation (8.20) is perfectly correct. Now, if we seek a curvature profile given by $\kappa_p(y)$, the actual curve $z(y)$ having this curvature would be obtained by solving the differential equation

$$\kappa_p(y) = \frac{z''(y)}{\left(1 + z'(y)^2\right)^{3/2}},$$

that is, finding the function $z(y)$ whose first and second derivatives satisfy the previous equation. If the surface contains this curve and it is umbilical, its curvature profile would be exact, and its exact main curvatures in the neighborhood of the umbilical line would be the same as those we obtain from a surface computed with equation (8.20), if we use equations (8.15) to (8.17) to compute the main curvatures with it. In any case, the surface behavior far from the umbilical line will only be sensible if the surface is quite flat.

Further Development of the Low-Curvature Model

In order to understand the main structure of the progressive surface and the consequences derived from them, we can just use the surface model represented by equation (8.21) with parabolic horizontal sections, as we did in the previous examples. For example, the relation between the slope of the curvature profile and the gradual build-up of astigmatism can be grasped, which leads to a basic statement of Minkwitz theorem. However, we have seen in the previous examples that the parabolic model predicts a steady growth of astigmatism along the horizontal direction, with the effect that a profile offering a curvature increase of just two diopters produces up to six diopters of cylinder at 20 mm from the umbilical line. This is not how modern progressive lenses behave. If we select at random a representative number of different progressive lenses from different designers, all of them with 2 D of addition, we can check that the average maximum unwanted astigmatism produced in these lenses is also around 2 D. Higher addition values yield average maximum astigmatism values slightly larger than the addition, and the other way around, smaller additions yield average maximum astigmatism values slightly smaller than the addition. No doubt, the $q(y)$ functions cannot be neglected in the design of progressive lenses and we will see in this section that they can be selected to restrain the growth of astigmatism imposed by Minkwitz's law, at least when we move far enough from the umbilical line. Then, let us consider the whole expansion given by equation (8.20). We can now write the second derivatives of the surface in terms of the main curvature profile and the q functions appearing in equation (8.20)

$$z_{xx} = \kappa_p + \sum_{n=3}^{N} n(n-1)q_n x^{n-2},$$

$$z_{yy} = \kappa_p + \frac{1}{2}\kappa_p'' x^2 + \sum_{n=3}^{N} q_n'' x^n,$$

$$z_{xy} = \kappa_p' x + \sum_{n=3}^{N} n q_n' x^{n-1}. \tag{8.25}$$

From them, we can directly give an expression for the mean curvature

$$H = \kappa_p + 3q_3 x + \left(\frac{1}{4}\kappa_p'' + 6q_4\right)x^2 + \frac{1}{2}\sum_{n=3}^{N}\left[(n+2)(n+1)q_{n+2} + q_n''\right]x^n.$$

The cylinder, once again, is much more complex, as it requires the computation of square roots of sums squared. We can still give a closed expression for the cylinder squared, $C^2 = \sum_{n=0}^{2N} c_n(y)x^n$, where the functions $c_n(y)$ are given by

$$c_n = \sum_{m=S}^{T} (\omega_m \omega_{n-m} + 4\zeta_m \zeta_{n-m}),$$

where $S = 0$ and $T = n$ if $n \le N$, otherwise $S = n - N$ and $T = N$, and where

$$\omega_0 = 0, \ \omega_1 = 6q_3, \ \omega_2 = 12q_4 - \frac{1}{2}\kappa_p'', \ \omega_{n\ge 3} = (n+2)(n+1)q_{n+2} - q_n'',$$

and

$$\zeta_0 = 0, \ \zeta_1 = \kappa_p', \ \zeta_{n\ge 2} = (n+1)q_{n+1}'.$$

The first coefficients for the expansion of the squared cylinder are

$$c_0 = 0,$$
$$c_1 = 0,$$
$$c_2 = 36q_3^2 + 4\kappa_p'^2,$$
$$c_3 = 144q_3 q_4 - 6q_3 \kappa_p'' + 24\kappa_p' q_3',$$
$$c_4 = -12q_4 \kappa_p'' + \frac{1}{4}\kappa_p''^2 + 4\left(36q_4^2 + 60q_3 q_5 + 9q_3'^2 + 8\kappa_p' q_4' - 3q_3 q_3''\right).$$
$$\cdots$$

From these coefficients we can draw some interesting conclusions. First, $c_0 = 0$ means there is no cylinder on the umbilical line, as it should be. Second, $c_1 = 0$ means the cylinder grows symmetrically to either side of the umbilical line, at least at close distance from it. Third, q_3 enters the second-order coefficient, so for small values of the x coordinate to either side of the umbilical line, the cylinder is given by

$$C \approx 2\kappa_p' |x| \sqrt{1 + 9(q_3/\kappa_p')^2}. \tag{8.26}$$

The most extended statement of Minkwitz's theorem is that of equation (8.24), where the cylinder grows proportional to the distance from the umbilical line, with the proportionality constant being twice the rate of growth of the curvature along this umbilical line. We see that, if the surface contains cubic terms in the variable x (that is, a nonzero q_3 function) then the cylinder will grow *faster* than predicted by the standard formulation of Minkwitz's theorem. In a real lens with a perfectly vertical umbilical line, we would expect a surface symmetrical with respect to the umbilical line, as Minkwitz assumed in his derivation [246]. However, as most modern lenses are asymmetrical (mainly to accommodate the required nasal inset of the umbilical line) a small amount of q_3 can be present in some parts of the umbilical line, giving rise to a slightly higher rate of growth of astigmatism. It is clear from the present deduction that, when designing progressive lenses, it is quite important to try keeping q_3 as close to zero as possible (even though the umbilical line may not be vertical, the present model can be applied locally, selecting the local Y axis as the local direction of the umbilical line).

Assuming q_3 is set to zero, then we get $c_3 = 0$ and simpler versions of the higher-order coefficients. Now we can see how we could overturn the cylinder growth in parabolic progressive surfaces. Assume we select a target value for the cylinder at distance x_t from the umbilical line, $C_t(x_t, y)$. Then we can try solving the differential equation for q_4

$$C_t^2 = 4\kappa_p'^2 x_t^2 + \left[-12q_4\kappa_p'' + \frac{1}{4}\kappa_p''^2 + 4\left(36q_4^2 + 8\kappa_p' q_4'\right) \right] x_t^4, \qquad (8.27)$$

and we could proceed stating further requirements on the cylinder and solving the resulting differential equations for the higher-order functions $q_{n>4}$. The procedure just outlined is interesting from the theoretical point of view, mainly to have a better grasp of the links and restrictions appearing in progressive surfaces; however, it is not a practical procedure. On the one hand, the differential equations are not linear and cannot be analytically solved for any interesting curvature profile κ_p. Second, the model gives us no clue about what could be sensible values for the target cylinder. Choose this target unwisely and the equations will not have solutions, or their solutions will be impractical, leading to surfaces with strong, undesirable warpings and wrinkles.

We may still use a simpler approach to seek how much cylinder we can expect in a well-designed surface: assuming we set $q_3 = 0$ and that we can neglect the fifth- and higher-order terms in the expansion of C^2, let us find what the coefficient c_4 should be for the cylinder to be zero at a distance d from the umbilical line. Under these assumptions, the squared cylinder is given by

$$C^2 = 4\kappa_p'^2 x^2 + c_4 x^4.$$

If the cylinder is zero at $x = d$, then $0 = 4\kappa_p' d^2 + c_4 d^4$ and solving for c_4,

$$c_4 = -\frac{4\kappa_p'^2}{d^2} = -\left(\frac{2\kappa_p'}{d}\right)^2.$$

This coefficient makes for a cylinder that grows linearly from the umbilical line (with the slope given by Minkwitz's theorem), reaches a maximum value, and then becomes zero at $x = d$. We can compute the maximum value and its position by equating the derivative of C^2 to zero,

$$\frac{d\left(C^2\right)}{dx} = 8\kappa_p'^2 x - 4\left(\frac{2\kappa_p'}{d}\right)^2 x^3 = 0.$$

From this equation we see that the maximum cylinder is located at $x = d/\sqrt{2}$ and the value of this maximum cylinder is $C_{max} = \kappa_p' d$. Let us now assume the progressive surface requires a distance L to increase the curvature from the value needed for far vision, κ_0, to the value needed for near vision, $\kappa_0 + \kappa_A$. The average slope would then be $\kappa_p' \approx \kappa_A/L$, and the value of the maximum cylinder $C_{max} \approx \kappa_A(d/L)$. In a typical progressive surface,

$L \approx 20$ mm, and if we try to get the cylinder down to zero at a similar distance from the umbilical, then we get the interesting result $C_{max} \approx \kappa_A$, that is, the maximum value of unwanted astigmatism will be similar to the addition we intend to produce.

The previous discussion does not offer complete proof of the final statement but a plausibility argument. The interested advanced reader may try to use the disclosed ideas to further explore the relation between addition, curvature profile and cylinder. Our argument has been applied to a single value of the coordinate y. If we try to do the same for all the values of y, we need to check that (8.27) is compatible with the values selected for d at each vertical coordinate. Also, we should consider that a well-designed progressive surface will need more terms, as we have to consider not only the cylinder but also other properties, such as the mean curvature, the overall smoothness, and the optical distortion produced by the lens. In any case, if we survey real modern progressive lenses and study their progressive surfaces, we will find that the average maximum value of astigmatism is quite close to the addition value [247, 248]. The designer can still decrease this astigmatism, but it will be at the cost of spoiling other optical properties, as we will see later.

Progressive Surfaces Without Umbilical Line

So far we have assumed the umbilical line is the plane curve $z(0, y)$, contained in the plane ZY. Minkwitz himself in 1965 and Schönhofer in 1976 extended the main result to nonplane curves and asymmetric surfaces [249, 250]. The use of nonplane umbilical lines is necessary to account for the eye convergence in near vision, which requires the umbilical lines to be tilted toward the nasal side. More recently, Esser and coworkers have published a generalization of Minkwitz's theorem for surfaces without an umbilical line [251]. They use a similar approach to the one presented above to compute the curvature of progressive surfaces, but instead of expanding the whole surface into a power series, they expand the exact expressions for the curvature components to explore the behavior of curvature along isolated horizontal meridians. Their main result is that the increase of astigmatism that appears in the direction perpendicular to the umbilical line can be slightly slowed down if we loosen the umbilicality restriction, letting some astigmatism grow along this line. It is not clear though how the neighbor horizontal meridians interact, and forcing some behavior of astigmatism along one of them may cause unwanted effects above or below it. A simpler derivation of this result based on a local analysis of the surface is Blendowske [252].

The model presented in this section can also accommodate astigmatism on the vertical meridian, in which case it would stop being an umbilical line. Let $\mathbb{C}_0(y)$ be the cylinder along the Y axis in matrix form, whose components are $c_{0x} = c_0 \sin^2 \alpha$, $c_{0y} = c_0 \cos^2 \alpha$, and $c_{0t} = -c_0 \sin \alpha \cos \alpha$, c_0 and α being functions of the y coordinate. We just need to modify the conditions stated in equation (8.18) to allow for the cylinder to be present,

$$z_{xx}(0, y) = \kappa_p(y) + c_{0x}(y),$$

$$z_{yy}(0, y) = \kappa_p(y) + c_{0y}(y),$$

$$z_{xy}(0, y) = c_{0t}(y). \tag{8.28}$$

The corresponding mean and Gaussian curvatures on the vertical meridian are $H(0, y) = \kappa_p + c_0/2$ and $K(0, y) = \kappa_p (\kappa_p + c_0)$, respectively. The same surface expansion that we previously used can be combined now with conditions (8.28), and from this combination we obtain the general expression for the surface,

$$z(x, y) = \iint \left(\kappa_p + c_{0y} \right) dy^2 + x \int c_{0t} dy + \frac{\kappa_p + c_{0x}}{2} x^2 + \sum_{n \geq 3}^{N} q_n x^n. \tag{8.29}$$

If we neglect cubic or higher terms in x, the Hessian of the surface at any point (x, y) is

$$\mathbb{H}(x, y) = \begin{bmatrix} \kappa_p + c_{0x} & c_{0t} + \left(\kappa_p' + c_{0t}' \right) x \\ c_{0t} + \left(\kappa_p' + c_{0t}' \right) x & \kappa_p + c_{0y} + x c_{0t}' + \frac{1}{2} \left(\kappa_p'' + c_{0x}'' \right) x^2 \end{bmatrix}, \tag{8.30}$$

and from it we can compute the cylinder squared, up to second order in x:

$$C^2 = c_0^2 + \left[8 c_{0t} \left(\kappa_p' + c_{0x}' \right) - 2 \left(c_{0x} - c_{0y} \right) c_{0t}' \right] x$$

$$+ \left[4 \left(\kappa_p' + c_{0x}' \right)^2 + c_{0t}'^2 - \left(c_{0x} - c_{0y} \right) \left(\kappa_p'' + c_{0x}'' \right) \right] x^2 + \cdots \tag{8.31}$$

The presence of astigmatism on the vertical meridian effectively changes the behavior of the astigmatism outside it. The cylinder squared now has a linear term, and the coefficient of the quadratic monomial also has new terms. But let us assume that the cylinder on the vertical meridian is constant, for example, a prescription cylinder. The additivity of curvature (valid for shallow surfaces) tells us that if we add a constant cylinder $\mathbb{C}_0$ to the surface (8.20), it will add *everywhere*, the umbilical line will turn into a line with astigmatism $\mathbb{C}_0$, and Minkwitz's theorem will be valid not for the total cylinder but for the *difference between the total and the prescription cylinders*. We can easily check this by equating to zero the derivatives of the components of the added cylinder in (8.30) and subtracting $\mathbb{C}_0$ from it: We just get the Hessian in equation (8.22), that of a surface with an umbilical line. In summary, Minkwitz's theorem is safe even for surfaces that combine progression and constant (prescription) astigmatism.

Now, let us glimpse at the other idea: allowing some – small – cylinder on the Y axis to try reducing the rate of growth of unwanted astigmatism outside it. The symmetry of the problem suggests the axis of this added astigmatism should be horizontal or vertical, and in that case c_{0t} and either c_{0y} or c_{0x} will be zero. To shorten the discussion, we will pick here the horizontal axis case, which is easier to handle. From (8.31), the cylinder squared, up to second order in x, would read

$$C^2 = c_0^2 + \left(4 \kappa_p'^2 + c_0 \kappa_p'' \right) x^2.$$

Then we could select the function $c_0(y)$ having a sign opposite to that of κ_p'', so that the product $c_0 \kappa_p''$ is negative, and automatically we would have a smaller growth of astigmatism than that predicted by Minkwitz's theorem for surfaces with umbilical lines.

We will not extend this analysis any further. Some patents have been filed proposing the presence of astigmatism on the vertical meridian (or its generalization, the curve tilting toward the nasal direction), because the progressive surface combines progression and prescription astigmatism, [253], because the astigmatism on the surface is intended to compensate for oblique astigmatism, [254], or to control the overall distribution of astigmatism [255, 256]. However, most progressives lenses without prescription astigmatism are designed with no or negligible astigmatism on the vertical meridian, even the majority of the latest and supposedly more advanced models. If the prescription is astigmatic, we usually encounter the prescribed astigmatism at the vertical meridian, and the astigmatic error grows from there following Minkwitz's theorem precisely. Introducing astigmatism on the umbilical line may reduce its growth in some small regions, but to the best of the authors' knowledge, it is not clear that overall lens quality could be improved by this technique.

8.4.3 Characteristics of Progressive Lenses

A progressive power lens (PPL) requires that, at least one of its two surfaces is progressive, such as those described in the previous section. There has been pioneering work on progressive lenses with aspherical and somewhat progressive surfaces at the front, back, or both lens surfaces since the beginning of the twentieth century. The first was the already mentioned invention of Owen Aves. Also worth mentioning is the invention of Poullain and Cornet, [257], which already describes a lens with a surface having an umbilical line, and from Gowlland, [258, 259], which constructs a progressive lens by combining a spherical or toric surface with an off-axis paraboloid, with the increased curvature of the paraboloid toward its center generating the required addition. In his patent of 1915, Gowlland considers using the paraboloid either as front or back surface. These and other inventions never entered the market in a significant way. In some cases the performance fell behind that of bifocals available at the time, but in most cases the manufacturing technology was not advanced enough to produce affordable lenses with enough quality. The first progressive lens with commercial success in Europe was invented by Maintenaz, [260, 261], and launched in 1961. In the United States, it was the lens invented by Volk and Weinberg [245, 262]. Both lenses were made of glass, with an anterior progressive surface and a posterior spherical or toric one. Curiously enough, the two lenses were successful despite having extremely opposite designs. While Maintenaz's lens progressed from the far to the near prescription in a relatively short distance, the American lens had a much longer progression. Likewise, the unwanted astigmatism was more concentrated in the former, while evenly distributed in the latter. None of the two inventions introduced radically new elements to the lens design arena: The idea of using continuous power variation for presbyopia, with or without regions with stable power, and the geometrical concepts and tools involved in progressive surfaces, such as umbilical lines and differential geometry, all dated back to previous patents and well-established developments in mathematics and geometry. The key factors that allowed these lenses to succeed were probably the companion inventions of generating machinery by the same authors, the technological maturity of the industry to build and run this machinery, and the vision and support of the companies that launched the lenses, the French company Essel (now Essilor), in the case of Maintenaz, and Univis Co. in the case of Volk and Weinberg. With respect to the optical design, we will see that Maintenaz charted the course for most of the "general purpose" designs that came later.

Next we will give a general overview of progressive lenses stepping through the four more important concepts needed to understand them: the location of the progressive surfaces, the reference points in a PPL, the characteristics of the umbilical line, and the management of the distribution of unwanted astigmatism. Dissertations on progressive lenses have traditionally been design-based: The "theory" of progressive lenses was constructed as new and "more advanced" designs were introduced by various lens manufacturers. As the geometrical and optical properties of PPLs are difficult to understand, and even more so their performance (by comparison with single vision or bifocal lenses), in most cases there is no other description of the technical properties of the lenses than that provided by the manufacturers, which over recent decades has grown complex and, in many cases, vague. So we can find many fine dissertations on PPLs that, beyond the description of the basic properties, consist of a more or less complete compilation of design descriptions. Nowadays, lens designers perfectly understand the geometrical and optical properties of PPLs; modern software and computers can optimize a customized lens using exact ray or wavefront tracing with complex lens-eye models in a matter of seconds, but even today the performance of PPLs is not fully understood. Our description will not be design-based, but property/performance-based, pointing out what is known and what is probably not known. A complete dissertation on progressive lenses according to this scheme would require a whole book on its own, so we will focus here on the essentials. The reader will find excellent complementary material in these references [203, 263, 264, 265, 266, 267, 268].

Reference Points and Marking

One of the main reasons for introducing progressive lenses was getting rid of the dividing lines typical in bifocals. This is, of course, a nice advantage, but curiously enough, one of the causes of adaptation difficulties. The reason is that the user cannot tell the near region apart; he does not know if he is looking through the right point for the given viewing distance. This problem is not trivial, and the fact the lens has regions with unwanted power values worsens it. Besides, the eye care professional (ECP) cannot easily tell from any power-related feature whether the lens is correctly mounted and well aligned with the user's eyes. It is for this reason that any progressive lens requires the identification of reference points and regions with a clear marking system that any ECP can understand. Marks and reference points are critical for understanding lens performance and for fitting lenses correctly. These marks and their relation with lens fitting and use are shown in Figure 8.16. There are temporary marks that are symbols and lines stamped with erasable ink and determine the location of the reference points. In addition to these temporary marks, there are also permanent marks, symbols subtly engraved at the lateral sides of the field of view

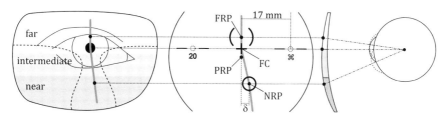

Figure 8.16 Marks and reference points in progressive lenses.

that provide information on the lens and also provide a reference system to relocate the reference points once the temporary marks have been removed. The main reference points are:

Fitting Point. Comonly referred as Fitting Cross (FC), it is the point that must be aligned with the pupil of the user at the main viewing position. In other words, when the user stands at the main viewing position, their line of sight should cross the lens through the fitting point. It is identified by a temporary cross (the fitting cross) or similar symbol.

Far Reference Point (FRP). Is the point at which the lens must provide the far vision prescription. It is typically located a few millimeters above the fitting cross. Its position is not indicated by an actual point, as it could interfere with the power measurement required for lens inspection. The temporary mark is usually a circle, parenthesis, braces, or any other symbol with the center at the FRP.

Prism Reference Point (PRP). This is typically located below the fitting cross. Progressive surfaces lack the required symmetries to define an optical center as we do with standard single vision lenses. We could choose to arrange a relative orientation of the two lens surfaces so that their normals are parallel at the fitting cross. This would guarantee zero eye deviation at the main viewing position. However the first PPL designers decided to locate a surface orientation control point (prism) a few millimeters below the fitting cross, at the geometrical center of the early, front-side progressive blanks. The thickness of a PPL can be reduced by tilting the back surface a positive angle around the X axis. This procedure introduces a base-down ground prism known as the *thinning prism*, which is the value we expect to find at the PRP. We should make clear that there is no reason the PRP must be at the geometrical center of blank. Depending on the prescription, and especially for positive ones, decentering the PRP may further optimize lens thickness. This was an option for front-side PPLs and now is almost a standard procedure for back-side PPLs.

Near Reference Point (NRP). At this point the lens should provide the near vision prescription. The NRP will be below the intermediate zone and displaced toward the nasal side. As in the case of the FRP, the NRP is not marked by a point but by some kind of hollow symbol whose center is the NRP.

The permanent marks or engravings consist of two symbols, the most popular being a hollow circle, symmetrically located at either side of the fitting cross. The distance between them is 34 mm and they are used to reconstruct the layout of temporary marks if necessary. For doing so, the practitioner locates the symbols, pinpoints them with an appropriate ink marker, and aligns these marks with the corresponding ones on a piece of cardboard with the lens layout provided by the lens vendor. Once the symbols are aligned, the practitioner may use the ink marker to copy the temporary marks on the lens, thus revealing the location of FRP, FC, PRP, and NRP. Below the alignment engravings and to the temporal side, we can find a number encoding the addition of the progressive lens. Below the nasal alignment symbol we may encounter other symbols that represent the manufacturer, the model, and the lens material, among others. We can also find lenses with no extra symbols at all on the nasal side. The permanent engravings are created in modern back-side PPLs by means of laser or needle engraving, with the latter typically made by the same lens generator after cutting the surface and before polishing.

On the line joining the alignment engravings and the fitting cross we usually see temporary line segments that are mainly used to help align the lens for the glazing process. It is also customary to stamp logos and lens model information as temporary marks.

Location of the Progressive Surfaces

From the days of the first successful PPLs up to the end of the twentieth century, progressive lenses were invariably made using a front progressive surface and a back spherical or toric surface. Front-side progressive blanks were made by a handful of companies all over the world. These companies would use their blanks to make their own products, or sell them to other lens manufacturers willing to offer progressive lenses. These blanks were cast with the progressive surface at the front side, and upon order from the ECPs with the prescription data, their back side would be resurfaced at the RX-lab with the spherical or toric surface that would provide the correct prescription. Computing and manufacturing a progressive surface is far more complex than computing and manufacturing spherical and toric surfaces, so molding the progressive surface at the front side of blanks was a good industrial strategy. There was, however, a limitation: The blanks had to be molded and stocked, so in order to keep costs low, the number of different stock-keeping units (SKUs), or different blanks should not grow too large. These lenses are known as front-side progressive lenses.

To describe the power balance in progressive lenses we will use $\mathbb{P}_{iF}$ and $\mathbb{P}_{iN}$ to designate the local refractive power of surface i at the FRP and NRP, respectively, as shown in Figure 8.17. In a front-side PPL the back surface is a sphere or a torus with constant refractive power $\mathbb{P}_2$. This back surface could also be aspherical or atoric, to provide compensation for oblique aberrations [269], in which case its curvature would not be constant, but the reader must differentiate between this aspherization, intended to provided constant power to the user at all gaze directions, and the aspherization leading to progressive surfaces, intended to produce different power at different points on the lens (more on this later). Now, if the prescription at the FRP is $[S, C \times \alpha] \equiv S\mathbb{I} + \mathbb{C}$, and the addition is A, then, within the thin-lens approximation,

$$S\mathbb{I} + \mathbb{C} = \mathbb{P}_{1F} + \mathbb{P}_2,$$
$$(S + A)\mathbb{I} + \mathbb{C} = \mathbb{P}_{1N} + \mathbb{P}_2.$$

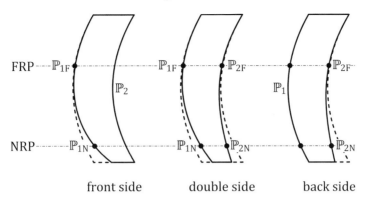

FRP

NRP

front side   double side   back side

Figure 8.17 Types of progressive lens according to the location of the progressive surface. The gray line stands for the progressive surface. The dashed line represents the curvature of a nonprogressive surface.

Subtracting the two equations we get $\mathbb{P}_{1N} - \mathbb{P}_{1F} = A\mathbb{I}$. Most front-side progressive blanks are locally spherical at FRP and NRP (in other words, the FRP and NRP are, at least, umbilical points). In this case we can write $\mathbb{P}_{1F} = B\mathbb{I}$ and $\mathbb{P}_{1N} = (A + B)\mathbb{I}$, were B is the *base curve of the progressive blank*. We see that, attending to the local curvature of the lens surfaces at FRP and NRP, a front-side PPL meets the same conditions as a front-side bifocal. Finally, let us remember that the refractive power of a front surface is given by $(n - 1)\mathbb{H}_1$, so in order to increase the refractive power to get positive addition, curvature must increase from FRP to NRP, that is, the front progressive surface bulges out at the near region (see Figure 8.17).

At the end of the nineties, some patents appeared claiming the combination of two progressive surfaces in such a way that one of them could subtract part of the unwanted astigmatism from the other one [270, 271][4]. Not that Minkwitz's theorem can be tricked, but using a second progressive surface could help redistribute the astigmatism of a given first progressive surface. Minkwitz's theorem can be applied to the wavefront refracted by the lens, and ophthalmic lenses are pretty much thin systems, [264, 265], so any progressive distribution of sphere and cylinder can be achieved by using either one of the surfaces of a lens or both. The balance of power in a double-side PPL (see Figure 8.17) is given by equations

$$S\mathbb{I} + \mathbb{C} = \mathbb{P}_{1F} + \mathbb{P}_{2F},$$

$$(S + A)\mathbb{I} + \mathbb{C} = \mathbb{P}_{1N} + \mathbb{P}_{2N}.$$

Now, we cannot infer any relation between the refractive powers at FRP and NRP for each surface. For example, the addition could be split between the two surfaces, so that

---

[4] At a fundamental level, this was the original idea of Owen Aves.

$\mathbb{P}_{1N} - \mathbb{P}_{1F} = \gamma A$ and $\mathbb{P}_{2N} - \mathbb{P}_{2F} = (1 - \gamma)A$, where γ is any number between 0 and 1. We could even consider making γ bigger than one or negative, so the partial addition at one of the surfaces should be negative (though that would not probably help to get a better lens). The mean sphere of the front surface at point F, $\mathrm{tr}(\mathbb{P}_{1F})/2$, can still be defined as the base curve of the double-side PPL.

Finally, the third type of PPLs, attending to the location of the progressive power, is the back-side progressive lens, represented to the right in Figure 8.17. Once again, some of the very first inventions reported back-side progressive surfaces, but then the configuration was abandoned as the industry focused on a RX-lab process that only generated spherical and/or toric surfaces on the back side of the lens blanks. This scenario changed dramatically when free-form technology burst into the ophthalmic industry at the end of the twentieth century. Free-form generators and polishers are machinery capable of shaping arbitrary surfaces on either side of a lens blank. As the RX-labs were already set up to process the back sides, they kept doing the same after incorporating the free-form machinery. The big difference comes from the fact that free-form technology can cut any progressive surface on the back side: Progressive blanks are no longer necessary, with large savings coming from much reduced stock (fewer SKUs are necessary) and cheaper nonprogressive blanks. From the optical point of view, the advantage of free-form technology is that the asphericity of each back surface can be tailored to minimize the oblique aberrations of the lens, as we will see later. For a back-side PPL we have

$$S\mathbb{I} + \mathbb{C} = \mathbb{P}_1 + \mathbb{P}_{2F},$$

$$(S + A)\mathbb{I} + \mathbb{C} = \mathbb{P}_1 + \mathbb{P}_{2N},$$

and subtracting the equations, $\mathbb{P}_{2N} - \mathbb{P}_{2F} = A\mathbb{I}$. The front surface is usually spherical, so $\mathbb{P}_1 = B\mathbb{I}$, B being the base curve of the lens. The back surface will also incorporate the prescription cylinder, $\mathbb{P}_{2F} = (S - B)\mathbb{I} + \mathbb{C}$, so in general, it will not be locally spherical at any point. Now, as the refractive power of the back surface is given by $(1 - n)\mathbb{H}_2$, getting a positive addition requires the curvature of the surface to be reduced, producing a flatter near region. It is also possible to account for different prescriptions at FRP and NRP, for example, accounting for lens effectiveness, as we saw in Section 8.2.2 (see for example [272]). In this case we need to differentiate between S_F and S_N, and between $\mathbb{C}_F$ and $\mathbb{C}_N$. The differences should be implemented in the refractive power of the back surface, $\mathbb{P}_{2F} = (S_F - B)\mathbb{I} + \mathbb{C}_F$, and $\mathbb{P}_{2N} = (S_N - B + A)\mathbb{I} + \mathbb{C}_N$.

An outcome from free-form technology is the proliferation of new designs. While front-side PPLs required a different SKU per addition value and design type, an infinite variety of designs and addition values can be made with free-form technology from a single semifinished blank. This is not a property of back-side PPLs, but the result of the technology used to make them. Free-form generators and polishers could be easily adjusted to shape the front side of the blank, in which case we could have the same infinite variety of designs as front-side PPLs. However, producing, storing, and processing back-side semifinished blanks would require new inversions, adding cost to the final product.

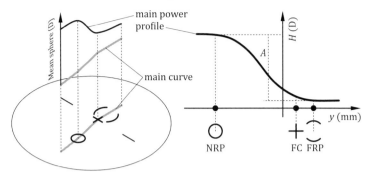

Figure 8.18 Principal curve and principal power profile of a progressive lens.

Structure of a Progressive Lens: Principal Curve and Power Profile

From now on we will give the name *principal curve* to the path joining the reference points FRP, FC, and NRP[5]. This curve can be an umbilical curve of the progressive surface or not. As we said before, in most cases, even for the majority of the most advanced designs available today, this principal curve is basically umbilical. We should avoid the label "meridian" for this curve, as "principal or main meridian" has a clear (and different) meaning in geometry that we continuously use in this book. Also, we should avoid using the label "vertical," as in a modern lens this path is curved toward the nasal side. A last remark about this main curve is that we should think of it as a property of the wavefront refracted by the lens (its *through-power*), rather than a property of the surfaces. For example, the combination of progressive cross-cylinders we analyzed in the introduction will produce a wavefront with an umbilical line, but the cylinders themselves do not have a single umbilical point.

We will give the name *principal power profile* to the power along the principal curve (see Figure 8.18). The pair constitutes one of the key elements determining the performance of a progressive lens for the three reasons we list next.

1. They link lens power and orientation of the line of sight. That is, the eye orientation needed to get the far prescription (far vision), partial addition (intermediate vision), or full addition (near vision) is determined by the power profile along the principal line. At first sight, this may seem obvious and of little importance, but it is not. The eye orientation required to maintain a visual task has a big impact on comfort and ergonomics [273, 274, 275]. Nonpresbyopic persons may tilt their heads and accordingly change their eye orientation while performing any visual task. This is a degree of freedom the user of multifocal lenses is going to lose, at least partially. The problem is shown in Figure 8.19. It shows a user's face in profile at four different situations. In subplot (a) the user is at the main viewing position: line of sight horizontal and passing through

---

[5] Depending on the design, it can also pass through PRP, but not necessarily.

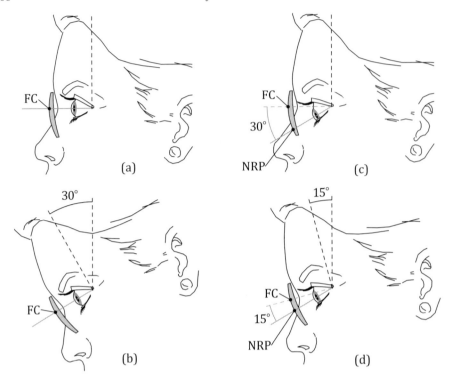

Figure 8.19 Relation among eye rotation, head inclination, and reference points in progressive lenses.

the fitting cross. In subplots (b) to (d) the fixation object forces the line of sight to form an angle of 30° with the main viewing direction. This inclination of the line of sight can be achieved by solely tilting the head (b), solely tilting the eye (c), or spliting the tilt between head and eye (d). A nonpresbyopic user may adopt any of the three positions (and any other in between), relieving the muscular stress derived from holding a position steady. On the contrary, the user of progressive lenses has to look through particular intervals along the principal line to get the extra power required by the object distance. For example, if the fixated object is at 40 cm or closer, advanced presbyopes will need the full addition, which the lens provides just around NRP. This requirement determines a fixed angle between the line of sight and the head axis, regardless of the vertical position of the fixation object. The effect becomes more acute as the amplitude of accommodation decreases with age. The ranges of clear vision that can be computed with (8.5) or (8.6) will become smaller with the progress of presbyopia, and the freedom to use different points of the lens along the principal line will be reduced, especially for intermediate and near vision.

2. The local slope of the power profile along the principal curve determines the width of the corridor, and this holds true as long as the principal curve is umbilical for the power distribution (in other words, for the wavefront refracted through the lens). The existence

of a corridor seems to be the main drawback and main limiting factor of progressive lenses, which is not true in general; the corridor is rather unlikely to be the cause of dissatisfaction for PPL (progressive power lenses) wearers. However, because of its bad name, it is not surprising that marketing material of lens manufacturers continuously addresses the width of the corridor as something that can be improved, and that is actually improved, by new "design technologies." However, geometry and optics dictate that the width of the corridor is uniquely linked to the local slope of the profile. The only "technology" that can create a wider corridor at some point along the principal line is that reducing the slope of the power profile at the same point. However, the power profile must increase its value from the far prescription to the near prescription (the addition) in the space between the FRP and the NRP, and this gives us an inferior bound for the slope. Reducing the slope requires increasing the distance from the FRP to the NRP, and in turn this demands a larger downward angle for the line of sight, which can be worse than the corridor itself: A balance between them must be sought. After this list we will see in greater detail the way to configure the power profile.

3. The third key component of the principal line is the lateral inset, that is, its displacement toward the nasal side to account for the convergence required to fixate near objects. Let us assume the umbilical line were perfectly vertical. When aiming an object located at intermediate or near distance on the sagittal body plane, eye convergence would cause the intersection point between the line of sight and the lens to move away from the umbilical line toward the nasal side, a region with unwanted astigmatism. To avoid this, and maximize the field of view, the umbilical line (and its associated corridor) should be tilted toward the nasal side. In some early designs the umbilical line was perfectly vertical on the progressive blank. These blanks should be rotated inward before blocking when surfacing the back sides. This operation would provide the required nasal displacement, but would also rotate the whole distribution of unwanted cylinder. Modern lenses are designed with displaced umbilical lines, allowing for better control of the symmetry of the design.

We will explore next in greater detail the properties of the principal curve and its associated power profile. A point on the principal curve will have coordinates $(\delta(y), y)$ for the right eye and $(-\delta(y), y)$ for the left eye, where $\delta(y)$ is a function giving the right inset at any given value of the y coordinate. We may locate the origin of coordinates at any point we want, for example the PRP. For positive y, the function δ is basically zero, as no inset is required for far vision. For negative y, $\delta(y)$ grows from zero to the maximum inset at the NRP. This growth can be linear or nonlinear depending on different design assumptions. Below the NRP the function $\delta(y)$ should stabilize, as more convergence is neither expected nor needed. We will assume that the principal curve is either umbilical or it has constant prescription astigmatism on it, so Minkwitz's theorem is at work. The power profile is then described by a scalar function $P(y)$ that assigns the mean sphere to each point on the principal curve. Let us also assume that the power profile is more or less stable at the far and near regions. Then $P(y)$ must be a decreasing function with the shape of a *sigmoid*, such as the curvature profile shown in Figure 8.15. For that example we used a piecewise defined polynomial, which is easy to integrate, but now it will be more convenient to use a single analytical function. A good candidate is the logistic function, suggested by Guilino in 1982, [276], which we will write in the form

$$P(y) = P_0 + \frac{A}{1 + \exp\left[\frac{2\ln 9}{L}(y - y_{10} + L/2)\right]}, \tag{8.32}$$

where P_0 is the prescription mean sphere at the FRP, A is the addition, and L and y_{10} are constants. When $y \ll y_{10} - L/2$, the exponent becomes negative and large, the exponential tends to zero, and $P(y)$ tends to $P_0 + A$. When $y \gg y_{10} - L/2$, the exponential tends to infinity and $P(y)$ tends to P_0. In between, the function changes smoothly between P_0 and $P_0 + A$. We can easily verify that $P(y_{10}) = P_0 + 0.1A$, so y_{10} is the vertical coordinate at which the power gets 10% of the addition. We can also define $y_{90} = y_{10} - L$, and verify that $P(y_{90}) = P_0 + 0.9A$, so y_{90} would be the coordinate at which the power gets 90% of the addition. As $L = y_{10} - y_{90}$, we conclude that L is the vertical distance between the points at which addition gets 10% and 90%. We will name L the *progression length* or *corridor length*.

A typical profile that could be used in a general-use, modern lens is shown in Figure 8.20. To the left we see a plot of the profile with the Y axis vertical and the P axis horizontal. We have chosen the point y_{10} to be coincident with the PRP (this is not necessary, but it is a reasonable selection). For the location of y_{90} we have chosen the top of the region surrounding the NRP. Once again, this is not necessary, but is quite common. The *minimum fitting height* (MFH) is defined as the minimum pupil height (see Appendix A) necessary for adequate fitting of the lens. If the distance from the fitting cross to the NRP were bigger than the pupil height for the frame selected by the user, the NRP would be cut out from the lens when glazing. It is clear that some allowance should be used below the NRP, and the MFH should be defined as the distance from FC to NRP plus this allowance. Unfortunately, the allowance is not defined in ISO or ANSI standards and each lens manufacturer uses its own criterion. Let us put some numbers to obtain a realistic relation between the corridor

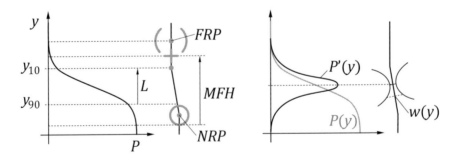

Figure 8.20 Relation between progression length and minimum fitting height, and corridor width in a sigmoidal profile (stable far and near regions).

length and the minimum fitting height: The fitting cross is typically located between 0 and 5 mm above y_{10} (the PRP in our example); let us take 3 mm. Let us assume now that the NRP is 2 mm below y_{90}, and we still pick 2 more millimeters below the NRP as allowance for the MFH. Then, $MFH = 3 + L + 2 + 2 = L + 7$. Let us assume now a typical MFH of 18 mm; the corridor length should then be $L = 11$ mm.

Let us explore now the relation between the profile and the corridor width. From Minkwitz's theorem, we know that for x close enough to the umbilical line, the cylinder grows linearly with x,

$$C(x) \simeq 2 \left| P'(y) \right| x.$$

Minkwitz's theorem was experimentally tested on actual progressive lenses by Sheedy et al. [277]. Their results were somewhat inconclusive, probably due to the averaging effect of the measuring instrument, which made the results very sensitive to the local rate of change of power. However, and according to the authors, Minkwitz's theorem seemed to be a good overall predictor of cylinder growth up to a few millimeters from the umbilical line. According to the experience of the authors measuring progressive surfaces with high accuracy profilers [278, 279], as long as the cylinder is below 0.1 D on the principal curve of the progressive lens, the linear growth predicted by Minkwitz's theorem holds true in the range between 4 and 6 mm from the umbilical line, depending on the local slope of the profile and the selected overall management of unwanted astigmatism. We can then use the previous equation to determine the distance from the principal curve (which should be umbilical or close to umbilical) at which the cylinder reaches some threshold value, C_{th},

$$x_{th}(y) = \frac{C_{th}}{2 \left| P'(y) \right|}.$$

The width of the corridor, $w(y)$, for the given cylinder threshold would be twice x_{th},

$$w(y) = \frac{C_{th}}{\left| P'(y) \right|}. \tag{8.33}$$

We can use our analytical profile and its derivative to compute the expected corridor width. To shorten the expressions, let us define $k = 2 \ln 9 / L$ and the midpoint of the profile, $y_{50} = y_{10} - L/2$. The derivative of the logistic profile is

$$P'(y) = -\frac{2A \ln 9}{L} \frac{\exp\left[k\left(y - y_{50}\right)\right]}{\left\{1 + \exp\left[k\left(y - y_{50}\right)\right]\right\}^2}, \tag{8.34}$$

and its width

$$w(y) = \frac{LC_{th}}{2A \ln 9} \frac{\left\{1 + \exp\left[k\left(y - y_{50}\right)\right]\right\}^2}{\exp\left[k\left(y - y_{50}\right)\right]}. \tag{8.35}$$

At any point along the corridor, the width is proportional to the corridor length and to the selected threshold for the cylinder, and is inversely proportional to the addition. This is true for any analytical profile with a sigmoidal shape. The derivative of the logistic function is bell-shaped, as shown in Figure 8.20, in the subplot to the right. The maximum of the derivative is located at y_{50}, the profile inflection point at which the slope is larger. It is straightforward to check that the minimum width for the logistic profile is

$$w_{min} = \frac{2LC_{th}}{A \ln 9}. \tag{8.36}$$

In the case of the previous example, where $L = 11$ mm, assuming $A = 2$ D, and choosing a cylinder threshold value of 1 D, we obtain $w_{min} = 5.1$ mm and $w(y_{10}) = w(y_{90}) = 13.9$ mm. Equations (8.35) and (8.36) are good estimators of the corridor width for progressive lenses with logistic-style power profile and umbilical principal curve, for cylinder thresholds not larger than 1 D and within the range $[y_{90}, y_{10}]$, though to compute the widths at the endpoints of the interval a smaller cylinder threshold would be better, as we will see later. The corridor of a logistic profile has a particularly small minimum width, w_{min}, because the slope at its inflection point is pretty large, $P'(y_{50}) = -0.217$ D/mm, assuming the same addition and progression length as before. If the profile were to grow linearly between y_{10} and y_{90}, the slope would be $0.8A/L = 0.160$ D/mm, which will yield a minimum width $w_{min} = 6.24$ mm. However, a linear profile would have a constant width as long as the slope is constant, while the logistic profile rapidly widens as we move up or down from the inflection point at y_{50}.

Designers may use other functions in order to have better control of the slope along the profile. For example, the function

$$P(y) = P_0 + \frac{A}{2} \left[1 - \frac{k(y - y_{50})}{(1 + k^s |y - y_{50}|^s)^{1/s}} \right],$$

is another sigmoid whose progression length is given by

$$L = \frac{1.6}{k (1 - 0.8^s)^{1/s}}.$$

The parameter s controls the slope at the inflection point. A value $s = 2.7$ yields a profile quite similar to the logistic one. Larger values will produce profiles that are more linear and wider at their inflection point, while smaller values of s (between 1 and 2.7) will generate profiles with even larger maximum slopes (smaller widths) that rapidly widen as we move away from the inflection point. The importance of this control over the slope is shown in Figure 8.21, where we present three $A = 2$ D profiles with y_{10} at zero and $y_{90} = -10$ mm, that is, $L = 10$ mm. The solid line corresponds to the logistic one, the other two to the generalized profile with $s = 1.5$ (dashed line) and $s = 10$ (dash-dotted line). At first glance the profiles seem pretty much alike, but obviously their derivatives and corridor structures differ a lot. We see that we can trade width between different places in the corridor. The $s = 10$ profile, highly linear, provides the wider corridor, opens up abruptly, but at y_0 and y_{90} is narrower than the other two. The profile with $s = 1.5$ has a very small minimum width at y_{50} but it is the widest at y_{10} and y_{90}. The logistic profile is in between. The user experience with the three profiles would be totally different. For example, we can move the fitting cross slightly downward with $s = 1.5$ to reduce the MFH, but we should not do so

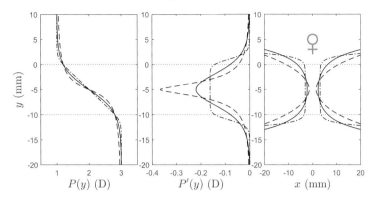

Figure 8.21 Comparison of three analytical profiles (left), their derivatives (center), and the corresponding corridors at $C_{th} = 1$ D (right).

with $s = 10$, as we would drastically reduce the field of view of the user 2 mm below the fitting cross.

Other profiles can be used that will distribute the power growth in different ways. For example, we may use asymmetric profiles, seeking a fast transition from the stable region at far vision to an approximately linear progression region, but then a slow transition from the center of the corridor to the near region. Or vice versa. Piecewise defined polynomials or spline functions can be used to achieve total control of the slope along the profile. In any case, the reader must be aware that in the end the profile must increase its power by A diopters in a limited vertical space. We can concentrate the growth in a small region, yielding a narrow corridor that quickly widens, or we can distribute the growth over the whole available space, which would produce a wider corridor that widens very little up to the stable regions. Either decision has its own drawbacks and advantages, and, more importantly, different users may prefer one option over the other according to their needs. We cannot *a priori* say that a corridor with a wider minimum width is better or worse (within certain limits); the rating will depend on the visual tasks and preferences of the user.

We will finish this section by considering another important fact regarding power profiles. Before the deployment of free-form technologies, the number of available designs was limited and basically the only quantifiable design characteristic available to the ECP was the MFH. A value of 18 mm was usually recognized in occidental countries as "standard," and PPL designs were labeled as "long" or "short" depending on their MFH being larger or smaller than 18 mm. At the same time, the idea that the smaller the MFH, the narrower the corridor permeated among lens manufacturers and ECPs. This fact was understandable, as the leading designs in occidental countries invariably had the y_{10} point 2 to 4 mm below the fitting cross and so MFH and corridor length (as we have defined it) would provide similar information. However, production of progressive lenses started in Japan in 1980, the first designs having a noticeable wider corridor than European or American lenses with the same MFH (see, for example [280]). In some cases, the feature was announced as a "new technology" that would provide this advantage. In fact, the technology consisted of displacing y_{10} upward with respect to the fitting cross. Some designs even had y_{10} above the fitting cross. Assuming y_{90} was kept at the same position and that the same allowance for the MFH below the NRP was provided, this new design decision automatically would provide a wider corridor with the same MFH. Of course, there was a price to pay: Some addition was noticeable at the fitting cross, and this would demand that users should tilt their gaze upward to get their far prescriptions "clean." Similarly, the fitting cross would be located well inside the corridor, with the horizontal field of view at FC being reduced as a consequence of the upward displacement of y_{10}. Experience demonstrates that this design decision is not necessarily better than the initial one, having y_{10} 2 to 4 mm below FC. Once again, the particular needs and adaptation history of the user will play a crucial role when choosing one design over the other. For the last 10 years or so there has been a tendency to produce designs with shorter MFH, with 16 mm becoming very popular. This shortening carries the need for squeezing FRP, FC, and y_{10} into a tighter vertical space. Still, having y_{10} below FC seems to be the preferred choice among European and American users.

Nowadays there is an explosion of design availability as a consequence of the spreading of free-form technology. The lesson from this example is that the ECP that wants better control of the selection of progressive designs should be aware not just of the MFH but also of some indicators about the start of the progression. Only the knowledge of the three values of y_{10}, L, and MFH will give the ECP a detailed idea about the performance of the progressive lens around the principal line.

Structure of a Progressive Lens: Distribution of Peripheral Power

The principal curve and its associated power profile can be considered the backbone of the progressive lens. Not only do they determine the gaze angles required for far, intermediate, and near vision, they also determine the distribution of power in the region surrounding the principal curve, which we will call the *Minkwitz region*. However, the lens designer still has a lot of work left. The remaining area outside the Minkwitz region, which we may call *peripheral region*, is weakly influenced by the profile; it rather depends on the $q_n(y)$ functions we defined in equations (8.20) and (8.29). This means there is a lot of freedom to decide how this distribution will be done.

The most effective method to attain a suitable distribution of power in the peripheral region seems to be *optimization*. It is possible to construct complete progressive surfaces in a fully parametric way [281], which is fast, but the type of possible power distributions is limited by the way the parametrization is constructed. Besides, parametrized surfaces are not guaranteed to be optimal with respect to unwanted astigmatism and mean sphere. Optimization techniques are based on the previous specification of target distributions of cylinder and mean sphere, $C_0(x, y)$ and $H_0(x, y)$. The progressive surface will be defined by a set of N parameters, $\mathbf{c} = \{c_n\}$, which will provide enough flexibility for its curvature to conform to any progressive distribution we want (see Section 2.4.2). Then we compute the cylinder and mean sphere of our surface, $C(x, y, \mathbf{c})$ and $H(x, y, \mathbf{c})$, and the merit function

$$\Phi(\mathbf{c}) = \sum_{k=1}^{M} \chi_k \left[C(x_k, y_k, \mathbf{c}) - C_0(x_k, y_k) \right]^2 + \eta_k \left[H(x_k, y_k, \mathbf{c}) - H_0(x_k, y_k) \right]^2, \quad (8.37)$$

where the sum is extended to a set of M points (x_k, y_k) that sample the whole lens area, and χ_k and η_k are point-dependent weights. Minimization algorithms are then used to find the set of coefficients $\mathbf{c}$ that will render Φ minimum, and hence the surface that will have distributions of cylinder and mean sphere as close as possible to the target functions. The problem is easily stated but not so easy to implement unless we have good candidates for the target functions and the weights. If any of these are not well chosen, the minimization process will render nonsense results. There are many different ways to tackle this problem, and each designer has its own set of tricks and tools. There are, however, two main approaches illustrated in Figure 8.22. In the first, the target functions C_0 and H_0 are created as realistic and close to the final result as possible. Of course, this approach requires a lot of experience on the part of the designer. For example, we could naively propose a target distribution C_0 with extremely low values of astigmatism, yet gathered in a small area. As there is no surface with such a distribution of astigmatism, there will be infinitely many surfaces equally distant from the target, the merit function will have many local minima, and convergence of the minimization algorithm to a useful solution would be impossible. A common trick here is starting with a feasible distribution of sphere and cylinder, making some modifications on them, and using the modified distributions as target. After running the optimization we check how close the result is from our expectations. If it is not what we wanted, we start over with a new modification, and so on. With this approach in which detailed target distributions are used, the weights are not very important. We can set them all to 1, or we can reduce its value in those regions where we are not certain of the quality of the targets.

The second approach to solve the minimization problem, illustrated in the left plot in Figure 8.22, consists of selecting very basic target functions and a more detailed distribution of weights. For example, we can set $C_0 = 0$ and $H(x, y) = \kappa_p(y)$. That is to say, we want no unwanted astigmatism at all, and we want the curvature all over the lens to follow the

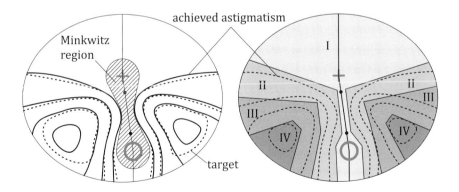

Figure 8.22 Distribution of peripheral astigmatism in a progressive lens.

power profile. Of course we know that is impossible. Then we create different regions to which different weights are assigned. In region I the weight will be the highest: We want the surface to strictly adhere to the targets in this region (which is possible). Then the weights are made successively smaller in regions II, III, IV, etc. This means the merit function allows the actual distributions to deviate more from the targets in these areas. The smaller the weight, the bigger the deviation. The way we configure the weight regions will determine the shape of the final distribution.

Detailed examples of the approach based on weights have been given by Loos [282], Hasenhauer, [283], Jonsson [284], Wang et al. [285], and Jiang, [286]. Isenberg [287] describes a method based on detailed target distributions of sphere and cylinder generated parametrically that are later improved through optimization. Dürsteler, [288], uses a hybrid method in which the points (x_k, y_k) are set as control points for which both target values and weights are specified. Other approaches to the problem of obtaining the peripheral power distributions may include more terms in the merit function to improve convergence or the smoothness of the final solution, for example those of Tazeroualti, [289] and Mendiola-Anda, [290]. Finally, it is worth mentioning that even though optimization seems to be the more widespread method to compute peripheral power distributions, other techniques have been proposed based on the resolution of differential equations, for example, the classical method used by Winthrop, [120, 291], based on the solution to the Laplace equation, and an improved version of the same by Tang et al., [292], which allows for more flexibility to produce different types of designs.

In Figure 8.23 we present four examples of astigmatism maps from actual progressive lenses, with the iso-lines representing astigmatism in steps of 0.5 Design (a) corresponds to a design from the beginning of the eighties in which the spread of astigmatism is relatively small. The consequence is that astigmatism grows to pretty high values, as explained previously. The distribution is symmetric with respect to the umbilical line, so the astigmatism lobes are tilted with respect to the vertical direction. Modern lenses are asymmetrical with respect to the umbilical line and symmetrical with respect to a vertical line, a feature that improves binocularity. Design (b) is asymmetrical and has a narrower near region. Although the build-up of astigmatism is fast (the first three iso-lines are closely packed), the growth stops at a lower value than in (a). Design (c) has even less astigmatism and the iso-lines are more spaced. This has been achieved by allowing the 0.5 iso-line to move above the fitting cross at the nasal and temporal sides and with an even narrower near region. Finally, design (d) has a longer power profile than the other three, and the 0.5 iso-line is allowed to keep a 40 degree inclination in the far region. This accounts for a lens with low astigmatism and spaced iso-lines, which means the spatial change of astigmatism is slow. We could, of course, present tens of other designs, some very similar, some very different. The reader should understand that the geometry allows for a great deal of flexibility to create different power distributions, with different degrees of spreading toward the near and far regions.

It has been a customary language in the ophthalmic industry to talk about *hard* and *soft* progressive designs. The terms were never precisely defined, but the general idea is that a lens design is hard if the spreading of the astigmatism distribution is small, or if the

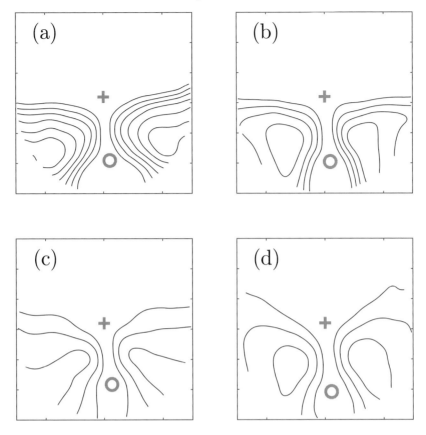

Figure 8.23 Examples of different distributions of astigmatism in PPLs. See the text for details.

corridor length is short. On the other hand, a PPL would be soft if the astigmatism is spread out and/or the length of the corridor is large. We have to consider that both properties are independent. We can construct a progressive lens with a short corridor and the astigmatism well spread out, And the other way around, we could think of a PPL with a long corridor but low spreading of astigmatism. The other combinations, long corridor and high-spread or short corridor and low-spread, are also possible. The latter two would be soft overall, and hard overall, respectively. We should think of softness and hardness in terms of the closeness of the iso-lines of sphere and cylinder (of course, we should use the same iso-lines to compare different lenses). Wherever the iso-lines are separated, the gradient of power will be small, and the lens will be soft. Wherever the iso-lines are closely spaced, the gradient of power is high and the lens would be hard.

8.4.4 Progressive Lens Performance

We have dealt so far with the description of the properties of progressive surfaces and lenses, and with the methods for obtaining such surfaces. The ECP aiming to correctly prescribe progressive lenses would probably raise the next question: *Among the many different types of PPL designs, which one is better for my patient? What are the rules for proper selection of PPL design?* This is a formidable question that on most occasions will lack a clear or unique answer, but regardless of its difficulty, it is a question that both industry and academy have to tackle. There are two aspects to the study of the performance of progressive lenses: On the one hand, the tools to test the user response to any type of lens, and in particular PPLs, are clinical trials. On the other hand, assessment tools able to quantitatively differentiate between different PPL designs are needed.

Following the description by Preston and Pope, [266, 293], clinical trials can be classified into four types: acceptance trials, preference of PPLs vs other solutions for presbyopia, comparison between different PPL designs, and finally objective assessment of different aspects of visual function when using PPLs.

PPL Assessment with Clinical Trials

Most of the available clinical trials of the first three types were conducted during the eighties and nineties, the period through which PPLs became the solution of choice for ECPs and presbyopes. For the last 15 years the main research focus has turned to the development and testing of new objective techniques that help us understand how PPLs affect different aspects of the visual function.

Progressive lenses are nowadays and by difference the preferred compensation for presbyopia among spectacle users, but as they struggled with bifocals, trifocals, and readers, a lot of studies took place to test acceptance of the "new" technology [294, 295, 296, 297, 298, 299, 300, 301]. All these were performed as field studies (except that by Augsburger) and reveal acceptation ratios over 90% and satisfaction ratios over 80%. Those studies also reveal certain patterns that could obscure the results of contemporary and later trials comparing different PPL designs: the satisfaction rate a user will give to the design strongly depends on

1. Being a previous PPL wearer,
2. Being a previous wearer of bifocals,
3. Addition level (the lower, the easier the adaptation),
4. Prescription (compulsory spectacle correction for both far and near vision eases the acceptation of PPLs),
5. Fitting accuracy. This was specifically pointed out in the study by Young and Boris [300]: acceptance ratios were clearly lower at practices with poorer or "more relaxed" fitting protocols. It will turn out that fitting accuracy is a critical factor for acceptance and adaptation to PPLs. From our point of view, even more important than PPL design.

Studies on the preference of PPLs vs other spectacle compensations for presbyopia are [302, 303, 304, 305, 306]. As Pope points out, in all these studies the PPLs were the solution of choice, though this result could be biased as most of these studies were sponsored by PPL manufacturers, and probably negative studies (for PPLs) would not be published. Preference of PPLs vs bifocals has grown from the study by Spaulding, in 1981, to above

90% for the latest studies. Now that spectacle compensation of presbyopia other than PPLs is becoming rare, our main interest focuses on the studies comparing different PPL designs. Interesting studies are those from Wittenberg, [307], Brookman, [308], Krefman, [309], Borish, [310], and, more recently, Rozas [311]. Once again, results from sponsored comparisons should be used with caution, as negative trials with different experiment designs would not be published, but more importantly, these studies compare lenses that can be quite different in any aspect: length and position of the corridor, power profile, and peripheral distribution of power, so it is almost impossible to decide which design characteristic could be responsible for the winning lens to score over the other. In that respect, studies comparing lenses that are identical except for a single design feature would be more interesting. For example, we could keep the same peripheral power distribution and change a single parameter of the power profile. Similarly, we could keep the profile constant and change a single parameter of the peripheral power distribution. The first study meeting this condition was conducted by Fisher [312]. He used six PPL prototypes, all of them identical except for the width of the near region at the NRP. This means that both the horizontal widths of the regions with cylinder below certain threshold and mean sphere above certain threshold changed uniformly from lens to lens. Of course, the wider the lens at NRP, the bigger the maxima of unwanted cylinder. But at least the distribution of this astigmatism was uniform across lenses. The far region was identical for the six designs. The objective of the experiment was to determine the subjective width of the region that the user perceived as "clear and comfortable" at either side of the NRP. Under well-controlled conditions including eye-tracking, Fischer determined that the edges of this region matched the points at which astigmatism reached 1 D, and as a consequence, the subjectively perceived widths correlated well with the physical width measured on the lens surfaces. Another enlightening piece of research was conducted by Preston in her doctoral thesis [293]. It comprises a preference study and a head movement study. For the whole research she accessed five PPL prototypes. Starting with an average model, there were another two with identical peripheral power distribution, but one of them shorter, the other one longer. The other two prototypes had identical profiles to the average model, but one was wider at the NRP, the other narrower. For the triad with varying corridor length, the distances from the fitting cross to the point having 95% of the add power (pretty much the distance between FC and NRP) were 14, 15.5, and 17 mm. For the triad with varying width at the NRP, the horizontal distances between 1 D cylinder iso-lines, 22 mm below the fitting cross, were 13.8, 15.8, and 22.7 mm. Of course, maxima of unwanted astigmatism was bigger for the shorter and wider designs, but the structure of the distribution of astigmatism around the maxima was the same for all the lenses.

The preference study was based on a typical double-masked protocol that presented the patients with two satisfaction questionnaires, one at the time of dispensing, the other after one week of wear. The results are summarized in Table 8.2. The main result is that, overall, the short corridor and the medium width were selected as better preferred (though the first one without statistical significance). All the patients had experience wearing PPLs, but segmenting the whole group into those with less and more than four years of experience,

Table 8.2 *Summary of results of the preference study by J. Preston. "P" stands for "preferred", "LP" stands for "Less preferred." For those groups presenting a significant preference ordering, "2nd" and "3rd" stands for the 2nd and last preferred options.*

			Preference			
Group	Long	Medium	Short	Wide	Medium	Narrow
Overall			P		P	
Experience < 4 yrs	P	2nd	3rd	3rd	2nd	P
Experience > 4 yrs	LP			3rd	P	2nd
Males	2nd	P	3rd	2nd	3rd	P
Females			P	3rd	P	2nd

the result is that the newer wearers prefer the long corridor and narrow width, while the experienced wearers prefer the medium width and pinpoint the long corridor as the less preferred (the other two equally chosen). Finally, there is also different behavior related to gender: Males prefer medium length and narrow width, while females choose short length and medium width. This preference study clearly reveals for the first time that the improved ergonomic obtained from short corridors may surpass the inconvenience of narrower corridors and higher levels of unwanted cylinder. It also reveals that, once the near region becomes wide enough, making it any wider will reduce user satisfaction (because of the subsequent increase of lateral astigmatism). The results from the segmentation according to user experience suggest that experienced users prefer medium/short corridors and balanced widths, while inexperienced users will prefer low gradient distributions, that is, longer corridors and narrow fields of view (softer lenses). The differences observed between genders suggest the average (larger) height of males may favor medium over short length corridors, while the average reading distance, which is correlated with Harmon distance (smaller in women, [313]), may favor narrower reading regions.

Preston's thesis also includes a head movement study made after the preference study with the triad of lenses having different widths at the NRP. By real-time measurement of the roll, pitch, and yaw angles of the head, control of the head position, and the reading material position and extension, Preston was able to measure the regions scanned by the user on the lens surface when reading text of three different type sizes. Her findings were:

1. Amplitude of lateral head movement depended on type size, but not on lens width.
2. The averaged effective width of the region used for reading was identical for the three lens widths (5.8 mm), but depended on the type size: 6.5, 5.8, and 4.6 mm for 10, 8, and 6 points, measured with the medium width lens.
3. The averaged vertical position selected for the users for reading (mm below fitting cross) changed very little with both lens width and type size; it was 11.6, 11.7, and 12.3 mm for type sizes 10, 8, and 6 points (all with the medium width lens), and 11.9, 11.7, and 12.0 mm for the narrow, medium, and wide lenses, respectively (all with 8 point type size).

The distance between 1 D cylinder iso-lines of the medium width lens at the vertical reading position was 7.4 mm. This means the wearers were using 87% of the lens area with astigmatism below 1 D for reading 10 point type, 78% for 8 point type and 62% for 6 point type, so the actual width of the reading region used by the reader strongly depends on the visual task being more or less demanding.

In recent years, preference studies seem to have given way to objective assessments of visual performance. Preference studies involve a very large set of variables, both on the side of the lens design and on the side of the wearers. These variables are compounded in complex ways, and it is difficult to separate their respective effects. In the work by Solaz et al., [314], a three-phase experiment was carried out comparing up to four different PPL designs. Their approach is different than that of Fisher or Preston; they intended to find which factors among lens design, refractive error, occupation, and viewing distance have a greater impact on user satisfaction. The first phase consisted of a field study involving 191 wearers who were asked to compare four different PPL designs, which are qualitatively described in the paper (the authors did not intend to relate particular features of the lens design to user satisfaction, but whether the design affects at all and how its influence compares with factors from the user or the visual task). The field study did not show correlation between PPL type and user satisfaction, although correlations appeared between overall satisfaction and satisfaction at different visual tasks when the data were clustered according to design, refractive error, or occupation. The second phase of the study was a controlled preference study using three out of the four initial designs, with the objective of determining the influence of design and viewing distance on user satisfaction. From the three possible comparison pairs, the study showed significant differences in just one pair. Finally, the third phase consisted of a laboratory study in which the three PPLs were compared for controlled visual tasks: far, intermediate, and near vision and a test combining the three distances. This last test offered enough resolution for the wearers to discriminate different discomfort perception between the three designs. Even so, not all the comparisons yielded significant differences. The whole study reveals the difficulty of finding correlations between subjective satisfaction and PPL design, especially when the designs are similar and more than one design feature is changed at the same time.

The last study we will review is that of Jaschinski et al. [315]. They compared the subjective performance of a general-use PPL and a lens specifically designed for computer vision. The latter belongs to a family of lenses intended for near and intermediate work. The power profile in these lenses is "stretched" upward, so the local slope is smaller and the corridor is wider. The downside is that far vision is affected to the point there is no sharp vision. At the fitting cross and above it there is still a clear amount of addition, so the far vision region is converted into a far-intermediate region. During the first part of the study, the wearers (not informed about the characteristics of the lenses they are going to use) underwent a two-week adaptation period (one week per lens) and a two-week test phase. This one consisted of the office use of a single design per week. In the second eight-week part the wearers were informed about the characteristics of the lenses and were set free to use them at their convenience. They found that wearers kept a more ergonomic head position (2.3 degrees lower) when using the computer lenses. At the end of the first part, 61% of the wearers preferred the computer lens, with the preference being clearer with larger adds (something to be expected, as advanced presbyopes have smaller ranges of clear vision and depend more heavily on the intermediate region of PPLs). Once the wearers were informed about the characteristics of the two lens models, the preference for the computer lens only dropped to 44%. This is an interesting figure, as the number is much higher than the market share for this type of specialized lenses, a mere 7%. No doubt, specialized designs will be the path to improve user satisfaction in the future, though a better understanding of the interaction between the parameters defining the lens design and the expected visual tasks is still required.

Objective Assessment of Ergonomic and Visual Performance

We have just seen how difficult it is to obtain statistically significant results in preference studies, because of the mixing of many different variables and how easily subjective bias can affect wearer selection (for example, the enthusiasm of the practitioner may trigger a positive answer from the wearer in field studies [310]). Another approach to the assessment of PPL performance consists of conducting objective tests that measure certain user parameters, or user performance in tasks that heavily rely on vision quality. Selenow and others undertook this research line and in a paper in 2000, [316, 317], they measured reading speed, reading comprehension, and the time needed to complete tasks involving alternating reading at different displays and spreadsheet operations, using single vision lenses and PPLs. They found differences in performance (in favor of the single vision lens) in all the tests except for reading speed, but the only one statistically significant was the test with alternate reading. In another experiment, [318], they found that reduction of contrast sensitivity correlated better than reduction of visual acuity when looking through the regions of progressive lenses having unwanted power. In unpublished work, the authors have confirmed the fact that subjective measurements of high-contrast visual acuity across the field of view do not correlate well with the PPL distribution of unwanted power. However, low-contrast visual acuity does correlate, and is easier and faster to measure than contrast sensitivity.

Another promising technique to assess progressive lenses is the measurement of eye and head movements [319, 320, 321, 322]. Saccadic eye movements, especially when reading low-contrast text, can easily discriminate not only between single vision and progressive lenses, but also between different designs of PPLs. Hutchings et al., [323], performed a study comparing eye and head movements before and after a four week adaptation period to two different PPLs (one week resetting period between the two designs). They found a clear increase in head movements, especially for reading and flash discrimination tasks, after the adaptation period with each PPL, but did not find differences between the two PPL designs. Mateo et al. [324], proposed a measuring set up for recording head, neck, and trunk position and movement during reading. They found significant differences between these parameters when comparing three different PPL designs, and thus they propose the tool to explore the match between lens design and user discomfort.

Another interesting assessment of PPLs is the one performed by Villegas et al. [181, 325]. In their first work, they developed a system that can be adapted to measure eye aberrations, local wavefront aberration of a single PPL lens, and the combined aberration of the lens-eye system. The main contribution to the RMS error at any point on the lens is defocus and astigmatism, as we should expect. After subtracting the second-order aberrations, the main terms in the corridor and the lateral regions of the PPL are coma and trefoil, with RMS magnitudes between 0.03 and 0.06 microns for an aperture of 6 mm. Without removal of second-order aberrations, RMS values are between 0.6 and 0.9 microns. In their second work they used a subjective extra channel to measure visual acuity, and compared it with objective metrics of retinal image quality. They found the interesting result that the interaction between eye and lens aberrations tends to equalize image quality between the central and the peripheral zones of the lens.

Scoring of PPLs

The conclusions drawn in the research works reviewed in the previous sections usually come from comparison of the power maps (typically mean sphere and cylinder) and whatever magnitude is measured, whether it is satisfaction, head movement, etc. Some authors point out that the maps of mean sphere and cylinder do not determine the "quality" of a lens [266, 326]. We would rather make a more precise assertion: The mean sphere and cylinder maps uniquely determine the progressive surface[6] (or, in the case of double-side progressive lenses, they determine the progressive refracted wavefront), but it is not possible, at least with the knowledge we have today, to infer user satisfaction from these maps. Anyway, the development of a parametrization system that could be applied to any progressive lens in order to make fair comparisons between their distribution of sphere and cylinder would seem necessary. Sheedy proposed such a system in 2004, [248], with a follow-up paper in 2006 [328]. The scoring system would be applied to plano, add 2 D lenses, assuming that the prescription would be a constant power background added to the power distribution of the plano lens. Sheedy uses the 0.5 D cylinder iso-line and the 0.25 and 1.75 mean sphere iso-lines to define distance, intermediate, and near regions and widths. He defines the next set of parameters that would be characteristic of the design:

1. Area of the distance region. The distance region is vertically enclosed by two horizontal lines, the upper one 1 mm above the FC, the lower at the vertical coordinate at which addition gets to 0.25 D. Laterally, the region is defined by the 0.5 D cylinder iso-line or the 0.25 mean sphere iso-line, whichever is more limiting. Four distance region widths are recorded at the FC, 1 mm above it, and 1 and 2 mm below it.
2. Three areas of the intermediate regions defined by the 0.5 D cylinder iso-line and vertical addition ranges [0.75, 1] D, [1, 1.25] D, and [1.25, 1.50] D. Four intermediate widths are defined at the edges of these addition intervals.

---

[6] A surface with its principal curvatures having the same sign in the domain of interest (as the majority of ophthalmic surfaces) is uniquely determined by the Gaussian curvature, and hence, by the two main curvatures [327].

3. Three areas of the near regions, which are determined by the 1.75 D means sphere iso-line of the 0.5 D cylinder iso-line, whichever is more limiting, and the horizontal lines located at 16.5, 18.5, and 20.5 mm below the FC. Four near region widths are computed at 14, 16, 18, and 20 mm below the fitting cross.
4. Highest value of unwanted cylinder.

Sheedy evaluated these parameters for 28 PPL designs whose power distributions were measured with a Class Plus mapper from Rotlex (see Chapter 10 for details of these instruments). Then he generated different rankings by normalizing areas and widths, balancing areas from different regions, and weighing the rankings with the values of maximum astigmatism. He proposes different rating combinations for different visual tasks, offering a quantitative way to select a design according to the user's need.

In another work, [247], Sheedy studied correlations between the parameters he defined in [248]. Now he introduces the slope along the profile, and finds the stronger correlation between slope and zone width, as could be expected from our model for progressive surfaces. Correlations between maxima of astigmatism and distance or near areas and/or widths is weaker or nonexistent, which confirms our previous statement that the distribution of astigmatism around the FC and the NRP can be managed in a quite independent way (at least outside of the Minkwitz region).

According to the authors' experience, Sheedy's scoring system fails to predict user satisfaction, both for general-use and task-oriented designs. Yet it is an important first step toward what should be a systematic and scientifically based PPL prescription aid. One of the main problems of Sheedy's scoring system is that it is based on mapper-measured power distributions and plano-add 2 lenses. The performance of a given design may and will change with addition, and the power perceived by the user may greatly differ from the one measured with a mapper. The last factor can be accounted for if we employ Sheedy's scoring system on user power maps [279, 329]. The former factor requires redefinition of the scoring system. Similarly, the correlations found by Sheedy can be refined and better implemented in the scoring system. For example, if a PPL design is displaced 1 mm up or down, its score would significantly change, and the changes would not be linearly distributed among the different lens regions, as a consequence of the structure of the power distribution. The same PPL fitted 1 mm higher or lower would change its performance, but probably not in the same way described by the scoring system. Another factor to be included is the distortion and swimming effects coming from the spatial variation of power. Some users are sensitive to these effects and although they are related to the maximum of astigmatism, they should be taken into account separately. In summary, we think Sheedy's work is pioneering and should be improved with the latest research on PPL performance.

8.4.5 Free-Form Technology and Custom-Designed PPLs

So far, we have been describing PPL power distribution in terms of surface curvature, that is, the local power of the lens is

$$\mathbb{P}(x, y) = (n - 1)\mathbb{H}_1(x, y) + (1 - n)\mathbb{H}_2(x, y),$$

if we assume it to be thin or

$$P(x, y) = \left(\mathbb{I} - \frac{t(x, y)}{n} \mathbb{P}_1(x, y)\right)^{-1} \mathbb{P}_1(x, y) + \mathbb{P}_2(x, y), \tag{8.38}$$

if we compute local back vertex power using the local thickness $t(x, y)$. However, we know from Chapter 6 that the power perceived by the user is not the paraxial back vertex power, but the curvature of the wavefront refracted by the lens at the vertex sphere, and we also know that this curvature is determined not only by the local curvatures at the points where the chief ray crosses the surfaces and the thickness between them, but on the obliquity angles that the chief ray makes with the surface normals. The curvature of the refracted wavefront at the vertex sphere is called the *user power*, power *as worn*, or *eye-point power*. The way obliquity affects progressive lenses and single vision lenses is similar. The base curve, the prescription, the tilt angles, and, to a lesser extent, the refractive index of the lens, determine the deviation of the user power from the paraxial power given by (8.38). The main conclusion is that we should be cautious when making an assessment on any type of progressive design over curvature-based maps, as they may differ significantly [180]. As advanced in the previous section, any attempt to create a scoring system for progressive lenses (or any other ophthalmic lens, actually) should be based on user power maps [279, 329]. In Figure 8.24 we can see the superimposed user power and lensmeter maps of two noncustomized PPLs, one of them front-side (a) and (b), the other one back-side (c) and (d).

User power, compensated power, customized design, free-form technology, and back-side PPL are terms usually mistaken in the sense that similar meanings are given to them, or rather, they are considered to be inseparable. Of course, they are related, but they are perfectly independent and separable.

User power. A power map, either sphere or cylinder, is a user power map if it represents the power perceived by the user. The X and Y axes can represent horizontal and vertical gaze angles, or Cartesian coordinates. The former case is clear, and the latter should be specified by the provider of the maps. For example, for any gaze angle we could compute the x, y coordinates of the impact point of the chief ray on the back surface of the lens. The user power map then could represent the power component (sphere, cylinder, or whatever) at each of these points. It is also possible to locate a plane perpendicular to the Z axis at the vertex of the back surface, then use the x, y coordinates of the impact point of each gaze direction in said plane. Selecting one representation or other would produce slight variations of the maps that should be taken into account. If a lens is designed and manufactured in such a way that its user powers at the FRP and NRP match the user prescription, we would say the lens is designed using user power as target. Obviously, when checked in a standard lensmeter, the measured power will not match the prescription. The idea of manufacturing lenses with user power as target is older than the standardization of free-form technology [269]. Computing user power requires a lens-eye model and exact ray/wavefront tracing.

Compensated power. If a user-power-designed lens is measured with a standard IOA (infinite on axis) lensmeter (see Chapter 10), the readings will not match the

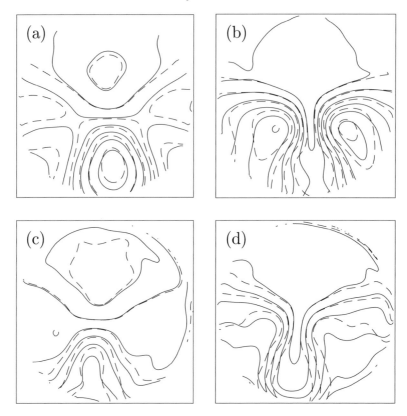

Figure 8.24 Comparison of user power (solid lines) and lensmeter power (dashed lines). Iso-lines of both sphere and cylinder start at 0.5 D and proceed in steps of 0.5 D. (a) and (b) Mean sphere and cylinder for a front-side PPL, base curve 3D, plano add 2. (c) and (d). Mean sphere and cylinder for a back-side PPL, base curve 4.5D, plano add 2.

prescription. Instead, we will have a different set of numbers for the sphere, the cylinder, the axis, and the addition. These numbers make up the compensated power. When computing a lens with user power as a target, it can be done at full field, meaning the oblique aberrations are removed or minimized all over the visual field, or can be done locally, adjusting the local power at the FRP and NRP to match the prescription, but ignoring the oblique aberrations elsewhere. Of course, full-field compensation is more challenging for the lens designer but a much better option for the user.

Customized design. The user power not only depends on the prescription and base curve. It also depends on the angular orientation of the lens with respect to the eye and the required eye convergence for the viewing distance. This means a lens can only by optimized with user power as a target if we know these user parameters. If we do not have access to these parameters we can use default values, and the option will, in general, be better than a curvature-based design. There is another way a lens can be customized, independent of the correction for oblique aberrations. The manufacturer could make a lens, PPL, bifocal, or single vision, with power distributions adapted to some specific need. This "custom design" could be computed with user power as target or not.

Free-form technology. Generating and polishing technology that allows for fast figuring of arbitrary surfaces. The standard process is tuned for plastic blanks, though it can be applied also to glass blanks, but with longer processing times. Most modern free-form generators could cut both the front and back surfaces, but the tooling and the machinery is fine-tuned for processing the back side of the blanks. Free-form technology can be used to make a spherical surface or the most advanced PPL design. In this sense, free-form is an enabling manufacturing technology. If the lens manufacturer does not have the means to optimize lenses designed with user power as target, it can still make standard lenses. Another aspect of free-form technology is that it allows for the manufacturing of an arbitrary large number of lens designs (whether single vision, progressive, or bifocals) at the same material cost of a standard single vision lens. Free-form technology then enables low-cost manufacturing of customized designs, or designs with user power as a target, or both.

Back-side PPLs. As free-form technology currently processes the blanks on the back side, PPLs made with this technology will be back-side PPLs. This is not particularly advantageous. It is said that because the progressive surface of a back-side lens is closer to the eye, its features subtend a bigger angle as seen from the rotation center of the eye. This suggests that a back-side PPL would seem "wider" than the corresponding front-side one. However, having the back surface closer to the eye also demands a bigger rotation angle to get the NRP from the FC. If we want to provide the same user experience, the back-side PPL should be a little bit shorter, spoiling the potential advantage just described. According to the authors' experience, cutting lenses on the back surface does not improve the lens performance design-wise. Another claimed advantage of back-side PPLs is that they create less geometrical distortion as a consequence of having smaller magnification at the near region than a front-side PPL. Distortion is produced whenever magnification is not constant, or whenever prismatic effect has a nonlinear dependence with position. In a front-side PPL the curvature of the progressive surface increases from the far region to the near region. As a consequence, the shape factor of the magnification grows from the DRP to the NPR. In back-side PPL the front surface is usually spherical, and the shape factor is constant. Then, it is true that a front-side PPL has a bigger change of magnification than a corresponding back-side PPL with the same design. However, we do not think this is a real advantage, at least, not always, and not for everybody. Wearers with good binocular vision adapt well to distortion, as the brain manages to compensate for it (while it cannot compensate for power error). On the other side, bigger magnification at the near region is a nice feature for presbyopes that is well rated in preference studies [330].

Optimization of Progressive Lenses at the Position of Use

The computation of progressive lenses corrected for oblique aberrations requires a different and more complex approach. To compensate for oblique astigmatism it is not enough to first compute the progressive surface and then construct the lens. Rather, a model of the complete lens along with the eye has to be created before optimization. The model takes into account base curve, lens material and position, and tilts of the lens with respect to the eye. Similarly, pupillary and working distance, along with the prescription, can be used to compute the lateral inset of the principal curve with more accuracy. All the information is collected in a merit function similar to (8.37), but now C and H will refer to power perceived by the user and the sum indexes will point to gaze angles instead of Cartesian coordinates. This process has been described by Haimerl and Baumbach [331] and in greater detail by Loos et al. [332].

Once again, a generic optimization could go as follows. We start with a good progressive design that we compute for the required addition and prescription plano. If we add the prescription to the local power of this design we would get the target lens power for any gaze direction. Of course, the addition must be performed in matrix form; we cannot just add cylinders without taking be the axis into account. Let $\mathbb{P}_T(u, v)$ be the target power for the gaze defined by the angles (u, v). As before, we create a temporary surface that depends on a set of coefficients $\mathbf{c}$. Then we model a lens comprising a predefined front surface, our temporary surface $z(x, y, \mathbf{c})$ at the back side and the desired refractive index, thickness, and orientation. For each gaze direction (u, v) we compute the power perceived by the user, $\mathbb{P}(u, v)$, and we construct the merit function

$$\Phi(\mathbf{c}) = \sum_{u,v=1}^{N} w(u, v) \, \|\mathbb{P}_T(u, v) - \mathbb{P}(u, v)\|^2 , \tag{8.39}$$

where $\|\ \|$ stands for the Frobenius matrix norm (see Appendix B), and $w(u, v)$ are gaze-direction dependent weights that help us control the amount of correction of oblique aberrations. These weights are necessary, as full compensation of oblique aberrations is impossible for all gaze directions. Finally, to this basic merit function we may add similar terms in which visual properties other than perceived power can be optimized, for example visual acuity, distortion, or even certain combinations of third- and higher-order aberrations.

We have described the process for optimizing a back-side PPL, as free-form technology is currently tuned for cutting and polishing the back surface. The process for a front-side one would be the same, changing the location of the surfaces. Double-side PPLs are a little bit more difficult, as the control of the overall lens bending requires additional regularization terms.

9

Low Vision Aids and High Power Lenses

9.1 Introduction

In previous chapters, we have studied ophthalmic lenses as compensating devices for ametropies. From a purely optical point of view, the problem of compensation of ametropies can be regarded as a problem of compensating defocus. However, in this chapter we will deal with another problem: the compensation of low vision. Although we will give a more precise definition later, we can consider that, in most cases, low vision is either a problem of low resolution or of a reduced field of view of the eye, which may be caused by a number of medical conditions. It is important to notice that an eye may present both defocus (ametropy) and lack of resolution/field of view (low vision). Therefore, in many cases we have to compensate both ametropy and low vision, which introduces some constraints to the design of optical aids for low vision, as we will see later. Obviously, given the complexity of the topic, our aim is somewhat limited to presenting the principles of the main optical systems designed for compensating, or, more accurately, for alleviating low vision problems. Therefore, we refer the reader for further information to specialized books on low vision (see, for example, [333], [334], and [335]).

Before proceeding to the study of low vision aids, it is important to notice the difference between compensating ametropies and low vision. Although, in both cases, we will employ imaging systems as compensating devices, they will be used in a different way. As we have already seen, the role of the ophthalmic lens is to form the image of an object within the focusing range of the ametropic eye. For example, in the case of distance vision for a myopic eye, the ophthalmic lens will form a virtual image of a distant object at the remote point of the observer's eye (i.e. the far point of its focusing range), which, in turn, will form a well-focused image on the retinal plane. Therefore, if the eye does not present any other condition in addition to myopia, the object will be perceived as sharp as if it were focused by an emetropic eye. It is important to take into account that compensation of ametropies with ophthalmic lenses has some limitations in the field of view and there is also a slight difference in the size of the perceived image but, in general, these differences are not noticed by the user of ophthalmic lenses once she has adapted to the usage of these devices (and, moreover, these problems do not arise in other compensating devices, such as contact lenses).

However, in order that a low vision patient might perceive correctly the image of a given object, it is not enough that a sharp image of the object is formed on the retina. Even in those circumstances, its vision impairment will not allow the patient to notice the fine details of the image. Thus, it is necessary to magnify the image in order to adjust the image size to the defective resolution of the low vision patient. To do so, the power of the optical system must be increased in such a way that, in many cases, a single lens will not yield the required magnification; use of an optical instrument formed by several lenses such as a telemicroscope is necessary. This has two main consequences. In the first place, the greater complexity and, consequently, greater size and weight of these systems make the task of adapting them more challenging than in the case of adapting lenses. Moreover, this adaptation process is further compounded by the fact that most users of these systems are elderly people who may present additional medical conditions that may impair their ability to hold and position those devices (for example, patients of Parkinson's disease may experience hand trembling that may make the correct positioning of a hand magnifier difficult). Second, the greater power (and magnification) of low vision aids is achieved through changing the values of other optical parameters, mainly field of view and depth of focus/field. In this context, it is quite important to notice that, as a rule of thumb, the greater the magnification, the lower the field of view and depth of field. This means that, contrary to what happens in the case of compensating ametropies, a user of low vision will not perceive the image of an object in the same way as a person with normal vision. In particular, for patients presenting subnormal resolution, the field of view perceived will be considerably lower than the one perceived by a person with normal vision. Moreover, in the case of nonplanar objects, reduction of the depth of field may lead to the impossibility of focusing the whole object at once.

Besides optical aids, other nonoptical devices are nowadays available to alleviate low vision problems. With the advent of powerful lightweight computers with planar displays, such as mobile phones, tablets, or e-readers, it is now possible to provide the low vision patient with portable devices equipped with cameras, which may increase his autonomy and independence by allowing him to perform a range of daily life tasks. For example, such devices can be used to read price tags in a supermarket or help the patient to fill out a form. However, the study of these nonoptical aids lies beyond the scope of this book. Finally, to end the chapter we will discuss briefly the role of high power ophthalmic lenses, particularly in the compensation of aphakia or for high myopia. We will discuss their differences with standard monofocal ophthalmic lenses, particularly in the employ of aspheric surfaces.

9.2 The Problem of Low Vision Compensation

9.2.1 Defining Low Vision

Colloquially, we may understand low vision as the inability to see normally or, more accurately, the suffering of visual impairment. When thinking about low vision, we may picture the image of a person struggling to read a book using a handheld magnifier. This image brings out the two foremost characteristics of low vision. First, it constitutes an impairment, that is, the person who suffers from low vision presents reduced visual capabilities and is unable to perform visual tasks in the same way that a person with normal vision does. Second, in many cases, a low vision patient needs an optical system (a magnifier, in our example) as an *aid* to allow her to perform tasks requiring vision, such as reading. However, it is quite important to introduce an accurate definition of low vision, particularly from a clinical and legal point of view because, as with any other impairment, it has legal consequences. For example, in many countries visual impairment entitles a person to a range of social benefits. Therefore, many countries have a *legal definition* of low vision [333, 334] and blindness, which may be understood as the complete absence of sight. Although there are some problems with this procedure, as is pointed out in [334], the legal definition of low vision is based on two visual characteristics: resolution and field of view. Resolution is characterized by the value of the best visual acuity measured using a test chart, normally adapted to the particular conditions of a low vision patient [334]. Best possible visual acuity is achieved when the eye is correctly focused, so it must be measured with the adequate correction of the individual's refractive error. The field of view is measured using a number of devices ranging from simple grids such as Amber's to more sophisticated perimeters [334].

The World Health Organization (WHO) has stated the criteria for classifying visual impairment and blindness, see Table 9.1 and reference [336]. According to the WHO, we may consider that a person presents low vision when the distance visual acuity (for the best correction) drops below 0.3 in decimal scale for the best eye. Table 9.1 is used by many

Table 9.1 *Different degrees of vision loss according to the table associated with category H54 Visual impairment including blindness (binocular or monocular) of the International Disease Classification IDC-10 given by the World Health Organization [336]. Those patients whose better eye presents a visual field lower than 10° in radius around the center of fixation are classified under category 3. Visual acuity must be measured using the best correction possible.*

| Category | Presenting distance decimal visual acuity | |
	Worse than	Equal to or better than
0 Mild or no visual impairment		0.3
1 Moderate visual impairment	0.3	0.1
2 Severe visual impairment	0.1	0.05
3 Blindness	0.05	0.02∗
4 Blindness	0.02*	Light perception
5 Blindness	No light perception	
9	Indetermined or unspecified	
	*or counts finger (CF) at 1 meter	

governments and organizations as a reference to determine whether a person is eligible for benefits or when he may join one of the national organizations that provides care and support for blind people. For example, in Spain, any Spanish national whose vision can be classified into WHO categories 2 to 5 is eligible for joining the *Organizacion Nacional de Ciegos Españoles* (National Organization of Blind Spaniards) ONCE [337], which entitles this person to receive counseling, training and education, etc. Similar benefits are entitled to visually impaired persons in other European countries through similar associations or organizations such as the *Royal National Institute of Blind People* in the United Kingdom, the french *Federation des Auvegles de France*, the *Deutsche Zentralbücherei für Blinde* in Germany, or the *Unione Italiana dei Ciechi e degli Ipovedenti* in Italy. In addition, many governments provide blind or visually impaired people with benefits, usually in the form of financial support, such as tax exemptions, allowances, grants, or reduced prices for public transportation. In the United States, at federal level, the benefits for those persons classified as legally blind range from tax exemptions to rehabilitation services [334], together with other benefits provided by the state in which the subject resides.

Although low vision can appear at any age, it is more common for people with advanced age. The more frequent conditions that lead to low vision for elderly persons are macular degeneration, glaucoma, diabetic retinopathy, and cataract, the usual causes of legal blindness in the United States [334]. For younger patients, in addition to the previously mentioned diseases, low vision may be due to other conditions such as retinitis pigmentosa, cone-rod distrophy, optic atrophy, or albinism [334]. Besides those pathological causes, low vision may also be the consequence of accidents such as solar burns or eye trauma. Regarding the spread of low vision, the WHO estimates [338] that around 285 million of people worldwide suffer vision impairment, of which around 39 million are blind and the remaining 246 million have low vision. It is important to notice that, according to WHO data, cataract is the foremost cause of low vision, particularly in low-income countries, so the organization considers that *"80% of all visual impairment can be prevented or cured"* [338]. It is also important to take into account that the prevalence of low vision problems will be higher in the decades to come, particularly in developed countries, due to higher life expectancy, which will lead to an increment of the elderly population. It is also worth noting that low vision constitutes a health problem with many social ramifications and, therefore, it is of paramount importance to aid people who are visually impaired so they can overcome their incapacity. In this context, it is the task of the practicioner to provide them with the tools and techniques that may be most useful to this end. Among these tools we find the visual aids that will be studied in this chapter, but before doing so it is necessary to understand the principles that guide the compensation of low vision.

9.2.2 Compensating Low Vision: Visual Magnification

As we have said before, in most cases compensating low vision implies use of an optical system (the low vision aid) in such a way that the image perceived by the eye is larger than the image of the object perceived without aid, so that the user would be able to distinguish

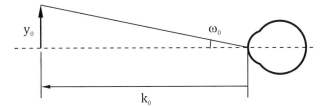

Figure 9.1 Observation of an object with the naked eye; the size of the image formed in the retina is proportional to the subtended angle ω_0.

better the fine details of the object. Let us suppose an emetropic eye looking at an object of height y_0 located at a distance k_0 from its vertex, as is depicted in Figure 9.1. In these conditions, the size of the image of the object formed in the retina is given by the angle ω_0 subtended by the object, so that, if this angle is equal or greater than the minimum angle of resolution (MAR) of the eye, η, then the person would be able to resolve the object. The problem with low vision patients is that, usually, their MAR is far greater than that of a person with normal vision, which is around 1 arcmin (equivalent to $AV = 1$ in decimal scale).

Therefore, in order to compensate a low vision patient, it is necessary to use an optical instrument to achieve greater resolution. In Figure 9.2, we have depicted an optical scheme representing the observation of an object through an optical system. It is important to notice that, although the object height, y_0, is the same of that of the object depicted in Figure 9.1, it is placed at a different distance, k, as measured from the eye. As we can see in Figure 9.2, the optical system forms an image of this object, which in turns acts as an object for the eye. The optical system not only forms this image with different size but also at a different location to that of the object. It is of paramount importance to notice that when we speak of "*magnification*," we think in terms of "*enlargement*" because what is perceived is an enlarged image of the object. This perception of enlargement is due to the greater angle ω' subtended by the image compared to that subtended by the object ω_0 when seen by the naked eye. Thus, we will define the so-called *visual magnification* of an optical system as the ratio between the angle subtended by the image formed by the system and that of the object seen with the naked eye. Mathematically:

$$M = \frac{\omega'}{\omega_0}. \tag{9.1}$$

In equation (9.1) we assume that the angles are measured in radians. If those angles are not too large, we can replace them by their tangents in equation (9.1). From Figures 9.1 and 9.2, we see that $\tan \omega_0 = y_0/k_0$ and $\tan \omega' = y'/k'$ so that

$$M \approx \frac{\tan \omega'}{\tan \omega_0} = \frac{k_0}{k'} \beta'. \tag{9.2}$$

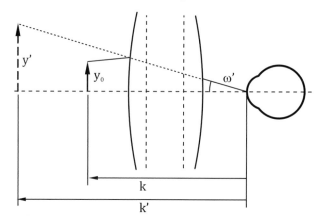

Figure 9.2 Observation of an object through an optical system. Notice that the system changes both the position and size of the image with respect to that of the object

Where $\beta' = y'/y_0$ is the linear magnification. Equation (9.2) implies that the sensation of "enlargement" associated with the usage of a magnifying optical device can be due to two causes, the enlargement given by the value of the linear magnification β' and the distance ratio k_0/k'. The distance ratio takes into account the fact that many optical systems produce the sensation of magnification rather by moving the image toward the eye than by producing an enlarged image of the object. This is a quite important property of the visual aids, particularly of telescopes, as they usually present a lateral magnification $|\beta'| < 1\times$, which is compensated by a closer location of the image to the eye and allows for a greater subtended angle. While we will see in the following sections some variations of equation (9.2) that are more convenient for the practitioner, the fundamental behavior of low vision aids is given by equation (9.2).

To finish this section, it is important to state the condition that must be accomplished in order to compensate a patient of low vision if his problem is manifested as a lack of resolution. To do so, let us consider again the situation depicted in Figure 9.2 but supposing that ω' is now the unaided eye MAR, that is $\omega' = \eta \equiv 1/AV$, where AV is the unaided visual acuity (also known as natural visual acuity) of the individual. In these conditions, and taking into account that the values of the involved angles are small, we could define the aided visual acuity as $AV_{ad} = 1/\eta_0$ η_0 being an angle given by

$$\eta_0 = \frac{\eta}{M}. \tag{9.3}$$

This angle η_0 could be interpreted as the "equivalent" MAR that would present a person able to observe the object with his naked eye. Equation (9.3) can be written as

$$M = \frac{AV_{ad}}{AV}, \tag{9.4}$$

which is the basic equation used to determine the magnification of an optical aid as a ratio between the aided (or target) visual acuity AV_{ad} and the natural visual acuity of the individual, AV. For example, if we would like to compensate a low vision patient with visual acuity $AV = 0.25$ using a target visual acuity $AV_{ad} = 1$, then we would need to employ an optical system with $M = 4\times$.

9.2.3 Role of Field of View, Depth of Field and Image Quality in Low Vision Compensation

Until now, we have presented the problem of compensating low vision as a magnification issue. However, the usage of a magnifying system introduces some limitations in aspects such as the perceived field of view, the depth of focus, or the quality of the image produced by the low vision aid.

Given the inverse relationship between field of view and magnification that holds for any imaging system, one of these issues is the field of view. As a rule of thumb, the greater the magnification the lower the field of view perceived through an optical system. This effect was not apparent in previous chapters due to the characteristics of the systems employed for compensating ametropies. For spectacle lenses, the low magnification of those devices combined with the relatively wide mounting frames make the field limitation unnoticeable by the user in most cases. Regarding other devices, neither contact lenses nor intraocular lenses are prone to limit the field of view. However, this is not the case with low vision aids, for which field limitation is always present and has an influence in their design and in the way in which they are employed. Consider, for example, the simple task of reading a newspaper using a handheld magnifier and how the user must scan the whole page of the newspaper with the magnifier in order to read the whole of it due to the limitation in the field of view that introduces the magnifier.

Another important optical parameter that must be taken into account when dealing with low vision aids is the depth of field (DOF). In general, for low vision aids, DOF decreases with magnification and this fact introduces some restrictions in their usage. The DOF reduction makes more complicated the accurate placement of the low vision aid in order to observe a given object, particularly when using handheld devices. By itself this can be a serious problem, but it is usually compounded by the fact that many low vision patients might present conditions, such as hand tremors, which make difficult or, even, impossible the use of handheld devices. As we will see later, this reduced DOF has an influence in the design of low vision aids, particularly for devising better ways to hold these systems accurately in the most ergonomic way possible.

Finally, another important general aspect in low vision systems optical design is the role played by aberrations. Compared to spectacle lenses, low vision aids have more optical power and this, by itself, indicates that they will present a higher amount of aberrations. This situation, which can be also considered a kind of "rule of thumb," is disadvantageous because the image quality will be reduced a priori. However, there are two factors that alleviate this situation. First, in many cases low vision aids have more degrees of freedom for balancing aberrations. A typical example will be the use of achromatic doublets as magnifiers allowing for a reduced chromatic aberration. Second, in many cases the reduction of optical quality is not noticed by the low vision patient as his vision is already degraded. In any case, great effort should be carried out by the designer of low vision aids in order to get the best image quality for these devices with the highest degree of comfort and ergonomy.

9.3 Low Vision Aids for Close Objects: Magnifiers

In this section we will discuss those low vision aids designed to help the user to perform tasks requiring near vision such as reading, writing, etc. For low magnifications, the optical instruments that will be employed are composed by a single lens, so they are known as *simple microscopes* or, more commonly, *magnifiers*. For high magnifications, simple microscopes will not work because of design constraints making necessary the use of a *compound microscope* (or simply "microscope") usually in the form of a *telemicroscope*. We will present first the relevant optical characteristics of a magnifier: visual magnification, field of view, and depth of focus. Afterward, we will give a brief description of microscopes to end the section studying the optical design of magnifiers.

9.3.1 Optical Characteristics of a Magnifier

Visual Magnification

Figure 9.3 illustrates the way in which a magnifier is used. As can be seen in this figure, the object is located between the object focus and the magnifier, which is represented as a thin positive lens. In these conditions, the image formed by the magnifier is a virtual image whose size y' is greater than that of the object, and which is located at a distance k' from the eye, so we can write the visual magnification of this system as

$$M = \frac{k_0}{k'}\frac{s'}{s} \equiv \frac{k_0}{k'}\frac{k'+d}{k+d} = \frac{k_0}{k}\frac{1+dK'}{1+dK}, \tag{9.5}$$

where K' and K are the vergences corresponding to distances k' and k, respectively. Equation (9.5) is widely employed in low vision [334] as it allows one to consider visual magnification as the product of two factors, ratio distance magnification (RDM) defined as $RDM = k_0/k$ and lens vertex magnification (LVM) given as $LVM = (1 + dK') / (1 + dK)$. According to Brilliant [334], the LVM is related to the change in the angular size of the image compared with that of the object, while the RDM indicates the ratio between the distance eye-object with and without aid. As can be seen in Figure 9.3, if the object is placed between the object focus and the magnifier, then an enlarged virtual image will be obtained so that $LVM > 0$. On the other hand, in order to achieve a RDM greater than 1, it is necessary that $k_0 > k$, and this can be achieved using high power lenses. Therefore, a magnifier is a high power lens forming a virtual enlarged image of a real object, as shown in Figure 9.3. The lens vertex magnification is named differently in other texts. For example, Fannin et al. [132] define LVM as *"angular magnification"* and visual

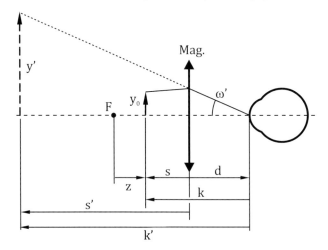

Figure 9.3 Image forming in a magnifier represented as a thin positive lens Mag. The object is located between the object focus F and the magnifier, so a virtual image is formed that is seen by the observer.

magnification as "*effective magnification*"; therefore, the reader should be aware of those different nomenclatures in order to identify the components of the magnification.

It is quite important to understand the relationship between visual magnification and the optical power of the magnifier. To do so, let us consider a particular case that is commonly found in practice: The object is placed at the object focus of the magnifier. In these conditions we see that $k = -f' - d$ and $k' = \infty$, so equation (9.5) becomes:

$$M_{k' \to \infty} \equiv M_N = \frac{-k_0}{f'} = -k_0 P, \tag{9.6}$$

where P is the lens power. We will define M_N as commercial magnification (although it receives different denominations in the literature, see, for example, [132, 334]), which is an important quantity as it confirms that the greater the lens power, the greater the magnification. In the literature [334, 333] we usually find normal magnification written as $M_N = P(D)/4$, where $P(D)$ stands for lens power measured in diopters. This simplification of equation (9.6) holds for emmetropic and not presbiopic patients, for which k_0 would be equal to the near point distance $k_0 = p = -250\,\text{mm} = -0.25\,\text{D}^{-1}$. This election of k_0 is arbitrary and may change among low vision aid manufactuers, so it is customary for many low vision aid manufacturers to give the value of the commercial magnification plus the power of the lens, for example $3.125 \times /12.5(D)$.

Example 40 A manufacturer sells a handheld magnifier with magnification (the manufacturers refer usually to M_N as "magnification" without further adjective) $M_N = 3\times$ and power $P = 8$ D, determine the value of k_0 used by the manufacturer.

From equation (9.6) we see that $k_0 = -M_N/P = -3/8\,\text{m} \equiv -375\,\text{mm}$. Notice that manufacturers tend to use values of k_0 greater (in absolute value) than the standard value of

250 mm, as this value corresponds to an accommodation amplitude of 4 D, which is a figure rather high when compared to the typical range of accommodation amplitude presented by low vision users. In this context, a value of $k_0 = -375\ mm$ would be equivalent to an accommodation amplitude of $\mathcal{A} = 2.67$ D, which is more reasonable for a middle-aged person.

Until now we have considered a nonpresbyopic subject but, usually, this is not the case, so that the low vision patient might have to wear a device (a contact or spectacle lens) for compensating his presbyopia together with the low vision aid. The easiest way to take into account the joint effect of the low vision aid and the ammetropy compensation is through the concept of *equivalent power*. To do so, let us consider the optical scheme depicted in Figure 9.4, in which we have depicted a magnifier and the lens used for compensating presbyopia, the so-called addition lens. As can be seen in this figure, it is assumed that the distance between the addition lens and the eye is negligible compared to the distance d between the magnifier and the addition lens. In these conditions, the equivalent power is defined as:

$$P_{eq} = P_{mag} + A - dP_{mag}A, \tag{9.7}$$

where P_{mag} is the magnifier power and A is the power of the addition lens. Therefore, if the object is located at the focus of the equivalent lens, as depicted in Figure 9.4, then the visual magnification observed by the subject would be $M_N = -k_0 P_{eq}$. The concept of equivalent power is quite useful to the practitioner because, in some cases, "sharing" the total power needed by the patient between two optical elements (magnifier and addition lens) is advantageous, while, in other circumstances, using a high power lens would

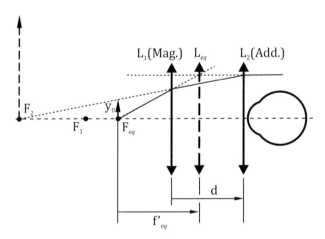

Figure 9.4 Optical scheme used to derive the concept of equivalent power. The magnifier (L_1) and the addition lens (L_2) form a compound optical system that is equivalent to the lens L_{eq} with equivalent focal length f'_{eq}. The ray tracing shows how to determine the position of the equivalent lens and its object focus F_{eq}.

allow the user to perform near tasks without the need to employ two different optical elements. More information about the concept and usage of equivalent power can be found in
specialized books such as references [333] or [334].

Field of View and Depth of Field

Although the most important optical parameter of a magnifier is the power (and consequently, its magnification), the practitioner must always take into account the fact that
prescribing a low vision aid requires the balancing of the advantage of an increment in
patient resolution with the disadvantages of a reduction in the field of view (FOV) and the
depth of focus.

We will study first the field of view, which, in the case of a magnifier, is similar to the
static field of view of a spectacle lens. In Figure 9.5, we have represented the optical scheme
for computing the semi-field of view. According to this figure, the image linear semi-field
y'_m can be computed as

$$y'_m = (d - s') \tan \omega'_m = \frac{D}{2}\left(1 - \frac{s'}{d}\right) \tag{9.8}$$

where D is the magnifier diameter and d is the distance between the magnifier and the eye.
In low vision, we are more interested in the object field of view y_m (that is, the extent of
the object that can be seen through the magnifier), which can be computed from the image
field of view as

$$2y_m = \frac{2y'_m}{\beta'} = \frac{s}{s'}2y'_m = Ds\left(\frac{1}{f'} + \frac{1}{s} - \frac{1}{d}\right). \tag{9.9}$$

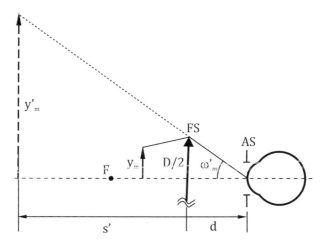

Figure 9.5 Field of view limitation in the magnifier-eye system. The magnifier, whose radius is $D/2$,
acts as a field stop (FS), while the eye's pupil is the aperture stop (AS) of the system. As both pupils
are located at the image space, they limit both the angular ω'_m and linear y'_m image semi-fields.

When the object is located at the focus of the magnifier, the frontal distance s is $s = -f'$ and equation (9.9) becomes

$$2y_m = \frac{Df'}{d} \equiv \frac{-k_0 D}{dM_N}. \tag{9.10}$$

This equation demonstrates the inverse relationship between the object field of view $2y_m$ and commercial magnification M_N, as stated before. As can be seen in equation (9.10), the field of view also depends on the diameter of the magnifier and the distance between the magnifier and the eye. It is interesting to note that the lower the distance d, the greater the FOV. Therefore, many low vision patients tend to approximate the eye to the magnifier as much as possible in order to get the maximum FOV.

Example 41 Compute the field of view of a $M_N = 4.5\times$ magnifier with diameter $D = 25$ mm and the eye located at 10 mm from the magnifier. Assume that $k_0 = -250$ mm and the object is located at the object focal plane of the aid. Would the user be able to observe the whole extension of a 7" screen e-reader without moving the magnifier?
   Substituting all data in equation (9.10) we see that

$$2y_m = \frac{-(-250) \times 25}{10 \times 4.5} = 138.9 \text{ mm}.$$

A 7" screen corresponds to a diagonal size of $x = 7 \times 25.4 = 177.8$ mm. As this length is greater than the object FOV, the subject would not be able to focus the whole extension of the e-reader screen.

   The depth of field (DOF) of a magnifier can be defined as the axial extension of the object space where the observer perceives a focused image of the object seen through the magnifier. To do so, the magnifier must form the virtual image of the object between the far and near points of the observer. To obtain a mathematical expression for the DOF let us consider an emmetropic observer whose near point is located at a distance k_0 from the eye. As the subject is emmetropic, the far point is located at infinity, so the far point of the DOF coincides with the object focus of the magnifier; therefore, the far object distance will be $s_{far} = -f'$. On the other hand, when the object lies at the near point s_{near} of the DOF, then the image is formed at the near point of the observer, so that $k' = k_0$. Therefore, the image distance is $s'_{near} = k_0 + d$, and applying Gauss equation we find that the near object distance is $s_{near} = (k_0 + d)f'/(f' - k_0 - d)$ and the DOF can be written as

$$\Delta s \equiv |s_{near} - s_{far}| = \frac{f'^2}{|f' - k_0 - d|}. \tag{9.11}$$

In order to relate the DOF with the magnification, in a similar way as we did with the field of view, let us now consider the particular case when the eye is located at the image focus of the magnifier, $d = f'$. In these conditions, equation (9.11) turns into:

$$\Delta s = \frac{f'^2}{|k_0|} = \frac{|k_0|}{M_N^2}. \tag{9.12}$$

Therefore, as stated by equation (9.12), the DOF also presents an inverse relationship with the magnification, but, in this case, the DOF is proportional to the inverse of the square of the magnification. This results in small values for the DOF for medium- to high-power magnifiers. For example, considering a value of $k_0 = -250$ mm, a $10\times$ magnifier would present a DOF of $\Delta s = 2.5$ mm. Such a small DOF makes the observation of nonplanar objects difficult and also has a big influence in the design of magnifiers, particularly of high-powered ones.

Microscopes

Equation (9.6) implies that to achieve large magnification values it is necessary to employ high power lenses. However, as power increases it becomes more and more complicated to manufacture a single lens with the required power because those high power lenses will present both low diameter and low image quality (the aberrations are greater for high power lenses). Thus, in order to get high values for magnification, a more complicated optical system is required. This system is the compound microscope, simply known as microscope in the literature. The name derives from the fact that the microscope is formed by two subsystems, objective and eyepiece, separated by a considerable distance. The objective is a high power positive system, which forms an enlarged image of the object that is, in turn, imaged by the eyepiece. As can be seen in Figure 9.6, the eyepiece acts as a magnifier of the intermediate image formed by the microscope objective. Therefore, it is usual to describe the visual magnification of a microscope as the product of the lateral magnification of the objective and the visual magnification of the eyepiece,

$$M = \beta'_{ob} M_{ep}, \tag{9.13}$$

β_{ob} being the lateral magnification of the objective and M_{ep} the visual magnification of the eyepiece. In optical instrumentation, it is usual to engrave the lateral magnification of the objective in its body in the form of a number followed by the multiplication sign. For example, an objective branded as $10\times$ presents a lateral magnification $\beta'_{ob} = -10\times$. The same happens with eyepieces that are marked with the value of the visual (commercial) magnification, usually taking $k_0 = -250$ mm. In the case of eyepieces, the mark is similar to that of the objectives and this marking is quite helpful in order to compute the microscope magnification easily. For example, if we have a microscope whose objective is labeled as $10\times$ and its eyepiece is $12.5\times$, then the visual magnification of the microscope will be $M = -10 \times 12.5 = -125\times$. It is important to point out the minus sign in the value of the visual magnification, which appears due to the image inversion.

As can be seen in Figure 9.6, where it has been depicted, the typical configuration of a microscope focused at infinity (therefore, it can be used by emmetropic or compensated observers), the lateral magnification of the objective can be written as $\beta'_{ob} = -t/f'_{ob}$ being f'_{ob} the focal length of the objective and t the so-called "tube length" or "mechanical tube length" [339]. On the other hand, as the intermediate image lies on the object focus of the

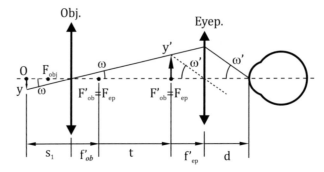

Figure 9.6 Schematic representation of a compound microscope formed by two subsystems (here depicted as positive thin lenses) named objective and eyepiece. The objective forms an enlarged intermediate image, which, in turn, acts as the object for the eyepiece. The eyepiece will act as a magnifier forming the final image to be seen by the observer's eye.

eyepiece, then its visual magnification is $M_{ep} = -k_0/f'_{ep}$, where f'_{ep} is the focal length of the eyepiece. Therefore, the magnification of the microscope can be written as

$$M = \frac{k_0 t}{f'_{ob} f'_{ep}} = \frac{-k_0}{f'_{eq}}, \tag{9.14}$$

where f'_{eq} is the equivalent focal length of the microscope, which coincides with the inverse of the equivalent power of the instrument.

Example 42 Compute the equivalent focal length of a $M = -125\times$ microscope if it is designed with $k_0 = -250$ mm.

According to equation (9.14), we see that $f'_{eq} = -k_0/M = -250/125 = -2$ mm, which is equivalent to a power of $P_{eq} = -500$ D. Obviously, this extreme value of the equivalent power cannot be achieved using a single lens (simple magnifier), and this is why two separate optical systems, the objective and the eyepiece, must be combined to form the compound microscope.

9.3.2 Optical Design of Magnifiers

The optical design of low vision aids has two main goals: optical quality and ergonomy. The second aspect is almost as important as the first because in many cases a low vision patient does also present additional illness that may impair her neuromotor capabilities. Thus, attending to ergonomy, we can classify magnifiers in three groups: handheld, mounted, and stand magnifiers.

A *handheld magnifier* consists of a lens (usually a low power one) mounted on a frame with a handler. These devices are employed mainly to carry out specific tasks, generally of short duration, which do not require great magnification. As can be seen in Figure 9.7, many models include an illuminating system in order to improve the contrast of the image perceived by the user. The advantages of handheld magnifiers are its simple design, low

Figure 9.7 Left: photograph of a handheld magnifier with an integrated illumination system. Right: photograph of a flat field magnifier; notice the difference in size between the letters in the central and external areas of the lens, which give an indication of the high magnification of the lens.

cost, ease of use, and discretion. On the contrary, they are not indicated for performing high precision tasks of long duration, and they will not be useful to those patients presenting psychomotor problems such as tremors, etc.

In a *stand magnifier* the lens is mounted in such a way that the distance between the lens and the object remains fixed. These are usually medium- to high-power lenses and typically they are used for observing planar or nearly planar objects (due to the low depth of focus). The simplest stand magnifier is the *flat field* magnifier depicted in Figure 9.7, which consists of a plano-convex lens with the plane surface located in contact with the planar object (usually they have a concave surface with very low curvature to avoid the formation of scratches). This kind of stand magnifiers are quite convenient for reading as the weight of the lens helps to keep the object in position and it is easier to maintain the working distance [333].

Due to their relatively high magnification, stand magnifiers are quite convenient for performing precision tasks. As they rest on the surface of the object they leave the hands of the user free, and they are also insensitive to involuntary hands movements. The stand magnifiers are found in spherical (for medium-power) and aspherical (for high-power) designs [333]. As in the case of the handheld magnifier, a built-in illuminating system can be incorporated in the mount [333].

The final type of magnifier that can be found is the *mounted magnifier*, which consists of a high power positive lens mounted on a spectacle frame (this arrangement is known in the specialized low vision literature as *"microscope"* [334]) or it may consist of a lens attached to the spectacle frame of the user by means of a clip-on mount or similar device [333, 334]. The greatest advantage of mounted magnifiers are that they maintain the binocular vision of the user provided the lenses are properly positioned in the frame. Other advantages are an improved cosmetic appeal, hands-free, and large field of view [334].

9.4 Low Vision Aids for Distant Objects: Telescopes

In this section, we will present low vision aids designed for looking at distant objects. Those aids are telescopes that can be found in two basic designs, the Keplerian telescope and the Galilean telescope. Those designs will be presented first in the so-called afocal configuration, as is customary in optics. However, in many cases, and certainly in low vision practice, most telescopes can be focused and they are not employed in the afocal configuration. Therefore, we will briefly study in this section the differences between afocal and nonafocal telescopes. Afterward, we will descibe other optical properties such as field of view and depth of focus, and we will finish by describing the use of telescopes in near vision, particularly the telemicroscope.

9.4.1 Afocal Telescopes

In Figure 9.8, we can see the usual scheme of a Keplerian telescope in afocal configuration. In this figure, the telescope is represented by two convergent lenses (objective and eyepiece) located in such a way that the image focus of the objective coincides with the object focus of the eyepiece. For an afocal telescope the distance between the objective and eyepiece (the length of the telescope) is equal to the sum of the focal image distances of the objective and eyepiece. Therefore, for an afocal telescope

$$d_\infty = f'_{ob} + f'_{ep} = \frac{P_{ob} + P_{ep}}{P_{ob}P_{ep}},$$ (9.15)

where d_∞ is the length of the afocal telescope, f'_{ob} is the focal length of the objective and f'_{ep} the image focal length for the eyepiece. With the afocal condition stated in equation (9.15) we can derive the properties of afocal telescopes. First, we see that the power of an afocal telescope is zero, that is, $P_\infty = 0$, which can be easily proved by substituting the length of the afocal telescope, given by equation (9.15) into the formula of the power of a compound system (9.7). As a consequence the focal distance of an afocal telescope is infinite, hence the name "afocal." Also, if two parallel rays enter the telescope, they will emerge parallel.

Afocal telescopes present more properties particular to this kind of system. The angular magnification (and, thus, the visual magnification for far objects) is constant. From Figure 9.8, we see that its value is

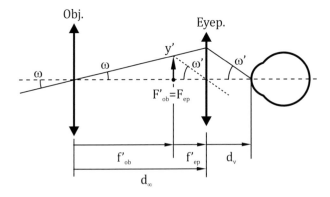

Figure 9.8 Keplerian telescope in afocal configuration.

$$\Gamma_\infty \equiv M_\infty = \frac{\tan \omega'}{\tan \omega} = \frac{-y'/f'_{ep}}{y'/f'_{ob}} = -\frac{f'_{ob}}{f'_{ep}} = -\frac{P_{ep}}{P_{ob}}. \tag{9.16}$$

Moreover, the lateral magnification is also constant, regardless of the object position, and its value is the inverse of the visual magnification, so that

$$\beta'_\infty = \frac{1}{M_\infty} = -\frac{P_{ob}}{P_{ep}}. \tag{9.17}$$

This inverse relationship presents an apparent problem, because if $|M_\infty| > 1$, then it means that $|\beta'_\infty| < 1$ and the size of the image is smaller than that of the object. However, although both statements are true, there is no contradiction, as the afocal telescope acts as a vergence multiplier. For a given object, the image through the telescope is always located nearer to the eye, in such a way that the angular size of the image is greater than that of the object, and the perceived (or apparent) size is bigger. In fact, when an afocal telescope is focused on an object located at a distance s from the objective and an image is formed at a distance s' measured from the eyepiece, image and object vergences are related by [334]

$$V' = \frac{M_\infty^2 V}{1 - d_\infty M_\infty V}, \tag{9.18}$$

where $V = 1/s$ and $V' = 1/s'$ are the object and image vergences. Equation (9.18) is known in the literature as the vergence amplification formula or Freid's equation [334], and it states that the vergence of the image can be approximated by the object vergence multiplied by the square of the visual magnification of the telescope. Therefore, although the image size is reduced by a factor M_∞, the image is M_∞^2 times closer to the object, which explains why, provided $|M_\infty| > 1$, the observer will experience the sensation of a magnified image, although we must properly speak of a closer image.

Example 43 A low vision patient uses a $M_\infty = -2.5\times$ telescope whose eyepiece has a power of 25 D. If the patient observes an object of 2 mm height located at a distance of 3 m from the objective of the telescope, compute the position of the image measured from the eyepiece.

First, we will compute the power of the objective from equation (9.16) so that $P_{ob} = -P_{ep}/M_\infty = -25/-2.5 = 10$ D. Afterward we will calculate the length of the telescope as

$$d_\infty = \frac{P_{ob} + P_{ep}}{P_{ob} P_{ep}} = \frac{10 + 25}{10 \times 25} = \frac{35}{250} = 0.14 \text{ m},$$

as the object vergence is equal to $V = 1/s = -1/3 = -0.333$ D, then we see that the image vergence is

$$V' = \frac{2.5^2 \times (-0.333)}{1 - 0.14 \times -2.5 \times -0.333} = -21.22 \text{ D}.$$

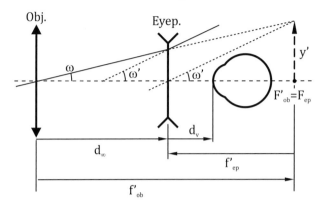

Figure 9.9 Optical scheme of a Galilean telescope. Notice that, compared with a Keplerian telescope, the eyepiece is a divergent (or negative) lens.

Thus, the image is formed at a distance $s' = (-21.22)^{-1} = -0.047$ m or 47 mm to the left of the eyepiece.

All of the properties of the afocal telescopes studied in this subsection can be applied to the Galilean telescope, which is formed by a convergent (or positive) objective combined with a divergent (or negative) eyepiece as shown in Figure 9.9. However, although the imaging properties of both types of afocal telescopes are similar, there are two important differences. First, the sign of the magnification for the Galilean telescope is positive, so the image has the same orientation as the object, avoiding the need to use an image inverter. A second difference is that for the same amount of magnification and power of the objective, the length of the Galilean telescope is lower than that of the Keplerian. These two facts have been conveniently used in low vision practice when the patient does not need high magnifications.

9.4.2 Nonafocal Telescopes

An afocal telescope allows an emmetropic user to focus an object located at the infinite. However, this is not the more usual case in the context of low vision for two reasons. First, the object can be located within a wide range of distances depending on the situation. For example, when looking at an object placed on a shelf in a supermarket, the distance from this object to the eye is around 1 m. According to the rule of vergence amplification given by equation (9.18), a telescope with $M_\infty = -2.5\times$ would form the image at a distance of, approximately, 160 mm from the eye, which is too close to be focused by most observers. Second, even if the object is located far enough away, the observer could not be emmetropic, so the image can be formed outside of his range of sharp vision.

A common solution for these drawbacks consists of making the telescope nonafocal (or focusable) by changing the distance between the objective and the eyepiece. If we give the notation $\Delta d = d - d_\infty$ to the difference between the length of the nonafocal telescope, d, and that of the afocal one, d_∞, then we can write the equivalent power of the nonafocal telescope P_{eq} as follows:

$$P_{eq} = -\Delta dP_{ob}P_{ep} = \frac{\Delta dP_{ep}^2}{M_\infty}. \tag{9.19}$$

Therefore, if a nonemmetropic individual with refractive error R observes a distant object with the telescope, she can move the eyepiece in order to focus the image of the object given by the telescope at her remote point. In these conditions, it can be demonstrated that the amount of displacement necessary to focus the image is

$$\Delta d = \frac{-R}{P_{ep}^2(1 + d_v R) - P_{ep}R}, \tag{9.20}$$

where d_v is the distance between the eyepiece and the eye. Equation (9.20) indicates the relationship between the refractive error of the observer and the displacement necessary in order to focus the image of a distant object with a nonafocal telescope, and it constitutes the basis of the graduate scales that can be observed in the focusable eyepieces.

Regarding magnification, for a general case of a near object and ammetropic observer, it is convenient to solve each particular case using the general equation (9.1) for computing the magnification, as shown in the following example.

Example 44 Compute the length and magnification of a nonafocal telescope whose objective and eyepiece have powers of 5 D and 25 D, respectively, if the telescope is used by a myopic user with -2.5 D of refractive error to focus an object placed 1.5 m away from the objective of the telescope. Note: take the vertex distance as $d_v = 14$ mm.

First, we will calculate the location of the intermediate image. According to the wording of the problem, $s_1 = -1500$ mm measured from the objective of the telescope. This distance corresponds to an object vergence of -0.67 D, so, by applying the Gauss law, we see that the image vergence after the objective is $V_1' = P_{ob} + V_1 = 4.33$ D, so the intermediate image is located at a distance $s_1' = 230.94$ mm from the objective of the telescope.

We will focus now on the location of the final image of the telescope, s_2'. In order that the myopic user could be able to focus it, the final image should be placed at the remote point of the observer. Taking into account the distance d_v between the eyepiece and the eye, and the remote distance of the observer $r = -400$ mm (which corresponds to a refractive error of -2.5 D), we have $s_2' = r + d_v = -386$ mm, the corresponding vergence being $V_2' = -2.59$ D. The vergence of the conjugate object is given again by the application of Gauss' equation $V_2 = V_2' - P_{ep} = -2.59 - 25 = -27.59$ D. Therefore, the intermediate distance should be located at a distance of $s_2 = -36.24$ mm measured from the eyepiece.

The computation of the intermediate and final images results in two conditions that must be simultaneously fulfilled: The intermediate image should be located at a distance $s_1' = 230.94$ mm measured from the objective and $s_2 = -36.24$ mm from the eyepiece. From the laws of geometrical optics, it can be demonstrated that these two conditions could only be satisfied if the length of the telescope is equal to $d = s_1' - s_2 = 230.94 - (-36.24) = 267.2$ mm. The length of the equivalent afocal telescope is $d_\infty = (P_{ob} + P_{ep}) / (P_{ob} P_{ep}) = 30/125 = 0.24$ m $\equiv 240$ mm, so the afocal telescope would be 27.2 mm longer. To compute the telescope magnification, we will use general equation (9.1). For this, we have to compute lateral magnification β', object distance k_o when observed without instrument, and the distance from the eye to the final image, k'. The lateral magnification is

$$\beta' = \frac{s_2' s_1'}{s_2 s_1} = \frac{-386 \times 230.94}{-36.24 \times -1500} = -1.63 \times .$$

For the object distance, we can assume that $k_o \simeq s_1 = -1500$ mm and, as mentioned previously, the final image is formed at the remote point of the object so that $k' \equiv r = -400$ mm. Thus, the magnification of the telescope is

$$M = \frac{k_0}{k'} \beta' = \frac{-1500}{-400} \times -1.63 = -6.15 \times .$$

The magnification of the equivalent afocal telescope (the one that would have the same objective and eyepiece but in afocal configuration) is $M_\infty = -5\times$. So we can see that the nonafocal telescope presents greater magnification (in absolute value) than the equivalent afocal one. However, this would not be necessarily true in all cases. In general, the values of the magnification for the nonafocal and afocal telescopes will be of the same order.

9.4.3 Other Optical Properties of Telescopes: Field of View and Depth of Focus

As a rule of thumb, telescopes present a limited field of view for both Keplerian and Galilean designs. The latter has a further constraint of the field of view because the exit window (image of the field stop) is virtual, which gives the characteristic "keyhole view" of those instruments. In the case of the Keplerian telescope, the optical scheme for computing the field of view is shown in Figure 9.10. As can be seen in this figure, the eyepiece of the telescope acts as a field stop and, simultaneously, as exit window, as there are no lenses

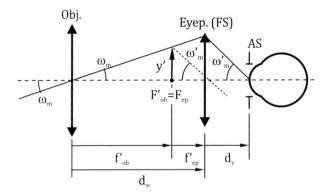

Figure 9.10 Optical scheme for computing the field of view ω_m of a Keplerian telescope. In this scheme FS stands for "field stop," while AS means "aperture stop."

behind the eyepiece. On the other hand, as in the case of the magnifier, we will assume that the entrance pupil of the eye acts as an aperture stop for the telescope. In these conditions, if the diameter of the eyepiece is D_{ep}, then the angular image semi-field is given by

$$\tan \omega'_m = \frac{D_{ep}}{2d_v},$$ (9.21)

where d_v is the distance between the eyepiece and the vertex of the eye (see Figure 9.21). For an afocal telescope, the computation of the angular object semi-field ω_m is straightforward as the angular and visual magnification are the same. Therefore, we have

$$\tan \omega_m = \frac{\tan \omega'_m}{|M_\infty|} = \frac{D_{ep}}{2d_v |M_\infty|}.$$ (9.22)

This equation illustrates the inverse relationship between field of view and magnification. It is important to notice that we have used the absolute value of the visual magnification as we are interested only in the extension of the FOV, and the sign of the angle is irrelevant in this case. As we will see in the following example, the inverse relationship between FOV and magnification leads to low values for the field of view even for moderate magnifications.

Example 45 Compute the field of view of a Keplerian telescope with magnification $M_\infty = -2.5\times$ if the diameter of the eyepiece is equal to 12.7 mm $(0.5'')$ and the vertex distance is 14 mm. What will the maximum observable extension of a poster located 6 m away from the observer be?

According to equation (9.22), the tangent of the semi-field of view is

$$\tan \omega_m = \frac{D_{ep}}{2d_v |M_\infty|} = \frac{12.7}{2 \times 14 \times 2.5} = 0.1814,$$

thus, the object semi-field is equal to $\omega_m = \tan^{-1}(0.1814) = 10.3°$ so the complete angular field of view is $2\omega_m = 20.6°$. If the poster is placed $s = -6$ m from the telescope, the lateral field that can be seen through the instrument will be

$$2y_m = 2 |s| \tan \omega_m = 2 \times 6 \times 0.1814 \equiv 2.17 \, \text{m}.$$

Although, in this case, the observer could see a wide area of the poster with the telescope, this is not what usually happens when telescopes with medium to large magnifications are used, as they tend to strongly limit the FOV.

Regarding the depth of focus, its computation is based on the same approach employed with the magnifier. For the afocal telescope, we will make use of equation (9.18) but, for the sake of simplicity, we will suppose that $d_v \simeq 0$ and the output vergence will be $V' = M_\infty^2 V$. Within this approximation, the image vergence V' is equal to the inverse of the distance k' between the eye and the final image. In these conditions, the limits of the DOF are given by the points in the object space for which V' coincides with the vergence of the remote

$V_1' = R$ and near $V_2' = P$ points of the observer. Thus, for the distance point, its position s_1 is given by the following equation

$$s_1 = \frac{M_\infty^2}{R},$$ (9.23)

while the location of the near point can be computed from this expression

$$s_2 = \frac{M_\infty^2}{P}.$$ (9.24)

Therefore, the depth of focus of the afocal telescope can be written as

$$\Delta s = |s_2 - s_1| = M_\infty^2 |p - r|,$$ (9.25)

where $r = R^{-1}$ and $p = P^{-1}$ are the distances from the eye to the remote and near points, respectively. From equation (9.25) we can deduce that, contrary to what happened for magnifiers and microscopes, in the case of the telescope, the DOF is greater than that of the naked eye as it is multiplied by the square of the magnification.

9.4.4 Telescopes in Near Vision: Telemicroscope

In principle, the vergence amplification, see equation (9.18), prevents the usage of telescopes in near vision as they would form the image of a near object so close to the eye that it could not be focused by the observer. However, this problem can be solved using an addition lens known in the context of low vision as "reading cap" [333, 334]. This is a convergent lens that can be attached to the objective of the telescope. If the object is located at the focus of the addition lens, it will form an image at infinity that can be focused by the telescope, as depicted in Figure 9.11. In this way, an afocal telescope can be focused to any object distance by choosing the proper power for the reading cap. The system composed by the reading cap and the afocal telescope is known as a *telemicroscope*.

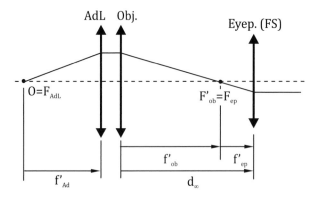

Figure 9.11 Optical scheme of a telemicroscope formed by the combination of an addition lens (AdL) and an afocal telescope.

As the telemicroscope is used to focus near objects, the magnification can be calculated from the focal distance of the telemicroscope. To do so, we can consider that the addition lens and the objective are so close that the distance between them is negligible, so the power of the set formed by the addition lens and the objective would be $P_1 = A + P_{ob}$. In these conditions the equivalent power of the telemicroscope will be

$$P_{TM} = P_1 + P_{ep} - d_\infty P_1 P_{ep}, \tag{9.26}$$

substituting $P_1 = A + P_{ob}$ and $d_\infty = P_{ob}^{-1} + P_{ep}^{-1}$ in equation (9.26) we have

$$P_{TM} = -A \frac{P_{ep}}{P_{ob}} = AM_\infty. \tag{9.27}$$

Thus, the power of the telemicroscope is equal to the power of the addition lens multiplied by the magnification of the afocal telescope. In these conditions, the magnification of the telemicroscope can be obtained directly from equation (9.6) as

$$M_{TM} = -k_0 P_{TM} \equiv M_A M_\infty, \tag{9.28}$$

where $M_A = -k_0 A$ is the normal magnification of the addition lens.

9.5 Low Vision Aids: Field Increasing Aids

Until now, we have studied classical optical systems (magnifiers, telescopes, and telemicroscopes) employed for enhancing the visual acuity of the user. However, as we have already seen, not all low vision problems can be compensated by increasing the size of the object. Some conditions exist that may reduce the field of view of an individual. To compensate those patients, a number of aids have been devised. Some of them are instruments like the reversed Galilean telescope, while others are formed by combination of mirrors and/or prisms, which are described in detail in the general references of low vision, such as the works of Dickinson [333] or Brilliant [334]. However, it is interesting to describe briefly the reversed Galilean telescope and how it helps to increase the field of view of the user.

In Figure 9.12, we show the optical scheme and the ray tracing used to compute the FOV of a reversed Galilean telescope. As we can see, the magnification of the reversed Galilean fulfills the condition $M_\infty < 1$, as the power of the objective is bigger that the eyepiece power. This is necessary because, as it is an afocal telescope, the FOV of the system can be computed using equation (9.22), so that the field of view is inversely proportional to magnification. Obviously, the usage of a device with magnification $M_\infty < 1$ implies some loss of visual acuity for the observer, but this loss is unavoidable given the inverse relationship between field and magnification. Curiously, the reversed Galilean is a fairly common optical system, as it can be found as door peephole viewers in many homes.

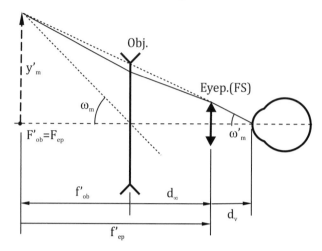

Figure 9.12 Field of view of the reversed Galilean telescope. Note that the absolute value of the objective power is bigger than the eyepiece power so that $M_\infty < 1$. Note that the eyepiece is the field stop (FS) of the system, while the pupil of the eye (not represented) would be the aperture stop.

9.6 High Power Ophthalmic Lenses

High power lenses are used in ophthalmic optics for two purposes. The first is to compensate high amounts of hyperopia or myopia, while the second is to compensate aphakic patients. Currently, the usage of intraocular lenses for compensating the aphakic eye is so extended that high power spectacle lenses for aphakia are only used when the implementation of an IOL is not possible. In any case, we will give some details on aphakic lenses when we discuss high power positive lenses.

In general, high power means stepper surfaces and, as a consequence, greater thickness, weight, magnification (in absolute value), aberrations, and prismatic effects. Therefore, as a rule of thumb, we may say that the user of high power lenses may experience a greater loss of image quality, and a reduction in comfort and esthetics, than the user of low power ones. We will see now in detail the characteristics of high power positive and negative lenses.

9.6.1 High Power Positive Lenses

For hyperopes, the refractive error seldom surpasses 10–12 D. For aphakic patients, the value of post-surgery refractive error depends on the value of refractive error before surgery through physiological parameters such as the length of the eye and the shape of the cornea. With IOLs, there are a number of empirical equations for predicting the value of the post-surgical refractive error. For example, Sanders et al. [340] state that the post-surgery refractive error is given by

$$R = 80.4 - 1.65l_o - 0.7K, \tag{9.29}$$

where l_o is eye length measured in millimeters and K is the refractive power of the cornea obtained through keratometry and measured in diopters. For example, an aphakic patient with an eye length of $l_o = 20.5$ mm and $K = 35.5$ D would present a refractive error $R = 21.72$ D according to equation (9.29). For this reason, the compensation of aphakia usually requires positive lenses with more power than those for high hyperopia.

For aphakic lenses magnification can reach values of approximately $1.33\times$ due to the high power and the change in the optics of the eye when the crystalline lens is removed, see [132] for further details. Conversely, the field of view is reduced due to the high power of the lens and the ring scotoma that appears at the periphery of the lens due to the high amount of prismatic deviation at this area [132]. Aberrations also have an effect on the reduction of the effective field of view of these lenses.

Regarding the design of high power positive lenses, besides the use of high refractive index materials for reducing thickness, it is very common to find aspherical designs. The goal pursued with the use of aspherical designs is to achieve a thickness reduction and to improve the optical quality of the lens. However, those two goals can only be achieved with complex surfaces with a high number of degrees of freedom. On the other hand, high power lenses usually present a limitation in the maximum diameter available due to the high curvature of the lens surfaces. This limitation can be solved with so-called lenticular designs. In a positive lenticular the front surface has a different radius of curvature at the center, which is steeper, than at the periphery, which is flatter (see Figure 9.13). A comprehensive review of lenticular lenses can be found in [56].

The practitioner should be extremely careful when fitting aphakic lenses because any undesired decentering could significantly impair the user's vision. It is recommended that, for aphakic patients, the naso-pupillary distance should be measured accurately in order to avoid undesired prismatic effects in the horizontal direction [132]. Vertically, the procedure recommended by Fannin and Grosvenor [132] is to fit the aphakic lens with the optical center at the same height as the boxing line and manipulate the frame so that the optical axis of the lens passes through the rotation center of the eye. Nowadays, it is possible to design the lens specifically for a given value of pantoscopic and facial tilts, and additional

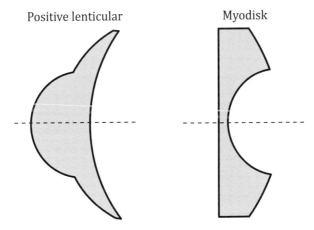

Figure 9.13 Lenticular designs for positive (left) and negative (right) high power lenses. The negative design with an anterior surface flat is known as myodisk.

parameters such as naso-pupillary distance, pupil height, shape of the spectacle frame rim, etc., optimizing, at the same time, thickness and aberration correction.

Finally, we must point out that after losing the crystalline lens, the retina of the aphakic eye loses the protection against low wavelength radiation, particularly UVA, UVB, blue, and violet ranges. Thus, it is mandatory to use a filter for low wavelengths in the UV and visible ranges. Usually, filters for aphakic lenses are designed with a cutoff frequency around 400 nm [132]. In addition, as the aphakic eye is more sensitive to glare [132], some of the radiation in the interval between 400 and 500 nm will also be filtered, which gives those lenses their characteristic yellowish appearance.

9.6.2 High Power Negative Lenses

High myopia is a common case in optometric practice. Epidemiologic studies, such as those of Ling in 1995 [341] and 2000 [342] show that around 15% of a young (eighteen years old) population presents a refractive error equal to or smaller than −6 D. This percentage of high myopia prevalence increases with age due to conditions such as Myopia magna. One solution for compensating high myopia is the prescription of high negative power spectacle lenses. They present some particularities, which will be described in the following paragraphs.

One of the most relevant problems is the large edge thickness presented by high power negative lenses. In Chapter 4, we presented an approximate equation for computing the edge thickness of a lens given by

$$t_e = t_c - \frac{D^2 P_{TL}}{8(n-1)},$$ (9.30)

where t_c is the center thickness, D is the lens diameter, P_{TL} is the thin-lens power, and n the refractive index. According to equation (9.30), there are two ways of reducing the edge thickness: (1) by increasing the refractive index, (2) by reducing the diameter. A third way is to employ aspheric designs (like with high power positive lenses), which can also reduce lens aberrations.

Regarding the optical properties, contrary to positive lenses, high power negative lenses have a wide field of view, but with reduced magnification [132]. Lenticular designs are very useful when the power of the lens is so high that the resulting lens diameter will be very small using a standard meniscus shape. One of the most popular of those designs is the myodisc shown in Figure 9.13, but there are other lenticular solutions available [56].

Finally, given the large edge thickness presented by high power negative lenses, there is an increased risk of internal reflections at the edge of those lenses. These internal reflections give rise to the appearance of ring images of the lens edge that are both unesthetic and annoying. There are several techniques for reducing such reflections [132], with the common goal of all of them being to change the edge surface in such a way that it becomes a diffuse reflecting surface instead of specular. Typically, tinting or grinding the edge gives very good results in reducing those unwanted edge reflections [132].

10

Lens Manufacturing and Measurement

10.1 Introduction

The process of ophthalmic lens manufacturing has unique characteristics in the context of optical technologies. These characteristics are associated with the type of product, its distribution, and commercialization. The most important factors are the variety of surfaces that must be generated, the materials employed to build the lenses, the manufacturing technologies used, the precision of the surface, and the processing time and cost.

With few exceptions, a spectacle lens has only two surfaces and a single refractive index. With this simple geometry, we must provide solutions for all refractive errors. In technical optics, the degrees of freedom required to minimize aberrations are achieved using multiple simple surfaces, typically spheres. For example, a single focal SLR camera lens can easily have more than 10 surfaces. Spheres are by far the easiest surfaces to manufacture, and mass production makes possible affordable prices. In the case of ophthalmic lens manufacturing, the degrees of freedom to deal with lens design lie in the two surfaces of the lens. Additionally, the front surface, the base, is chosen from a reduced set of fixed surfaces, leaving the back surface as the main responsible for dealing with the prescription, power distribution and aberration handling. Therefore, generators and polishers must be able to produce a large variety of surfaces, including flats, spheres, conicoids, astigmatic surfaces, and, in general, free-form surfaces [343].

Historically, the standard material for lenses has been glass. Properties like stability, hardness, and refractive index make glass the first choice for lens manufacturing. However, its inherent fragility and difficulty of production using molding techniques has promoted the development of polymeric materials, such as CR-39, that are appropriate for molding, easy to surface, and compatible with standard tinting and coating processes. For these reasons, today in ophthalmic optics the more common materials are plastics. For an introduction to optical materials the reader is referred to Chapter 1, where we present a comprehensive presentation of the subject.

With respect to manufacturing, the ophthalmic industry combines in a unique way the molding and generating processes. In the optics industry, mass production applications tend to use only one technology. For example, molded optics is used in focusing lenses of the optical disc readers, while the majority of precision glass lenses are surfaced by generation and polishing. Meanwhile, in the ophthalmic lens sector the usual process uses a plastic molded blank for which one of the sides is finished by a generating and polishing process.

Materials and manufacturing are tightly coupled with the power tolerances of an ophthalmic lens. Current norms (ISO and ANSI [344, 345]) imply that overall surface deviation can be quite large, up to 20 microns, while local errors should be smaller than 1 micron. In technical optics, the precision is typically an order of magnitude less, and some special applications, like lithography imaging lenses, require even more, in the order of 1 nm. This relatively low surfacing tolerance is what makes it possible for ophthalmic lens generators to produce lenses at a much higher rate than precision optics systems. The typical surfacing time in ophthalmic lens manufacturing is under $5'$, to be compared with at least $60'$ in precision optics. The surfacing time should be framed in the more general process of manufacturing a spectacle lens. Ophthalmic lenses must be produced and delivered very fast. From the lens order to its delivery there should be no more than two days. Taking into account the high number of processes associated, from surfacing to tinting and coating (that imply idle curing times), this requirement imposes a high degree of organization and logistics in the manufacturing process.

Closely related to the manufacturing process is lens production quality control. In ophthalmic lens manufacturing we consider two main types of defects, cosmetic and refractive. In this chapter, we are going to center our discussion on refractive quality, in other words how to check if the lens actual power distribution matches the expected one. The most common way to verify the refractive properties of a lens is the measurement of the through power, that is, refractive power measured in a transmission configuration as in the object-lens-eye system. For a homogeneous lens material with constant refractive index, refractive power is determined by surface curvature; therefore, for spherical surfaces we can calculate the surface power by measuring the radius of curvature. This direct measurement can be done by using a spherometer calibrated in diopters. However, for a general surface the curvature is no longer a scalar and the spherometer is of very limited use. In this case the surface sag can be measured using a perfilometer, and from the resulting point cloud we can calculate the refractive power in the whole lens by calculating the curvatures (see Appendices C and D). Although sag is a direct measurement, the process of measuring the whole lens can be time consuming. An alternative to sag measurement are focimeters and lens mappers. These systems permit measurement of through power by observing a test or pattern through the lens, making possible a very fast through power measurement.

An important question arises related to the very definition of power and the different technologies that can be used to measure it. The back vertex power of standard single vision lenses is the one that should match the prescription. However, local back vertex power is not uniquely defined in multifocal and/or customized lenses. Different measuring technologies will yield different types of local power, and in general the measured power does not match the user perceived power, as described in Chapter 8. Although ISO and ANSI regulations are clear about measuring paraxial back vertex power at one or two points on the lens surface (optical center or DRP and NRP), even today there is not a clear consensus on how local power should be measured and assessed in a progressive lens.

With all these ideas in mind, in the first two sections of this chapter we are going to review the main spectacle lens surfacing technologies and the lens manufacturing process in a modern free-form fabrication line. After this, we will review lens power measurement tools and techniques to finally discuss the issue of the differences between user power and the measured power. We will finish this chapter with a discussion about the use of ISO test procedures in the case of progressive power lenses (PPL).

10.2 Lens Surfacing

Lens surfacing is the process of producing optical elements with polished surfaces. In this section we are going to present the two main technologies used in the ophthalmic industry, grinding and polishing, and casting. In grinding and polishing a material blank is first shaped by a subtractive process, grinding, that leaves an optically rough surface that must be finished to achieve optical performance, polishing.

10.2.1 Lapping

Historically, the grinding and polishing processes were carried out by a process called lapping [25, 346]. In lapping lens surfacing is achieved through a series of grinding operations in which the lens is held against a metallic tool of opposite curvature as shown in Figure 10.1. The final step is a polishing process using a soft lap tool and the finest grain abrasive powder. In the lapping process, the lens starts as a glass blank consisting of a thick disk of material that is blocked and fixed to a shaft [347]. The lap tool is moved in a random way from side to side of the lens while vertical pressure is applied to the lap. The glass material is removed by the use of an abrasive slurry inserted between the lap tool and the lens. From the flat to the spherical surfaces, the process goes through a series of finer grain abrasive slurries. A typical process would pass through roughing (0.3 mm grain size), truing (75 μm), and smoothing (10 μm). The last polishing step uses a cloth fixed in the lap tool with an abrasive of 1–2 μm grain size.

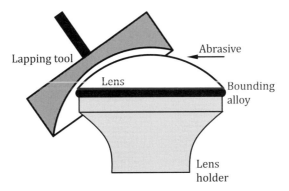

Figure 10.1 Lapping process for lens surfacing.

   This process not only surfaces the lens but also slowly degrades the lap tool, traditionally an iron-cast mold. For this reason these iron tools were substituted by cast aluminum tools in which the grinding is done using interchangeable adhesive pads that are fixed to the lap tool. The abrasive pads can be easily replaced as necessary and so the working life of the lap tool is increased. Although the slurry is no longer necessary, a coolant is used to dissipate the heat and remove away the glass material.

   The process of lapping is still used for mass-produced spherical surfaces, and the lapping process is extensively used to manufacture flat surfaces with optical quality.

10.2.2 Sphero-Torical Generators

In lapping, the first grinding steps to generate a rough surface are time consuming. The solution to this problem was to use a generation process. The first spectacle lens generators consisted of a milling machine with a ring tool in which the rim was impregnated with sintered diamond and were designed to work with glass. In a spherical generator the lens blank is blocked in a vertical spinning axis and the ring tool is presented to the blank tilted and displaced with respect to the vertical as shown in Figure 10.2. To generate the desired curve, the rotating ring tool is displaced vertically while the blank is spinning. If the generator is well calibrated, in the final position the rim of the ring tool should be aligned with the vertical spinning axis and the ring tool axis should point to the center of curvature of the generated sphere [25, 344]. If d is the diameter of the ring tool and α is the angle between the ring tool axis and the spinning axis, the radius of curvature of the generated sphere is given by

$$R = \frac{d}{2\sin\alpha}. \tag{10.1}$$

When the generator axis rotates about a point other than the center of curvature of the spherical surface, one meridian of the lens will have the radius of curvature of the generator circular sweep while the other will have the radius of curvature as given by equation (10.1). This is the base of the universal toric generator, and this has been the main process for

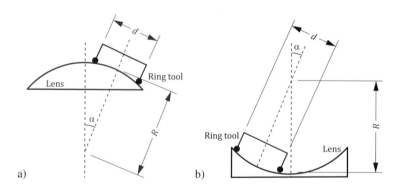

Figure 10.2 Spherical surface generation using a ring tool a) Convex and b) concave case.

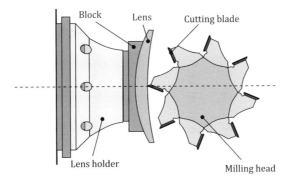

Figure 10.3 Diagram of a milling head used in cut to polish generation.

manufacturing toric surfaces. These toric generators produced rough surfaces, and they required, before polishing, an intermediate finishing step.

With the appearance of new plastic materials, tools designed for glass tend to *load*, decreasing their cutting capability [344]. This issue led to the last evolution in sphero-torical generators before the appearance of free-form technology: The "cut to polish" process. This technology consisted of the use of a crown-shaped milling tool for lens surfacing, such as the one shown in Figure 10.3. These generators can produce toric surfaces with main radii depending on the diameter, tilt angle, and rotation axis of the tool. However, the main advantage is that in plastic materials cut-to-polish generators can generate surfaces ready for the last polishing step, saving the finishing step. With cut-to-polish technology, the last polishing step is performed using aluminum tools with detachable pads.

Lap and ring tools are calibrated to produce a surface with a given refractive power; usually they are stocked in steps of 0.25 D. However, the power of a spherical surface on air depends on the refractive index of the material, which means the tools generate the labeled power only for a reference refractive index. If the tool is designed to generate a spherical surface of radius R, for a material with index n, the generated refractive power is

$$P = \frac{n-1}{R} \; ; \tag{10.2}$$

however, the power labeled in the tool is

$$P_{ref} = \frac{n_{ref} - 1}{R} \; . \tag{10.3}$$

If $n_{ref} \neq n$ then the tool must be selected with a labeled refractive power of

$$P_{ref} = \frac{n_{ref} - 1}{n - 1} P \; . \tag{10.4}$$

This process is called tool power conversion. For example, in the United States $n_{ref} = 1.523$. In other countries you will find two sets of tools, one for glass with $n_{ref} = 1.523$ and the second for CR-39 with $n_{ref} = 1.498$ [348].

10.2.3 Free-Form Surfacing

Lapping and sphero-torical generators can only produce monofocal lenses. Progressive power lenses (PPL) need nonrotationally-symmetrical surfaces in at least one of the two sides of the lens. The traditional approach to this problem is to use front surface progressive blanks and then generate a torical surface in the back side to compensate user refractive errors in the near and far zones of the PPL. These progressive blanks are usually fabricated in plastic materials using a casting process. However, the use of progressive blanks imposes limitations in the geometry and the number of available lens designs. For this reason, these types of PPL do not allow for a full personalization of the spectacle lens taking into account aspects as the frame shape, interpupilar distance, and facial and pantoscopic angles, for example. Moreover, two lenses made with the same progressive blank but with different back spherotoric surfaces (in order to take account of two different prescriptions) may present noticeable differences in image quality.

The free-form solution to this issue is the generation of back surface PPLs using front surface spherical blanks. Free-form is a manufacturing technology that allows for the generating and polishing of arbitrary surfaces. A lens is free-form if at least one of its surfaces is made with free-form technology, and that surface is not necessarily spherical neither torical. Free-form machines may generate standard surfaces, but as they can also be generated with standard methods, the corresponding lenses should not be named free-form lenses. Free-form surfacing uses numerically controlled generators and polishers.

In free-form the front surface is blocked and the blank is fixed to a lathe. Then, the back surface is generated using diamond turning with an oscillating tool head. In this context diamond turning means that the lens blank is mounted in a lathe and the cutting tool has a diamond single point (see Figure 10.4(a)). In this configuration, while the lens is spinning the cutting tool traces a spiral path shaping the surface. The generation of nonrotational symmetric surfaces is accomplished by mounting the cutting tool on an oscillating tool head that is synchronized with the lathe, so that the tool moves forward or backward as the piece rotates. The surface finish provided by modern free-form generators minimizes the polishing needs and is usually a cut to polish procedure. In free-form surfacing, polishing takes place by means of a flexible pad mounted in a numerically controlled tool as shown in Figure 10.4(b). Due to the single point generation, free-form is not well suited for brittle

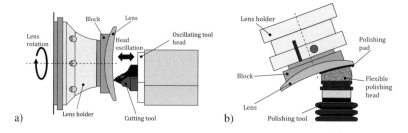

Figure 10.4 Free-form surface generation a) single point diamond turning b) flexible polishing pad.

materials such as optical glass. Free-form able materials include nonferrous metals like copper, aluminum, or brass, and plastics like PMMA or CR-39 [343].

The combination of the spiral path and the oscillating head permits the creation of highly nonsymmetrical surfaces with very few limitations in shape and curvature. Also, the surface does no longer need any symmetry and can be defined in a sag file, making the free-form manufacturing a CAD-CAM process. That means that the sag file, and therefore the lens design, is decoupled from the surfacing. Different lens design software (LDS) can interact with the free-form generators and the laboratory management software (LMS) as we will explain later.

European manufacturers of free-form machinery include Schneider, Satisloh, Optotech, and Comes. In the United States the main players are Gerber Coburn and DAC. Most free-form generators are cut to polish, although there are exceptions like Optotech that uses a two-step figuring process. With respect to precision, free-form manufacturers claim 0.01 D, although this is very difficult to achieve in plastic because of the thermo-mechanical stresses that tend to deform the lens as it is blocked, surfaced and unblocked. However, a well-tuned free-form process with well-maintained cutting tools, soft pads, and cooling slurries guarantees better precision than standard grinding and polishing manufacturing.

It is important to remark that these free-form advantages do not mean necessarily that a free-form lens is optically better than a traditional lens. Free-form is a manufacturing process, not a design method or an optical design software. However, free-form decouples manufacturing from design. Almost independently of the selected base, the degree of freedom on the back lens surface makes possible a degree of aberration correction and personalization not achievable in traditional manufacturing. A proper optical design tool combined with a well-tuned free-form line is what can make spectacle lenses with better optical qualities for the end user.

Another big advantage of the free-form process is that the PPL manufacturing is no longer tied to base-curve restrictions. The traditional solution for PPL manufacturing was the use of front surface progressive blanks for which the prescription was completed by cutting a sphero-torical surface in the back side. These progressive blanks come typically with 5–6 fixed progressive surfaces optimized for each addition. In free-form, the progressive surface is cut in the back side of a spherical blank, and each PPL back surface can have its own shape. Moreover, the same optical performance can be achieved for most prescriptions using different spherical bases. This characteristic allows the ECP to choose the best personalized tradeoff between base curve and progressive surface based on the frame, personal parameters, and life-style of the final user.

In Figure 10.5, we show an example of how to build a PPL with prescription $[2, 1 \times 30°]$ add 2 using the classical and free-form approaches. In the first row of Figure 10.5 we show the classical method. In this case, the PPL is built using a front side progressive blank with 2 D addition. The blank has 5 D and 7 D in the far and near zones, respectively. For this progressive blank the desired prescription can be achieved surfacing a torus in the back side with cross cylinder prescription $(-3 \times 30°)(-4 \times 120°)$. In the final lens, the far and near powers are denoted by an ellipse with the appropriate orientation and semi-axes

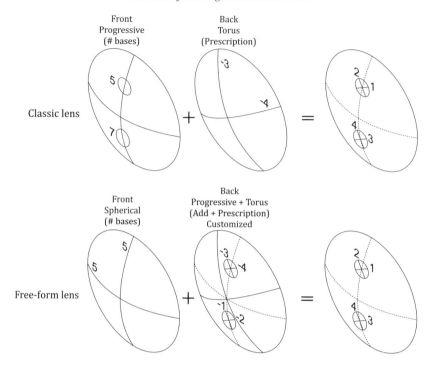

Figure 10.5 Classical vs free-form lens design. A PPL with prescription [2, 1 × 30°] add 2 is manufactured using classical and free-form techniques. In the first row, we show the classical method. In this case we use a front side progressive blank with 5 D and 7 D in the far and near zones. For this blank the desired prescription can be achieved using a torus in the back side with cross cylinder prescription (−3 × 30°)(−4 × 120°). In the resulting lens the far and near powers are denoted by an ellipse with the appropriate orientation and semi-axes values. In the second row, we present the free-form solution to the same problem. In this case we use a base 5 front side spherical blank where the back side is a progressive surface with the appropriated far and near main powers, denoted by the ellipses in the back surface. As in the classic lens of the first row, the final PPL prescription is also [2, 1 × 30°] add 2. However, in the classical case, corridor length, field width, and maximum cylinder are determined by the blank. In the case of the free-form lens all these elements can be designed in a per-user basis making possible personalized designs using the same blank.

values. For the far zone we have a cross cylinder prescription of (2 × 30°)(1 × 120°) and for the near zone (4 × 30°)(3 × 120°). In the second row of Figure 10.5, we present the free-form solution to the same problem. In this case we use a base 5 front side spherical blank where we have surfaced a progressive surface in the back side. The back surface has the appropriated far and near main powers with cross cylinder prescription (−3 × 30°)(−4 × 120°) and (−1 × 30°)(−2 × 120°) (these powers are denoted by the ellipses in the back surface). As in the classic lens example, the final PPL prescription is also [2, 1 × 30°] add 2. Although the final prescription is the same, in the classical case the PPL parameters, such as corridor length, field widths, maximum cylinder, etc., are mainly determined by the blank, and there is no room for lens personalization. In the case of the free-form lens all these parameters can be designed in a per-user basis making possible the use of multiple personalized designs using the same blank. This possibility is one of the main advantages of the free-form manufacturing: You do not need all the bases with all the additions; the addition and the lens design are "included" in the surfacing process.

The final reason why free-form personalization makes better lenses is asphericity. In optics, asphericity is the way to correct for lens aberrations when you cannot use many lens elements. For optical systems working in the paraxial region, aspheres do not present any advantage. However, off-axis systems like a spectacle lens can be highly improved using aspheric free-form surfaces. A classic example is the reduction of spherical aberration with aplanatic lenses for which the surface is a cartesian oval. Another example is discussed in Section 6.3.1, where the use of aspheric surfaces makes possible the design of flatter lenses. In ophthalmic optics, compensation must be made using just one lens per eye, for this reason the capability for generating aspheric designs is very important. Each prescription, each lens position, each user life-style (that defines the user working object space), requires a particular asphericity to get the best possible compensation for all sight directions. Classical PPL lenses manufactured with progressive blanks use the same asphericity for thousands of different situations, and the asphericity may become useless. In summary, the free-form capability to make a different surface for each lens, combined with the appropriate lens design tool, allows the optimum compensation of lens aberrations for each user, including:

- Lens optimization for the power that the user perceives for each sight direction
- Lens optimization for any position of use (for example, facial and pantoscopic angles, corneal and interpupilar distances)
- Lens optimization for any spherical base curve (different frame wraps can use different bases maintaining lens optical performance)
- Optimize the lens for each prescription (for example, flatter lenses for big prescriptions and use of different corridor lengths in PPL)
- Optimize the lens for different object spaces, which are determined by the user's life-style (for example, an office vs a sport design)
- Better control of lens thickness.
- Off-centering without losing optical quality.

Despite these advantages, free-form manufacturing and the best optical design tool cannot violate some fundamental laws of optics that make impossible some achievements like:

- Progressive lenses that violate Minkwitz's theorem. The main consequence is that it is not possible to have PPLs with different corridor widths while keeping the same slope along the corridor.
- Lenses without aberrations. Off-axis perfect imaging is not possible, although with proper design the main aberrations affecting the visual system can be controlled, especially in constant power areas.
- Third-order eye aberrations cannot be compensated simultaneously for all directions of sight. By their nature, third-order aberrations can be compensated only within a very

narrow field. For example, keratocone-induced coma can be corrected only for a fixed sight direction

- PPLs without adaptation period. The varifocal character of the PPL introduces lateral aberrations that need an adaptation period. For example, the transition between the far and near zones in a PPL induces distortion all over the lens. The amount of distortion depends on the prescription, but in any case it is necessary to always have an adaptation period.

As we discussed in detail in Chapter 8, the position of the progressive surface in a PPL, back vs front side, is not that important. The relevant thing is that the progressive surface must be free-form and well designed. The optical performance of any front side PPL can be reproduced by a back side PPL and the opposite is also true. This is true as long as the back surface stays convex enough; this translates into smaller obliquity angles and a better control of oblique aberrations through asphericity. Therefore, in positive lenses with a flatter back side, ideally the free-form surface should be the front one. On the other hand, in negative lenses with a flatter front side, the ideal free-form surface should be the back surface. However, the manufacturing advantages of surfacing the back side using a few spherical bases together with a good design make back side PPLs the choice for the industry.

10.2.4 Injection Molding and Casting

Injection molding and casting are processes designed for plastic materials. As explained in Chapter 1, polymeric materials can be divided in two big families, thermosets and thermoplastics. The main difference is that thermosets are cross-linked polymeric materials that do not liquefy upon heating. Instead, if they are heated at enough temperature they will decompose. On the other hand, in thermoplastics the polymeric chains are not cross-linked and they can become liquid with temperature [347]. This physical behavior determines the appropriate technique for lens manufacturing using plastic materials: injection molding for thermoplastics and mold casting for thermosets.

In injection molding a glass or metal mold is produced with several lens cavities and sprues (the channels connecting the cavities). Pellets of thermoplastic material are fed into a molding machine and pumped by a screw through a series of heaters that liquefy the plastic and fill the lens cavities as shown in Figure 10.6. Traditionally, injection molding optics had low quality; however, currently injection molding can produce elements with an optical quality up to 1λ. The technique is well suited for low cost production of the same element in high volumes. Molds can be expensive, but low material cost and fixed costs like labor and energy make injection molding very competitive for quantities exceeding the hundreds. Injection molding is the technique of choice for mass producing prismatic elements, aspheres, and lens arrays [25, 347]. Due to the thermo-mechanical stresses and physical behavior of thermoplastics, shrinking and birefringence are big issues for this manufacturing technique. Thermoplastic materials of interest in the ophthalmic industry include polymethyl methacrylate (PMMA) and polycarbonate [343]. Polycarbonate is a very interesting thermoplastic because it can also be cut in free-form generators.

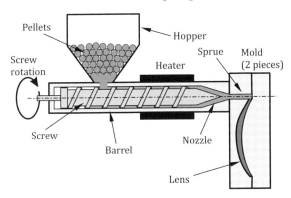

Figure 10.6 Injection molding lens manufacturing.

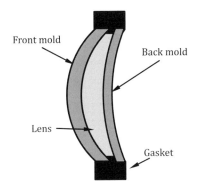

Figure 10.7 Diagram of a glass mold used in casting.

The second main technique for manufacturing plastic lenses is casting. Here thermoset materials like CR-39 are molded to its final shape and surface quality using a glass mold. First, the mixture of liquid monomer and initiator that will polymerize is injected in a glass mold. The mold is built with front and rear molds separated by a flexible gasket designed to handle the shrinkage as shown in Figure 10.7 (for example, CR-39 shrinks almost 14%.) Once filled with the monomer, the mold is heated to initiate the polymerization process, with typical polymerization times and temperatures of 6 h at $80°$ C. To obtain a precise shape with minimal shrinkage it is necessary to use specially designed temperature cycles. Also, the addition of extra components to the monomer-initiator mix, such as prepolymers or additional initiators, is usual. Although the only difference between thermal and light-induced polymerization is the initiator, fotopolymerization is not recommended in ophthalmic lens casting. The reason is that the material thickness involved (about 1 cm) makes control of the homogeneity of the final polymeric product very difficult. Casting is very well suited for mass production of blank lenses and aspheres. Today almost all semifinished blanks are made in plastic using casting.

10.3 Free-Form Lens Manufacturing

Modern lens manufacturing using stock semifinished blanks was established in the 1950s. Since then, the manufacturing process has consisted of two stages. The first step is the manufacturing of a plastic molded blank using a set of predefined spherical front surfaces: the base curves. Sometimes, the blanks are referred to as semifinished lenses because they are only half finished; only the front surface is definitive. These blanks are kept in stock in different bases and materials at the surfacing laboratory, making blank inventory management a key part of the manufacturing process. The second step consists of the finishing of the back surface of the blank by a grinding and polishing process. Alternatively, blanks can be prepared with both surfaces already finished, and in this case we will talk of stock lenses, normally single vision, which do not require additional surfacing. For both cases there is a last glazing step to fit the lens to the spectacle frame.

The free-form process entered the ophthalmic industry in the early 2000s. Production-wise, its main advantage is the capability to manufacture any custom-designed lens out of standard spherical blanks. Additionally, progressive blanks can also be used in a free-form process to produce double-sided PPLs.

The advantages of free-form technology have made a quick transition possible and nowadays the ratio of free-form to traditional sphero-torical generators is about 70:30. In this section we are going to describe the free-form manufacturing process in detail. For a good description of classical lens manufacturing technologies the reader is referred to [11] and [349].

The manufacturing of a personalized free-form lens must pass though all the traditional steps of building a prescription lens from a blank: job preparation, blocking, surfacing, polishing and engraving, tinting, coating, and quality control [350]. Among them, job preparation, surfacing, and quality control have to be adapted to a modern free-form manu-facturing process. In Section 10.2.3 we introduced free-form surfacing, and in this section we are going to explain the specifics of the job preparation and quality control steps for a free-form laboratory. In this context a job is an order for prescription ophthalmic lenses or spectacles. For the other manufacturing steps the reader is refereed to [344] for blocking, and [351] for tinting and coating.

Modern free-form lens manufacturing makes possible the fabrication of personalized PPLs designed specifically for each user and life-style. This is only possible using a specific process in which surfacing technology is as important as the LMS for laboratory production control and the lens design software (LDS) for the calculation of the surface for each personalized lens.

In a free-form laboratory, the LMS is mainly responsible for the laboratory integration and the interface with both the LDS and the eye care professionals (ECP). For example, three well-known LMS providers are DVI [352], Ocuco [353], and CCSystems [354]. The LMS integration task consists of the inter-connection of all lab equipment (genera-tors, polishers, glazing machines, computers, mappers) that makes possible full automation of the lens manufacturing process. For this, the LMS, the LDS, and the lab equipment must follow a communications standard. In the optical industry this standard is the Data

Communications Standard (DCS) of The Vision Council (TVC) [355] (some times this is referred to as Optical Industry Association – OMA – compliant systems). In this context, a job is represented in the LMS and LDS as a file or set of files, managed by the LMS, that describes the single lens or pair of lenses that are being manufactured. Also, the LMS allows the automation of some tasks like selection of base curve and job validation. One example is validation of the surfaces calculated by the LDS, checking if they can be surfaced by the laboratory generators. Together with the Data Communications Standard, the Vision Council has created a Lens Description Standard [356] to provide a consistent and precise method for describing the geometry of the lens surfaces of a job, including aspheric and progressive surfaces. For example, the use of the Lens Description Standard ensures accurate control of lens thickness during job processing. In addition, the Lens Description Standard provides a consistent method and nomenclature for describing the specifications and attributes of both finished and semifinished spectacle lens blanks as supplied by the lens vendors.

The second leg of a free-form manufacturing process is the lens design software (LDS). The LDS is a calculation utility for designing aspherical single-vision lenses and/or PPLs using free-form technology. Three well-known suppliers are Essilor [357], Zeiss [358], and IOT(Indizen Optical Technologies) [359]. Typically, in a free-form process the LDS calculates the concave side of a lens while the convex is determined by the base curve of the blank. Modern LDS are TVC/OMA compliant and use the TVC Data Communications and Lens Description Standards for integration with existing LMS systems. In this way the LMS and the LDS applications are decoupled and can be provided by different vendors. Even, a laboratory running a given LMS can use several LDS packages from different providers. After the LMS sends a job file with the prescription and design, the LDS calculates the lens surface and lens parameters and completes the job files with the necessary data to produce a free-form lens. Among these data are the concave surface point cloud and the lens thickness and layout. Additionally, the LDS can calculate the best suitable front curve for a job. The functionality included in modern free-form LDS packages includes:

- Calculation of the free-form aspheric back surface for single-vision and progressive lenses.
- Design of flatter, thinner lenses with a better optical quality than their classical equivalents.
- Automatic selection of the optimal base.
- Use of Standard input & output formats TVC/OMA compliant
- A calculation time of a few seconds
- Design based on user-perceived power, with improved off-axis performance.
- Compensation for pantoscopic and wrapping angles
- Personalization based on frame, interpupillary distance and life-style.

After the introduction of the LMS and LDS functionalities we are going to describe the steps for a job manufacturing in a free-form laboratory. Figure 10.8 shows a diagram of the information flow for this task.

1. First, the ECP sends the prescription data to the laboratory. Then a laboratory technician reviews and introduces the job order in the LMS. Every parameter needed to calculate a PPL is contained in that order. The order includes prescription, assembly, and customization data of the lens such as the pantoscopic and facial angle, the horizontal boxed lens size of the frame, the lens design, and the refractive index (the complete set of parameters is described in [355]).

2. The LMS sends the job order to the LDS in a LDS file. At this point, the LDS calculates the back surface of the PPL and determines the optimal base curve. All this information is sent back to the LMS in two files, an LMS file with records related to the actual designed lens and a surface definition file (SDF). All the label definitions of an LMS file and the format for the surface definition file can be found in the TVC Data Communications Standard [355]. The surface definition file can include information of both eyes in the same file and also admits surface normal values at the edges.

3. The LMS file created is sent to the LMS to review the results of the calculated lens. Meanwhile the surface definition file is sent to the server and stored in a predefined location. Once the lab technician has checked and accepted the parameters of the calculated lens, the LMS file is sent to the server too. From the server, the LMS and surface definition files are distributed to the free-form line (blocker, generator, and polisher) for surfacing the lens. The free-form line, indicated in Figure 10.8, can be an independent system of the server or part of the LMS system. As long as the elements of Figure 10.8 are TVC/OMA compliant, the free-form line, the LMS, and the LDS will be able to work together.

The use of TVC/OMA compliant systems can be extended to the quality control of free-form lenses with big advantages. Here, quality control consists of the evaluation of the quality of the manufacturing process of a free-form lens using a free-form line. The

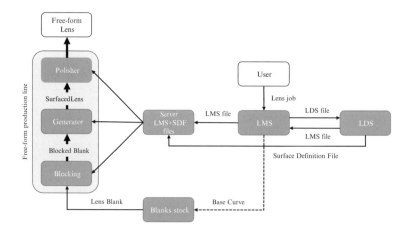

Figure 10.8 Free-form manufacturing work flow.

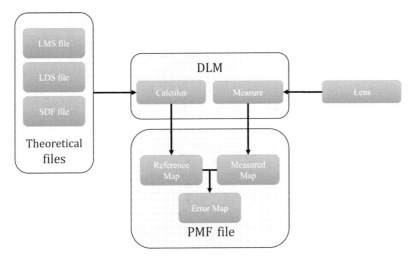

Figure 10.9 TVC compliant quality control workflow.

free-form manufacturing process has many factors that influence the final quality of the lens. For example, the polishing pads can became worn or the generator speed might not be appropriate for the material being used or curvatures being cut. For this reason continuous motorization of the free-form production is highly advisable and the use of a TVC/OMA compliant lens mapper facilitates this task. In Figure 10.9, we show the typical quality control work flow using a TVC/OMA compliant lens mapper. On one side, the files that describe the theoretical (expected) lens, the LDS, LMF, and SDF files, are fed to the mapper system. If the LDS, LMS, and mapper are TVC compliant, the mapper system can use the production files to calculate the theoretical power matrix, $\mathbb{P}_T$, that one will obtain if the lens is flawless: the reference map. On the other hand, the mapper can measure the actual power matrix of the manufactured lens, $\mathbb{P}_M$: the measured map. From these two results, the mapper can calculate the error map $\Delta\mathbb{P} = \mathbb{P}_T - \mathbb{P}_M$ and the corresponding mean sphere and cylinder error maps. For standard mappers the output of the measurement and the comparison with the reference is given in a Power Map Datasets file (PMF) [355], which can be used by an external automatic quality control assessment system. In the PMF format, all the power maps of a given job can be included, including the measured, theoretical, and error maps. In the PMF format, the power matrix is described providing the maps for the mean sphere, cylinder, and axis. Additionally, modern mappers can measure the lens in transmission and in reflection. The reflection map is very useful for checking the base curve, and the transmission map is used to calculate the through-power matrix. The PMF format contemplates this possibility, including the four possible combinations of measurement method (reflection, transmission) and map type (theoretical or measured). For example, the IOT Futura LDS from IOT [359] and the Dual Lens Mapper from A&R [360] can work together in this way.

10.4 Lens Measurement

The rest of this chapter is dedicated to lens measurement techniques and their application in the quality control process. We are going to start with a discussion about the spherometer, the most straightforward method to check for the curvature of spherical and torical surfaces. Next, we will present the technology behind focimeters and lensmeters, the most used systems in production and clinical environments. Both methods measure power at a single point, so we will continue with lens mappers that can obtain the power matrix over a set of positions of the lens. Next we will discuss the differences between the user power and the focimeter power and its implications for the quality control of a PPL. We will close this section with a presentation of how the use of the power map of a lens can be combined with the standard ISO test to improve the quality control process of PPLs using machine learning recommendation methods.

10.4.1 Lens Gauges and Spherometers

Lens gauges and spherometers are tools for measuring surface curvature of spherical and torical surfaces. If the refractive index is known, the curvature can be translated into refractive power. A lens gauge is a mechanical device with two lateral pins and a central measuring needle with a dial gauge driven by a clockwork (for this reason sometimes this device is denominated a lens clock). The two lateral pins are fixed and the central probe is spring loaded. When the pins are placed on a spherical surface the central needle will be depressed or raised relative to the fixed pins depending on the central sag s (see Figure 10.10). If the distance between the central probe and the fixed pins is known, the radius of curvature can be determined. As shown in Figure 10.10, if d is the distance between the central probe and

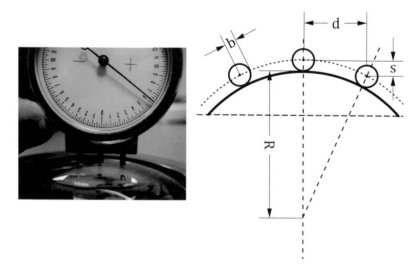

Figure 10.10 a) Lens gauge b) effect of the tip ball size.

the fixed pins, b is the radius of curvature of the tip of probe and pins, and s is the measured sag by the probe, the radius of curvature is given by

$$R = \frac{d^2 + s^2}{2d} \mp b .$$ (10.5)

Where the plus or minus sign is used for convex or concave surfaces. Most devices have a very small needle tip size that makes $b \approx 0$; however, for soft surfaces a ball at the needle tip is necessary and its radius must be taken into account.

As in the case of lapping tools, the dial of a lens gauge is directly scaled in diopters, usually with two scales for concave and convex surfaces to take into account the sag probe tip radius. Commercial lens gauges are calibrated for a reference refractive index, n_{ref}, typically 1.5 or 1.53. As in the case of lapping tools, if the refractive index of the material, n, is different from the reference we must correct the refractive power reading by

$$P = \frac{n - 1}{n_{ref} - 1} P_{dial} ,$$ (10.6)

where $P_{dial} = (n_{ref} - 1)/R$ is the power read in the lens gauge dial.

It is very important for the central probe to be perpendicular to the surface. Otherwise the measured sag (and therefore the power) will be incorrect. For example, for a convex surface, if the central probe is not normal to the surface, the measured sag and the radius of curvature will be greater than expected, and therefore the measured refractive power smaller. The lens gauge can be used for measuring cylindrical and torical lenses by rotating the lens gauge and aligning the fixed pins with the principal meridians.

The spherometer is similar to the lens gauge but uses a stable base that guarantees the normal position of a central sag probe with respect to the lens surface. This is usually done using three fixed pins instead of two. The three pins form an equilateral triangle and the sag probe is located at the geometrical center. Other spherometer models use a circular rim that rests on the surface with the sag probe at the center. For this kind of spherometer, the distance d of equation (10.5) is the radial distance from the central probe to the fixed pins or to the circular rim.

10.4.2 Focimeters (Optical or Digital)

The image focal length of a lens can be referred to the image principal plane, (effective focal length, or EFL) or to the vertex of the back surface (back focal length, or BFL). In Figure 10.11 H and H$'$ are the object and image planes, respectively, and F$'$ is the back focal point. The BFL f_B' is measured from the lens back vertex, while the EFL, f', is measured from H$'$. Therefore, for lenses with the same EFL but different shapes the focal point will be located at different positions with respect to the back vertex [47], giving different values for the BFL.

Ophthalmic lenses compensate refractive errors by imaging the infinity on the eye's far point, but according to the former discussion the image position with respect to the lens back vertex depends on the lens shape. In consequence, lenses with the same EFL

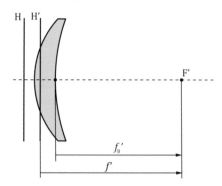

Figure 10.11 Effective focal length versus back focal length for a thick lens. In the figure, H and H′ are the object and image principal planes, F′ the back focal point, f'_B is the BFL, and f' is the EFL.

and vertex distance but different shapes will compensate different refractive errors. For this reason, ophthalmic lenses are specified in terms of three parameters, back focal length (BFL), base curve and central thickness. Two lenses with the same BFL, using the same vertex distance to the eye, will compensate the same refractive error, although they can have different EFL (and different equivalent power).

The focimeter is an optical instrument to measure vertex focal length and its inverse, vertex power. In ophthalmic optics, the main use of the focimeter is the measurement of the BFL of ophthalmic lenses. The basic focimeter setup is shown in Figure 10.12(a). A test target T is placed at the front focal point, F_c, of a collimating lens, L_c. If there is no test lens, the collimator produces a collimated beam and the target image can be observed by an unaccommodated user by means of an afocal telescope formed by the objective L_1 and the ocular L_2. In Figure 10.12(a) the observed image is represented by the small inset below the focimeter diagrams. As we show in Figure 10.12(b), if a test lens, L, is placed between the collimator and the telescope, the output beam will be defocused and the telescope will produce a blurred image. To recover a sharp image, the target must be axially displaced a distance z away from the collimator frontal focus, F_c, until the image T′ of the target through the collimator lies at the test lens focal point. In this case, the lens will generate a collimated beam and the observer will perceive a focused image through the telescope.

Distances z and z' are the conjugate object and image distances of the target measured from the collimator focal points. In air the relation of these two distances and the collimator EFL, f_c', is given by Newton's formula

$$zz' = -f_c'^2 . \tag{10.7}$$

If the test lens is placed with its back vertex at the collimator image focal point, by definition $z' = -f'_B$ where f'_B is the BFL of the test lens. Therefore, in this case we obtain that $f'_B = f_c'^2/z$ or

$$P_V = zP_c^2 , \tag{10.8}$$

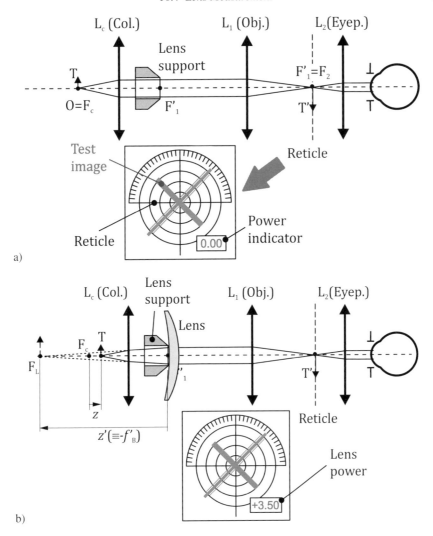

Figure 10.12 Focimeter diagram, a) without test lens b) with the test lens inserted (see text for details).

where P_c is the power of the collimator and $P_V = 1/f'_B$ the lens back vertex power (BVP). If we express P_c in D, and z in mm, equation (10.8) becomes

$$P_V (D) = \frac{z\,(mm)}{1000} P_c^2 (D) .$$ (10.9)

The front vertex power can be measured if the test lens is placed with its front vertex on the collimator frontal focal point. In this case the distance z' will be the lens front focal length (FFL) and from it we can determine the front vertex power using equation (10.8).

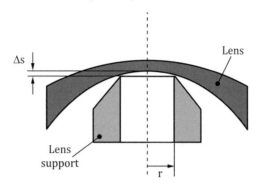

Figure 10.13 Sagittal height error in the vertex position of a lens.

From Equation (10.8) the relative error in the BVP is given by

$$\frac{\Delta P_V}{P_V} = \frac{\Delta z}{z} + 2\frac{\Delta P_c}{P_c} , \tag{10.10}$$

where we have taken into account the measurement errors in the target displacement, z, and the collimator power, P_c. However, to complete the analysis of the main error sources we must take into account the sagittal height error. This error appears because we have assumed that the lens vertex is located at the collimator front focal point, but in standard focimeters, the lens is placed on an annular support of radius r as shown in Figure 10.13. Due to the surface curvature, the lens vertex position will be axially displaced from the plane of the lens support by a quantity Δs given by

$$\Delta s = R - \sqrt{R^2 - r^2} , \tag{10.11}$$

where R is the smaller radius of the back surface. If the center of the annular support is located at the frontal focal point of the collimator, then $z' \equiv f'_B$ and $\Delta s \equiv \Delta f'_B$. Therefore, the sagittal height error will induce an error in the lens BFL because what we are measuring is $\bar{f}'_B = f'_B - \Delta f'_B$ and the back vertex power will be

$$P'_V = \frac{1}{\bar{f}'_B} = \frac{1}{f'_B - \Delta f'_B} = P_V + \Delta s P_V^2 , \tag{10.12}$$

and

$$\Delta P_V = P'_V - P_V = \Delta s P_V^2 , \tag{10.13}$$

will be the error in the BVP due to the sagittal height error. Using equation (10.13), the total relative error of P_V including the sagittal height error will be

$$\frac{\Delta P_V}{P_V} = \frac{\Delta z}{z} + 2\frac{\Delta P_c}{P_c} + \Delta s P_V . \tag{10.14}$$

We can see how the main error sources weigh in focimeter measurement accuracy. The first contribution is the error in the position of the target image, $\Delta z/z$. In principle, the target image position z will be determined by the precision of the linear stage used to displace the target. However, the main dependence will be related to the depth of focus imposed by the annular lens support, which in the standard focimeter becomes the aperture stop in target image formation. The smaller the annular lens support the bigger will be the depth of focus. Therefore, small annular lens supports will increase the target image positioning error $\Delta z/z$.

The sagittal height error contribution ΔsP depends on the sagital error and the power of the lens. For a given base curve, the first grows larger with power for negative lenses, and the opposite for positive lenses. Also, a small annular support is preferable to reduce the sagital error.

Finally, the collimator contribution $\Delta P_c/P_c$ depends on the collimator aberrations. The most important is the spherical aberration that translates as a direct error of the collimator power error ΔP_c. In this case, the use of a small annular support limits the aperture and therefore the contribution of the collimator lens aberrations. Another possible source of error could be the chromatic aberration of both ophthalmic lens and collimator. To remove this source of error, many manufacturers employ a band-pass filter or a quasi-monochromatic light source like an LED.

Another important property of the focimeter is its measuring range. For positive lenses, the test, initially at the collimator front focal point, should be displaced toward the collimator. Thus the range of positive target image distances, z, is limited by the collimator EFL so the maximum value is $z_{max} = 1/P_c$. From equation (10.8), the maximum positive BVP we can measure is $P_V = P_c$. For standard focimeters, this limit is typically around 10 D. For negative lenses, the test should be displaced away from the collimator and its value will be negative, $z < 0$. Therefore, for negative lenses there is no limit other than that imposed by the focimeter housing.

Up to now we have assumed that the lens is centered or that there is no surfaced prism in the lens. As we show in Figure 10.14, if we place on the annular support of the focimeter a decentered monofocal lens, with no ground prism, with a BVP P_V and displaced vertically d, a prism, p_y, appears that according to Prentice's law is given by

$$p_y = -P_V d . \tag{10.15}$$

This induced prism displaces vertically the intermediate image by a quantity

$$\Delta y = -f'_{ob} p_y = f'_{ob} P_V d , \tag{10.16}$$

where f'_{ob} is the telescope objective EFL.

To measure the prismatic deviation, a reticle is located at the focal plane of the objective lens of the telescope. Concentric circles are engraved in the reticle, their radii calibrated in prism diopters. The polar coordinates of the center of the target image, which are the components of the prismatic effect, are determined with this engravings. In Figure 10.14, the prismatic power is 2Δ at $270°$. Another possibility to measure the prism is the use of a variable prism device (Risley's prism) that is positioned between the lens and the

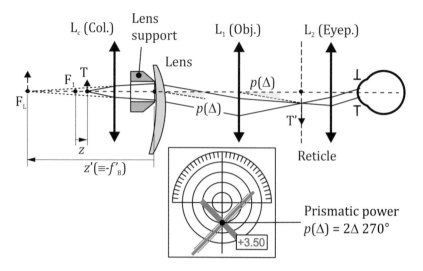

Figure 10.14 Prismatic effects in the focimeter.

objective lens of the telescope. When the lens to be measured produces prismatic deviation, the Risley's prism is set to counteract it, its setting providing the desired measurement.

Despite its simple operation, the classical focimeter has some disadvantages [361]. Prismatic effects and power measurement at different points of the lens requires re-adjustment. This is especially problematic in the case of PPL lenses with no specific symmetry and spatially variable spherical power and astigmatism. Also, as the focimeter is a subjective instrument, automation of the focusing process is not easy. As a subjective instrument, a noncompensated refractive error, or the accommodation of the user, may introduce significant errors in the measurement. One alternative to the subjective character of the focimeter is to transform it into an objective instrument, for example using a projection lens to generate a real image on a screen; however, all the operative problems associated with complex lenses like PPLs will remain.

For all these reasons, modern automatic lens testing systems do not use the imaging system described in Figure 10.12. Instead they use a set of collimated beams and measure the deflection for each beam using a Shack-Hartmann measuring system [79], and for this reason they are usually called lensmeters. Single point lensmeters are based on the deflection of a collimated beam as it goes through the lens. If a beam hits a lens at a lateral position $\mathbf{d} = [d_x, d_y]^T$ with respect to the local optical center located at position (x, y), the prismatic deviation of the beam $\mathbf{p} = [p_x, p_y]^T$ is given by the generalized Prentice's law as [362]

$$\mathbf{p} = \mathbf{p}_0 - \mathbb{P}\mathbf{d}, \qquad (10.17)$$

where $\mathbf{p}_0 = [p_{0x}, p_{0y}]^T$ is the local ground prism and $\mathbb{P}$ the local dioptric power matrix at position (x, y). From Equation (10.17), using a single beam measurement we will produce

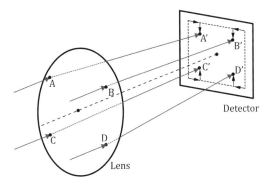

Figure 10.15 Lensmeter measurement setup using four beams. From the measured deflection of the four beams at points A, B, C, and D, it is possible to calculate the local dioptric power matrix and the grounded prism.

two equations and five unknowns given by the three independent components of the local dioptric power matrix and the two components of the local ground prism. To solve for all unknowns, we need a minimum of three measurements around location (x, y). One possible arrangement involves four measuring points at the corners of a square, and the power and prism are measured at the center of the square as depicted in Figure 10.15. In this way, a lensmeter allows for the automatic measurement of the local diopter power matrix and the local ground prism, from which local spherocylindrical prescription and local prism are easily derived.

Let us see an example of the application of Prentice's law for the case of spherical monofocal lenses. If we want to measure the power at the optical center of the lens, we can use just two parallel beams aligned with the optical axis and separated a distance $2d$. Their incidence coordinates on the lens with respect to the optical center will be $\mathbf{d}_1 = [0, d]^T$ and $\mathbf{d}_2 = [0, -d]^T$ as shown in Figure 10.16. In this case, equation (10.17) becomes

$$\begin{bmatrix} p_x \\ p_y \end{bmatrix} = -\begin{bmatrix} P_V & 0 \\ 0 & P_V \end{bmatrix} \mathbf{d}, \tag{10.18}$$

where P_V is the BVP of the spherical lens and $\mathbf{d}$ the lateral displacement of the measuring beam with respect to the optical axis. For beams 1 and 2 of Figure 10.16 the prismatic deviations are

$$p_{y1} = -P_V d$$
$$p_{y2} = P_V d. \tag{10.19}$$

On the other side, from the lateral position of the two beams measured at the sensor we have

$$p_{y1} = -\frac{d - h}{z}$$
$$p_{y2} = \frac{d - h}{z}, \tag{10.20}$$

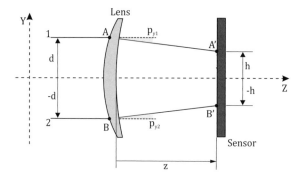

Figure 10.16 Use of Prentice's law for the measurement of the power of a monofocal lens.

from which we can calculate the lens BVP as

$$P_V = \frac{d - h}{zd}. \tag{10.21}$$

Automatic lensmeters use a monochromatic light beam with a given wavelength; however, lens power should be given for a specific wavelength that does not necessarily match the lensmeter wavelength. For example, a typical reference value is the Fraunhofer d line ($\lambda_d = 587.56\,\text{nm}$). Due to lens material chromatic dispersion we must take into account the variation in refractive index between the wavelength used by the lensmeter and the d line [363]. In air, the lens power is approximately proportional to $n - 1$, where n is the index of refraction for a given wavelength. If we want to determine the power for the d line from the lensmeter measurement, the corrected BVP for the d line, P_d, will be

$$P_d = \frac{n - 1}{n_d - 1} P_V, \tag{10.22}$$

where P_V is the BVP measured by the focimeter, and n_d and n are the refractive index for the d line and the lensmeter wavelength. Chromatic dispersion can be an issue for materials with high index and low Abbe number. As an example, let us consider two well-known glass materials used for spectacle lens manufacturing: the Hoya Crown C3 and the Schott SF4 dense Flint [364]. In Table 10.1 we show the Abbe number, the refractive index for each material and the correction factor we should apply to compute P_d from the measured power, P_V, in the case that the lensmeter uses the Fraunhofer C line ($\lambda_C = 656.28$ nm).

From Table 10.1, if a lens manufactured with C3 glass were found to have a power of 10 D for the C line, we will have a power of 9.95 D for the d line. Likewise, if a lens made of SF4 glass were found to have a power of 10 D for the C line it will have a power of 9.9 D for the d line. In both cases the chromatic dispersion error is very close to the ISO standard for lens measurements of 0.12 D and should be corrected. Most modern lensmeters either use a green or a red monochromatic lights. For the green light case, as the wavelength is very close to the d line, no correction is necessary and we can assume that the power measured corresponds to P_d. For the red wavelengths it is not necessary to

Table 10.1 *Lensmeter power correction for the C line using index dispersion and Abbe number.*

Material	ν	n_d	n_C	$\frac{n_C-1}{n_d-1}$	$\frac{2\nu-1}{2\nu}$
Hoya C3	58.96	1.5182	1.5156	0.995	0.991
Schott SF4	27.58	1.7552	1.7473	0.990	0.982

know the index dispersion; instead we can use the material Abbe number to correct for the chromatic dispersion. The Abbe number is defined as

$$\nu = \frac{n_d - 1}{n_F - n_C} \, , \qquad (10.23)$$

where n_F is the refractive index for the Fraunhofer F line ($\lambda = 486.134$ nm). If we assume that the lensmeter wavelength is close to the C line, given the halfway position of the d line between the F and C lines, and we use the approximate relation $n_F - n_d \approx n_d - n_C$, we can correct the measured power for the C line. From equation (10.22), if we approximate $n \approx n_C$ then the corrected BVP for the d line, P_d, is given by

$$P_d = \frac{2\nu - 1}{2\nu} P_V \, . \qquad (10.24)$$

For this reason, lensmeters using a red light source usually have the possibility to specify the Abbe number as a parameter of the measurement software. Coming back to the former example of Hoya C3 and Schott SF4 glasses, in Table 10.1 we also show the correction factor using only the Abbe number. As can be seen, the correction factor is very similar to that obtained from the two refractive indexes for the d and C lines.

We would like to end this section with a mention of the spherical aberration effect in the lensmeter measurement. In equation (10.17), we are assuming that the power at the four measuring points and the paraxial power at the center of the square defined by the four beams is the same, as shown in Figure 10.15. In the case of strong spherical aberration this will no longer be true. For spectacle lenses, spherical aberration is not an issue due to the small curvatures of the surfaces and the fact that the lens is measured and used in air. For contact lenses, in wearing conditions, the air interface at the back surface of the contact lens is missing. However, when we measure a contact lens with a focimeter, the back surface interface is in air. With the introduction of this interface, the lens becomes meniscus-shaped with higher spherical aberration. Following Campbell et al., [363] the BVP measured from the four measurement points at a radial distance r from the center of the square pattern can be modeled as

$$P_V = P_0 + \alpha \, (P_0 r)^2 \, , \qquad (10.25)$$

where P_0 is the paraxial power at the center of the square pattern defined by the measuring points and α is the spherical aberration coefficient. To solve this issue, in [363] a lensmeter with two square patterns with radial distances from the center of r_1 and r_2 is proposed. For this instrument the paraxial power can be estimated as

$$P_0 = \frac{P_1 r_2^2 - P_2 r_1^2}{r_2^2 - r_1^2},$$

(10.26)

where P_1 and P_2 are the BVP measurements for each square pattern.

10.4.3 Lens Mappers

Lens mappers are systems that can measure the power matrix for each point of a lens, $\mathbb{P}(x, y)$. Normally, the power matrix is represented by depicting the mean sphere, the cylinder, and the axis for each location. The different representations of the power matrix in terms of the main powers, sphere, cylinder, and axis are summarized in Table 4.2. As an example, Figure 10.17 shows a PPL measurement. In this figure, we can see the typical outcome for the spatial distribution of the power of a PPL as measured by a mapper. For instance, the corridor of the cylinder and the near zone for the mean sphere is clearly visible. The maps in Figure 10.17 have been measured with a Dual Lens Mapper from A&R. In these maps we can see the theoretical maps calculated from the LDS and LMS files side by side with the measured maps, allowing for a quick visual comparison. Also, superposed to the power maps we can see the PPL reference points: distance, near and prism reference points, fitting cross, and alignment reference markings. These reference points are relevant for lens quality control because the comparison between the theoretical and measured results can be used to automate quality control of a PPL [345]. The capability for producing this kind of information is what makes a mapper an invaluable tool for the quality control of PPLs. In the next three sections we are going to present the technologies behind the main commercial mappers: Shack-Hartmann systems, deflectometers, and coordinate measuring machines (CMM) profilometers.

Shack-Hartmann Mappers

Shack-Hartmann systems [343, 365] are sensors that can measure the output wavefront of a lens, W, from which we can extract the power maps. The main characteristic of a Shack-Hartmann mapper is the use of a lenslet array that consists of a matrix array of several hundred microlenses, all with the same focal length, focused on the imaging sensor. Each lenslet has a small square aperture ranging from 50 to 450 μm in width. For instance, this technology is used by the Visionix mappers in what they call "Visionix wave front technology " [366].

In Figure 10.18(a), we show the working principle of the Shack-Hartmann mapper. When the input wavefront reaches the array plane, each lenslet will focus the local wavefront at a displaced spot of the imaging plane. For every lenslet, the reference point is the intersection of its optical axis with the focal plane. For an incident axial plane wavefront, all lenslets will focus on their respective reference points. If the wavefront is not plane but curved, each lenslet will receive a tilted wavefront that will produce a transverse displacement of the image point. Following Figure 10.18(b), if f is the focal length of the lenslet,

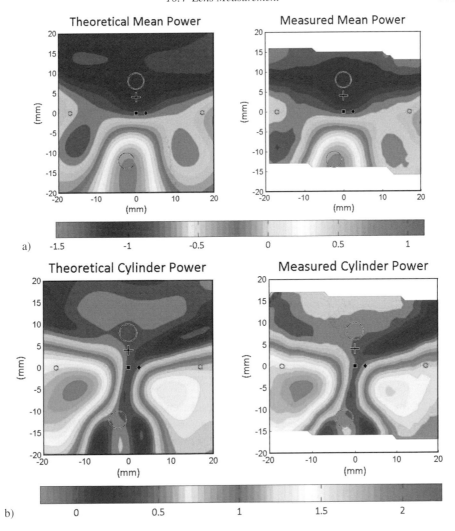

Figure 10.17 Example of a PPL measurement with a lens mapper a) theoretical maps for the sphere as obtained from the LDS and LMS files (right) and as measured with a DLM (Dual Lens Mapper) (left), b) equivalent maps for the cylinder. (Courtesy of Indizen Optical Technologies).

the displacement vector at the image plane between the reference and displaced points is given by

$$\mathbf{d} = [d_x, d_y] = [-f \tan \delta_x, -f \tan \delta_y], \qquad (10.27)$$

where $[\delta_x, \delta_y]$ are the direction angles of the wavefront normal, as shown in Figure 10.18(b). If the tilted input wavefront can be approximated as a plane for the lenslet aperture, the wavefront gradient at the center of the lenslet is given by

$$\nabla W = [\tan \delta_x, \tan \delta_y]. \qquad (10.28)$$

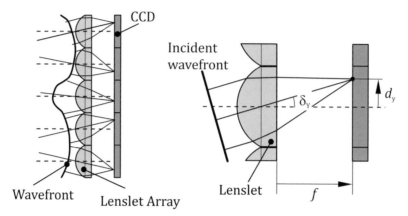

Figure 10.18 Shack-Hartmann mapper working principle. a) The incident wavefront generates a displaced spot for every lenslet of the array, and the reference point is the intersection of the lenslet optical axis with the focal plane. b) Assuming a locally plane wavefront, the displacement of the focal point with respect to the reference point is a measure of the local slope $\partial W/\partial y = \tan \delta_y$.

Finally, from equations (10.27) and (10.28), the measured wavefront gradient at each lenslet position is given by

$$\nabla W = -\frac{\mathbf{d}}{f} . \qquad (10.29)$$

From the Shack-Hartmann measurement we can calculate the power matrix at each lenslet location using two methods. In the first technique, we can use Prentice's law [362]. For each lenslet, we have five unknowns, the three power matrix elements, and the ground prism, $\mathbf{p}_0$. If we assume that the power matrix and the ground prism varies slowly across lenslets, from each set of four-connected lenslets we can set up eight linear equations

$$\left[\delta_x, \delta_y\right]_i = \mathbf{p}_0 - \mathbb{P}\mathbf{d}_i, \quad i = 1\ldots 4. \qquad (10.30)$$

From this set of linear equations, the three elements of the power matrix $\mathbb{P}$ and the two components of $\mathbf{p}_0$ can be obtained. This technique allows for the measurement of discontinuous power maps as in the case of bifocal lenses.

The second method uses a Zernike expansion of the wavefront W [343]. In this case, if we assume that the wavefront is continuous across the lens aperture, and $Z_n^m(x, y)$ are the Zernike polynomials, we can always model the lens wavefront by a Zernike expansion of the wavefront $\overline{W}$ given by

$$\overline{W}(x, y) = \sum_{n,m} a_{n,m} Z_n^m (x, y), \qquad (10.31)$$

where $a_{n,m}$ are the Zernike coefficients. However, what we measure with the Shack-Hartmann sensor is the wavefront gradient, ∇W. In this case we can fit the measured derivatives to the gradient of the estimated wavefront $\overline{W}$ obtaining the coefficients $a_{n,m}$ by

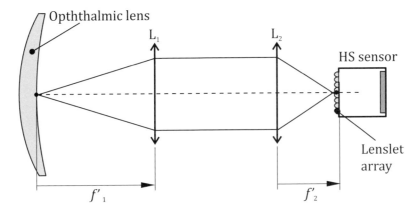

Figure 10.19 Diagram of a relay system for the imaging of a input wavefront on the lenslet array in a Shack-Hartmann mapper. The wavefront at the lens output is imaged onto the Shack-Hartmann array by the relay system formed by the lenses L_1 and L_2. The magnification will be $m = f'_2/f'_1$.

minimizing the functional $U\left(a_{n,m}\right) = \Sigma_{x,y} \left\| \nabla W - \nabla \overline{W} \right\|^2$ [367]. Finally, as explained in Appendix D the power matrix can be calculated as

$$\mathbb{P} = \mathbb{H}\left(\overline{W}\right), \tag{10.32}$$

where $\mathbb{H}$ is the Hessian operator. If the input beam is collimated (plane input wavefront), equation (10.32) will produce a paraxial approximation for the lens power, which for the axial direction will match the BVP of the lens. That is, at the optical axis, a Shack-Hartmann mapper will give the same power as a lensmeter.

A Shack-Hartmann sensor measures the wavefront arriving at the lenslet array plane. However, the back surface of the lens cannot be in direct contact with the lenslet array plane, because of the curvature of the lens surface, and mechanical constraints from the housing of the sensor and the lens support. Also, the lens is usually bigger than the lenslet array. These problems can be easily solved by means of a relay system that transports the wavefront emerging from the back surface of the lens to the lenslet array plane, as shown in Figure 10.19.

Deflectometric Mappers

Deflectometric mappers are based on the measurement of the deflection of a light beam as it passes through the lens or is reflected by the lens surface. Transmission deflectometric mappers can measure the effects of both surfaces and thickness, making possible measurement of the lens power matrix for each point. Reflection deflectometric mappers measure only the power of the corresponding specular surface. This is useful, for example, for checking the lens base curve or measuring the progressive side power distribution of a PPL independently of the base curve.

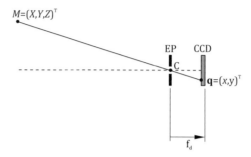

Figure 10.20 Pinhole camera model. The image of 3D point **M** is the 2D point **q**, located at the intersection of the principal ray with the focal plane. The camera center, **C**, is at the center of the entrance pupil of the camera objective.

For commercial mappers, there are two main technologies available: fringe deflectometry and moiré deflectometry. The working principle in fringe deflectometry is the observation of the distortion of a pattern when it is observed through a lens or reflected by a lens surface. From observation of the distortion suffered by the pattern, measurement of the deflection angles and the power matrix for all lens locations is possible. In moiré deflectometry, the moiré pattern formed by two grids is observed with and without the lens. In this case, the change in the moiré fringe pattern allows determination of the lens deflection and the power matrix.

In this section we are going to assume that the camera of the mapper can be modeled as a pinhole system, [368], see Figure 10.20. This imaging model is named the perspective projection model. In this model the 2D image $\mathbf{q} = (x, y)^T$ of a 3D point $\mathbf{M} = (X, Y, Z)^T$, is formed at the intersection of the principal ray with the image plane. The principal ray passes through the camera center **C** and the object point **M**. The image plane is placed at distance f_d from the camera center. This distance is referred to as the camera focal distance in the computer vision literature, not to be confused with the objective lens focal distance. Using this imaging model, image **q** of point **M** is calculated as

$$
\begin{bmatrix} u \\ v \\ w \end{bmatrix} = \begin{bmatrix} -f_d & 0 & 0 & 0 \\ 0 & -f_d & 0 & 0 \\ 0 & 0 & 1 & 0 \end{bmatrix} \begin{bmatrix} X \\ Y \\ Z \\ 1 \end{bmatrix}, \tag{10.33}
$$

with

$$
\mathbf{q} = (x, y)^T = \frac{1}{w}(u, v)^T. \tag{10.34}
$$

If the aperture is small, camera objectives can be modeled as a pinhole camera, with the camera center located close to the entrance pupil center. In general, the more symmetrical the layout of the objective of a camera, the better the approximation of the camera objective as a pinhole camera.

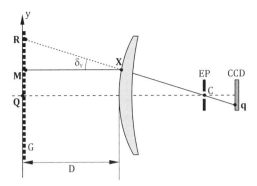

Figure 10.21 Transmission fringe deflectometry setup. A grid G is observed by the camera. Without the lens, point **R** of the grid will image on point **q** of the camera. With the lens inserted, point **M** will image on **q**. From the displacement **d** = **R** − **M** it is possible to calculate the deflection $\delta = (\delta_x, \delta_y)$.

The perspective projection model is very convenient for metrology purposes. For example, given the focal distance f_d and the image coordinates $\mathbf{q} = (x, y)^T$, determination of the ray vector **r** pointing to the object point **M** and the angle subjected by **r** with the optical axis is immediate. Another example where the pinhole model is very useful is in the measurement of 3D coordinates. With two calibrated cameras, it is possible to determine the two rays $\mathbf{r_1}$ and $\mathbf{r_2}$ that point to a 3D point **M**. Measurement of the X, Y, Z coordinates of **M** can be achieved by finding the intersection of rays $\mathbf{r_1}$ and $\mathbf{r_2}$.

The basic setup for transmission fringe deflectometry is shown in Figure 10.21. A grid G (or a fringe pattern in general) is observed with a (pinhole) camera. In transmission fringe deflectometry, the grid pitch is big enough to be resolved by the camera. Without the lens, the grid point $\mathbf{R} = (R_x, R_y)^T$ is imaged on the camera point **q**. In this case, following Figure 10.21, point **q** will appear black. When we place the lens between the grid and the camera, point $\mathbf{M} = (M_x, M_y)^T$ will be imaged on **q** because of refraction. In this example, point **q** will appear white. The use of a grid, or a pattern in general, allows the determination of the position of point **M** in the grid reference frame and its distance to the reference point **R**. The fringe pattern can be static like a printed screen [369] or dynamic, in the sense that several fringe patterns can be displayed in the screen to help in the precise location of points **R** and **M** [370].

Through calibration of the deflectometric setup, it is possible to know the separation between the lens and the grid, D. Also, using a square grid it is possible to calculate the position of points **R** and **M** in the grid reference system. Finally, from the displacement $\mathbf{d} = \mathbf{R} - \mathbf{M} = (d_x, d_y)$, the deflection $\delta = (\delta_x, \delta_y)$ at point **X** of the lens is given by

$$\tan \delta_x(\mathbf{X}) = \frac{d_x}{D}, \quad \tan \delta_y(\mathbf{X}) = \frac{d_y}{D}. \tag{10.35}$$

From this equation, the sensitivity of the transmission fringe deflectometry method is inversely proportional to the screen-lens distance D. The bigger this distance, the more precise the measurement of the deflection. On the other hand, the smaller the pitch of the grid G, the better is the measurement of the displacement **d**. As the rays are perpendicular to the wavefront, the gradient of the wavefront W at the lens output is given by

$$\nabla W = \left(\tan \delta_x, \tan \delta_y \right). \tag{10.36}$$

As in the case of the Shack-Hartmann mapper, we can fit this gradient to the gradient of a Zernike expansion of the wavefront $\overline{W}$ and from it calculate the power matrix as

$$\mathbb{P} = H\left(\overline{W} \right). \tag{10.37}$$

Once you have the deflection map δ, another possibility is the use of the generalized Prentice's law [362] to calculate the power matrix as

$$\mathbb{P} = \begin{bmatrix} \delta_{xx} & \delta_{xy} \\ \delta_{yx} & \delta_{yy} \end{bmatrix}, \tag{10.38}$$

where $\nabla \delta_x = \left(\delta_{xx}, \delta_{yx} \right)$ and $\nabla \delta_y = \left(\delta_{yx}, \delta_{yy} \right)$.

This method is one of the more frequent techniques used by commercial mappers. For example, the Dual Lens Mapper from A&R [360] uses transmission fringe deflectometry for the through power measurement mode. Some mappers from Lambda-X [371] use a variant of the technique that is denominated "Phase-Shifting Schlieren" [372].

Polished surfaces can also be measured by observing the reflection of a fringe pattern, and in this case we talk of "reflection fringe deflectometry" [373, 374]. In Figure 10.22, we show a typical setup for reflection fringe deflectometry. A grid G is observed in reflection with a camera. As in the transmission case, the grid pitch must be big enough to be resolved by the camera. As we show in Figure 10.22(a), using a reference mirror, the grid point $\mathbf{R} = \left(R_x, R_y \right)$ is imaged on the camera point **q**, through the reflection in a small patch with normal vector $\mathbf{N_0}$. Point **q** of the captured image will appear black when **R** is on a black bar of the fringe pattern. When we place the lens, we can think that it behaves locally as an inclined reflecting patch with normal vector **N**, as shown in Figure 10.22(b). Now, the lens reflects point $\mathbf{M} = \left(M_x, M_y \right)$ on camera point **q** via lens point **X**. Now, image point **q** may

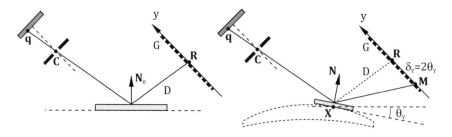

Figure 10.22 Reflection fringe deflectometry setup. The reflection of grid G is observed by the camera a) using a reference flat specular surface, with normal $\mathbf{N_0}$, point **R** form image on **q**. When the lens is placed with normal **N**, point **M** will form an image on **q**. From the displacement $\mathbf{d} = \mathbf{R} - \mathbf{M}$, it is possible to calculate ray deflection $\delta = \left(\delta_x, \delta_y \right)$ and local slope of the lens $\theta = \left(\theta_x, \theta_y \right)$.

appear white whenever **M** is on a white bar of the fringe pattern. As in transmission fringe deflectometry, the use of a grid, or a fringe pattern, allows determination of the positions of points **M** and **R** in the plane of the grid, and from them calculation of the displacement **d** = **R** − **M** is possible. In a similar way to the transmission case, the fringe pattern can be static like a printed screen or dynamic, in the sense that several fringe patterns can be used. From Figure 10.22, if the normal of the lens reflecting patch **N** forms an angle $\theta = (\theta_x, \theta_y)$ with the reference to normal **N**$_0$, then the angular displacement between **R** and **M**, $\delta = (\delta_x, \delta_y)$, is given by

$$\delta_x = 2\theta_x, \ \delta = 2\theta_y . \tag{10.39}$$

Through calibration, it is possible to determine the distance between the reference mirror and the grid D, and from the displacement **d** we can calculate the tilt angle θ at point **X** of the lens as

$$\tan 2\theta_x(\mathbf{X}) = \frac{d_x}{D}, \ \tan 2\theta_y(\mathbf{X}) = \frac{d_y}{D} . \tag{10.40}$$

As in the transmission case, sensitivity in the reflection fringe deflectometry method is inversely proportional to the screen-mirror distance D. The greater is D, the more precise will be measurement of the slopes $\theta = (\theta_x, \theta_y)$. Once we have the local tilt, it is straightforward to measure the normal. If $z(x, y)$ is the sag of the reflecting surface, the local gradient is given by

$$\nabla z = \begin{bmatrix} \tan\theta_x \\ \tan\theta_y \end{bmatrix}, \tag{10.41}$$

and for small slopes the normal vector will be

$$\mathbf{N} = \begin{bmatrix} \tan\theta_x, \tan\theta_y, 1 \end{bmatrix}^T . \tag{10.42}$$

As in the case of the Shack-Hartmann measurement, we can fit the measured sag gradients to the Zernike expansion of the sag, $\bar{z}$, and from it calculate the power matrix of the reflecting surface as (see Appendix D)

$$\mathbb{P} = (n-1)\,\mathbb{H}\,(\bar{z}), \tag{10.43}$$

where n is the lens refractive index and $\mathbb{H}$ is the Hessian. Another possibility is the direct derivation of the surface sag gradient, equation (10.41), and then calculation of the power matrix as

$$\mathbb{P} = (n-1)\begin{bmatrix} \nabla z_x \\ \nabla z_y \end{bmatrix} = (n-1)\begin{bmatrix} z_{xx} & z_{xy} \\ z_{yx} & z_{yy} \end{bmatrix}, \tag{10.44}$$

where $\nabla z_x = (z_{xx}, z_{yx})$ and $\nabla z_y = (z_{yx}, z_{yy})$. Therefore, by reflection fringe deflectometry it is possible to measure only the refractive power of one surface of the ophthalmic lens. To measure the whole lens power matrix one needs to measure the frontal and back surfaces z_1 and z_2 to compute the thickness function $z_{12}(x, y) = z_2(x, y) - z_1(x, y)$ and from it calculate the power matrix as $\mathbb{P} = (n-1)\,\mathbb{H}\,(z_{12})$. For this reason, ophthalmic lens mappers use reflection fringe deflectometry to measure only one side, for example the frontal surface, to check for the base curve.

In reflection fringe deflectometry, if the system uses one camera and one screen, there is a fundamental indetermination in the slope measurement. The reason is that for any position along the view line for image point **q**, any surface patch with the appropriate slope will reflect point **M** on point **q**. This indetermination can be solved by using more than one camera or more than one fringe pattern [373]. A commercial mapper that uses this principle is the SpecGAGE3D from 3D-shape work [375]. However, if the lens is supported in a lens holder and always placed in the same spatial position with respect to the camera and screen, distance D is almost fixed, and with just one camera and one pattern screen the measurement can be performed without indetermination. This is the case for the Dual Lens Mapper from A&R [360]. For this mapper, a dynamic pattern on the back side of the lens allows measurement of the power matrix in transmission, and a second pattern in the side of the camera allows for measurement of the front surface of the lens. In the Dual Lens Mapper, the reflection pattern is formed by an array of LED (light-emitting diodes) point sources, a configuration known as the Hartmann test [376].

The last deflectometric technique that we are going to discuss is moiré deflectometry. In fringe deflectometry, the measuring principle was modification of the observed fringe through the lens. In moiré deflectometry, what we observe are the changes in the moiré pattern that is formed by the superposition of two gratings. The moiré fringe pattern is measured before and after the lens is placed in the mapper, and from the changes of the moiré fringe pattern, the lens power matrix can be obtained. The main advantage of moiré deflectometry is that the gratings do not need to be resolved by the camera, allowing for good sensitivities using compact designs. However, the small grating period means that diffraction effects must be taken into account, especially for the moiré fringe contrast. Moiré deflectometry exploits two difractive dual effects: Talbot and Lau moiré fringe formation.

The Talbot effect [377] is a diffraction phenomenon: When a collimated monochromatic wavefront is incident on a grating, the geometrical image of the grating appears at regular distances from the incidence plane. These distances are multiples of the Talbot distance D_T given by

$$D_T = 2p^2/\lambda \, , \qquad\qquad (10.45)$$

where p is the grid period and λ the wavelength. If the two gratings are separated a Talbot distance, the propagation analysis can be performed using geometrical optics principles, although we must always remember we need diffraction theory to fully explain the results.

The typical setup for Talbot moiré deflectometry is depicted in Figure 10.23. Two grids G_1 and G_2 are separated by the Talbot distance, D_T, and diffusing screen A is placed just behind the second grating. Without a lens, when a collimated beam incides on G_1, a geometrical image of the grid is formed on the plane of the second grating G_2, and the diffusing screen allows the observation of the moiré pattern formed by the superposition of G_2 and the image of G_1. Using a geometrical analysis, an axial ray incident on point

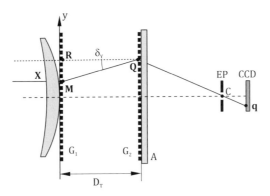

Figure 10.23 Talbot moiré deflectometry setup, see text for details.

R on G_1 will propagate axially until hitting point **Q** on G_2 (see Figure 10.23), and we will register the moiré fringe at point **q** of the camera. For example, if both gratings are rotated, the resulting moiré pattern is a set of straight lines perpendicular to the bisectrix of the grating directions. When we place a lens before grid G_1, the axial ray incident on point **X** of the lens will be deflected, and point **M** of G_1 is imaged on point **Q** of G_2. The overall effect of the deflection is that a distorted image of the grating G_1 is projected on G_2. This distorted projection produces a change in the moiré pattern observed in point **q** of the camera. Using square grids, it is possible to calculate displacement $\mathbf{d}(\mathbf{X}) = \mathbf{R} - \mathbf{M}$ [369]. The deflection angles at point **X** of the lens $\boldsymbol{\delta}(\mathbf{X}) = (\delta_x, \delta_y)$ will be

$$\tan \delta_x(\mathbf{X}) = \frac{d_x}{D_T}, \quad \tan \delta_y(\mathbf{X}) = \frac{d_y}{D_T}. \tag{10.46}$$

As the incident wavefront is plane, the gradient of the output wavefront will be

$$\nabla W = (\tan \delta_x, \tan \delta_y). \tag{10.47}$$

As in the case of transmission fringe deflectometry, from deflection $\boldsymbol{\delta}$ we can either estimate the wavefront and compute the power matrix from the Hessian (see equation (10.37)) or calculate the power matrix from the gradients of the deflections (see equation (10.38)). In commercial systems, Talbot moiré deflectometry is the technique used by the Rotlex Class Pluss mapper [378] and the NIMO TF from Lambda-X [371].

The other diffraction effect used in deflectometry is the Lau effect [379, 380]. The Lau effect consists of the formation of moiré fringes at infinity when two equal gratings separated by a multiple of the Talbot distance, D_T, are illuminated by a monochromatic extended light source. In Figure 10.24, we show a typical setup for a Lau moiré deflectometer. Two equal gratings G_1 and G_2, separated by the Talbot distance D_T, are illuminated by an extended incoherent source L. As in the case of Talbot moiré deflectometry, when the grids are separated by the Talbot distance, we can make a geometrical analysis of the measuring principle. Without a lens, the superimposed images of the grids will be imaged

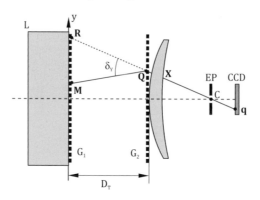

Figure 10.24 Lau moiré deflectometry, see text for details.

on the camera sensor, with the corresponding scale change, producing a moiré pattern. In this case, the depth of field of the pinhole camera makes imaging of the propagated grids as if they come from infinity possible. Without the lens, points $\mathbf{R}$ of G_1 and $\mathbf{Q}$ of G_2 will both form images on point $\mathbf{q}$ of the camera. If we introduce a lens, the ray coming from $\mathbf{q}$ will be deflected in point $\mathbf{X}$ of the lens. In this case, point $\mathbf{M}$ on G_1 will pass through $\mathbf{Q}$ on G_2 and will form an image on $\mathbf{q}$. This change in the imaging conditions of the grating G_1 produces a change in the moiré pattern observed in the camera image plane. From this change in the moiré pattern, using crossed fringe patterns for G_1 and G_2, it is possible to determine displacement $\mathbf{d}\,(\mathbf{X}) = \mathbf{R} - \mathbf{M}$ in the plane of G_1 and calculate the deflection at point $\mathbf{X}$ of the lens using equation (10.46) and the lens output wavefront gradient using equation (10.47). Following the same procedures as in Talbot moiré deflectometry, the power matrix can be calculated from the Hessian of the wavefront or from the gradient of the deflections.

   Although they seem very different, Talbot and Lau deflectometric set-ups are dual by ray path inversion. In the Talbot moiré deflectometer, a collimated beam, coming from infinity, illuminates the two grids, and the moiré image is formed on a diffusing screen, which behaves as a secondary extended source. On the other hand, in the Lau moiré deflectometer, an extended light source illuminates the two grids and the moiré image is formed at infinity. So, the two systems are a ray-path reversed version of each other. This is evident also in the equations that relate deflection with grid pitch and grid separation. The main advantage of the Lau moiré deflectometer is that the set-up can be very compact, sharing the sensitivity of the moiré techniques. One commercial system that uses the Lau set-up is the Free-Form Verifier from Rotlex [378].

CMM Profilometers

In the last two sections, we presented optical systems able to measure the output wavefront of a lens. They are well suited for calculating the power matrix, are fast and precise, and can have a big spatial resolution. However, all lens mappers we have presented are based on measurement of the deflection of a light beam as it is refracted or reflected by a lens. Therefore, all these mappers measure the wavefront gradient, not the wavefront

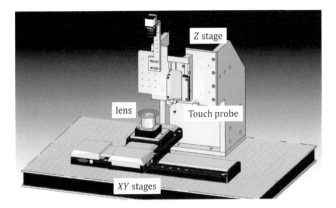

Figure 10.25 Cartesian CMM configuration; the *XY* linear stages move the lens while the touch probe measure the sag (courtesy of Indizen Optical Technologies).

directly. When we need the topography of a lens (for ray tracing, for example) very small curvature errors quickly accumulate as we integrate these measured gradients over the lens surface. For this reason, coordinate measuring machines (CMM) profilometers are used as an alternative to lens mappers when we want to measure surface topography directly. One of the simplest setups is the column type or cartesian CMM. In this configuration, two linear actuators move the lens in the *XY* plane, and a vertical touch probe directly measures the sag $z(x, y)$ of the lens, see Figure 10.25; the vertical touch probe is mounted in a *Z* stage to make possible the loading and unloading of the lens. The CMMs used in ophthalmic lens profilometry have a small working volume (around 1000 cm^3) with a volumetric error below $1 \ \mu\text{m}^3$.

The measurement principle is very simple; as the lens is displaced by the *XY* stage, the touch probe in contact with the lens surface measures the sag $z(x, y)$ for every location. Once we have measured the sag, the surface power matrix is given by

$$\mathbb{P} = (n - 1) \mathbb{H}(z),$$
(10.48)

where n is the lens index and $\mathbb{H}$ the Hessian (see Appendix D).

Coordinate measuring machines profilometers are the most direct and simple way to obtain lens surface sag. However, there are some important aspects that must be taken into account for correct use. First, the touch probe does not end in a punctual tip. Rather, it ends in a sphere, see Figure 10.26. This means that when we measure a surface with a touch probe we do not measure the surface directly, but a convoluted version given by the tangential contact of the sphere with the surface. Following Figure 10.26, if point $\mathbf{q} = (u, v, w)$ at the sphere center is the position returned by the touch probe, to obtain the corrected position on the surface $\mathbf{r} = (x, y, z)$, we must compute the normal of the convoluted surface at $\mathbf{q}$, $\mathbf{N}(u, v, w)$ and calculate $\mathbf{r}$ as

$$\mathbf{r} = \mathbf{q} - \rho \mathbf{N}(\mathbf{q}),$$
(10.49)

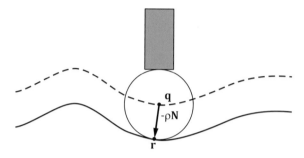

Figure 10.26 CMM touch probe ball radius correction, see text for details.

where ρ is the ball radius. For this reason, when the tip ball of a CMM profilometer is changed, it is very important to check its radius for a correct compensation for the ball size. Another important aspect of the tip ball of the touch probe is its material. Using the Z stage, the touch probe can jump from measuring point to measuring point; however, to increase measurement speed it is usual that the touch probe stays in contact with the surface as it is dragged on the surface while the XY stage moves the lens. In this case, residues may accumulate on the ball, depending on the affinity between the ball and surface materials, and altering both, the tip ball and the surface. To avoid this, the ball tip material should be chosen according to the surface material. For example, ruby and zirconia are recommended for plastics and glass, for steel molds the better choice is zirconia and for aluminum molds use of silicon nitride is advised, see [381] for a list of possible materials for CMM touch probes.

The final issue with CMM profilometers is the calibration of the geometrical setup. In a CMM machine, it is very important to ensure the right orthogonal orientation of the CMM linear stages and the touch probe to minimize the squareness error. While it is necessary to take into account 21 error parameters to fully calibrate a column-type CMM, the squareness error alone accounts for more than 50% of the total accumulated error [382]. Small squareness errors can have a dramatic effect on the measured sag and the matrix power of a lens surface [382], and it is very important to take them into account, especially if the CMM is assembled from off-the-shelf parts. For example, in Figure 10.27(a), we show the sag residues for a spherical lens surface of radius 104,45 mm. Figure 10.27(a) shows the sag residues in microns with respect to the ideal sphere before calibration of the CMM profilometer. From calibration, the dominant error was a misalignment between the Z and the Y axis of about 1°. Figure 10.27(b) shows the sag residues in microns after calibration. As shown in the figure, the squareness error can induce sag measurement errors above 5 microns, which after calibration can be reduced to 0.1 microns. This kind of highly structured sag error will inevitably be translated to the power matrix (see equation (10.48)). Figure 10.28 shows the spherical power of the same lens as in Figure 10.27, before and after correcting the squareness error. As the results show, bad CMM calibration may have dramatic effects on the measurements, where, rather than a large absolute error, we should expect spurious spatial distributions of sag and power.

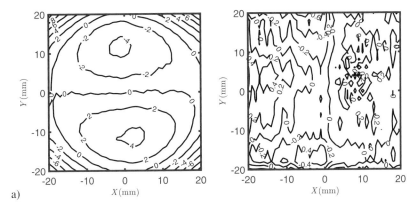

Figure 10.27 Squareness error. Sag residues in microns for a spherical lens, a) before and b) after calibration.

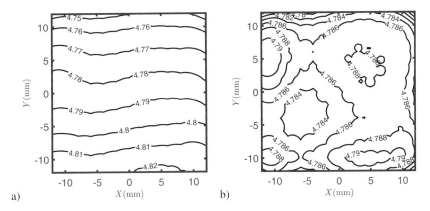

Figure 10.28 Squareness error. Spherical power in D for the same spherical lens as Figure 10.27, a) before and b) after calibration.

User Power and Focimeter Power

In paraxial conditions, measuring in the optical axis, standard focimeters and lens mappers measure the paraxial BVP of the lens; this is what we will call the focimeter power. As the measuring point moves away from the paraxial zone, the focimeter power changes in a different way for every instrument, in a form that depends on the measuring technology and the ray tracing [383, 384]. The paraxial BVP depends only on the lens surface curvatures, refractive index, and central thickness, and for its calculation one always uses a collimated input wavefront. However, the actual optical power perceived by an user, the user power, depends additionally on eye-lens position (the distance between the cornea and the lens vertex), lens orientation (pantoscopic and facial angles), and obliquity of the aiming direction. To better understand the differences between the focimeter and the user power we are going to recall the BVP calculation for a thick lens (see Section 4.3). Referring to

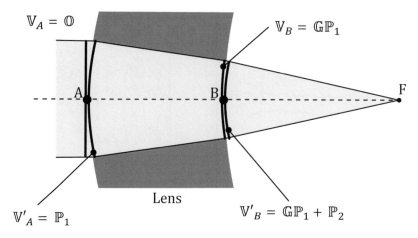

Figure 10.29 Back vertex power of a thick lens.

Figure 10.29, for an object point at infinity (a collimated input beam) and normal incidence, the paraxial back vertex power of a thick lens is given by

$$P_V = V'_B = gP_1 + P_2 \,, \tag{10.50}$$

where $g = \left(1 - \frac{t_c}{n}P_1\right)^{-1}$ is the lens shape factor, P_1 and P_2 are the refractive powers of both surfaces, n the refractive index, and t_c the lens central thickness. The BVP is defined as the inverse of the distance between the back vertex of the lens and the focal point; thus the BVP for the thick lens is given by the output vergence V'_B. In our example, we are assuming spherical surfaces, and therefore, in normal incidence for an on-axis object, the refractive powers, the vergences, and the lens shape factor are scalars.

For far vision, the ray tracing of Figure 10.29 is the same as the one used to describe the focimeter in Figure 10.12; therefore, for the axial view direction, the power measured by a focimeter is the BVP power of the lens.

The nominal or prescribed power, or simply the prescription, is the BVP necessary to compensate for a refractive error referred to the spectacle plane.

Before the introduction of free-form technology, single vision lenses were calculated so that the nominal power was equal to the power measured with a focimeter or similar instrument in paraxial conditions. In this case, it is said that the lens design is classical or optimized in curvatures. With the appearance of free-form technology, the manufacturing of a lens in which the user power would match the nominal power for all viewing directions is a possibility. In this case, the relative position of the eye and the lens and the user's personal parameters must be known beforehand. For this reason, a lens designed using user power is known as a personalized lens.

Now we are ready to define user power more precisely. So far, we have assumed that the incidence angle is close to zero and that we are in the paraxial region. Therefore, if the lens

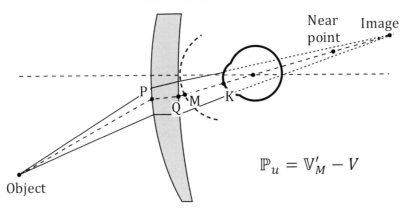

Figure 10.30 When the user aims obliquely, the perceived power depends on the object location and the aiming direction. If V is the object vergence and $\mathbb{V}'_M$ is the output vergence measured from the vertex spheres, the user perceived power is $\mathbb{P}_u = \mathbb{V}'_M - V$.

is well designed in curvatures, for an eyesight direction passing through the optical center of the lens, the user perceived power and the focimeter power should match.

However, in wearing conditions, the eye gaze, the surface normals, and the incidence angles may move out of the paraxial region. For monofocal lenses, this is especially important at the extremes of the field of view. Also, this is the case if the pantoscopic or facial angles are large. But maybe the most relevant case is the near zone of a PPL, where the eye must aim obliquely with an angle that can reach $30°$.

As we show in Figure 10.30, when we aim obliquely through an ophthalmic lens, surface normals and beam axes do not have the same direction and this produces two main effects. First, lens thickness variation with respect to the axial direction appears. In Figure 10.30, this effect is clearly seen as the dependence of the distance $\overline{PQ}$ with the incidence angle. In positive lenses, the effective lens thickness for oblique sight directions is smaller than in the center. On the other hand, for negative lenses, the effective lens thickness for oblique sight directions can be much larger than in the center. Second, for oblique directions, the distance from the lens back vertex Q to the eye vertex changes. Moreover, when the user aims obliquely, the reference point for measuring the output vergence is no longer the lens back vertex Q. Instead, the correct reference point for measuring the output vergence is point M, situated on the vertex sphere (see Section 6.3). For a personalized lens design, once we set up the object space for every sight direction, from the object point we have to trace the main (or chief) ray, which passes through the object and the rotation center of the eye. This ray tracing gives us the incidence angles on each surface and the path traveled in each medium. With this information, we can compute the object vergence, V, and then start the vergence propagation. First, we have refraction on the front surface at point P; this is followed by a propagation to the back vertex Q, and then we have refraction at Q. Finally, there is propagation to the vertex sphere point M. The vergence at this point $\mathbb{V}'_M$ minus the object vergence at point P, V, gives us the actual oblique power perceived by the user for any viewing direction, the user power $\mathbb{P}_u$, given by

$$\mathbb{P}_u = V'_M - V. \tag{10.51}$$

From its definition, point M is at the same distance from the cornea as the axial back vertex of the lens. Therefore, if we assume that the lens is designed to compensate the nominal refractive error, $\mathbb{R}$, the user power for all aiming directions, $\mathbb{P}_u$, must match the nominal prescription,

$$\mathbb{P}_u = \mathbb{R}. \tag{10.52}$$

In summary, for oblique directions, lens thickness and back vertex position changes add up, and there is a difference in user power compared to paraxial BVP. For example, assume a spherical lens designed for focimeter power, for which user power increases with obliquity. In Figure 10.31(a), this effect produces a lens image focus lying before the near point for the eye (to simplify, we are assuming the eye is using maximum accommodation). While at normal incidence the image focus and the near point will coincide, for oblique incidence they do not superpose.

Correct personalized design of the lens for near sight compensation must take this into account, generating the image at the near point for all aiming directions, and not only for

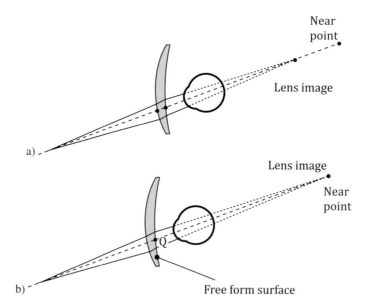

Figure 10.31 Near vision ray tracing of a beam entering the eye at oblique aiming a) with a spherical lens b) with an optimized free-form surface. For the spherical lens, oblique power does not match the nominal power and the lens image does not focus on the near point. On the other hand, the use of free-form surfaces makes possible the compensation of obliquity at some gaze directions.

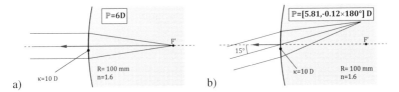

Figure 10.32 User power versus focimeter power, a) in normal incidence oblique power only depends on surface curvature, b) in oblique incidence, the power also depends on the incidence angle.

the main viewing direction. In other words, for a well-designed personalized lens, user power and oblique power will coincide.

To illustrate the change in user power with obliquity, in Figure 10.32, we show an example of the change in optical power with obliquity for a spherical diopter. In Figure 10.32(a), we depict a spherical diopter of radius $R = 0.1$ m, and curvature of $\kappa = 10$ D, separating air and a media with index $n = 1.6$. For an axial point, the refractive power of the diopter is $P = (n - 1)/R = 6$ D. In Figure 10.32(b), we show the same surface when the incidence angle is $\theta = 15°$. Although the curvature remains the same, the refractive power changes to $\mathbb{P} = [5.81, -0.12 \times 180°]$. In general, oblique power of a spherical surface is a function of surface curvature κ, index n, and obliquity angle θ

$$\mathbb{P} \equiv \mathbb{P}(\kappa, n, \theta),$$ (10.53)

and for normal incidence, $\theta = 0$ the refractive power becomes the refractive power of a spherical diopter

$$\mathbb{P}(\kappa, n, 0) = \kappa (n - 1).$$ (10.54)

Now we can consider oblique refraction through a thick lens to calculate oblique power. Formally, the process is the same as we used to calculate paraxial BVP. First, there is refraction at the first surface, followed by propagation to the second surface, which is taken into account as a shape factor, and finally there is refraction at the second surface. However, for oblique incidence, both refractive powers and the shape factor depend on the angles involved. Referring to Figure 10.33 if θ and φ are the incidence angles on the first and second surface, respectively, and we denote with A and C the input and output points, the oblique power for the lens shown in Figure 10.33 will be

$$\mathbb{P} = \mathbb{G}\mathbb{P}_{1\theta} + \mathbb{P}_{2\varphi},$$ (10.55)

where $\mathbb{P}_{1\theta}$ and $\mathbb{P}_{2\varphi}$ are the oblique powers for surface 1 and 2, respectively, and

$$\mathbb{G} = \left(1 - \frac{\overline{AC}}{n}\mathbb{P}_{1\theta}\right)^{-1},$$ (10.56)

is the shape factor.

Now we are ready to answer the question of what power we are measuring when we use a lensmeter at an off-axis point as Q in Figure 10.30. At this point, we must consider

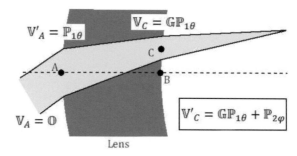

Figure 10.33 Oblique power calculation for a thick lens.

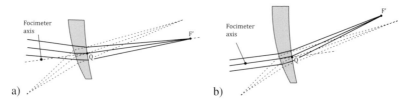

Figure 10.34 Ray paths for a) focus on axis (FOA) and b) infinite on axis (IOA) focimeters.

what type of focimeter we are using. There are two types of commercial focimeters, which differ in the orientation of the incoming beam: infinite on axis (IOA) and focal point on axis (FOA) focimeters [345].

Let us give the notation Q to the point at the back surface of the lens we locate at the center of the focimeter support ring. As we press the lens against the support, the normal to the back surface at Q will be coincident with the instrument axis.

In the FOA focimeter (see Figure 10.34(a)) the light comes from point F' (the image of the test target in a standard focimeter) and refracts backward through the lens, emerging as a parallel beam from the front surface. The power measured by the FOA focimeter is characterized by having $\varphi = 0$ and $\theta \neq 0$.

In the IOA focimeter, (see Figure 10.34(b)), the normal to the back surface at Q is still coincident with the instrument axis. However, a beam of parallel rays parallel to the instrument axis refracts forward through the lens, with $\theta = 0$, emerging at Q with some obliquity $\varphi \neq 0$. The lensmeter described in a previous section is IOA-type. In both cases, the nonzero angles are small, as they are nothing other than the prismatic deviation produced by the lens at Q.

We may compare these ray tracings with the user power ray tracing also shown in Figure 10.34 with dashed lines. For these beams, both θ and φ may become quite large (up to 40 degrees in a standard frame). For each type of power, the thickness traveled by the chief ray through the lens is different. The input vergence giving the two focimeter powers is always zero, while the input vergence giving user power is zero for far objects only. Finally, the focimeter power, whether FOA or IOA will be the vergence of the output beam at Q, while the user power matches the output vergence at the vertex sphere, as previously explained.

The differences between the IOA or FOA focimeter power and user power are well known in the literature. The standard ISO 8980-2 [345] advises on the differences between measuring a lens with IOA and FOA focimeters. For example, the ISO standard warns that lens regions with prism may lead to differences between FOA and IOA power because of the different obliquity of the ray paths. Unfortunately, the ISO 8980-2 is too vague and imprecise when speaking about measuring addition. It recommends measuring the near power by placing the surface in which the addition is ground on the lens support. Then, front side PPLs should be measured at the near point by the front side, and back side PPLs should be measured on the back. In any case, the norm ISO 8990-2 [345] only acknowledges that there are differences if the addition is measured by the front or the back, and that there are also differences if IOA or FOA focimeters are used, and that the manufacturer should state the measurement method.

As said before, an ophthalmic lens can be designed for focimeter power or curvatures, a classical lens, or can be designed for user power including the user parameters, a personalized lens. In classical lenses, the ECP will read the correct prescription on a FOA focimeter (the most extended type), but this design strategy has important drawbacks and some minor advantages. The drawbacks are that due to obliquity, the user power at the reading point will not be as expected, astigmatism perceived by the user at the reading point will be higher, and, when there are tilts, user power will not be correct for the whole lens. In parallel, customization based in the position of use and user parameters is not possible. The advantages of classical lenses are that when there are no tilts, user power is correct at the center of the lens, and at the reading point the focimeter should read nominal near power (far power plus addition) within ISO tolerances.

On the other hand, for personalized lenses the user power will coincide with the nominal power at the position of use. The user will find important optical advantages while the ECP and the manufacturer may find some minor drawbacks. The advantages are that with this kind of lenses, visual acuity is higher, tilts can be taken into account, and the visual field in the near region is wider. However, for personalized lenses, the focimeter will not read the nominal power at the far region when there is tilt. Also, the focimeter will not read nominal power at the near region. For this reasons, quality control of personalized lenses requires the manufacturer to stamp the expected focimeter power as well as prescription on the lens envelop, so that the ECP can check that the lens power is correct to tolerances when using the focimeter.

The intrinsic differences between the user power and focimeter power not only affect the way we should interpret the power measured at a reference point. Also, these differences affect the way we can infer the performance of a PPL from its power maps. The visual field perceived by the user in the near region is usually smaller than the one measured with a mapper if the lens has been optimized for focimeter power. In the same situation, the perceibed astigmatism lobes are typically larger [329].

As a final consideration, understanding the difference between user and focimeter power and the role of obliquity allows for the comprehension of the differences in optical performance between classical and personalized lenses. Obliquity is the most important factor that differentiates user and focimeter power. Aspherical surfaces that compensate for oblique errors are the final reason why personalized free-form lenses provide a better user experience than classical lenses. This is a very important point that the ECP should understand when performing quality control on a lens with a lensmeter. As an example, in Figure 10.35, we show a comparison of the user power and the focimeter power for a spherical (classical) and an aspherical (personalized) solution for the same nominal prescription of $-5\,$D. For this purpose, we have calculated the user power and the (FOA) focimeter power for a 60 mm diameter lens with base 1, index $n = 1.5$ and center thickness $t_c = 1$ mm. In both cases, the back surface curvature is calculated so that the nominal BVP is $P_V = -5\,$D. For the aspherical case, we have chosen a conicoid with asphericity parameter $p = 1.5$ (see Section 2.2.2) and have used the vertex curvature as a free personalization parameter. Figure 10.35 shows the resulting power profiles for the vertical diameter of the lens.

In the first row of Figure 10.35, we show the spherical lens calculation for the mean sphere and the cylinder. For the axial viewing direction ($y = 0$), user power and focimeter power coincide. However, as we depart from the axial direction, obliquity produces differences between user and focimeter power. In particular, at the border of the lens ($y \approx 30$ mm), while the focimeter power is still within tolerances, the user power is almost 1 D different in sphere and cylinder. In consequence, although focimeter power is good at the axis, the user will perceive a totally different lens at the lens border.

In the second row of Figure 10.35, we show how the use of a conicoid in the back surface allows for a better lens in terms of user power, although the focimeter power readings will be worse than in the spherical case. For example, we can see that the difference between nominal power and user power will be below 0.5 D for sphere and cylinder for the whole lens. Meanwhile, the focimeter cylindrical power will give a 1.5 D error with respect to the nominal power at the lens border, although the user-perceived cylinder will be close to 0.25 D. This is a very simple case in which we could have improved the performance of the spherical lens by selecting a steper base curve, but still makes the point in stablishing the differences between focimeter power and user power.

10.4.4 PPL Measurement and ISO Standards

We are going to finish the chapter with a discussion about quality control in PPLs. The standard ISO 8090-2 [345] specifies the requirements for the optical and geometrical properties for uncut finished PPLs. In the standard, additional to establishing the absolute tolerances to the mean sphere, cylinder, and prismatic powers, the test procedure to perform these measurements is also described. The test method is based on measurement with a lensmeter at three control points: distance, near, and prism reference points (DRP, NRP, and PRP, respectively). Therefore, any PPL with the correct power values at the three mentioned

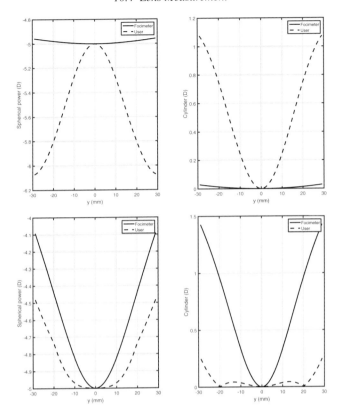

Figure 10.35 User and focimeter power for a spherical lens (first row) and an aspherical back surface lens (second row) (see text for lens details).

points will pass the ISO test. However, in a free-form generated PPL with complex power maps, this is not enough to decide whether the lens is correctly manufactured or not. The reason is that free-form generation is a complex and delicate process that can produce complex error patterns in the power matrix that can alter lens performance significantly [385]. A given power map can have small errors with respect to the theoretical values at the control points, but, outside the control points, the error makes the lens useless, especially if the error affects the near, distance, or corridor zones. In [386] there is a collection of the most common error maps arising in free-form surfacing. The white-paper also included a list of possible error causes. For example, it is well known that a centering error in the free-form generation process produces a central dot pattern in the sphere and cylinder error maps. Another example is the use of excessive polishing speed, which gives place to the appearance of ring patterns in the spherical and cylinder error maps.

To illustrate the problems of quality control in PPL free-form surfacing, in Figures 10.36 and 10.37, we show the measurements of two identical PPLs with prescription $[0, 0 \times 0°]$ add 2.25. In Figure 10.36, we show the sphere and cylinder error of the first lens with

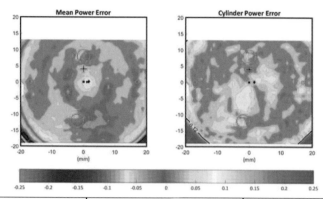

DRP				NRP				PRP			
Theoretical		Measured		Theoretical		Measured		Theoretical		Measured	
H	*C*	*H*	*C*	*H*	*C*	*H*	*C*	Prism	Axis	Prism	Axis
0.04	−0.01	0.07	−0.04	2.17	−0.35	2.17	−0.35	0.48	270	0.61	261

Figure 10.36 Example of a PPL lens that is well manufactured and that also will pass the ISO test. The two maps above are the residues with respect to the theoretical mean spherical power and the cylinder (the lighter grey the better). Below we can see the measured and theoretical values at the DRP, NRP, and PRP (courtesy of Indizen Optical Technologies).

respect to the theoretical values. The light grey colors and greyscale bar indicate low error. Below the maps, we have included the lensmeter measurements and the theoretical expected values for the DRP, NRP, and PRP in D. In this case, the ISO test result and the error map coincide, and both methods will classify the lens as correctly manufactured.

In Figure 10.37, we show the second lens, that was manufactured with an error of about 0.2 D in the lens center. The error can be seen as a central dot clearly visible in the error maps. As in the former case, we show the error maps with respect to the theoretical power, and below the maps we include a table with the spherical, cylinder, and prismatic powers measured at the three ISO control points in D. As we can see, the lens is well manufactured according to the ISO test, but the error map shows an error zone in the central portion of the lens, in the connection between the far region and the corridor. In this case, this error affects an important vision zone, and the PPL should be discarded.

With these two examples, we have shown how the ISO test is not general enough for testing the quality of a free-form surfacing process. Quality control test methods based on the measurement of power in a discrete set of control points should be used with care, and whenever possible the error map should be included in the decision-making process. For this reason, it is advisable to develop a quality control metric that also includes the error map. The simplest choice will be the calculation of the RMS error between the measured and the theoretical power maps. However, this global measure does not take into account the importance of the different zones of the PPL. A surfacing error such as the one shown in Figure 10.37 will have a small RMS error, because its spatial extension is small, but it is

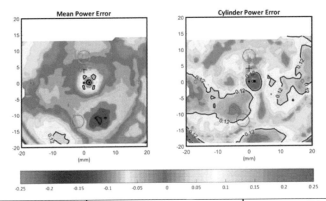

DRP				NRP				Prism			
Theoretical		Measured		Theoretical		Measured		Theoretical		Measured	
H	C	H	C	S_H	C	H	C	Prism	Axis	Prism	Axis
0.04	−0.01	0.00	−0.04	2.17	−0.35	2.12	−0.37	0.48	270	0.46	277

Figure 10.37 Example of a PPL lens that has a manufacturing error but will pass the ISO test. The two maps above are the residues with respect to the theoretical mean spherical power and the cylinder. The manufacturing error produces a central defect in the mapper measurement. Below we can see the measured and theoretical values at the DRP, NRP, and PRP (courtesy of Indizen Optical Technologies).

located on the connection between the far zone and the corridor, producing an unexpected distortion each time the user lowers their eyes from the far to the near zones of the PPL, and making it potentially a big usability problem.

Therefore, it is advisable to use a decision-making process that includes multiple factors. For example, the three RMS values of the error map of the lens power at the near, far, and corridor zones, combined with the ISO test, can provide a better view of the quality of the PPL. This multiple factor procedure makes machine learning methods very well suited for the automatic classification of a PPL job using the error maps as input for the classification process. For example, IOT [359] has developed a Quality Control assistant app for the automatic classification of a PPL using machine learning techniques. This tool is based on the use of a set of PPL lenses that have been labeled as "good" or "wrong" by IOT. This labeled set of lenses are used as a training set for a supervised learning classification system. The use of machine learning techniques does not only allow a "good"/"wrong" classifica-tion but also calculates the feasibility of the classification. This kind of tool also helps the temporal tracking of the production line and establishes a common ground for all staff of an organization involved in the quality control process, especially for big laboratories with several facilities. Moreover, the spatial structure of the error maps carry information about the error source, making possible the determination of the most probable cause for the surfacing error, such as generation misalignment, blocking issues, or polishing problems.

In Figure 10.38, we show a screenshot of the QCApp quality control assistant software from IOT. This application uses two error maps, sphere and cylinder, to classify the lens

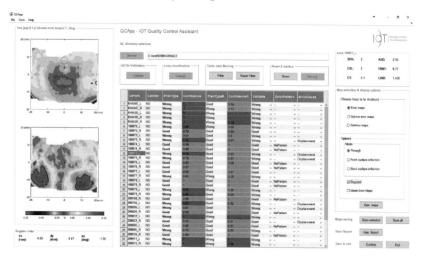

Figure 10.38 Screenshot of the IOT QCApp assistant software. The job #168875_L (highlighted in the LensId column) is classified as good, with a confidence of 0.92 over 1. The error maps for the mean sphere and the cylinder are shown in the left panel. After spatial registration, there is a big error in the cylinder map, but it is located in the astigmatism lobes. The localization of the cylinder error together with the good quality of the sphere enables the system to classify the lens as "good" (courtesy of Indizen Optical Technologies).

as "good" or "wrong" and gives a confidence level for the classification. In Figure 10.38, the error maps of a PPL job #168875-L (highlighted in the "LensId" column) are shown in the left panel. These error maps were measured with a dual lens mapper from A&R [360] in transmission mode. As can be seen in the column "ConfidenceR," after the spatial registration of the theoretical and measured maps, the PPL is classified as "good," with a confidence of 0.92 over 1. This is a good example of the importance of manufacturing errors in the different zones of the PPL. Although there is a big error in the cylinder error map (in dark gray), it is located in the astigmatism lobes, and its importance for PPL usability is small. The location of the error together with the good quality of the sphere enables the system to classify the lens as "good" with a high confidence level.

11

Filters and Coatings

11.1 Introduction

In the modern ophthalmic industry, coatings have gained a growing importance, as they allow lens manufacturers to change many properties of lenses, improving their overall quality. Thanks to coatings, it is now possible to reduce the amount of light reflected by a lens, or make lens surfaces more resistant to abrasion just by adding thin coatings of a few microns depth.

Coatings may also be used to filter out radiation that is potentially harmful to the human eye. The goal is to reduce the amount of harmful radiation that reaches the eye after passing through the lens. Even sunlight may be potentially dangerous to the ocular tissues, particularly the short wavelengths in the ultraviolet (UV) range. An example of this is the condition known as cataracts, whose development is certainly related to the absorption of sunlight UV radiation throughout life. Besides sunlight, a number of potentially hazardous artificial light sources, such as lasers, are employed in many activities, making protective equipment, such as safety eyeglasses, necessary to prevent ocular damage.

The goal of this chapter is to present the main ocular problems associated with electromagnetic radiation and to describe the optical filters, particularly sunglasses, used for reducing this radiation hazard. Additionally, we will introduce the main characteristics of the coatings employed nowadays by the ophthalmic industry. We will begin this chapter with a discussion of the hazards associated with electromagnetic radiation, with special focus on sunlight. Afterward, we will present the main optical properties of the filters employed in the ophthalmic industry, with particular attention to the relevant international standards. Then, we will discuss some general principles on prescribing optical filters and a brief survey of other safety eyeglasses. Next, we will summarize the main properties of the optical coatings employed in the ophthalmic industry. Finally, we will close the chapter by studying the physical properties of anti-reflective coatings.

11.2 Ocular Hazards Due to Electromagnetic Radiation

It is well known that electromagnetic radiation may inflict ocular damage when it is absorbed by eye tissues. The resulting damage depends on two main factors: the amount of energy absorbed and the wavelength of the radiation. Due to the preventive mechanisms of the human visual system against damage produced by visible light, it is more usual to find damage caused by infrared and/or ultraviolet radiation. However, elevated but harmless amounts of visible light incident on the eye may induce glare, which, in some cases, may lead to a temporal incapacitation of the visual system.

In order to quantify the amount of energy that may cause damage to the eye, it is necessary to use some concepts of radiometry. When electromagnetic radiation (note that visible light is, indeed, electromagnetic radiation with wavelength between 380 and 780 nm) passes through a given surface, the radiant flux Φ_w is the amount of energy by unit of time carried by this radiation. Radiant flux is measured in watts (W), the power unit of the International System of units (IS). The spectral radiant flux is the radiant flux per unit of wavelength, and it is measured in watts per meter (Wm^{-1}) in the IS. For a given point over a surface, the amount of incident radiant flux by unit of area is known as the irradiance E and it is measured in watts per square meter (Wm^{-2}). If the surface is exposed to radiation during a time t, then the amount of radiant energy received (known as dose or exposure) is $H = Et$ and is measured in Joules (J). Exposure is the prime magnitude when speaking of radiation damage, as only the radiant energy absorbed by a given material or tissue may produce damage to this material or tissue (Drapper-Gotthus law).

When radiation passes through a surface, part of the incident energy is usually transmitted and some is reflected back. Therefore, if Φ_{wo} is the incident radiant flux, Φ_{wt} the transmitted flux, and Φ_{wr} the reflected one, we can define two coefficients that will quantify the transmission and reflection of radiant energy: transmittance and reflectance. Transmittance T is the ratio between the transmitted radiant flux and the incident one, so

$$T = \Phi_{wt}/\Phi_{wo}, \tag{11.1}$$

while reflectance is defined as

$$R = \Phi_{wr}/\Phi_{wo}. \tag{11.2}$$

It is worth noting that the conservation of energy requests that $R + T = 1$. Transmittance and reflectance may also be defined as percentages, so $T\,(\%) = 100T$.

Radiation may cause damage to ocular tissues through thermal and photochemical [387] processes. In the thermal case, the tissue is heated by the absorbed radiation, producing a burn if the increment in temperature is enough to induce protein denaturalization (this threshold may be as low as an increase of 2° C for some tissues [388]). Thermal damage is usually linked to medium or high wavelengths, typically in the infrared range. One example, is the so-called eclipse sunburn, which may appear at the retina when looking at a solar eclipse without any kind of ocular protection.

Photochemical damage is due to short wavelength radiation. In this case, the damage is produced by photochemical reactions, which lead to changes in the proteins that form the tissues of the eye. Two main mechanisms of photochemical damage to the retina have been described in the literature. The first is photoreceptor damage caused by UV light. This problem is associated with the rodopshin absorption spectrum. The second damage mechanism is the depigmentation of the retinal pigmented epithelium caused by short wavelength radiation ranging from UV to the blue visible spectrum.

Whatever the damage mechanism may be, we can always define, for a given tissue or anatomical structure, a threshold level of exposure, known as maximum permissible exposure (MPE). The MPE is defined as the maximum amount of energy received by an ocular structure (cornea, lens, or retina) before the appearance of damage [387]. Due to the complicated interaction between radiation and living tissues, MPE also depends on the exposure time, and MPE values are given as a function of wavelength and exposure time. It is important to keep in mind that damage is usually accumulative, so the effect of small amounts of energy absorbed at different times is accumulated. When the combined dose is greater than the MPE, then the ocular tissues will be damaged. An example of this accumulative effect is the UV-induced cataract presented by elderly people after lifelong exposure of the eye lens to low doses of UV radiation. It is also quite important to consider the way in which the light arrives at the eye. For example, the effect of a collimated laser beam and that of an extended object are quite different (see Figure 11.1). For a collimated beam, radiation is concentrated upon the retina, while for extended objects, exposure levels are greater in the eye lens, as each point of this lens receives radiation coming from all the object points. In some cases, ambient light is an important factor in the development of radiation damage, particularly for the corneal tissues. Therefore, besides exposure, there are other parameters that are useful in order to assess the potential of radiation damage, such as wavelength, light collimation, etc.

The human eye is exposed to natural and artificial sources of radiation such as incandescent lamps, discharge lamps, arc lamps, LED (light emitting diode), or lasers. In the context of ophthalmic optics, we are mostly interested in protecting the eye from solar radiation with sunglasses, although we will describe briefly some guidelines of ocular protection from artificial sources. The Sun radiates in a wide range of wavelengths, as it behaves as a thermal radiator that follows the black body radiation law [389]. The maximum spectral flux for the solar spectrum in the upper atmosphere is reached for the wavelength of $\lambda = 555$ nm, which corresponds to a black body with a temperature of $5778°$ K, and the interval of emitted wavelengths extends from 250 nm to 2500 nm, that is from UVB to the near infrared. The different gaseous molecules present in the atmosphere absorb some of this radiation, particularly in the band from 250 to 315 nm (UVB), which is absorbed by the ozone (O_3) at the ozone layer of Earth's stratosphere. Other absorption bands occur at mid-infrared and are mainly due to atmospheric water vapor. Note that the amount of radiation observed is not the same at sea level as at high altitude, because, in this latter case, the thickness of the atmosphere is reduced.

The human eye has a degree of natural protection against elevated levels of solar radiation. Eyebrows and eyelashes create a shadow that protects the cornea from vertical radiation. Pupil diameter is reduced when high levels of visible radiation are present. In sunny latitudes, dark iris colors due to an increased amount of melanine are quite common in order to absorb the maximum amount of ambient light. The aversion reflex and the sensation of glare prevents the eye from looking at intense light sources, such as the Sun, avoiding in

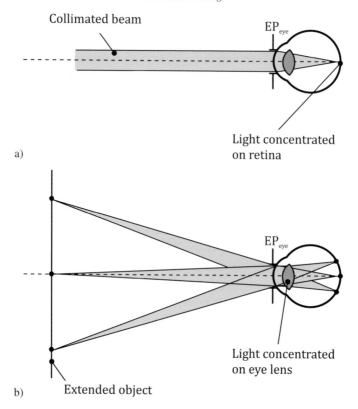

Figure 11.1 Effect of a collimated beam (a) and an extended object (b) on the concentration of light over eye structures. While collimated light is concentrated mainly upon the retina, the light coming from the points of an extended object passes through the lens, so each point of the lens receives light from all the object points, and the energy is concentrated within the eye lens.

this way retinal damage. Finally, to avoid accumulated UV damage to the retina, the eye lens becomes absorbent for blue wavelengths with age.

Despite the protective mechanisms described in the preceding paragraph, the eye is vulnerable to tissue damage due to radiation. We will describe now the main conditions/trauma associated with radiation damage to the eye. In this description, we will separate the effect of radiation on the three main structures of the eye: the cornea, the lens (and iris), and the retina.

Radiation Damage to the Cornea/Conjunctiva

Due to its position, the cornea/conjuctiva is the part of the eye that is more exposed to ambient radiation, especially for high altitudes due to less atmospheric absorption. Therefore, people who dwell in mountain places are more prone to develop chronic damage to the cornea. In winter, there is an additional risk factor because at high altitudes the earth is usually covered with snow or ice, and the amount of ambient radiation over the cornea is increased, which makes necessary the use of protective equipment for the eye. Another situation in which ambient radiation may be higher than usual is at summer in or near large water reservoirs (such as the sea, lakes, rivers, etc.) due again to light reflection at the water surface. In general, solar radiation only produces corneal damage by absorption of UV radiation, while infrared damage may be caused by artificial sources [132].

Corneal damage caused by solar ultraviolet light may be temporary or permanent. In the first case, we find the photoqueratitis that is caused by absorption of UV radiation by the corneal ephithelium. Photoqueratitis has the same symptomatology as conjunctivitis and usually disappears 24 hours after exposure [132]. Permanent corneal damage is caused by repeated exposure to low levels of UV radiation and it takes different forms. The most common are pinguecuale and pterygium. Pingueculae are small and yellowish elevations of the surface of the conjunctiva [132], while pterygium are a kind of translucent sheet that partially covers the corneal surface, usually in zones located close to the corneal limb due to the growing of vascular and convective tissue over the corneal surface. In some cases, pterygium may grow close enough to the corneal apex to interfere with the subject vision and it should be removed surgically. Accidental exposure to intense sources of infrared radiation (IR) (such as fire, flame, welding torch, etc.) may produce traumatic damage to the corneal tissues due to burning injuries.

Radiation Damage to the Eye Lens

As with the cornea/conjunctiva, the lens may suffer radiation damage when absorbing UV and infrared radiation, although infrared damage is mostly associated with accidental exposure to high radiance artificial sources. For young individuals, the eye lens is transparent to visible radiation; however, with aging the lens becomes more absorbent for blue-violet visible light, acquiring a characteristic yellow/brown tint. The accumulative effect of a lifelong exposure to UV radiation may favor a condition known as cataracts. When cataracts are present, the lens experiments a loss of transmittance accompanied by an increment in the scattered radiation. As a result, contrast sensitivity and, ultimately, visual acuity decrease. Cataracts usually worsen with time, and they may lead to a complete loss of vision. Therefore, when a highly developed cataract is present, the only solution is the surgical extirpation of the lens and its substitution with an intraocular lens (see Chapter 8). It is important to notice that UV radiation is not the only cause of cataracts. Other processes such as the use of certain drugs or some trauma may also lead to the development of this condition.

Radiation Damage to the Retina

Retinal tissues may suffer both thermal and photochemical damage when the image of a high power light source is formed upon the retina. In the case of the Sun or other visible light sources, the aversion reflex prevents the user from looking directly at an intense light source for enough time to cause damage. However, we must point out the existence of highly collimated light sources operating outside the visible range of the spectrum, which may cause retinal injuries without the user knowing. This is particularly important with laser radiation, due to the high degree of collimation and monochromaticity of this kind of radiation sources. The average person is not usually exposed to large amounts of laser energy, but given the numerous applications of lasers, it is fairly common to find those kind of devices at work. In those cases, the international standards establish the security precautions and the characteristics and specifications of the individual protective equipment that should be used when high amounts of laser energy are present.

As we have described the potential hazards that radiation may cause in the ocular tissues and structures, we will next discuss the ophthalmic filters used to protect the eye from potentially harmful radiation.

11.3 Filters for Ocular Protection

In this section, we will study the foremost optical characteristics of ophthalmic filters. These characteristics are highly regulated by standards such as the international standards ISO 8980-3 [390], ISO 14889 [20], or the American ANSI Z80.3. These standards classify the ophthalmic lens into different classes attending to the value of the visible transmittance, although other important aspects, such as chromatic reproduction, should also be considered. There are many ways to reduce the transmittance of ophthalmic optics, but nowadays the most employed are organic tints and photochromatic layers. Finally, to end this section, we will briefly discuss some aspects of prescription of ophthalmic filters, and we will describe occupational protective goggles such as laser protection eyeglasses.

11.3.1 Optical Characteristics of Ophthalmic Filters

Before discussing the regulations established by the standards, it is necessary to define some magnitudes of interest. In the last section, we defined transmittance as the ratio between transmitted and incident radiant fluxes. In general, the amount of radiant energy transmitted will depend on the wavelength. Therefore, we will define spectral transmittance as

$$T(\lambda) = \frac{\Phi_{wt}(\lambda)}{\Phi_{wo}(\lambda)}, \qquad (11.3)$$

where $\Phi_{wt}(\lambda)$ is the spectral flux transmitted for the wavelength λ and $\Phi_{wo}(\lambda)$ is the incident spectral flux for the same wavelength. Following standard ISO 8980-3, we will define the luminous (or visible) transmittance as

$$T_v = \frac{\int_{380}^{780} T(\lambda)V(\lambda)D_{65}(\lambda)d\lambda}{\int_{380}^{780} V(\lambda)D_{65}(\lambda)d\lambda}, \qquad (11.4)$$

with $D_{65}(\lambda)$ being the spectral distribution of CIE's D65 (illuminant D65 from the Commission Internationale de l'Eclairage) standard illuminant and $V(\lambda)$ the spectral sensitivity function of the human eye. Thus, in the context of the norm ISO 8980-3, luminous transmittance represents the transmittance (clarity) of the lens as perceived by an average human

Table 11.1 *Categories of ophthalmic filters according to international standard ISO 8980-3. Visible transmittances are expressed as percentages.*

Category	Range of T_V	Max $T_{SUV A}$	Max $T_{SUV B}$
0	[80%, 100%]	T_V	T_V
1	[43%, 80%]	T_V	$0.125 T_V$
2	[18%, 43%]	T_V	$0.125 T_V$
3	[8%, 18%]	$0.5 T_V$	$0.125 T_V$
4	[3%, 8%]	$0.5 T_V$	< 0.001

observer when looking at the sky. Notice that equation (11.4) represents the particular case when the incident spectral flux coincides with that of the illuminant D65, while in the general definition of luminous transmittance $D_{65}(\lambda)$ is substituted by the incident radiant flux $\Phi(\lambda)$. The values of the product $D_{65}(\lambda)V(\lambda)$ for the spectral range between 380 and 780 nm are tabulated in appendix A of norm ISO 8980-3 [390].

In equation (11.4), if we use different spectral regions we can define other integrated transmittances. For example, the standard ISO 8980-3 defines the solar UVA and UVB transmittances substituting the visible spectrum range [380, 780] nm for the corresponding UVA [315, 380] nm or UVB [280, 315] nm spectral ranges and replacing the standard illuminant $D_{65}(\lambda)$ and the spectral sensitivity function $V(\lambda)$ for the solar radiation function $E_{S\lambda}(\lambda)$ and the spectral effectiveness function $S(\lambda)$, respectively. Those functions are tabulated in appendix B of standard ISO 8980-3 [390] and are labeled $T_{SUV A}$ and $T_{SUV B}$ in Table 11.1.

Another important parameter is optical density, defined as

$$D = -\log_{10} T = \log_{10} \frac{1}{T}, \tag{11.5}$$

where T is the transmittance (in many practical cases the optical density is defined for the luminous transmitance). Optical density acts as the inverse of transmittance and represents the opacity of a given material. Due to its logarithmic nature, optical density is well suited to specify filters with very low transmittance, and it is a common specification for the protective eye goggles.

In Table 11.1 we can see the five categories in which the international standard ISO 8980-3 divides the ophthalmic filters. Thus, we will find lenses from category 0, which corresponds to lenses with visible transmittance greater than 80%, to category 4, which corresponds to highly opaque lenses. The standard also specifies the values of the UV transmittance in the UVA and UVB regions, labeled $T_{SUV A}$ and $T_{SUV B}$ in Table 11.1. This is extremely important because the reduction of visible transmittance should be accompanied by a corresponding reduction of UV transmittance or ocular damage may result: If the luminous transmittance is reduced, then the amount of visible light that passes through the lens diminishes. In order to compensate for this loss of light, the eye pupil diameter will be increased. But, if the pupil diameter is greater, then the amount of nonvisible radiation, particularly UV radiation, which enters in the eye will be greater. Thus, if UV radiation is not filtered, the effect of a visible-filter before the eye would increase the risk of eye injury due to UV radiation. It is necessary to keep in mind that the photochemical damage associated to UV light is accumulative, so a slight increment of the daily exposure could have long-term negative consequences for the eye. Therefore, in order to avoid this problem, ISO 8980-3 standard, and all the other standards that refer to ophthalmic filters, such as the American Z80.3, requests that the amount of UV radiation transmitted by the lens should not be greater than the luminous transmittance. For example, according to Table 11.1, a category 3 filter with luminous transmittance of $T_v = 14\%$ should have a solar UVA transmittance $T_{SUVA} = 7\%$ and a solar UVB transmittance $T_{SUVB} = 1.625\%$.

Another important optical property for ophthalmic filters is the color reproduction. An ophthalmic filter may alter the color perceived of a given object. In the first place, a filter usually presents a different transmittance for each wavelength. Indeed, some filters are colored, typically in a green or brown shadow. Also, even when a gray (sometimes called "neutral") filter is used, the color intensity is lowered and the color is perceived as "darker." The change of perceived color when looking through a filter may be quite relevant in some activities, such as driving, which require a flawless identification of objects (traffic signals or lights) in order to avoid accidents. Thus, all standards that apply to ophthalmic filters regulate the color reproduction properties of the filter for a filter to be suitable for driving. For example, the international standard ISO 14889 makes the following requirements for an ophthalmic lens for driving:

1. Spectral transmittance $T(\lambda)$ for any wavelength comprised between 500 and 600 nm should be greater than $0.2T_v$.
2. Luminous transmittance should be 8% or better in daylight conditions.
3. Luminous transmittance should be 75% or better in night conditions.
4. Relative visual attenuation coefficient, Q, should be greater than 0.8 for red, 0.8 for yellow, 0.6 for green, and 0.4 for blue lights.

The red, yellow, green, and blue lights are defined as the product of the spectral distribution of CIE's illuminant A, $S_A(\lambda)$, the spectral sensitivity function of the human eye $V(\lambda)$ and the corresponding spectral transmittance distribution T_{RED}, T_{YELLOW}, T_{GREEN}, and T_{BLUE} for each light. Those functions are tabulated in annex I of standard ISO 8980-3. With this, the relative visual attenuation coefficient for red light is defined as

$$Q_{RED} = \frac{\int_{380}^{780} T(\lambda)T_{RED}(\lambda)S_A(\lambda)V(\lambda)d\lambda}{T_v}, \tag{11.6}$$

with $T(\lambda)$ being the spectral transmittance of the filter and T_v its luminous transmittance for the D65 illuminant.

When a certain amount of radiation passes through a bulk material, absorption and/or dispersion may decrease the amount of radiant flux that is propagated. Let us now consider a collimated beam propagating along Z axis between two planes, 1 and 2, separated by

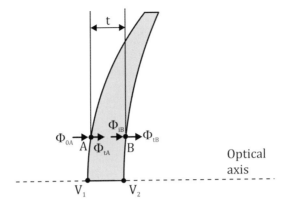

Figure 11.2 Transmittance of an ophthalmic lens. The incident flux Φ_{0A} arriving at point A exits the lens at point B after passing through a thickness t. The incident radiation is partially reflected and absorbed by the lens, resulting in the final transmitted flux $\Phi_{tB} = T\Phi_{0A}$.

distance z. If Φ_1 is the amount of radiant flux at the plane 1 and Φ_2 is the radiant flux transmitted to the second plane, then the Lambert-Beer law establishes that

$$\Phi_2 = \Phi_1 e^{-\alpha z}, \tag{11.7}$$

where α is the so-called coefficient of absorption. For transparent materials, α is close to zero and its values are greater for absorptive materials. The coefficient of absorption is also wavelength dependent, which reflects the fact that many materials present selective spectral absorption.

We can now compute the spectral transmittance of an ophthalmic lens manufactured in a material with refractive index $n(\lambda)$ and coefficient of absorption $\alpha(\lambda)$. In the following discussion, we will drop the wavelength dependence for the sake of clarity. As we can see in Figure 11.2, where we have represented a section of an ophthalmic lens, an incident radiant flux Φ_{0A} arrives at point A of the first surface of the lens. As predicted by the Fresnel equations, due to the difference of index between the surrounding media, n_0, and the lens material, n, part of the incident light is transmitted and part is reflected. If the incident beam incides parallel to the normal of the surface at point A, then we see that Fresnel equations can be written as

$$R = \frac{(n - n_o)^2}{(n + n_o)^2}, \quad T = \frac{4nn_0}{(n + n_0)^2}. \tag{11.8}$$

If the incident medium is air ($n_0 = 1$), then we see that the light transmitted into the lens at point A is

$$\Phi_{tA} = \frac{4n}{(n + 1)^2}\Phi_{0A}. \tag{11.9}$$

This flux is propagated through the lens until it arrives at point B. In this propagation, part of the light is absorbed, so, according to Lambert-Beer's law, the flux incident at point B is

$$\Phi_{iB} = \Phi_{tA}e^{-\alpha t} = \frac{4n}{(n+1)^2}e^{-\alpha t}\Phi_{0A}, \tag{11.10}$$

where t is the lens thickness between points A and B. Finally, applying again Fresnel's equations at point B, we find that

$$\Phi_{tB} = \frac{16n^2}{(n+1)^4}e^{-\alpha t}\Phi_{0A}. \tag{11.11}$$

Thus, the general formula for the transmittance of a lens in air is

$$T = \frac{\Phi_{tB}}{\Phi_{0A}} = \frac{16n^2}{(n+1)^4}e^{-\alpha t}. \tag{11.12}$$

Example 46 An ophthalmic lens is manufactured in a material with refractive index $n = 1.6$ and absorption coefficient $\alpha = 0.35$ mm^{-1} for the wavelength $\lambda = 555$ nm. If the lens power is $P = +3.50$ D, the lens diameter $D = 65$ mm, and the center thickness is $t_c = 4.5$ mm, compute the spectral transmittance for 555 nm at the center and at the edge.

   Let us compute first the edge thickness using equation (4.37)

$$t_e = t_c - \frac{PD^2}{8(n-1)} = 4.5 - \frac{0.0035 \times 65^2}{8 \times (1.6-1)} = 1.42 \text{ mm.}$$

We can now calculate the requested transmittances using equation (11.12)

$$T_c = \frac{16 \times 1.6^2}{2.6^4} \exp(-0.35 \times 4.5) = \textbf{0.185 (18.5\%)},$$

$$T_e = \frac{16 \times 1.6^2}{2.6^4} \exp(-0.35 \times 1.42) = \textbf{0.545 (54.5\%)}.$$

Note the considerable variation of the transmittance between the edge and the center of the lens.

11.3.2 Types of Ophthalmic Filters

In this section, we are going to discuss the main types of ophthalmic filters. We will begin with solid tinted filters, generally made of glass. Then, we will study immersion tinted lenses, which are one of the most popular types of filter that we can find nowadays. For example, immersion tinting is very well suited for producing solar protection lenses made of plastic. The third category, photochromatic filters, is quite important in modern ophthalmic optics, and it constitutes one of the more active fields of research in the ophthalmic industry. Finally, we will briefly study polarizing filters and low transmission coatings.

Solid Tinted Lenses

As is well known, adding certain metallic oxides, such as Fe_2O_3 and CuO, to molten glass will produce a glass with selective wavelength absorption. Using different concentration of oxides, we can change the absorption coefficient and set the final transmittance. Mass tinting is the preferred method for producing low cost filters in optics manufacturing. However, its usage in the ophthalmic industry is rather limited, as the resulting lens transmittance will depend on lens thickness. Moreover, these filters require huge stocks of semifinished blanks with different optical densities, which is both impractical and quite expensive.

Tinted Lenses

Nowadays, lens tinting is the most common method for producing plastic ophthalmic filters. This technique consists of dipping the finished plastic lens in a hot bath containing organic molecules. These organic molecules will act as "color centers" that absorb the incident light in the same way as metallic oxides do. The tint molecules will penetrate the surface by diffusion due to the temperature and the concentration gradient between the liquid and the lens. The physics of this diffusion process prevents the tint molecules from traveling long distances within the lens. Therefore, instead of a uniform distribution of color centers over the whole volume of the lens, as in the case of solid tinted lenses, we will have a thin absorbing layer located just beneath the lens surface (see Figure 11.3). Therefore, the transmittance of a tinted lens is $T = T_l^2$, where T_l is the transmittance of this absorbing layer. This layer transmittance term includes absorption by the tinted layer and light reflection at the air-lens surface and the tinted layer-clear lens one. Light reflection between tinted layer and the lens is relevant only when the tint changes significantly the refractive index of the material. Thus, the transmittance does not depend on lens thickness, contrary to what happened with mass tinted lenses.

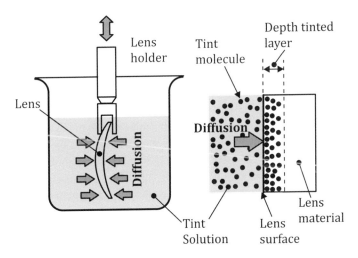

Figure 11.3 Manufacturing process of a tinted lens. The lens is immersed in a solution containing the tint molecules, which are diffused into the lens surface. The penetration of the tint molecules is limited to a thin layer beneath the lens surface, as can be seen in the drawing on the right.

The resulting spectral transmittance of a tinted lens is due to several factors, such as the concentration of tint molecules, the bath temperature, and, foremost, the immersion time. The longer the lens is immersed in the bath, the greater the concentration of tint molecules and the lower the transmittance. This property allows the manufacturing of gradient tinted lenses by lifting the lens slowly from the bath, so that the lower part of the lens, which has been submersed for longer, becomes darker than the upper one. Tinted lenses have uniform transmittance and the manufacture of these lenses does not require the stocking of tinted semifinished lenses (as with mass tinted lenses) because the coloration is applied after lens manufacturing. The main disadvantage of tinted lenses is the stability of the transmittance throughout the lifetime of the lens, because with the wear of continuous use, the lens tends to lose optical density (fading).

Photochromic Lenses

Certain chemical substances, particularly silver halides, react with UV radiation in such a way that they absorb visible light when exposed to UV radiation. This photochemical reaction is, in most cases, reversible. When the source of UV radiation is removed, the material fades. Photochromic materials are those materials that become darker in the presence of UV radiation and transparent when no UV radiation is present. Therefore, photochromic materials present variable (visible) transmittance with two stationary states: a dark (or activated) state with a typical transmittance of about 15%, which is reached when the material is exposed for a certain time to UV radiation; and a clear (or fade) state (typical transmittance of 80%), which happens when UV radiation is absent for a given amount of time. Ophthalmic lenses made of photochromic material are quite convenient for their users, as they become darker under the sun and lighter indoors. Therefore, when the user is outdoor, the photocromic lens will protect the eye against an excess of solar radiation. Moreover, a photochromic lens will provide good transmittance inside a building, where the lens fades because of the lack of solar UV radiation and because the artificial sources used in lighting are poor UV emitters. In Figure 11.4, we show a semifinished photochromic lens whose left part has been exposed to solar radiation, while the right one has been covered with a sheet of paper. As we can see in this figure, the lens part exposed to UV radiation is considerably darker than the other part. The spectral transmittance depicted in Figure 11.5 shows a considerable diminution of the visible transmittance in the dark state of the lens. It also shows the good filtering properties of photochromic lenses in the UV region either in the activated or deactivated state. As we can see in Figure 11.5, a photochromic lens filters out almost all radiation below 400 nm. Therefore, photochromic lenses are a good solution for providing UV protection. Indeed, Corning introduced its family of photochromic lenses in 1981, known as CPF, which filter UV and blue visible radiation. Those lenses were intended for subjects suffering from a range of conditions, including macular degeneration, glaucoma, and retinitis pigmentosa, among others. More details about CPF lenses can be found in [132].

Despite the above-mentioned advantages, there are also some drawbacks. As mentioned previously, the activation mechanism depends on the amount of UV radiation received,

Figure 11.4 Photograph of a photochromic semifinished lens. The left side has been exposed to UV radiation, and thus it is darker than the unexposed right side.

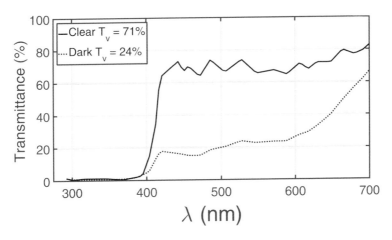

Figure 11.5 Spectral transmittance (expressed as percentage) of the lens depicted in Figure 11.4 for both activated (dark) and deactivated (clear) states. Notice the very low transmittance for the UV region (below 400 nm).

which typically comes from sunlight. If the sky is clear, the presence of UV radiation makes the lens darker, reducing the luminosity as intended. However, it may be that on a cloudy day a considerable amount of UV radiation will reach the earth's surface, while visible radiation is filtered out by clouds. On these days, the photochromic lens may reduce the brightness of the objects perceived by the user, although overall luminosity is not high.

Another problem is related to the difference between activation and deactivation times. Photochromic activation is considerably faster than fading. A typical photochromic lens may pass from fade (80% transmittance) to fully dark (15% transmittance) states in about 15 minutes, but it is noticeably darker in less than one minute. Conversely, fading takes longer. For example, a monochromatic lens with transmittance 85/15 (maximum transmittance 85%, minimum transmittance 15%) starting in the dark state needs around 10 minutes to get 70% of the transmittance and another 15/20 minutes to reach the maximum transmittance. Indeed, one of the most active areas in optical materials research is the production of photochromic materials with low fading times [391]. Another factor that influences the final transmittance of a photochromic lens is temperature. In general, high temperatures make the photochromatic reaction less efficient [132]. As a consequence, photochromic lenses exposed to UV radiation become darker in winter. This may be an inconvenience if the sky is overcast because, as mentioned previously, the lens is activated by UV sunlight, while the amount of visible light is low. The first photochromic lenses were made of glass with an addition of silver halides [132]. One may think that, given the uniform distribution of silver halides within the lens volume, glass photochromatic lenses would present a variation of transmittance between the lens center and the edge, like solid tinted lenses. However, this is not the case, because activation depends on the presence of UV radiation, and this radiation is absorbed by the outermost layers of the lens when they activate. Thus, the only part of the lens that becomes darker is a thin layer beneath the anterior surface of the lens. As a result, the transmittance variation along the lens surface is small. The first photochromic lens was Bestlite from Corning, Inc., which was first manufactured in 1965. Further developments by Corning were lenses with greater transmittance variation such as Photogray (1968), Photogray Extra (1978), and Photobrown Extra (1981), or faster lenses such as Photogray II introduced in 1984. Corning also developed high index photochromic lenses. Other manufacturers of photochromic lenses made in glass were Chance-Pilkington Ldt (Reactolite) in Britain and Deutsche Spezialglass AG in Germany [132].

The development of photochromic lenses in plastic materials was considerably more difficult. The first attempt to develop a photochromic plastic lens was made by Americal Optical Corp. in 1983 with the Photolite. Photochromaticity was achieved by embedding silver halides within the polymeric matrix. However, there were problems with the stability and kinetics of the photochromic cycle. An important research effort was carried out with the aim of producing polymer-compatible photochromic compounds. As a result, Pittsburgh Plate Company (PPG) presented in 1989 a new lens named Transitions Comfort. This lens introduced a new photochromic technology, compatible with plastic lenses, based on organic compounds such as spiropyrans or spirooxacines [392]. A year later, PPG and Essilor International created a joint venture company named Transitions Optical Inc. for the development and production of plastic photochromatic lenses [393]. As with Corning's glass photochromatic lenses, Transitions has launched several generations from Transitions One in 1991 to the current Transitions Signature and Transitions Vantage. Also, a number of specialized lenses for tasks such as driving or outdoor sports have been produced [393]. Other companies that share the plastic photochromatic market at the beginning of the twenty-first century are Mitsui Chemical (Sun Sensors, originally from Corning and acquired by Mitsui in 2014), Rodenstock (ColorMatic), Signet Armorlite (PhotoViews), VisionEase (LifeRx), and Hoya (Suntech). It is interesting that most of the photochromic technologies available – with exceptions like Sun Sensors, which are manufactured en masse – are applied to clear lenses as a thin photochromic layer. This method has several advantages. First, there is no need to stock special semi finished lenses of photochromatic material. Also, high index photochromic lenses may be produced (if the photochromic material is compatible with high index resins), and finally the transmittance is uniform along the lens surface. However, there are also some drawbacks, especially lower durability. Dip-coating and embedding (Transitions), spin-coating (Suntech), or film (LifeRx) are the current technologies employed to produce thin photochromic layers.

High Reflective Coatings

In the following sections, we will see in detail the properties of anti-reflective coatings. However, we may say now that such coatings are formed by several thin layers of different materials. If the arrangement of materials (refractive index) and thickness are correct, the different reflected beams interfere destructively. The final result is a considerable reduction of the light reflected by the lens surface, with the consequent increment of the overall transmittance of the lens. By changing the arrangement of layers, we can achieve just the opposite effect. In this case, instead of having low reflectance, we will obtain highly reflective coatings. As $R + T = 1$, an increment of reflectance implies a decrement of transmittance. Therefore, we can produce a filter by deposing a highly reflective coating over a given lens surface. Usually, only the front surface of the lens is coated. In this way, we avoid the appearance of ghost images caused by reflections at the back surface of the lens. Obviously, as the coating is uniformly deposed over the lens surface, transmittance is also constant along the whole surface of the lens.

As we will see later, the reflectance of an anti-reflective multilayer (or highly reflective, as in the present case) is highly dependent on wavelength. This will produce the so-called residual colors. Residual colors are the highly colored images of intense objects (usually primary sources such as lamps) that are formed after reflection over the lens surfaces. Due to the relation between transmittance and reflectance, the color transmitted by the lens is the complementary of the color reflected. For example, if we have a white source and the residual color image from a high-reflectance filter is green, the source will be perceived as purple when looking at the source through the lens.

An alternative to highly reflective multilayer coatings is the deposit of a thin layer of a single metallic oxide. The resulting lenses have uniform transmittance across the lens and form a specular image of the scene in front of the lens. The reflective layer is usually deposited over the back surface of the lens with an anti-reflective layer superposed in order to reduce the amount of reflected light. The main drawbacks of high reflective coatings is their durability, as they may degrade during the lifetime of the lens for different reasons. This problem is not presented by metallic oxides coatings, which are, usually, more resistant to wear.

Polarizing Filters

Natural sunlight is not polarized. This means that the direction of the electric and magnetic vectors, over a plane transverse to the light propagation direction, is random. Certain devices may change the state of polarization of light. For example, a linear polarizer transforms natural light into linear polarized light: After passing through the device, the tip of the electric (and magnetic) vector follows a linear path over the transverse plane. This is possible because a linear polarizer presents two privileged directions (or axes): the transmission and the extinction axis. These axes are perpendicular between them and they form the transverse plane. Therefore, when light passes through a linear polarizer, its electric field vector can be decomposed into two perpendicular components parallel and perpendicular to the principal axis of the polarizer. The component parallel to the extinction axis is absorbed, while the other is transmitted. As a result, a linear polarizer transforms the incident light into linearly polarized light, regardless of its initial polarization state. This means that, unless the incident beam is linearly polarized, with the electric vector parallel to the transmission axis, part of the light is lost at the exit of a linear polarizer. If the incident light is depolarized, such as sunlight, then the transmittance of the device will be $T = 0.5$, and half of the incident light will be absorbed.

In nature, we can find partially polarized light after reflection on water, snow, or ice [389]. In this case, the direction of the electric vector is mostly parallel to the reflecting surface. If sunlight, reflected on a smooth horizontal surface, passes through a vertical linear polarizer an almost complete extinction will result. This is quite useful in certain activities such as driving or fishing, where the usage of a vertically polarized filter helps in reducing glare and improving the contrast perceived by the user. Therefore, polarized sunglasses are a good option for outdoor use and they are quite popular among certain professionals for the above-mentioned advantages.

The polarizers employed by the ophthalmic industry are, usually, thin films that can be adhered over the lens surfaces. Typically, as with highly reflective coatings, the polarized film is only attached to the front surface of the lens. Polarizer films can be combined with other treatments. For example, by combining a polarizing film and one or more photochromatic layers, a versatile lens with variable transmittance results, as the lens Transitions Drivewear from Younger Optics. Those combined photochromic-polarizer lenses can be used in many applications. For example, when driving, a conventional photochromatic lens is barely useful, as it will not activate behind car windscreens that are UV filters. However, the combined photochromatic-polarizer lens can be used for driving due to the polarizer, and also for outdoor usage, when the photochromic activation is added to the polarized light filtering.

11.3.3 Prescription of Ophthalmic Filters

Types of Ophthalmic Filters

Ophthalmic filters have different applications in ophthalmology and optometry. According to Fannin et al. [132], we find the following types of filters:

1. Therapeutic filters used in the treatment of several conditions such as aphakia, macular degeneration, etc.
2. Sunglasses, or filters intended for protecting the eye against the excess of sunlight radiation in outdoor activities.
3. Lenses with selective absorption of visible radiation.
4. Lenses with UV filters.
5. Photochromic lenses.

A therapeutic filter is intended for use in the treatment of several conditions. For example, a filter with relatively high visible transmittance (around 80%) may be useful in the management of photo phobia at indoor locations. Other typical examples are blue-filtering lenses, such as Corning's CPF family, used by patients affected by macular degeneration, retinitis pigmentosa, cataracts, or diabetic retinopathy. These filters reduce the amount of potentially harmful short wavelength radiation, and they are helpful in preventing further worsening of those conditions. There are also additional benefits of these filters such as the improvement of visual acuity and contrast sensitivity. However, this enhancement of visual acuity and contrast sensitivity might not happen for all patients [394], so the practitioner must be aware of this.

Sunglasses are general purpose filters devised for outdoor use. As a rule of thumb, visible transmittance of a sunglass should always be lower than 67% [132]. However, the actual value of this parameter depends on its intended use. For example, on clear winter days, when snow or ice covers the ground, the amount of ambient light received by the eye may be quite high. In this situation, a ISO 8980-3 type 4 filter with transmittance lower than 8% should be necessary to avoid ocular discomfort. As mentioned previously, industrial standards require that sunglasses must filter out UV radiation to avoid potential damage to the eye. In many countries, current legislation regards sunglasses as ocular protection devices. Therefore, sunglasses must fulfill the same requirements (such as impact resistance, etc.) as any other ocular protective equipment. See Chapter 1 for further details.

Another kind of filter are those lenses with selective absorption of the visible spectrum. For persons without ocular pathologies, these filters may improve visual comfort, as they filter out the blue/green light that is more efficiently scattered within the eye. However, this subjective sensation strongly depends on the individual. As yellow filters reduce the amount of blue light reaching the retina, they could be helpful in preventing diseases such as macular degeneration. Also, it is believed that yellow filters are good for relieving ocular fatigue associated with an excess of ambient blue light [395]. However, it should be noted that blue light does also have benefits for human health such as sleep regulation, so blue light filtering is still subject to research among the optometric community.

Filtering out UV radiation reduces the potential risk of ocular injuries, which is of paramount importance for people dwelling at locations exposed to high UV levels, such as coasts or mountains. Also, we must realize that most plastic materials do not block UV radiation completely. For example, in Chapter 1, we saw that the UV cutoff wavelength for CR-39 is $\lambda_{UV} = 350nm$, so part of the UVA portion of the spectrum passes through lenses made in CR-39. To avoid this problem, many manufacturers of plastic lenses employ UV filters. There are two ways of applying UV filters to plastic lenses. The first consists of the addition of UV absorbers to the monomer before polymerization, so the cutoff wavelength rises to 400 nm. It is also possible to filter out UV radiation with a properly designed anti-reflective coating. In this case, the coating acts as a wavelength selective filter that lets pass the visible portion of the spectrum but reflects UV radiation.

Photochromic lenses protect the eye from excessive ambient visible light when the user is outdoors. They are also very good UV filters, as, in the activated state they absorb this radiation to become darker. However, the optometrist must be aware that photochromic lenses are not a substitute for sunglasses because they do not activate indoors, as this is prevented by the absorption of UV radiation by the glass panes of the windows, or windshields in the case of vehicles.

Prescribing Sunglasses

Most of all, sunglasses must protect the eye against glare and other discomforts caused by an excess of visible light. Outdoors, the amount of ambient light depends mainly on geographical and meteorological factors. Ambient luminance is quite variable; it can reach values higher than $15,000$ cd/m^2 in snow mountains on clear winter days [132], while the comfort level of ambient luminance ranges between 250 and $2,000$ cd/m^2. In Table 11.2, we present the ranges of maximum and minimum luminances for the different categories of filters as defined in the standard ISO 8980-3. We have added a column indicating the typical usage of each category. We must carefully interpret this column. For example, we have indicated that the typical usage for category 0 lenses is "indoor." However, it is obvious that these lenses can also be worn outdoors, provided the ambient conditions are not extreme, as the eye is adapted to a wide range of ambient luminances. As we can see in Table 11.2, the filters belonging to category 3 and 4 are suited for very particular situations. In the case of category 3, they are indicated for the practice of aquatic sports, which are usually practiced in summer when the amount of ambient luminance is quite high, and it is augmented by reflection of light on the water surface. Type 4 are intended for outdoor mountain activities such as skiing or climbing, particularly at high altitudes and in winter.

Glare protection is not the only requirement of sunglasses. They must also block UV radiation, particularly when the transmittance of the filter is low. Indeed, it is very important

Table 11.2 *Comfortable luminance levels for the different categories of filters described in the international standard ISO 8980-3 and typical usage for the different categories.*

Category	Range of T_v	L_{min}(cd/m^2)	L_{max}(cd/m^2)	Typ. Usage
0	[80%, 100%]	$250 - 315$	$2000 - 2500$	indoor
1	[43%, 80%]	$345 - 580$	$2500 - 4650$	urban outdoor
2	[18%, 43%]	$580 - 1390$	$4650 - 11100$	rural outdoor
3	[8%, 18%]	$1390 - 3125$	$11100 - 25000$	aquatic sports
4	[3%, 8%]	$3125 - 8330$	$25000 - 66600$	winter sports

to raise awareness in the population of the need of acquiring only sunglasses manufactured according to the requirements of the relevant standards, such as ISO 8980-3 or ANSI Z80.3. The use of a visible filter, which is transparent to UV radiation, increases the risk of ocular injuries and conditions caused by excess of UV radiation. Moreover, as sunglasses are considered an eye protection device, a lens manufactured according to the standards provides the user with protection against mechanical impacts and other potential risks for the eye.

Therefore, in the process of sunglass prescription, the optometrist must first know the activities the user intends to participate in, particularly outdoors activities such as trekking, climbing, snowboarding, etc. Afterward, the practitioner should inform the patient which range of transmittance she considers optimum for the intended use and environment. She may also guide the user in selecting the color of the filter, and she must inform the patient of the existing regulation regarding the use of sunglasses when driving vehicles both in terms of transmittance and color reproduction.

11.3.4 Occupational Eye Protective Devices

Some professional activities involve working with or in the presence of highly radiant sources. As with other work-related risk, when the subject's eyes are exposed to high amounts of radiation, the normative demands the use of ocular protection goggles. From the many examples of these protective eyeglasses, we are going to present two of them: welding lenses and laser protection glasses.

Welding Lenses

Welding is an activity that requires heating metals beyond the point of fusion. Several techniques are employed to do so, and some of them produce a high temperature flame, while others use high voltage electric sparks to weld the metal. Whatever the technique employed, a high temperature plasma is produced in the welding process. This plasma emits high amounts of energy, which could be dangerous for the eyes of the welding operator. It is important to note that the hot metal at the welding junction is also a potent emitter of energy due to its high temperature. Therefore, in the process of welding, the eye is exposed to high amounts of radiation in the visible, UV, and infrared bands.

In order to protect the eye from this radiation, welding glasses have a very low transmittance, with values as low as $T = 0.00027\%$ [132], which is equivalent to an optical density of $D = 5.6$. Indeed, welding lenses are classified according to the shade number N_s, which is related to optical density by

$$N_s = \frac{7}{3}D + 1.$$

The shade number lies between $N_s = 1$ and $N_s = 14$. Selection of the proper shade number depends on multiple factors, such as the type and power of welding machine, whether the user is the welding operator, or whether she works close to a welding station.

Finally, regarding the spectral transmittance of welding lenses, the normative requires total protection against nonvisible radiation, so these lenses are, effectively, band pass filters that reject radiation below 420 nm and above 780 nm.

Laser Protection Goggles

Laser radiation is used in many applications such as quality control, machining, or surgery. A laser beam is highly collimated, so when it enters the eye, it is focused on the retina. Due to the high power of some industrial lasers, damage to the eye may result even for small exposures. The problem is compounded by the fact that many lasers operate in the UV or infrared, and they are highly monochromatic, so they cannot be detected by the human eye. This means that if there is a risk of ocular damage due to laser radiation, protective equipment should be worn according to the relevant standards (EN207, EN208 in Europe and ANSI Z136.1 in the United States).

Lasers may operate in several modes: continuous emission, pulsed, Q-switched, etc. For certain wavelengths and pulse frequencies, a pulsed laser could be more dangerous than a continuous one. Therefore, the European standard EN208 establishes several laser categories: D is a continuous laser, I a pulsed one, Q a Q-switched pulsed laser, and M for mode-locking pulsed lasers. The norm also fixes the MPE level depending on wavelength and type of laser. For example, according to EN207 standard, for the range of wavelengths between 350 and 1400 nm, the MPE for a D laser is 10 W/m^2 for emission durations higher than 5×10^{-4} s. The level of protection required for a given exposure is computed as $L = \log_{10}(H/\text{MPE})$ where H is the exposure. For example, a continuous laser operating in the wavelength of 780 nm with a protection level $L = 1.5$ protects the eye against a maximum irradiance of 315 W/m^2 according to the standard EN207.

Usually filters absorb radiation, so they can be damaged if the amount of incident radiation is greater than the damage threshold of the filter. A damaged laser protection lens

Figure 11.6 Photograph showing a detail of a laser protection eyeglass. The marks over the lens indicate the optical density for different wavelength intervals. Notice the lateral screen for protecting the eye against glazing beams.

is no longer useful as a protective device and should be discarded. The European standard EN208 requires the manufacturer to mark both the maximum power and maximum energy that could withstand the lens. The second is important, as it indicates the useful lifetime of the laser protection goggle. EN208 also requires marking of the wavelength (or wavelength interval) for which the lens is intended to protect, the laser type marked as D, I, Q, or M, the letter R followed by the level of protection (L) indicated as a number, the name of the manufacturer, and a letter indicating ambient resistance against scratches, impacts, etc. As can be seen in Figure 11.6, the mounts of laser protection glasses present, usually, wide sides and screens to protect the eye against glazed beams that could impinge laterally on the eye.

11.4 Anti-Reflective Coatings

11.4.1 Ghost Images

As we have seen in the previous section, even for completely transparent lenses, a certain amount of light is reflected at the lens surfaces. Given the high grade of smoothness of the surfaces of ophthalmic lenses, this reflection is mostly specular. Therefore, if we consider a spherical surface of radius r and a bright object located at a distance s of the surface, then the location of the reflected image s' is given by Abbe's invariant for reflection

$$\frac{1}{s'} + \frac{1}{s} = \frac{2}{r}.$$

The amount of light reflected by the surface of a lens is quite low. According to Fresnel's equations, the reflectance of the surface of a lens in air is

$$R = \left(\frac{n-1}{n+1}\right)^2,$$

where n is the refractive index of the lens. For example, the surface of a lens made of a material with $n = 1.5$ presents a reflectance $R = 0.04$, so less than 5% of the light is reflected. This means that lens reflections would only be noticeable for objects that present high luminance. Those high luminance objects are usually primary light sources such as the Sun or lamps.

In the context of ophthalmic optics, we will define a ghost image as the image formed when the light coming from an object is reflected by, at least, one of the surfaces of the ophthalmic lens. In Figure 11.4, we can see a variety of images produced by the reflection of light coming from a window and ceiling luminaries. Notice the differences in brightness and size of these images. Ghost images may be appreciated by the user of the ophthalmic lens or by surrounding people. In the latter case, there is just an aesthetic problem. If the user perceives ghost images superposed to the field of view, they may interfere with normal vision. This will happen if the following conditions are met:

1. The ghost image is formed between the remote and the near points.
2. The intensity of the ghost image is high enough so it can be distinctively perceived.
3. The lateral position of the ghost image is not very far from the sight direction.

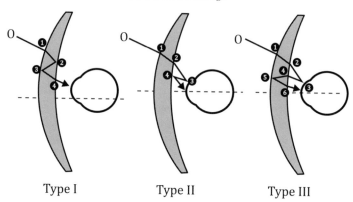

Type I   Type II   Type III

Figure 11.7 Ghost images of type I, II, and III, formed by the ophthalmic lens-eye system when the object is located in front of the lens. The numbers indicate the different refractions or reflections experienced by the ray coming from the object.

Therefore, the most important characteristics of ghost images are their location (both axial and lateral) and intensity. If we consider the lens-eye system, there are only five situations in which a ghost image may be perceived by the lens user. Of these situations, the object that forms the ghost image is located in front of the lens in three cases (see Figure 11.7). In the other two cases, the object is located behind the subject's eye, as shown in Figure 11.8. In this book we will follow the classification scheme of Jalie [56], so we will consider the following categories of ghost images.

- *Type I.* As we show in Figure 11.7, this image is formed when light coming from the object passes through the front surface of the lens; then it is reflected on the back surface of the lens, and again on the front one. The light then bounces back and, after refraction at the back surface, it arrives at the eye.
- *Type II.* As can be seen in Figure 11.7, in this case, the light coming from the object passes through the lens, and then it is reflected at the subject's cornea, so it is directed now toward the lens. When the light arrives at the back surface of the lens, it is reflected again and, finally, it enters the eye.
- *Type III.* This case is similar to the former, but the light reflected by the cornea is reflected on the front surface of the lens.
- *Type IV.* In this case, the object is located behind the eye, and the light is reflected on the back surface of the lens.
- *Type V.* In a quite similar way as with type III, the light coming from an object located after the lens is reflected on the front surface of the lens and directed toward the eye, as we can see in Figure 11.8.

For thick lenses, computation of the location of ghost images is quite cumbersome. However, if the thin lens approximation holds, this calculus can be greatly simplified, as we will see in the following example.

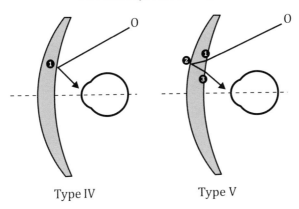

Type IV   Type V

Figure 11.8 Ghost images of type IV and V, formed by an ophthalmic lens when the source is located after the lens. The numbers indicate the different refractions or reflections experienced by the ray coming from the object.

Example 47 Compute the axial location of the ghost image of type I formed by a thin ophthalmic lens of total power P, base P_1, and refractive index n if the object vergence is L. Consider the lens as a thin one.

We will compute first the curvatures of the front and back surface of the lens as

$$K_1 = P_1/(n-1), \quad K_2 = (P_1 - P)/(n-1).$$

Afterward, we will follow the ray path corresponding to the type I image, as depicted in Figure 11.7. First, we have a refraction at the front surface, so that $nL'_1 = L + P_1$. As the lens is thin, we see that $L_2 = L'_1$, so the vergence after reflection on the back surface of the lens is

$$L'_2 = 2K_2 - L_2 = \frac{2(P_1 - P)}{n-1} - \frac{P_1}{n} - \frac{L}{n}.$$

The light is then reflected on the front surface of the lens. Thanks again to the thin lens approximation, we see that $L_3 = L'_2$, and the image vergence

$$L'_3 = \frac{2P_1}{n-1} - L_3 = \frac{2P}{n-1} + \frac{P_1}{n} + \frac{L}{n}.$$

Finally, the light is refracted at the back surface of the lens, so, considering that $L_4 = L'_3$, we have

$$L'_4 = P_2 + nL_4 = P_1 + P_2 + \frac{2nP}{n-1} + L = \frac{(3n-1)P}{n-1} + L.$$

Thus, the image vergence of the type I ghost formed by a thin lens is

$$L'_I = \frac{3n-1}{n-1}P + L.$$

Table 11.3 *Reflectance for the different ghost images for the general case and for a lens in air. T_1 and T_2 are the transmittances of the front and back surfaces of the lens, R_1 and R_2, are the reflectances of the front and back surfaces of the lens, R_c is the corneal reflectance, n is the refractive index of the lens, and n_c is the refractive index of the cornea.*

Reflectance	Type I	Type II	Type III
General	$R_I = R_1 R_2 T_1 T_2$	$R_{II} = T_1 R_c R_2$	$R_{III} = T_1^2 R_c T_2 R_1$
Lens in air	$R_I = \frac{16n^2(n-1)^2}{(n+1)^8}$	$R_{II} = \frac{4n(n-1)^2(n_c-1)^2}{(n+1)^4(n_c+1)^2}$	$R_{III} = \frac{64n^2(n-1)^2(n_c-1)^2}{(n+1)^8(n_c+1)^2}$

$$\cdots$$

Reflectance	Type IV	Type V
General	$R_{IV} = R_2$	$R_V = R_1 T_2^2$
Lens in air	$R_{IV} = \frac{(n-1)^2}{(n+1)^2}$	$R_V = \frac{16n^2(n-1)^2}{(n+1)^6}$

Note that this expression does not depend on the shape of the ophthalmic lens due to the thin lens approximation.

The axial locations of the other types of ghost images can be found in [56]. Regarding the intensities, we can compute a reflectance associated with each type of ghost image in a similar way as we did when we computed the transmittance of an ophthalmic lens. In Table 11.3, we present the values of the reflectances for the different types of ghost images for both the general case and the particular case of a lens in air.

Some of the solutions for dealing with ghost images are either changing the fabrication base or the pantoscopic angle of the mount [132]. However, using anti-reflective coatings is more effective. For example, considering an organic lens made of a material with refraction index $n = 1.6$, the value of the reflectance for a ghost image of type I is, according to the formula of Table 11.3, $R_I = 0.0076$. If both surfaces of the lens are covered with an anti-reflective coating with surface reflectances $R_1 = R_2 = 0.001(0.1\%)$ then $T_1 = T_2 = 0.999$ and the ghost image reflectance is $R_I = (0.001)^2 (0.999)^2 = 9.99 \cdot 10^{-7}$. Thus, the amount of reflected light has been lowered by three orders of magnitude.

11.4.2 Thin Film Anti-Reflective Coatings

A thin film coating is a structure formed when one or more materials, typically metallic oxides, are deposited over an optical surface as a succession of thin layers. By a proper selection of the thickness and refractive index (material) of each layer, it is possible to change the reflectance and transmittance of the surface through an interference process (see Appendix F). Thin film coatings are used in optics for manufacturing interferometric filters, high quality mirrors, and, finally, anti-reflective coatings. In this latter case, the

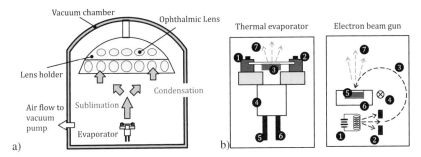

Figure 11.9 a) Drawing of a vacuum deposition chamber, where we can see its main elements: evaporator, lens holder, and vacuum pump. b) Detail of a thermal evaporator and an electron beam gun and its parts. Thermal evaporator key: 1) Anode, 2) Cathode, 3) Boat and material, 4) Holder, 5) Electric wire, 6) Water pipe, and 7) Evaporating material. Electron beam gun key: 1) Filament, 2) Collimation slit, 3) Trajectory of the electron beam, 4) Magnetic field (pointing toward the paper), 5) Material, 6) Holder, and 7) Evaporating material.

objective is to reduce the surface reflectance beyond the limit given by Fresnel's equations. In practice, if the reflectance of the surface without coating is about 4%, it is possible to achieve reflectances as low as 0.01% even with quite simple anti-reflective coatings.

Thin film anti-reflective coatings are manufactured in a process known as vacuum deposition. In Figure 11.9(a), we show a drawing of a vacuum chamber. This is a pressure chamber that contains a rotating bell-shaped holder, which holds several ophthalmic lenses for coating. The other element of the drawing is the evaporator. In Figure 11.9(b), we have depicted the scheme of the two most common types of evaporators: the thermal evaporator and the electron beam gun. The deposition process starts with creating the necessary degree of vacuum. This usually requires the consecutive action of two pumps: first, a standard rotary or centrifugal pump diminishes the pressure within the chamber to a level in which the second pump, usually a turbomolecular one, can operate to achieve the requested operating vacuum of the machine. Once the vacuum is made, the evaporation process begins. There are many ways of achieving this evaporation, ranging from simple heating to more sophisticated techniques such as bombardment with an electron beam.

In thermal evaporators, the material is placed in a special holder, known as a boat, which is located between two electrodes – see Figure 11.9(b). When a high current passes through the boat, the temperature of the material rises above the sublimation point, and the coating material sublimates. The hot vapor produced in the evaporator rises to the top of the chamber, where the lens holder is placed. Then, a condensation process takes place and the material is deposited over the surface of the lenses. In order to produce a uniform deposition, the lens holder rotates slowly through this process. The thickness of the layer is controlled by using a quartz oscillator balance, which measures the quantity of material deposited and, hence, the layer thickness. Once the first layer is produced, other materials are evaporated until the final coating is achieved. For electron beam guns, the process is quite the same, but the evaporation is achieved by bombarding the material with a beam of electrons accelerated through electric and magnetic fields, as shown in Figure 11.9(b).

In Appendix F, we present the general theory of multilayer films, in the form set out by Abelès. This theory describes the phenomenon of multiple interferences, which happens at a multilayer structure. Abelès theory is able to predict the values of optical parameters, such as reflectance, of a given multilayer film, regardless of the values of incidence angle and wavelength, and the polarization status of the incident light. However, in its general form, this theory of multilayer films is quite complicated. In order to ease the study of thin layer films, we will restrict ourselves to conditions of normal incidence, and we will consider structures with a limited number of layers. Therefore, we will now study three examples of anti-reflective coatings formed by one, two, and three layers, respectively. These examples are based on the classical treatment of multilayer films given in reference [389], so we refer the reader to this book for additional information.

Single Layer Anti-Reflective Coatings

Let us consider now a single layer, as depicted in Figure 11.10, with thickness t_1, and made in a material whose refraction index is n_1. The refractive indexes of the substrate and the external medium are n_s and n_0. According to Abelès theory, this layer is characterized by the following transfer matrix (see Appendix F)

$$M_1 \equiv \begin{bmatrix} m_{11} & m_{12} \\ m_{21} & m_{22} \end{bmatrix} = \begin{bmatrix} \cos \delta_1 & \frac{i}{\gamma_1} \sin \delta_1 \\ i\gamma_1 \sin \delta_1 & \cos \delta_1 \end{bmatrix}, \tag{11.13}$$

where $\delta_1 = \frac{2\pi}{\lambda} t_1 n_1 \cos \theta_1$ is the phase difference, λ the wavelength of the incident light, and θ_1 the angle formed by the ray inside the layer with the normal. In our case, $\theta_1 = 0$, so $\cos \theta_1 = 1$. The term γ_1 is given by $\gamma_1 = \frac{n_1}{c} \cos \theta_1$, where c is the speed of light in vacuum. Again, in conditions of normal incidence, $\gamma_1 = n_1/c$.

From the elements of the transfer matrix, it is possible to obtain the value of the reflection coefficient as

$$r = \frac{\gamma_0 m_{11} + \gamma_0 \gamma_s m_{12} - m_{21} - \gamma_s m_{22}}{\gamma_0 m_{11} + \gamma_0 \gamma_s m_{12} + m_{21} + \gamma_s m_{22}}, \tag{11.14}$$

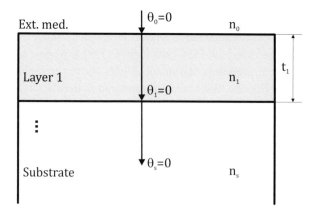

Figure 11.10 Schematics of a single layer film in conditions of normal incidence.

with $\gamma_0 = n_0/c$, $\gamma_s = n_s/c$ in conditions of normal incidence. By substituting the elements of the transfer matrix given by equation (11.13) in equation (11.14), we get, after some algebra

$$r = \frac{(n_0 - n_s)\, n_1 \cos \delta_1 + i\left(n_0 n_s - n_1^2\right) \sin \delta_1}{(n_0 + n_s)\, n_1 \cos \delta_1 + i\left(n_0 n_s + n_1^2\right) \sin \delta_1}. \tag{11.15}$$

Ideally, the reflection coefficient must be zero for a perfect anti-reflective coating. As it is a fraction of complex numbers, both real and imaginary parts of the numerator must be null. This means that the first condition for obtaining a single layer anti-reflective coating is

$$\cos \delta_1 = 0.$$

Unfortunately, as δ_1 varies with the wavelength, it is not possible to accomplish this equation for every wavelength. Instead, a single wavelength, known as design wavelength, λ_d, is chosen, so that

$$\delta_1 = \frac{2\pi}{\lambda_d} n_1 t_1 = \frac{\pi}{2}.$$

By rearranging terms, we get a condition on the layer thickness

$$t_1 = \frac{\lambda_d}{4 n_1}. \tag{11.16}$$

A layer manufactured with this thickness is known as a quarter wave layer or a $\lambda/4$ layer. The reflection coefficient of a $\lambda/4$ layer for the design wavelength is

$$r = \frac{n_s n_0 - n_1^2}{n_s n_0 + n_1^2},$$

and the corresponding reflectance

$$R = |r|^2 = \frac{\left(n_s n_0 - n_1^2\right)^2}{\left(n_s n_0 + n_1^2\right)^2}. \tag{11.17}$$

It is very important to realize that equation (11.17) gives only the reflectance for the design wavelength. The general equation for the spectral reflectance of a $\lambda/4$ layer is

$$R(\lambda) = \frac{(n_s - n_0)^2\, n_1^2 \cos^2\left(\frac{\pi \lambda_d}{2\lambda}\right) + \left(n_s n_0 - n_1^2\right)^2 \sin^2\left(\frac{\pi \lambda_d}{2\lambda}\right)}{(n_s + n_0)^2\, n_1^2 \cos^2\left(\frac{\pi \lambda_d}{2\lambda}\right) + \left(n_s n_0 + n_1^2\right)^2 \sin^2\left(\frac{\pi \lambda_d}{2\lambda}\right)}. \tag{11.18}$$

Note that for $\lambda = \lambda_d$, we get the reflectance given by equation (11.17).

In Figure 11.11, we have plotted the spectral reflectance for a single quarter wave layer made in Cryolite ($n_1 = 1.30$) deposited over a Crown glass substrate $n_s = 1.523$. The external medium is air $n_0 = 1$, and the layer has been designed for a wavelength of $\lambda_d = 550$ nm, so the thickness of the layer is

$$t_1 = \frac{\lambda_d}{4 n_1} = \frac{550}{4 \times 1.30} = 105.8\,\text{nm}.$$

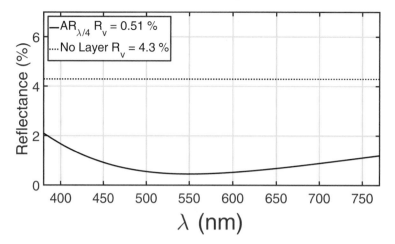

Figure 11.11 Reflectance of a single layer of $\lambda/4$ thickness. The values chosen in this plot are: substrate crown $n_s = 1.523$, $\lambda/4$ layer of Cryolite $n_1 = 1.30$ for a design wavelength of $\lambda_d = 550$ nm. The reflectance of the substrate without coating is also plotted in order to compare the visual reflectance with or without coating.

The reflectance for the design wavelength is

$$R \equiv \frac{\left(n_s n_0 - n_1^2\right)^2}{\left(n_s n_0 + n_1^2\right)^2} = \frac{\left(1.523 - 1.3^2\right)^2}{\left(1.523 + 1.3^2\right)^2} = 0.0027 \ (0.27 \%).$$

Therefore, we get a reflectance of 0.27 % for normal incidence and design wavelength. As we can see in Figure 11.11, the reflectance is greater for other wavelengths, as expected, but it is always lower than 2% in the visible spectrum. Moreover, as we have selected the design wavelength at the center of visible spectrum, the visible transmittance is quite low $R_v = 0.51 \%$, especially when compared to the visual reflectance of the uncoated substrate $R_v = 4\%$. In theory, it is possible to obtain a $\lambda/4$ layer that presents null reflectance for the design wavelength. This happens when the following condition is accomplished

$$n_1 = \sqrt{n_0 n_s}. \tag{11.19}$$

In practice, however, it is quite difficult to find a suitable index matching. For example, a perfect anti-reflective coating for a crown substrate $n_s = 1.523$, with air, $n_0 = 1$, as external medium requires a material with a refraction index $n_1 = \sqrt{1.523} = 1.234$, which is excessively low. If it is not possible to get an exact index matching, we have to chose the value of n_1 as close as possible to the one predicted by equation (11.19). This is why we selected Cryolite in our previous example, as Cryolite is one of the coating materials with the lowest refraction index.

   A final point regarding the spectral reflectance shown in Figure 11.11, is the color of the reflected images. As the spectral reflectance is not uniform, the reflected image of a white object will be colored. This phenomenon is known as residual color, and it is characteristic of multilayer coatings. In the case represented in Figure 11.11, the reflectance is lower at the central part of the spectrum and greater at the extremes. Therefore, as a rule of thumb, the residual color would be an additive mixture of red and blue wavelengths, resulting in a purple/violet shade.

Two-Layer Anti-Reflective Coatings

Let us consider now a two-layer $\lambda/4 - \lambda/4$ film. This means that the coating will be composed of two layers made in materials with refractive indexes n_1 and n_2, respectively. According to the convention of Appendix F, we will choose layer 2 (n_2) as the medium closer to the substrate, so layer 1 (n_1) is deposited over layer 2. For the design wavelength λ_d, the thickness of the layers are $t_1 = \lambda_d/(4n_1)$ and $t_2 = \lambda_d/(4n_2)$, respectively.

The transfer matrix of layer 1 for normal incidence is

$$M_1 = \begin{bmatrix} \cos \delta & i\frac{c}{n_1} \sin \delta \\ i\frac{n_1}{c} \sin \delta & \cos \delta \end{bmatrix},$$

where $\delta = \frac{\pi \lambda_d}{2\lambda}$, with λ the wavelength of the incident light. The transfer matrix of the second layer is

$$M_2 = \begin{bmatrix} \cos \delta & i\frac{c}{n_2} \sin \delta \\ i\frac{n_2}{c} \sin \delta & \cos \delta \end{bmatrix},$$

so the transfer matrix of the whole structure is

$$M = M_2 M_1 = \begin{bmatrix} \cos^2 \delta - c\frac{n_2}{n_1} \sin^2 \delta & ic \sin \delta \cos \delta \left(\frac{1}{n_1} + \frac{1}{n_2} \right) \\ \frac{i}{c} \sin \delta \cos \delta (n_1 + n_2) & \cos^2 \delta - c\frac{n_1}{n_2} \sin^2 \delta \end{bmatrix}.$$

When $\lambda = \lambda_d$, we have $\delta = \pi/2$, so the transfer matrix becomes

$$M = \begin{bmatrix} -c\frac{n_2}{n_1} & 0 \\ 0 & -c\frac{n_1}{n_2} \end{bmatrix}.$$

Substituting the elements of the transfer matrix in equation (11.14) we get the reflection coefficient for the design wavelength

$$r = \frac{n_s n_1^2 - n_0 n_2^2}{n_s n_1^2 + n_0 n_2^2},$$

and the corresponding reflectance

$$R = \left(\frac{n_s n_1^2 - n_0 n_2^2}{n_s n_1^2 + n_0 n_2^2} \right)^2. \tag{11.20}$$

Thus, the mathematical condition that must be met in order to get a $\lambda/4 - \lambda/4$ coating with no reflectance for the design wavelength is

$$\frac{n_2}{n_1} = \sqrt{\frac{n_s}{n_0}}. \tag{11.21}$$

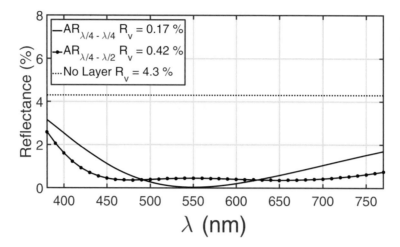

Figure 11.12 Spectral reflectance for a two-layer coating in two configurations: $\lambda/4 - \lambda/4$ (continuous line) and $\lambda/4 - \lambda/2$ (continuous line with black circles). In both cases, the refractive indexes of the layers are $n_1 = 1.3$ (Cryolite) and $n_2 = 1.60$ (Al$_2$O$_3$), respectively, with a Crown substrate $n_s = 1.523$, and $n_0 = 1$. The design wavelength is $\lambda_d = 550$ nm in both cases. We have also depicted the reflectance of the substrate when no coating is applied (dotted line).

In Figure 11.12, we have represented the spectral reflectance of a $\lambda/4 - \lambda/4$ coating made of Cryolite ($n_1 = 1.3$) deposited over Al$_2$O$_3$ ($n_2 = 1.60$), and this latter over a Crown substrate. If the external medium is air, the reflectance for the design wavelength is

$$R = \left(\frac{n_s n_1^2 - n_0 n_2^2}{n_s n_1^2 + n_0 n_2^2} \right)^2 = \left(\frac{1.523 \times 1.3^2 - 1.60^2}{1.523 \times 1.3^2 + 1.60^2} \right)^2 = 7.3 \times 10^{-6}.$$

Thus, as we can check in Figure 11.12, the reflectance becomes almost zero for the design wavelength. This is due to good matching between the indexes of the materials and substrate. For our example $n_2/n_1 = 1.60/1.3 = 1.23076$, while $\sqrt{n_s/n_0} \equiv \sqrt{1.523} = 1.234$. Thus, the condition stated in equation (11.21) is almost fulfilled. As in the former case, when the wavelength does not correspond to the design wavelength, reflectance increases, producing a residual color.

An alternative to $\lambda/4 - \lambda/4$ coating is setting the inner layer as a $\lambda/2$ one, which means setting the thickness of this layer as $t_2 = \lambda_d/(2n_2)$. As we can see in Figure 11.12, a $\lambda/4 - \lambda/2$ film presents a flatter spectral reflectance than the equivalent $\lambda/4 - \lambda/4$ coating. In the example of Figure 11.12, the value of the visible reflectance is greater for the $\lambda/4 - \lambda/2$ coating, as we have kept the same materials and design wavelength. However, by a proper choice of those parameters, it is possible to obtain low reflectance coatings with a flatter spectrum than the one produced by the $\lambda/4 - \lambda/4$ solution. This could be convenient in some cases, and this shows that two-layer coatings are more flexible than single layer ones.

Three-Layer Anti-Reflective Coatings

The last of our examples of anti-reflective coatings is a three-layer structure. As in the previous case, we have two possible designs: $\lambda/4 - \lambda/4 - \lambda/4$ and $\lambda/4 - \lambda/2 - \lambda/4$. Let us consider first the former one. This coating is formed by three layers made of materials with refraction indexes n_1, n_2, and n_3, enumerated in an inverse order of that of deposition. Therefore, n_1 corresponds to the most external layer, while n_3 is the layer closer to the substrate. In the $\lambda/4 - \lambda/4 - \lambda/4$ configuration, the thickness of the layers are $t_1 = \lambda_d/4n_1$, $t_2 = \lambda_d/4n_2$, and $t_3 = \lambda_d/4n_3$, respectively. If we illuminate the layer structure with monochromatic light, with $\lambda = \lambda_d$, in normal incidence, the transfer matrix of the whole coating is

$$
M = \begin{bmatrix} 0 & i\frac{c}{n_3} \\ i\frac{n_3}{c} & 0 \end{bmatrix} \begin{bmatrix} 0 & i\frac{c}{n_2} \\ i\frac{n_2}{c} & 0 \end{bmatrix} \begin{bmatrix} 0 & i\frac{c}{n_1} \\ i\frac{n_1}{c} & 0 \end{bmatrix} = \begin{bmatrix} 0 & -i\frac{n_2 c}{n_1 n_3} \\ -i\frac{n_1 n_3}{n_2 c} & 0 \end{bmatrix}.
$$

Substituting the elements of the transfer matrix in the general equation for the coefficient of reflection, we get, after some calculation

$$
r = \frac{\left(n_0 n_s n_2^2 - n_1^2 n_3^2\right)}{\left(n_0 n_s n_2^2 + n_1^2 n_3^2\right)}.
$$

Therefore, the reflectance of a three-layer structure $\lambda/4 - \lambda/4 - \lambda/4$ in normal incidence, and for the design wavelength is

$$
R = \left(\frac{n_0 n_s n_2^2 - n_1^2 n_3^2}{n_0 n_s n_2^2 + n_1^2 n_3^2} \right)^2. \tag{11.22}
$$

From this equation, we can derive the mathematical condition that should be accomplished in order to get an anti-reflective layer

$$
\frac{n_1 n_3}{n_2} = \sqrt{n_s n_0}. \tag{11.23}
$$

As in previous examples, we have plotted in Figure 11.13 the spectral reflectance calculated for a $\lambda/4 - \lambda/4 - \lambda/4$ layer made with the following materials: magnesium fluoride (MgF_2) with $n_1 = 1.38$ for the first layer, neodymium oxide (Nd_2O_3) with $n_2 = 2.0$ for the middle one, and, finally, thorium oxide (ThO_2) with $n_3 = 1.80$ for the innermost layer. The index matching is also good as $n_1 n_3/n_2 = 1.242$, while $\sqrt{n_s n_0} \equiv \sqrt{1.523} = 1.234$. This means that the value of reflectance for the design wavelength will be quite small. Effectively, by substituting the values of the refraction indexes we have

$$
R = \left(\frac{1.523 \times 2^2 - (1.38 \times 1.8)^2}{1.523 \times 2^2 + (1.38 \times 1.8)^2} \right)^2 = 4.07 \times 10^{-5}.
$$

However, the main difference between the three-layer and the coatings studied before is the broad region of the spectrum for which the reflectance presents almost negligible values. This translates into a very low value of the visible reflectance, with $R_v = 0.08\%$

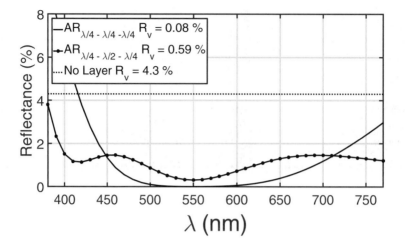

Figure 11.13 Spectral reflectance for a three-layer coating in two configurations: $\lambda/4 - \lambda/4 - \lambda/4$ (continuous line) and $\lambda/4 - \lambda/2 - \lambda/4$ (continuous line with black circles). In both cases, the refractive indexes of the layers are $n_1 = 1.38$ (MgF$_2$) and $n_2 = 2.0$ (Nd$_2$O$_3$), and $n_3 = 1.80$ (ThO$_2$), with a Crown substrate $n_s = 1.523$, and air as external medium $n_0 = 1$. The design wavelength is $\lambda_d = 550$ nm in both cases. We have also depicted the reflectance of the substrate when no coating is applied (dotted line).

according to Figure 11.13. Therefore, by increasing the number of layers, we multiply our design options. Finally, as with the two-layer coatings, we can choose a value of thickness different to the usual $\lambda/4$. For example, in Figure 11.13, we have also represented a $\lambda/4 - \lambda/2 - \lambda/4$ coating made of the same materials, and with the same design wavelength as the $\lambda/4 - \lambda/4 - \lambda/4$. In this case, the change of thickness results in a flatter reflectance spectrum, as in the case of the two-layer coating.

11.5 Other Coatings

Nowadays, most ophthalmic lens are provided with several coatings to improve not only their optical properties, but also the chemical and mechanical properties of their surfaces [396]. These coatings or treatments are applied by layers using different techniques, such as evaporation in a vacuum chamber, dip or spin coating, etc. Besides the now ubiquitous anti-reflective coatings, it is very common to find in modern ophthalmic lenses two additional coatings: hard and hydrophobic. We will give now a brief explanation of these coatings, and we will end this section with a short discussion on the general arrangement of the different coatings over the lens surface and the chemical and mechanical compatibility requirements that must be achieved.

11.5.1 Hard Coatings

As we saw in Chapter 1, glasses are harder than plastics, so we need to apply more pressure to glass in order to get a permanent deformation of the material. Indeed, in the case of

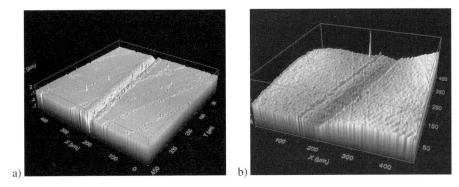

Figure 11.14 Confocal topography of a) a thin scratch and b) a thick one over the surface of an ophthalmic plastic lens.

glasses, it is more common to produce the breaking of the material rather than a perma-nent deformation. However, plastics are softer and tend to exhibit permanent deformations even when moderate pressures are applied. This greater softness causes the appearance of so-called esthetic defects, mainly scratches, over the surfaces of the lens. The physical mechanism leading to the formation of scratches is abrasion, which is caused by small and hard objects rubbed against the lens surface. Usually, this is due to the mishandling of the spectacles. A typical example is the resting of the spectacles upside down against a table so that the front surface of the lens is in touch with the surface of the table. In this case, the superficial irregularities of the table or hard particles trapped within the table surface may cause abrasion, resulting in the generation of scratches.

When examining the scratches formed over the surface of an ophthalmic lens using an electronic or confocal microscope, we usually find two main types of scratches, known as thin and wide. In Figure 11.14, we show the topographic map of these two types of scratches measured with a confocal microscope with nanometric resolution (Sensofar's Plμ). The topographic profiles of this couple of scratches are also shown in Figure 11.15(a) and (b), respectively. These profiles allow us to determine the physical dimension of the scratches, with the width and depth of the thin scratch being ≈ 50 μm and ≈ 10 μm, respectively, as we can see in Figure 11.15(a). The corresponding figures for the wide scratch (Figure 11.15(b)) are ≈ 170 μm and ≈ 320 μm, respectively. The dimensions of the scratches are important, as they determine their optical behavior. As a rule of thumb, thin scratches, although barely visible with the naked eye, produce a considerable amount of dispersed light due to diffraction and scattering. Conversely, wide scratches are visible to the naked eye (and they may even be annoying for the lens user), but they do not disperse light in the same way as thin ones do.

Therefore, depending on the mechanical properties of the lens material, it might be advisable, even mandatory, for some very soft materials such as polycarbonate to deposit a hard coating over the lens surface. The first attempt to increase the hardness of plastic lens surfaces happened at the beginning of the 1970s with the so-called quartzsage process

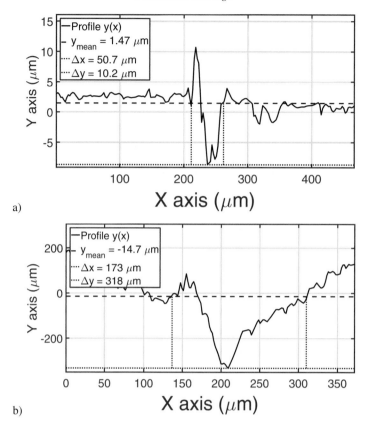

Figure 11.15 a) Confocal topography of a wide scratch over the surface of an ophthalmic lens made of plastic. b) Transverse profile across the scratch, showing its main dimensions.

[396]. This technique consisted of coating the lens surface with a layer of quartz (SiO_2), protecting the lens from the small particles that cause the abrasion. However, there were two problems with this first generation of hard treatments. First, thermal and chemical compatibility between the quartz and a polymeric substrate is poor and, second, quartz, although very hard, is also fragile and tends to fracture easily, resulting in the detachment of entire flakes of the coating.

As a result of the shortcomings shown by quartz coatings, new hard coatings based on organic molecules were developed [396]. These coatings were formed by organic compounds containing silicon (Si), belonging to the family of the polysiloxanes. The presence of Si atoms within the polysiloxane molecules increases the hardness of the surface, while, as both coating and substrate are organic materials, thermal and chemical compatibility are also greater. Unfortunately, issues still remained, in this case, regarding the compatibility between hard and anti-reflective coatings [396]. The problem was that the polysiloxane layer was too flexible for depositing over it a fragile anti-reflective layer.

Thus, although the lens was more resistant to abrasion, the durability of the anti-reflective coating diminishes, resulting in a loss of useful life for the lens.

The solution to the problem of compatibility between anti-reflective and hard coatings constitutes the third and current generation of composite coatings [396]. A composite material is formed by the mixture of two different materials (such as fiberglass and a polymer) with different mechanical properties, resulting in a material with enhanced mechanical properties.

In the case of hard coatings, the composite is formed by nanoparticles of silicon embedded in a liquid with a chemical structure similar to that of the polysiloxane molecules [396]. After coating the lens with the mixture, and carrying out a thermal polymerization process, a hard layer is formed over the lens substrate. Lens manufacturers mainly employ dip coating and spin coating in order to deposit a uniform layer of varnish over the surface of a lens. Dip coating consists of dipping the lens into the varnish and, afterward, lifting it slowly. In this way, a uniform layer of varnish is formed over the lens by surface tension. The other process consists of pouring a certain quantity of varnish over the lens surface while the lens is spinning at high speeds. The centrifugal forces due to the lens rotation ensure the formation of a uniform varnish layer over the lens surface. In both cases, after deposing the liquid, the lens is heated up to $100°$ C to polymerize the varnish. The composite mix yields the required mechanical properties: abrasion resistance due to the embedded nanoparticles of silicon, and enough toughness for supporting the fragile anti-reflective coating without breaking. Several mechanical tests are carried out by lens manufacturers in order to check that these properties are effectively present in their hard-coated lenses [396].

11.5.2 Hydrophobic Coatings

Unlike other optical devices, which are more protected, ophthalmic lenses are directly exposed to the environment. As a consequence of this exposure, lenses are usually befouled by many different substances. Some of them are solid, but many are liquids, particularly water and oily solutions. This accumulation of dirtiness may impair the optical properties of the lens. For example, a deposit of grease over the lens surface may produce undesired effects such as light dispersion, scattering, diffraction, and even interference with the anti-reflective coating [396]. Moreover, if a liquid solution is in contact with the lens surface, it is possible that chemical reactions arise between the substances dissolved in the solution and those present on the lens surface. Any chemical reaction involving the substances that form the anti-reflective layer is of particular concern, as it might damage the layer. Finally, it is obvious that the more often the lens is soiled, the greater the number of times that the user must wipe the lens surface, so the risk of surface abrasion increases, particularly if the user is not careful enough in this cleaning.

Therefore, the ophthalmic industry has developed hydrophobic treatments in order to ease the problems caused by liquids deposited over the surfaces of the lenses. The goals of a hydrophobic coating are to reduce the formation of liquid deposits over the lens surface and to act as a barrier protecting the inner layers of the surface treatment, particularly the anti-reflective coating. Modern hydrophobic treatments [396] act in two ways. First, they considerably reduce the surface roughness by completely covering all the micropores and other surface irregularities present on the lens surface. As a consequence, the coefficient of friction of the surface is greatly reduced. The second mechanism of action of the hydrophobic treatment is chemical. The composition of the hydrophobic layer is rich in fluoride, which acts as a repellent for water and oils [396]. In the end, the combined effect of a smoother surface and the repellant effect of fluoride-based compounds reduce the surface of contact between the liquid droplets and the hydrophobic layer. As a result, the liquid droplets slide more easily over the lens surface, which eases lens cleaning considerably. Hydrophobic coatings can be applied either as a varnish or by vacuum deposition [396].

11.5.3 Order of Deposition and Compatibility

To end this chapter, we are going now to briefly discuss two important aspects of ophthalmic lens coatings: the order of deposition and the compatibility. The various coatings that form the whole layer structure of the lens are deposited in a very particular order. This is due, in part, to dependencies between the different coatings. For example, one of the missions of the hydrophobic layer is to act as a barrier between the external medium and the AR (anti reflection) layer. Therefore, the hydrophobic layer must be deposited over the anti-reflective layer. In Figure 11.16, we show the order of deposition of the different layers that form the lens coating. It is important to notice that not all the treatments may be available at the same time for a particular lens. However, we can safely say that most of the ophthalmic lenses manufactured nowadays have, at least, an anti-reflective coating.

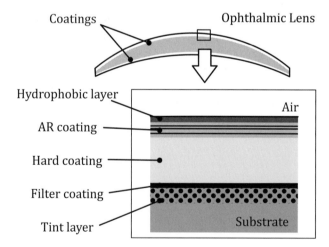

Figure 11.16 Schematic drawing showing the order of deposition of the different coatings over the lens surface. The thickness of each layer is not drawn to scale.

The order of deposition shown in Figure 11.16 is as follows:

Hydrophobic layer. This is the outermost layer to ensure that liquids are not retained by surface irregularities. The thickness of the hydrophobic layer is very small, so that this coating does not interfere with the anti-reflective coating.

Anti-reflective coating. This is laid just after the hydrophobic coating. Otherwise the inner layers would produce interferential effects, which may reduce its effectiveness.

Hard coating. This coating is located under the anti-reflective layer in order to provide a mechanical basement with enough flexibility and resistance. Its mission is to prevent the break up of the anti-reflective coating due to stress. The thickness of this layer is greater than the others in order to ensure the desired mechanical properties of abrasion resistance.

Filter coating. If the lens is provided with a polarizing, photochromic, and/or absorbing filter, it is placed between the hard coating and the substrate. In this way, the gap created by the hard layer prevents any interference with the anti-reflective coating.

Substrate and tint layer: If the lens has been tinted by immersion, then the tint molecules penetrate under the lens surface and form a thin layer that provides uniform absorption of the incoming light along the whole surface of the lens. It is important to remark that coatings are always applied after tinting.

When an ophthalmic lens is manufactured, the treatments are applied in the order inverse to deposition. Usually, between the application of each layer, the lens passes through a cleaning process in order to ensure the highest possible quality of the final product. As a consequence, the application of surface treatments is a lengthy process that requires a great amount of the time necessary for manufacturing an ophthalmic lens.

Finally, all the treatments applied to a lens must be compatible among them. This compatibility is measured attending to both physical and chemical parameters. From the point of view of physics, the layers must be optical, mechanical, and thermally compatible, which means that: (1) they do not introduce undesired optical effects such as interference between layers, (2) they do not cause other layers to break under stress, and (3) their thermal expansion coefficients are similar enough to ensure that the layers do not break or detach after successive cycles of heating/cooling. In turn, chemical compatibility implies that the layers are inert, so they do not react either among each other or with environmental substances.

Appendix A

Frames

A.1 Historical Notes

Eyeglasses, *specta eyeglasses*, or *spectacles* are devices intended for supporting a pair of ophthalmic lenses in front of the eyes, so that they can accomplish their role as compensating devices for ametropies. There is no conclusive evidence on the subject of the invention of eyeglasses. The most robust evidence points to Italy at the end of the thirteenth century [241]. Indeed, some references to "glasses for seeing" are found in decrees from Venice [398, 399], while in a famous sermon preached by Giordano da Rivalto in Florence in 1305, the first clear mention of spectacles was recorded. In the sermon, Giordano acknowledged he had met the inventor, and that spectacles had been around for at least 20 years. The invention of eyeglasses has been associated with names such as Roger Bacon and Salvino d'Armati [400, 401], but these associations seems to be refuted [398]. Allexandro della Spina, a friar at the Dominican convent of St. Catherine in Pisa who died in 1313, was credited in the *Chronicles* of the convent with having seen spectacles made by the inventor and then managing to understand and manufacture them [398, 399, 401]. Pictorial evidence and archeological records show that the first spectacles consisted simply of two rims joined by a bridge piece for holding the lens over the nose. This kind of spectacle is known as a "pierce-nez" from its French name [132].

In the sixteenth and seventeenth centuries, eyeglasses, still in the form of the "pierce-nez," were held against the user's head with the help of a cord or a ribbon attached to the lateral ends of the mount, in the same place where the lug is located in modern spectacles. Throughout these centuries, they gained popularity despite some early detractors [402]. Pictorial and sculptural evidence of this era, particularly the "Portrait of Cardinal Fernando Niño de Guevara" by the Spanish painter El Greco [403], suggests the use of temples for holding spectacles. However, it is not until the beginning of the eighteenth century that conclusive evidence of the use of temples or sides in eyeglasses appeared. Indeed, a British optician, Edward Scarlett, showed, in 1730, the first illustration of a spectacle frame with sides in his trade card [401]. A curious point is that the sides of Scarlett's eyeglasses ended in a spiral, which was a somewhat unusual form given the posterior development of spectacles. Subsequent spectacle frames were made with sides ending in rings, purportedly for carrying a ribbon or cord to ensure better fitting of the eyeglasses to the user's head.

The development of spectacle frames continued through the nineteenth and twentieth centuries by the use of new materials such as acetate, nylon, aluminum, stainless steel, and titanium, and by enhancing the design and manufacturing process. It is interesting to point out the ever increasing role that fashion is playing in the design of eyeglasses, which are in some cases regarded not only as a compensating device for the eye but also as a fashion complement.

A.2 Frame Materials

Throughout history, spectacle frames have been produced using different materials, ranging from wood to precious metals [132, 404]. After the Industrial Revolution, mass production was applied to spectacle frame manufacturing, with two main types of materials, plastics and metals. A typical example of the first is nitrocellulose, while metallic spectacle frames were produced in alloys, usually containing nickel. With the arrival of modern plastics such as polymers based on cellulose acetate, nylon, or epoxies, the number of materials available for plastic frames has been considerably enlarged. On the other hand, new alloys such as stainless steel and the popularity gained by tough metals such as titanium have also created a wider number of options for metallic mounts. Finally, we must not forget the importance of composite materials in the modern ophthalmic industry and the role that new manufacturing processes will play, particularly 3D printing, in the foreseeable future. Therefore, we will now describe briefly the main properties of the materials employed in the manufacturing of spectacle frames, and we will group them in three main categories attending to their nature.

A.2.1 Plastics

As happened with ophthalmic lens materials, the rapid expansion of the chemical industry at the end of the nineteenth and beginning of the twentieth century made a number of plastics for the spectacle frame manufacturing industry available. In general, plastics present good mechanical properties for spectacle mounts; they are tough enough to resist wear, they can be tinted, and, as should be the case of materials used in the ophthalmic industry, they had good biocompatibility. We will describe now the main plastics used in frame manufacturing.

Cellulose Nitrate

Cellulose nitrate, nitrocellulose, "gun cotton," or celluloid is a material that appeared in the nineteenth century. It was made by combining nitric and sulfuric acids with cotton fibers. The resulting material, which was highly flammable and even explosive, was employed in a number of applications, ranging from films to smokeless powder. Nitrocellulose for spectacle frames was first produced by heating a mixture of cotton fibers and nitric acid. Afterward, it was mixed with camphor to obtain the desired plastic properties, and in this way, a material that can be kneaded and rolled into sheets is obtained. Finally, by joining the sheets together, a resulting rectangular block of nitrocellulose is obtained (see reference [404] for further details on cellulose manufacturing).

Despite its good properties as spectacle frame material such as toughness and the ability to be easily polished, cellulose nitrate is extremely flammable, which nowadays prevents its use for spectacle frame production.

Cellulose Acetate

In the first half of the twentieth century, a substitute for celluloid was developed. Although the name of the new material was cellulose acetate or xilonite, it is universally known in the ophthalmic industry as "acetate." The basis of cellulose acetate is the mixture of cellulose, usually extracted from cotton, with acetic anhydride $(CH_3CO)_2O$. Commonly, a phthalate compound is added as a plasticizer [404]. Being a thermoplastic material, it can be easily molded, and this property can also be used in the fitting process to dilate a frame made with acetate by applying heat. However, care should be taken because too much heat can damage the material. For this reason, acetate is not common for manufacturing spectacle mounts that must be worn in extreme hot climates such as tropical or subtropical ones. There are other two undesirable properties of cellulose acetate. It is quite hygroscopic, so water absorption may cause deformations in the mount (this also make its use in tropical zones impractical). In addition, cellulose acetate tends to wear out easily and can also react with human sweat and chemicals such as deodorants, cosmetics, etc. One of the signs of this deterioration is the filtering of the phthalic compounds to the surface of the mount, thus creating a white coating that is both quite noticeable and unesthetic.

Cellulose Propionate

Cellulose propionate is another alternative to cellulose nitrate. It is obtained from mixing cellulose flakes and propionic acid. It is usually molded by injection, so the cellulose propionate flakes are first heated and then injected into a mold. Compared to acetate, cellulose propionate is more stable in hot climates and it is quite inert, so it does not cause skin irritation or allergies. For spectacles frames made of this material, the lens can be inserted into the rim by heating, but care should be taken because if too much heat is applied the mount can be deformed. Cleaning of cellulose propionate mounts should be done with warm water and never with solvents such as alcohol or acetone [404].

Nylon (Polyamide)

Nylon is the commercial name for a series of plastics made of polyamides (a macro molecule presenting amide bonds). Discovered in 1935, nylon was one of the first synthetic fibers ever produced. Nylon can also be manufactured in film format, and it is possible to produce mechanical parts through extrusion, casting, or even injection molding, because Nylon is a thermoplastic material. Nylon is also suitable for use in 3D printing [405], which is an interesting feature given the range of possibilities offered by this new manufacturing technique. Nylon has good properties as a material for spectacle frames, as it is hypoallergenic, lightweight, and tough. The main problem presented by nylon is that it only admits opaque shadows, so the range of colors available is somewhat reduced [404]. Besides the classical nylon, some manufacturers have produced polyamides with enhanced properties such as Silhouette's SPX [404].

Polycarbonate

The use of polycarbonate in spectacle frames is usually restricted to protective eyeglasses made by injection molding. Those frames are lightweight and they present high impact resistance. They are also flexible enough for allowing the fitting of the lens, although they cannot be heated. Polycarbonate mounts are usually transparent and can include nosepads made of rubber or silicon in order to protect the user. Regarding the maintenance of those mounts, it is important to notice that polycarbonate can be dissolved by acetone, so this product should be avoided in the cleaning of those spectacle frames.

Optyl

Optyl is a thermosetting plastic belonging to the family of epoxy resins. The main characteristic of epoxy resins is the presence of an epoxide group (two carbon atoms joined by one of oxygen), which confers the final polymer its mechanical properties, in particular its toughness, allowing, for example, its use as high strength adhesives. Optyl frames are produced by casting and curing and can be heated up to $200°$ C without losing their form [404]; therefore, they are suitable for heat manipulation in the glazing process. Spectacle frames made of optyl can be polished in a polyurethane bath, and they can be tinted by immersion, in a similar way as with CR-39 lenses [404].

A.2.2 Metals

Nickel Alloys

Nickel is a common component of a wide range of alloys used in ophthalmic mount manufacturing. Mixed with copper and zinc, it forms the so-called "nickel silver" or "German silver." The composition of nickel silver is around 20% nickel, 60% copper, and 20% zinc. Spectacle mounts made of this alloy can be easily manipulated in order to adjust them to the user.

Another widely employed alloy that contains nickel is "monel," which is a nickel copper alloy with a higher proportion of nickel (around 67%) than nickel silver. Monel is highly resistant to corrosion, and it admits many colorations [132], although its high nickel content makes it more expensive than nickel silver.

Stainless Steel

It is well known from antiquity that iron (and hence steel) is quite susceptible to corrosion, which made its usage in spectacle frames difficult because contact with the fatty acids of the skin accelerates corrosion. Although some metallurgical techniques were available for avoiding this problem (such as the chromium-coated tips on the weapons of the renowned terracota army in China [406]), it was not until the discovery of stainless steel in the last decades of the nineteenth century that corrosion-resistant steel was produced. Besides the basic components of steel such iron and carbon, stainless steel has, in addition, a proportion of nickel and chromium. Stainless steel frames are tough and lightweight and admit many colorations through enameling [132].

Aluminum

Aluminum alloys are widely employed in applications that need a metal with both great strength and light weight, and hence its use in spectacle frames. Aluminum is resistant to corrosion and can be coated by anodization, creating a layer of aluminium oxide. Anodization makes the metal more resistant to rusting and also allows for a better application of paints.

Precious and High Value Metals

Traditionally, gold has been employed as a material for luxury eyeglasses, and details of the use of this metal in spectacle frames can be found in the books of Fannin [132] and Wakefield [404]. It is important to be aware of the legal provisions that affect this kind of spectacle frame, particularly those regarding the marking of gold [132].

Among the high value metals used in the ophthalmic industry, titanium is undoubtedly the most popular given its intrinsic properties. With a greater strength than steel but lightweight, titanium mounts are both light and tough. Unfortunately, titanium is not only expensive by itself, but its manufacturing process is complicated by its high fusion temperature, which makes spectacle frames made of titanium even more expensive. Other high value metals employed in the ophthalmic industry are beryllium and cobalt [132, 404].

A.2.3 Composite and Special Materials

A composite material is formed when two or more materials with different properties are combined in such a way that the resulting composite has better properties than either of their components. Composite materials are widely used in high technological areas such as the automotive or aerospace. The use of composite materials allows frames with enhanced strength and light weight to be employed in demanding situations such as outside sports. On the other hand, in recent years, 3D printing has appeared as a new manufacturing technique, allowing the fast and cheap production of a wide range of parts. We may expect that in the coming years, more and more products will be made using 3D printing methods, including spectacle frames. For this reason, we will give a brief description of the materials that would be susceptible of being employed in the production of spectacle frames with 3D printing.

Carbon Fiber

Carbon fiber is formed when a thread of graphite wires is embedded in a thermosetting resin monomer and then polymerized. The resulting material is both tough and light, and in some applications it has replaced light metals such as aluminum. Carbon fiber mounts are quite durable and thin, and they resist temperature well (due to the thermosetting resin that forms them). If a carbon fiber spectacle frame is not provided with joints in the rim, the practitioner should be extremly accurate in cutting the lens to the required form.

Kevlar

It is a widely employed material, particularly in military application as a basic component of helmets and body armor. In its origins, Kevlar was developed as a synthetic fiber of high tensile strength discovered and manufactured by DuPont company in the 1960s. When Kevlar fibers are combined with an epoxy resin, the resulting composite presents great resilience and strength. In the ophthalmic industry, Kevlar is widely employed as protective eyeglasses, especially for sports, although it is somewhat limited by the small range of colors available.

Materials for 3D Printing

Although in the beginning of development, the materials for 3D printing were limited to thermosetting and thermoplastics resins, nowadays even metallic pieces can be produced by this technique. For their application in the production of spectacle frames, extrusion deposition and photolithography are two technologies that present the greater prospects for industrial production. In the first, a thermoplastic material is extruded through an extrusion nozzle, which is positioned accurately with the help of step motors. In the second process, the 3D piece is formed by successive polymerization of 2D layers by photolithography. Some of the materials employed in spectacle frame production are suitable to be used in 3D printing, such as nylon. Although there are still problems with 3D printing (like the layer binding), it constitutes a technology that may change the way in which spectacle frames are manufactured.

A.3 Elements and Dimensions of Frames

In this section, we will describe the main elements that form a spectacle frame. It is important to point out that, although some elements such as sides are common for all ophthalmic frames, there are important differences depending on the particular kind of spectacle frame. According to the international standard ISO 7998:1984 [407], four basic types of spectacle frames are defined: plastic, metallic, rimless, and combined. We will give now a brief description of the first three classes of spectacle frames, indicating their main elements. Regardless of the particular kind of frame, we will distinguish the front and sides of the spectacle frame. The front contains the elements necessary to hold the pair of ophthalmic lenses together with a piece for supporting the frame on the user's nose. The role of the frame sides is to hold the spectacle frame against the head. For this reason, the end parts of both sides are placed between the head and the user's ear lobes. Therefore, every spectacle frame is held by three support points located on the nose and both ear lobes of the user. All the specialized terms introduced in this section are in agreement with the international standard ISO 7998:1984 [407], and we refer the reader to this standard or to books on ophthalmic lens prescription [404] for more information about this topic.

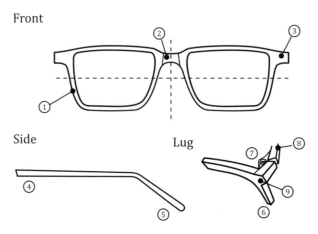

Figure A.1 Elements of a plastic spectacle frame. 1) Rim (left eye), 2) bridge, 3) lug, 4) knuckle, 5) side (or temple) end, 6) part of rim, 7) joint, 8) side knuckle, 9) lug.

A.3.1 Elements of a Plastic Frame

Usually, a plastic frame comes with three different parts: the front, and two sides also known as temples. As can be seen in Figure A.1, the front is formed by two rims for supporting the lenses plus a bridge between them. The temporal extremes of the front finish in two lugs (see Figure A.1), whose mission is to support the joints, which are small hinge-like metallic pieces that attach the sides of the frame to the front. Those joints allow the sides to be folded up when the spectacle frame is not used. The lens is inserted in the plastic frame mechanically, albeit on some occasions the frame is previously heated in order to ease this process. In this case, the practitioner must be careful to avoid any damage to the frame by excessive heat. In order to fix the lens, there is a groove in the inner part of the rim that matches with the lens bevel (the same applies for metallic frames).

The most common type of side used in plastic spectacle frames is the drop end side, which is depicted in Figure A.1. As can be seen in the detail of the lug shown in the aforementioned figure, the joint is placed in the inner side of both front and temple of the frame. It is quite common to find a reinforced wire placed within the frame side in order to strengthen it.

A.3.2 Elements of a Metallic Frame

Metallic frames (see Figure A.2) differ from plastic ones in some important aspects. First, the rims are not continuous and they have a separable closing block that allows for the placement of the lens. The two parts of this closing block are kept in place by a screw so that the practitioner can change the strength exerted by the rim over the lens. In this way, the rim can adapt itself better to the shape of the lens. It is, however, quite important to be extremely careful in not setting the rim too tight, as it would result in high pressures over the lens. High pressures over the lens contour may cause deformations on plastic lenses

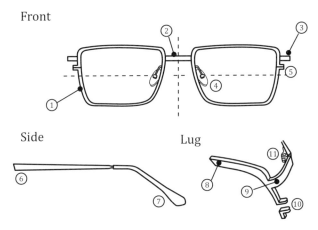

Figure A.2 Elements of a metallic spectacle frame. 1) Rim (left eye), 2) bridge, 3) lug, 4) pad and pad arm, 5) closing block, 6) side knuckle, 7) side end, 8) rim (part), 9) lug, 10) closing block, 11) joint.

that can be severe enough to produce changes in the optical properties of the lens [27]. In the case of glass lenses, if the lens border is not properly finished, the stresses induced by the rim can even cause breakage of the lens, with the corresponding risk of injuries to the user.

Another element that differentiates metallic frames from plastic ones is the support of the nose. In metallic frames, this is usually done by a pair of pads or nose-pads made of silicon, or other similar material, and they can be fitted to the particular user by means of adjustable pad arms.

A.3.3 Elements of a Rimless Frame

Rimless and semi-rimless spectacle frames are quite popular due to both esthetic and comfort factors. In the case of rimless frames (see Figure A.3), the lenses are no longer held by the rim, and the supporting elements of the spectacle frame (bridge, lugs, and sides) are directly attached to the lens by screws that pass through holes drilled in the lens. Due to the difficulties of drilling in glass and the consequent risk of glass breaking, only plastic lenses are usually used with rimless frames.

On the other hand, semi-rimless spectacle frames retain part of the rims, usually the upper part, and so the lenses are fixed to this upper rim through screws or, more commonly, through a retaining cord, usually made of nylon. For this reason, these frames are popularly known as "nylon supra mounts." Generally, for both types of spectacle frames, the elements of the frame are metallic in order to give both strength and lightness to the frame.

In order to fit a lens in a rimless or semi-rimless spectacle frame, it is necessary to perform a drilling operation after cutting the lens using a flat edge (obviously, a bevel is no longer needed if there is no rim). This is a delicate process that must be done with high accuracy in order to avoid mounting errors. Indeed, many modern cutting machines have

Rimless mount front

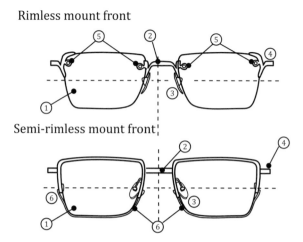

Figure A.3 Elements of a rimless (up) and a semi-rimless (bottom) frame front: 1) Rim (left eye), 2) bridge, 3) pads, 4) lug, 5) strap and 6) retaining cord.

a drill accessory incorporated, so that they can drill automatically after cutting the lens. Similarly, when fitting lenses in a nylon supra mount with retaining cord, it is necessary to groove the edge of the lens with high accuracy. The most usual position for the groove is centered in the edge; however, depending on the lens (for example, high negative power lenses), it may be advisable to locate the groove closer to the edge's frontal or rear end. This work is usually undertaken with a small machine that centers the edge properly while making the groove using a wheel whose cutting edge contains small pieces of diamond over its surface. As happened in the case of drilling, modern cutting machines have incorporated a groove wheel as an accessory.

A.3.4 Spectacle Lens Frame Dimensions and Markings

It is of paramount importance to have a system for referring and measuring the dimensions of a spectacle lens frame. Nowadays, the international standard ISO 8624 [408] establishes a unified system for measuring a spectacle frame. In Figure A.4, we have depicted a typical plastic frame, indicating its main dimensions for both front and sides according to the above-mentioned international standard. Although we have represented a plastic spectacle frame, the same applies for any other type of frame.

    As can be seen in Figure A.4, the measuring system defined by the ISO 8624 standard is based on two rectangles that circumscribe the inner part of both rims. For this reason, this system is known as the "box system." The geometrical centers of both boxes define two points known as boxed centers. The line that joins both centers is known as the horizontal center line (HCL). The horizontal center line acts as the x axis of the frame as we will see in the following section. Another auxiliary line is defined, parallel to HCL and located 5 ± 0.5 mm below it. In these conditions, the following dimensions are defined:

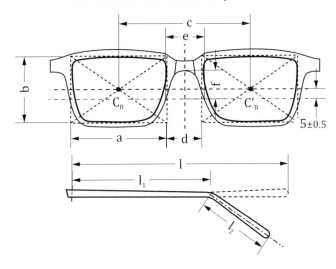

Figure A.4 Main dimensions of a spectacle frame according to the ISO 8624 standard. The name of each dimension is given in the text. Notice the boxes that circumscribe the inner part of the frame rims. All numerical dimensions are given in millimeters.

- *Horizontal lens size*: Is the width of the rectangle that circumscribes each rim; it is labeled as a in Figure A.4.
- *Vertical lens size*: Marked as b in Figure A.4, it is the height of the rectangle that circumscribes each rim.
- *Distance between lenses*: Is the horizontal distance between the two rectangles, d in Figure A.4, and it corresponds to the minimum horizontal distance between the lenses.
- *Distance between rims*: Or bridge width, is the horizontal distance between both rims, measured, as shown in Figure A.4, along the horizontal line located 5 mm below the HCL. It is labeled as e in Figure A.4.
- *Bridge height*: Is the vertical distance, marked as f in Figure A.4, between the midpoint of the lower edge of the bridge and the auxiliary line placed 5 mm below the HCL.

Notice that, according to Figure A.4, the *horizontal distance between centers, c* corresponds to the sum of the horizontal lens size and the distance between lenses, so $c = a + d$, an equation that will be quite useful in the next section.

Regarding the side of the spectacle lens frame, the *overall length of the side, l* in Figure A.4, is measured from the knuckle of the temple to the tip of the end of the side, but by straightening first the whole temple, as shown in Figure A.4. Two more lengths are defined: the *length to bend* (l_1 in Figure A.4), defined as the horizontal distance between the knuckle and the bend point of the frame side, and the *length of drop* (l_2 in Figure A.4), which is the distance between the bend point and the tip of the temple's end.

Spectacle lens frames are marked by the manufacturer following a standard. In Europe, the marking of spectacle frames is regulated by the international standard ISO 9546 [20], which establishes a number of guidelines for marking a spectacle frame, some of them mandatory. Among the mandatory information, we find the horizontal lens size and the distance between lenses, which are engraved or painted on the frame surface as two numbers (corresponding to the value of those distances in millimeters) separated by the symbol □. Thus, if we have a frame marked as 52□19, the horizontal lens size would be $a = 52$ mm and the distance between lenses $d = 19$ mm. Another dimension that should be marked in the frame is the overall length of the side as a number representing the value of this length in millimeters. It is also usual to mark in the frame the bridge width with a value in millimeters marked with a special symbol such as $//$ or $\frown$, for example $\widehat{17}$ would indicate a bridge width of $e = 17$ mm.

Other pieces information that must appear on the frame are the name of the manufacturer, model, color, and the material used in manufacturing the spectacle frame. All of these compulsory marks are placed on the inner surface of the frame. Usually, the horizontal size of the lens and the distance between lenses is marked at the front of the frame, while the remaining information is written on the temples. Besides the mandatory markings, the frame's manufacturer must provide on request additional data, such as the lens area in square millimeters, mass of frame, frame height, etc. For further information about this topic, see [409] and [404].

A.4 Lens Centering

Lens centering is the process of placing the lens in the proper position so it can fulfill its intended role as a compensating element for the nonemmetropic eye. To do so, it is necessary to align the frame, the lens, and the eye of the user. Once the frame is fitted to the head of the patient, the distance between the lens and the eye becomes fixed, so the alignment is made in the plane that contains the rims of the frame. To perform the centering of the lens, the practitioner has to carry out some operations. Some of them imply the taking of measurements over the spectacle lens frame and patient's face. This is necessary in order to locate three characteristic points, one for each of the elements involved in the centering procedure. We will start with the reference point of the eye, and then we will describe the corresponding points for both the spectacle frame and the lens.

The natural reference point of the eye is the center of the pupil. However, there is a problem, as neither the eye nor the subject's head remain stationary for long. Therefore, it is necessary to specify a reference position for both head and eyes. In most cases, we are interesting in compensating the vision of the user for distant objects. So we will adopt the primary gaze position [410] as a reference for the head and the eyes of a given individual. In order to define accurately this position, we need some anatomical terminology. Reid's reference line is defined as the line coming from the inferior edge of the left eye socket to the top of the bony ear hole. Thus, we will suppose that in the primary gaze position, Reid's reference line is perpendicular to the frontal plane of the individual. In other words, the head of the subject is erect and facing forward. Moreover, in this reference position,

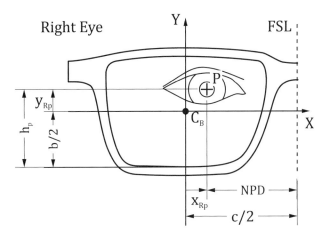

Figure A.5 Location of the right eye pupil P, in primary gaze position, referred to the boxed center C_B of the spectacle frame. Distances x_{Rp} and y_{Rp} correspond to the horizontal and vertical pupil coordinates.

the eyes are looking at a far object, which is placed directly in front of the subject, so the optical axes of both eyes are also parallel to Reid's reference line.

In these conditions, we will define the location of the pupil as the normal projection of the pupil center upon the plane defined by the rim of the spectacle frame. This point is represented by a cross (point P) in Figure A.5. As can be seen in this figure, the position of the pupil center is given by two distances. The first is the naso-pupillary distance (NPD), which is the horizontal distance between the pupil and a vertical line equidistant to the two boxed centers of the frame. This line is usually known as the frame symmetry line (FSL), and we will suppose that this line is contained in the median or sagittal plane of the individual, so it is also a symmetry line for the subject's face. The vertical distance between the pupil and the lower end of the inner rim is known as the pupil's height. Both NPD and pupil's height must be measured by the practitioner. For this, there are a number of techniques ranging from the manual measurement using a simple rule to the most sophisticated apparatus based on image capturing and processing. In any case, an accurate measurement of both anthropometric distances (NPD and pupil's height) is key to successful fitting, and extreme care should be taken in measuring those distances.

The second point that must be aligned is related to the spectacle frame. In this case, we speak of two reference points, one for each frame's rim. If we are working within the framework of the ISO 8624 standard, the obvious reference points are the boxed centers. Therefore, for each rim we will define a Cartesian coordinate system centered on the corresponding boxed center, with the x-axis parallel to the HCL and its y-axis parallel to the FSL, as depicted in Figure A.5. If we know both the spectacle frame dimensions and the anthropometric distances, it is easy to compute the pupil coordinates in those Cartesian systems. For the right eye, those coordinates are

$$x_{Rp} = \frac{c}{2} - NPD = \frac{a+d}{2} - NPD,$$

$$y_{Rp} = h_p - \frac{b}{2}.$$

(A.1)

where a, b, and c are the vertical lens distance, horizontal lens distance, and the distance between lenses, respectively.

For the left eye, the pupil coordinates are identical but for a global sign change in the horizontal coordinate, which is computed as $x_{Lp} = NPD - c/2$. This is due to the reversed orientation of the left eye rim of the spectacle frame. In those cases where the eyes are located symmetrically with respect to the frame in the horizontal direction, the pupillary distance can be substituted by the interpupillary distance (IPD) divided by two in equation (A.1). This is a usual practice in ophthalmic lens dispensing, but the reader must be aware that using NPD is always more accurate.

Example 48 The antropometric measurements of a individual are: Right eye $NPD = 31$ mm, $h_p = 21$ mm, Left eye $NPD = 32.5$ mm, $h_p = 22$ mm. This individual wears a spectacle frame marked as 52□18. If the vertical lens size of this frame is 36 mm, compute the pupil coordinates for both eyes.

According to the frame's data, we see that $a = 52$ mm and $d = 18$ mm, so $c = 70$ mm and we know also that $b = 36$ mm. Therefore, the coordinates for the right eye will be

$$x_{Rp} = \frac{c}{2} - NPD = \frac{70}{2} - 31 = 4 \, \text{mm},$$

$$y_{Rp} = h - \frac{b}{2} = 21 - \frac{36}{2} = 3 \, \text{mm},$$

while the corresponding ones for the left eye are

$$x_{Lp} = NPD - \frac{c}{2} = 32.5 - \frac{70}{2} = -2.5 \, \text{mm},$$

$$y_{Rp} = h_p - \frac{b}{2} = 22 - \frac{36}{2} = 4 \, \text{mm}.$$

According to these results, both right and left eye pupil's are closer to the nose than the boxed centers, and they are placed higher than the boxed centers of the frame, which is a quite common situation in practice.

We will discuss now the location of the third reference point, which is located on the lens. There are two main cases, either the lens has been marked with a focimeter or the lens manufacturer has provided a "fitting cross" marked over the lens surface. The first case occurs for standard monofocal lenses with or without prismatic decentration, while the second case is common for progressive free-form lenses or when the lens comes with a ground prism. For example, this is the case with pure cylindrical lenses when the orientation of the base of the prism is closer to that of the axis of the cylinder. Therefore, we will give different centering rules for each of those cases.

Centering Rule for Monofocal Lenses with Prismatic Decentration

In this case, the goal of the fitting process is to place in front of the eye (when looking at the primary gaze position) the point of the lens that produced the prismatic effect requested by the patient's prescription. For this, the practitioner can mark the desired point over the lens using the focimeter. Alternatively, he can mark the optical center of the lens and then proceed to calculate the proper position of the optical center of the lens with respect to the pupil and the boxed center of the mount. We have represented this situation in Figure A.6, where we have represented the position of the optical center relative to the boxed center, given by vector $\mathbf{r}_O$. Once this vector is known, it is possible to fix a blocking piece to the lens using a projection marker. Nowadays, this process is usually made automatically by the edger itself or by a centering device connected with the edger.

In order to compute the location of the optical center, $\mathbf{r}_O$, for a given prescription prism, $\mathbf{p}$, it is necessary to know the position of the pupil (given by vector $\mathbf{r}_P$) and the vector that gives the relative position of the pupil as measured from the optical center, vector $\mathbf{r}_{OP}$. According to Figure A.6, those vectors are related by $\mathbf{r}_O = \mathbf{r}_P - \mathbf{r}_{OP}$. Vector $\mathbf{r}_P$ is known through the anthropometric measurement and frame dimensions, while vector $\mathbf{r}_{OP}$ is determined by the prescription prism through the generalized Prentice's rule (see Chapter 5), $\mathbf{p} = -\mathbb{F}_d \mathbf{r}_{OP}$. The matrix $\mathbb{F}_d$ is related to the lens power and the viewing distance for which the prismatic effect is sought. For distant objects and thin lenses, $\mathbb{F}_d$ is well approximated by the lens power matrix $\mathbb{P}$, so we have

$$\mathbf{r}_{OP} = -\mathbb{P}^{-1}\mathbf{p}. \tag{A.2}$$

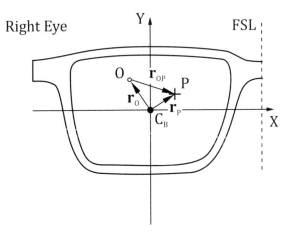

Figure A.6 Relative position of the optical center of the lens (marked as O) with relation to the pupil (marked as P) and the boxed center (marked as C_B) within a frame.

Here we are assuming the inverse matrix $\mathbb{P}^{-1}$ exists, which basically means $\mathbb{P}$ is neither zero nor a pure cylinder (see discussion in Section 5.4.6). Also, the amount of spherical power in $\mathbb{P}$ should be large enough that $\mathbf{r}_{OP}$ is a relatively small vector and the decentration can be achieved with the available lens diameter.

Once we know how to compute $\mathbf{r}_{OP}$, the location of the lens optical center relative to the frame center is given by

$$\mathbf{r}_O = \mathbf{r}_P + \mathbb{P}^{-1}\mathbf{p}. \tag{A.3}$$

Let us see an example of how this equation is used in an actual situation.

Example 49 The prescription for the right eye of a patient is $[-2.50, 1.25 \times 90°]$, $2.5\Delta120°$. If the subject wears a spectacle frame marked $54\square19$, with a vertical lens size of 34 mm, and her anthropometric measurements are $NPD = 33$ mm and $h_p = 19$ mm, compute the coordinates of the optical center of the lens referred to the boxed center of the frame.

We will first compute the prismatic power in Cartesian coordinates

$$\mathbf{p} = \begin{bmatrix} 2.5 \times \cos 120 \\ 2.5 \times \sin 120 \end{bmatrix} = \begin{bmatrix} -1.25 \\ 2.16 \end{bmatrix}\Delta.$$

We also get the dioptric power matrix of the lens from Long's equations

$$\mathbb{P} = \begin{bmatrix} -1.25 & 0 \\ 0 & 2.50 \end{bmatrix}D,$$

and its inverse from matrix algebra

$$\mathbb{P}^{-1} = \begin{bmatrix} -0.8 & 0 \\ 0 & 0.4 \end{bmatrix}D^{-1}.$$

Thus, we can now compute the vector $\mathbf{r}_{OP}$ from equation (A.2) as

$$\mathbf{r}_{OP} = -\begin{bmatrix} -0.8 & 0 \\ 0 & 0.4 \end{bmatrix}\begin{bmatrix} -1.25 \\ 2.16 \end{bmatrix} = \begin{bmatrix} -1.0 \\ -0.86 \end{bmatrix}\text{cm} \equiv \begin{bmatrix} -10 \\ -8.6 \end{bmatrix}\text{mm}.$$

Notice the change of units after application of the inverse form of Prentice's rule. The next step is the computation of the pupil coordinates from equation (A.1) as

$$x_{Rp} = \frac{c}{2} - NPD = \frac{54 + 19}{2} - 33 = 3.5\,\text{mm},$$

$$y_{Rp} = h_p - \frac{b}{2} = 19 - \frac{34}{2} = 2\,\text{mm}.$$

Finally, the coordinates of the optical center are given as

$$x_O = x_p - x_{OP} = 3.5 - (-10) = 13.5\,\text{mm},$$

$$y_O = y_p - y_{OP} = 2 - (-8.6) = 10.6\,\text{mm}.$$

It is interesting to note that the obtained solution is not entirely satisfactory, as the distance between the center of the frame and the optical center of the lens is more than one centimeter. Such long decentering would make necessary to use a high diameter lens. In such cases, ordering a lens with ground prism is more advisable, but to check this it is necessary first to compute the location of the optical center as it has been done in this example.

Centering Rule for Monofocal Lenses without Prism Decentration

This is one of the most common cases found in practice. The optical center of the lens is marked using the focimeter, and now we have to align this point with the references for the spectacle frame and eye. One may think that, if no prism is required, $p = \mathbf{0}$, the direct application of equation (A.3) would indicate that $\mathbf{r_O} = \mathbf{r_P}$. In other words, the position of the optical center and the pupil would coincide. This situation is known as pupil centering. However, although this is true in the horizontal direction, we will see that due to the pantoscopic tilt the position of the optical center should lie below the pupil.

In Figure A.7, we have represented a lateral view of the spectacle frame and the eye in primary gaze position. As we can see in this figure, the line of temple is parallel to the primary gaze position, but, due to the shape of the frame, the rim plane is tilted with respect to the sagittal plane of the head, which is perpendicular to the primary gaze position. This tilt of the rim's plane is known in the literature as *pantoscopic tilt*, and the angle that defines it is known as the *pantoscopic angle*. For most spectacle frames, the value of this angle ranges between $0°$ and $10°$, depending on the frame design.

If the lens is centered on the pupil, we have the situation depicted in the left side of Figure A.8, where the lens is tilted with respect to the primary gaze position. The effect of this pantoscopic tilt is the introduction of the so-called central oblique astigmatism. This is an oblique astigmatism produced by the inclination of the lens. Even at the the optical center, the user will experience an unwanted amount of astigmatism. Moreover, as we can see in the left side of Figure A.8, the optical axis of the lens given by line $\overline{PQ}$ in Figure A.8 no longer passes through the rotation center of the eye, point C_R. Thus, the design conditions are not fulfilled, and this may impair the quality of vision for the lens user. As can

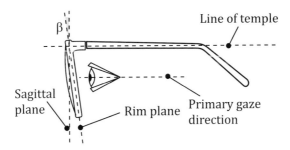

Figure A.7 Pantoscopic angle β is the angle formed by the plane containing the spectacle frame rims with the sagittal plane of the user head when he looks at a distant object in the primary gaze position.

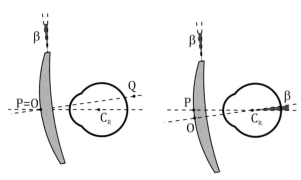

Figure A.8 Effect of pantoscopic angle without (left) and with (right) vertical decentration of the lens. The vertical decentration of the lens allows fulfillment of the design conditions of the lens, as the optical axis of the lens passes through the rotation center of the eye.

be seen in the right side of Figure A.8, if the lens is slightly decentered in the vertical direction, it is possible to find a situation where the optical axis of the lens passes through the rotation center of the eye, albeit not for the primary gaze position. In this case, the user will still experience some oblique astigmatism when looking through the lens, but, as the design conditions are now fulfilled, this unwanted astigmatism will lie under the tolerance level of the eye.

The proper amount of vertical decentration can be computed from the situation depicted in the right side of Figure A.8. From simple trigonometry, we see that $\overline{PO} = \overline{PC_R} \sin \beta$. If we assume that $\overline{PC_R} \cong l_2'$, where l_2' is the distance between the lens back vertex and the rotation center of the eye, then the centering rule for monofocal lenses can be stated as follows

$$x_O = x_p$$
$$y_O = y_p - l_2' \sin \beta. \tag{A.4}$$

where $\mathbf{r}_O = (x_O, y_O)$. So, the horizontal position of the optical center of the lens and the pupil are the same, while the vertical position of the lens is located below the pupil an amount that depends on the value of pantoscopic angle according to equation (A.4). In many practical situations, if we take $l_2' = 27$ mm, we can compute the vertical decentration of the optical center of the lens as $\delta y = y_O - y_p \cong -\beta \, (^\circ) / 2$, which is the amount of pantoscopic angle measured in degrees divided by two. This approximation is widely used in practice.

Centering Rule for Lenses Marked with Fitting Marks

The last case is that of the lenses that come from the manufacturer marked with a number of impressions over the lens surface to help the practitioner in the process of lens fitting. There are two kinds of fitting marks: The "fitting cross," a point usually marked with a cross, and some longitudinal marks that indicate the proper lens orientation. The fitting cross plays the

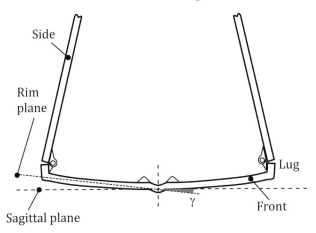

Figure A.9 Definition of facial angle, γ, as the angle formed by the projection of the sagittal and rim plane over the plane defined by the three points in which the spectacle frame is held against the head.

same role as the optical center marked with the focimeter. It is the reference point over the lens that should be properly located in relation to the pupil and boxed center of the frame. The longitudinal marks are a number of lines marked to help the practitioner find the proper orientation of the lens when using the lens blocker. Nowadays, there are three main kinds of lenses marked by the manufacturer, monofocal lenses with a ground prism, free-form monofocal lenses, and progressive addition lenses. For all of these lenses, the centering rule is quite simple; the fitting cross should be located in the same point as the projection of the pupil over the rim's lens in the primary gaze position. Moreover, the direction of the orientation marks should be parallel to the horizontal center line of the frame.

Despite its simplicity, extreme care should be taken in fitting this kind of lens, because it requires high accuracy in both the fitting process and the measurement of anthropometric distances. Moreover, many modern lenses can be designed specifically for a given user, which means that the practitioner should provide the manufacturer with all the relevant data regarding the spectacle frame and user. This includes the accurate measurement of all the relevant magnitudes, including pantoscopic and facial angles. The facial angle is defined as the horizontal angle formed by the rim's plane and the sagittal plane as shown in Figure A.9.

Appendix B

Introduction to Matrix Algebra

B.1 Introduction

In this appendix we are going to review the matrix algebra concepts necessary for the understanding of the matrix methods used in the book, such as the power matrix of a lens or the transfer matrix in multilayer films. For a comprehensive introduction to the subject the reader is directed to [411].

B.2 Matrices

A matrix is a collection of numbers arranged in rows and columns

$$\mathbb{A} = \begin{bmatrix} a_{11} & \cdots & a_{1m} \\ \vdots & \ddots & \vdots \\ a_{n1} & \cdots & a_{nm} \end{bmatrix}, \tag{B.1}$$

where a_{ij} are the matrix elements. The notation a_{ij} is used to design an element corresponding to the row i column j. The number of rows and columns determines the matrix dimensions and is usually written as $n \times m$, where n is the number of columns and m is the number of rows. In the book the most common matrix dimension is 2×2, and in this case, with only four elements we can avoid the use of indexes and can write the matrix as

$$\mathbb{A} = \begin{bmatrix} a & b \\ c & d \end{bmatrix}. \tag{B.2}$$

In this book, a matrix always represents a geometrical or optical concept that cannot be represented by a single number, and therefore, the elements of a matrix are always related between them. For example, if we change the reference system in which we have defined a geometrical concept represented by a matrix, the elements will not change independently.

With respect to vectors, a row vector with p components is a $1 \times p$ matrix. A column vector with p components is a $p \times 1$ matrix. A scalar is a real number that can be interpreted as a 1×1 matrix.

For example, temperature, pressure, lens thickness, and lens weight are all scalar magnitudes. Velocity, acceleration, electrical field, and position of a point are all vector magnitudes. Finally, the power of an astigmatic lens and the curvature of a surface without rotational symmetry must be described by a matrix.

B.3 Definitions

Square Matrix Matrix with the same number of rows and columns.

Main diagonal In square matrices, the main diagonal describes the elements in the diagonal that starts in the upper left corner and ends in the lower right corner. For a 2×2 matrix, the main diagonal is $[a\,d]$.

Symmetrical matrix A square matrix for which the entries are symmetric with respect to the main diagonal. A symmetric matrix is a square matrix that is equal to its transpose, $a_{ij} = a_{ji}$. For a 2×2 matrix, the matrix is symmetric if $b = c$.

Trace The trace of a square matrix, $\text{tr}\,(\mathbb{A})$, is the sum of the elements of the main diagonal. For a 2×2 matrix, $\text{tr}\,(\mathbb{A}) = a + d$. For a square matrix, of dimension n, $\text{tr}\,(\mathbb{A}) = \sum_{i=1}^{n} a_{ii}$.

Determinant The determinant is a number that can be computed from the elements of a square matrix with important applications in linear algebra such as the solution of a linear set of equations. For a 2×2 matrix, the determinant is

$$\det (\mathbb{A}) = |\mathbb{A}| = \begin{vmatrix} a & b \\ c & d \end{vmatrix} = ad - bc .$$

For a 3×3 matrix $\mathbb{A} = \begin{bmatrix} a & b & c \\ d & e & f \\ g & h & i \end{bmatrix},$

$$\det (\mathbb{A}) = \begin{vmatrix} a & b & c \\ d & e & f \\ g & h & i \end{vmatrix} = a \begin{vmatrix} e & f \\ h & i \end{vmatrix} - b \begin{vmatrix} d & f \\ g & i \end{vmatrix} + c \begin{vmatrix} d & e \\ g & h \end{vmatrix}.$$

Identity matrix A square matrix with main diagonal elements equal to 1 and zero otherwise. The identity matrix is denoted as $\mathbb{I}$. For example, the 2×2 identity matrix is $\mathbb{I} = \begin{bmatrix} 1 & 0 \\ 0 & 1 \end{bmatrix}.$

Diagonal matrix A square matrix with only main diagonal elements. For dimension 2, a diagonal matrix will be $\mathbb{D} = \begin{bmatrix} a & 0 \\ 0 & d \end{bmatrix}.$

Transpose matrix Transposition is an operator that switches the row and column indices of the matrix, producing another matrix denoted as $\mathbb{A}^{T}$. For example, the transpose of a square matrix of dimension 2 is

$$\mathbb{A}^{T} = \begin{bmatrix} a & c \\ b & d \end{bmatrix},$$

and the transpose of a square matrix of dimension 3 is

$$\mathbb{A}^\mathsf{T} = \begin{bmatrix} a & d & g \\ b & e & h \\ c & f & i \end{bmatrix}.$$

For vectors, the transpose of a row vector is a column vector and vice versa, that is, if $\mathbf{r} = \begin{bmatrix} x, y \end{bmatrix}$ then

$$\mathbf{r}^\mathsf{T} = \begin{bmatrix} x \\ y \end{bmatrix}.$$

B.4 Operations with Vectors and Matrices

Some of the basic arithmetic operations with scalars can be extended to matrices, such as addition. However, there are operations like multiplication that must be defined in a different way. In the next examples we will use arbitrary square matrices of dimension 2 and column and row vectors of dimension 2,

$$\mathbb{A} = \begin{bmatrix} a & b \\ c & d \end{bmatrix}; \; \mathbb{B} = \begin{bmatrix} s & t \\ u & v \end{bmatrix}; \; \mathbf{r} = \begin{bmatrix} x \\ y \end{bmatrix}; \; \boldsymbol{\rho} = \begin{bmatrix} \xi & \eta \end{bmatrix}. \tag{B.3}$$

Scalar multiplication. The multiplication of a scalar λ by a matrix $\mathbb{A}$ is performed element by element. This is a commutative operation

$$\lambda \mathbb{A} = \mathbb{A}\lambda = \begin{bmatrix} \lambda a & \lambda b \\ \lambda c & \lambda d \end{bmatrix}.$$

Sum of matrices. The sum of two matrices is performed element by element, and is only valid for matrices of the same dimension. This is a commutative operation

$$\mathbb{A} + \mathbb{B} = \mathbb{B} + \mathbb{A} = \begin{bmatrix} a+s & b+t \\ c+u & d+v \end{bmatrix}. \tag{B.4}$$

Dot product. For vectors $\mathbf{a}$ and $\mathbf{b}$ of the same dimension, the dot product, $c = \mathbf{a} \cdot \mathbf{b}$, is a scalar equal to the sum of the products of the corresponding elements of the two vectors and is denoted by a "dot." Algebraically, we will arrange the first vector as a row and the second vector as a column, so the number of columns of the first equals the number of rows of the second, and then multiply each component in order and finally sum all the products. For dimension 2 vectors, this operation is given by

$$\mathbf{r} \cdot \boldsymbol{\rho} = \underbrace{\begin{bmatrix} x & y \end{bmatrix}}_{1 \times 2} \underbrace{\begin{bmatrix} \xi \\ \eta \end{bmatrix}}_{2 \times 1} = x\xi + y\eta. \tag{B.5}$$

If θ is the angle between the vectors $\mathbf{r}$ and $\boldsymbol{\rho}$, the dot product can be calculated as

$$\mathbf{r} \cdot \boldsymbol{\rho} = \|\mathbf{r}\| \, \|\boldsymbol{\rho}\| \cos \theta, \tag{B.6}$$

where $\|\mathbf{r}\| = (\mathbf{r} \cdot \mathbf{r})^{1/2}$ is the norm of $\mathbf{r}$. If the vector ρ is a unit vector, $\|\rho\| = 1$, then the dot product $\mathbf{r} \cdot \rho$ is the projection of vector $\mathbf{r}$ along direction ρ. If two vectors are perpendicular, $\theta = 90°$ and $\mathbf{r} \cdot \rho = 0$.

Cross product. The cross product of two vectors $\mathbf{a} = \begin{bmatrix} a_1 & a_2 & a_3 \end{bmatrix}$ and $\mathbf{b} = \begin{bmatrix} b_1 & b_2 & b_3 \end{bmatrix}$ is a vector $\mathbf{c} = \mathbf{a} \times \mathbf{b}$ that is perpendicular (orthogonal) to both with a direction given by the right-hand rule and a magnitude equal to the area of the parallelogram spanned by $\mathbf{a}$ and $\mathbf{b}$. If $\mathbf{e}_x$, $\mathbf{e}_y$, and $\mathbf{e}_z$ are the standard basis (unit vectors and oriented along the XYZ Cartesian axes) then the cross product is defined as

$$\mathbf{c} = \mathbf{a} \times \mathbf{b} = \begin{vmatrix} \mathbf{e}_x & \mathbf{e}_y & \mathbf{e}_z \\ a_1 & a_2 & a_3 \\ b_1 & b_2 & b_3 \end{vmatrix} = \begin{vmatrix} a_2 & a_3 \\ b_2 & b_3 \end{vmatrix} \mathbf{e}_x - \begin{vmatrix} a_1 & a_3 \\ b_1 & b_3 \end{vmatrix} \mathbf{e}_y + \begin{vmatrix} a_1 & b_2 \\ b_1 & b_2 \end{vmatrix} \mathbf{e}_z, \quad \text{(B.7)}$$

that expressed in components will be

$$\mathbf{c} = \left(\begin{vmatrix} a_2 & a_3 \\ b_2 & b_3 \end{vmatrix} \quad - \begin{vmatrix} a_1 & a_3 \\ b_1 & b_3 \end{vmatrix} \quad \begin{vmatrix} a_1 & b_2 \\ b_1 & b_2 \end{vmatrix} \right). \quad \text{(B.8)}$$

Matrix multiplication. Two matrices can be multiplied if the number of columns of the first equal the number of rows of the second. When this condition is met, the element i, j of the product matrix is calculated as the dot product of the ith row of the first matrix by the jth column of the second matrix. The dot product is actually the product of a $1 \times n$ by a $n \times 1$ matrices.

$$\mathbb{A}\mathbb{B} = \begin{bmatrix} a & b \\ c & d \end{bmatrix} \begin{bmatrix} s & t \\ u & v \end{bmatrix} = \begin{bmatrix} as + bu & at + bv \\ cs + du & ct + dv \end{bmatrix}. \quad \text{(B.9)}$$

This operation is noncommutative, $\mathbb{A}\mathbb{B} \neq \mathbb{B}\mathbb{A}$. An important case of matrix multiplication is the product of a matrix by a vector

$$\mathbb{A}\mathbf{r} = \underbrace{\begin{bmatrix} a & b \\ c & d \end{bmatrix}}_{2\times 2} \underbrace{\begin{bmatrix} x \\ y \end{bmatrix}}_{2\times 1} = \underbrace{\begin{bmatrix} ax + by \\ cx + dy \end{bmatrix}}_{2\times 1}, \quad \text{(B.10)}$$

$$\mathbf{r}^{\mathsf{T}}\mathbb{A} = \underbrace{\begin{bmatrix} x & y \end{bmatrix}}_{1\times 2} \underbrace{\begin{bmatrix} a & b \\ c & d \end{bmatrix}}_{2\times 2} = \underbrace{\begin{bmatrix} ax + by & cx + dy \end{bmatrix}}_{1\times 2}. \quad \text{(B.11)}$$

as we can see, the product of a matrix by a vector is also noncommutative. The product of a matrix or vector by the identity matrix produces the same initial matrix or vector

$$\mathbb{A}\mathbb{I} = \mathbb{I}\mathbb{A} = \mathbb{A}; \quad \mathbb{I}\mathbf{r} = \mathbf{r}; \quad \mathbf{r}^{\mathsf{T}}\mathbb{I} = \mathbf{r}^{\mathsf{T}}. \quad \text{(B.12)}$$

Matrix inversion. The division operation is not defined for matrices. However, it is possible to calculate the inverse of a matrix. The inverse of matrix $\mathbb{A}$, denoted by $\mathbb{A}^{-1}$, is a matrix such as

$$AA^{-1} = A^{-1}A = \mathbb{I}. \tag{B.13}$$

This is very useful for solving linear equations. If we have a known column vector **b** and matrix $\mathbb{A}$, the solution to the set of linear equations

$$A\mathbf{x} = \mathbf{b}, \tag{B.14}$$

can be obtained by multiplying both sides by $\mathbb{A}^{-1}$, obtaining the solution as a column vector given by

$$\mathbf{x} = \mathbb{A}^{-1}\mathbf{b}. \tag{B.15}$$

For high dimension matrices, the inverse calculation can be complicated; however, for dimension 2, the inverse is given by

$$\mathbb{A}^{-1} = \frac{1}{\det(\mathbb{A})} \begin{bmatrix} d & -b \\ -c & a \end{bmatrix}. \tag{B.16}$$

For dimension 3 matrices, a useful formula is the Cayley–Hamilton decomposition given by

$$\mathbb{A}^{-1} = \frac{1}{\det(\mathbb{A})} \left\{ \frac{1}{2} \left(\left[\mathrm{tr}\,(\mathbb{A})^2 - \mathrm{tr}\,(\mathbb{A}\mathbb{A}) \right] \mathbb{I} - \mathbb{A}\,\mathrm{tr}\,(\mathbb{A}) + \mathbb{A}\mathbb{A} \right) \right\}. \tag{B.17}$$

Pseudo inverse. In the definition of the matrix inverse, we have assumed that the matrix $\mathbb{A}$ is square and that the linear set of equations $A\mathbf{x} = \mathbf{b}$ has the same number of equations (the number of rows of $\mathbb{A}$) as unknowns (the number of columns of $\mathbb{A}$). It is common to find problems for which there are more equations than unknowns. In this case, we can also write the linear set of equations as $A\mathbf{x} = \mathbf{b}$, but $\mathbb{A}$ is no longer square (it has dimension $n \times m$) because the system has n equations and m unknowns with $n > m$. In this case, it is possible to find a solution in the least squares sense given by

$$\mathbf{x} = \left[(A^T A)^{-1} A \right] \mathbf{b}, \tag{B.18}$$

where the matrix $\mathbb{B} = (A^T A)^{-1} A$ is called the pseudo inverse of $\mathbb{A}$.

Matrix norm. The Frobenius norm of matrix $\mathbb{A}$ is defined as the square root of the sum of its components squared,

$$\|A\| = \sqrt{a^2 + b^2 + c^2 + d^2}. \tag{B.19}$$

B.5 Axis Rotation

Matrices and vectors usually represent physical magnitudes with a directional character. Therefore, their elements depend on the coordinate system. If we change the coordinate system, all elements of a matrix or vector do change when expressed in the new system.

One common change of coordinate system is the rotation of the coordinate axes. Following Figure B.1, given a vector **r** with coordinates $\begin{bmatrix} x & y \end{bmatrix}^T$ in the system XY, if we

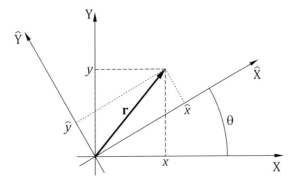

Figure B.1 Rotation of a coordinate system.

rotate the axes by an angle θ, we will have a new coordinate system given by $\hat{X}\hat{Y}$, and the elements of $\mathbf{r}$ in the new coordinate system will be given by

$$\hat{x} = x\cos\theta + y\sin\theta,$$
$$\hat{y} = -x\sin\theta + y\cos\theta.$$

This linear relation between $\hat{x}, \hat{y}$ and x, y can be expressed as a matrix product

$$\begin{bmatrix} \hat{x} \\ \hat{y} \end{bmatrix} = \begin{bmatrix} \cos\theta & \sin\theta \\ -\sin\theta & \cos\theta \end{bmatrix} \begin{bmatrix} x \\ y \end{bmatrix}, \tag{B.20}$$

where the matrix is known as the rotation matrix and is usually given the name $\mathbb{R}(\theta)$. In compact notation, the transformation of coordinates reads $\hat{\mathbf{r}} = \mathbb{R}\mathbf{r}$, where $\hat{\mathbf{r}}$ is the same vector as $\mathbf{r}$, but with its coordinates corresponding to the rotated system. The inverse relation would be $\mathbf{r} = \mathbb{R}^{-1}\hat{\mathbf{r}}$, where the inverse matrix always exists and is given by

$$\mathbb{R}^{-1} = \begin{bmatrix} \cos\theta & -\sin\theta \\ \sin\theta & \cos\theta \end{bmatrix},$$

as it represents a rotation by the same angle but in the opposite direction.

The components of a matrix also transform under rotation of the coordinate system, and the rotation matrices are also involved in the transformation. Assume the matrix $\mathbb{A}$ transforms an arbitrary vector $\mathbf{r}$ into vector $\mathbf{s}$, $\mathbf{s} = \mathbb{A}\mathbf{r}$. In the rotated system, we would have $\hat{\mathbf{s}} = \hat{\mathbb{A}}\hat{\mathbf{r}}$. Now, applying equation (B.20) to both vectors, we have

$$\mathbb{R}\mathbf{s} = \hat{\mathbb{A}}\mathbb{R}\mathbf{r}.$$

Left-multiplying both sides of the equation by $\mathbb{R}^{-1}$, we have $\mathbb{R}^{-1}\mathbb{R}\mathbf{s} = \mathbb{R}^{-1}\hat{\mathbb{A}}\mathbb{R}\mathbf{r}$, that is, $\mathbf{s} = \mathbb{R}^{-1}\hat{\mathbb{A}}\mathbb{R}\mathbf{r}$. If we compare the latter with the initial equation $\mathbf{s} = \mathbb{A}\mathbf{r}$, and recall $\mathbf{r}$ is arbitrary, we must conclude that $\mathbb{A} = \mathbb{R}^{-1}\hat{\mathbb{A}}\mathbb{R}$, or equivalently,

$$\hat{\mathbb{A}} = \mathbb{R}\mathbb{A}\mathbb{R}^{-1}, \tag{B.21}$$

which is the equation that gives us the components of the matrix $\mathbb{A}$ in the rotated system.

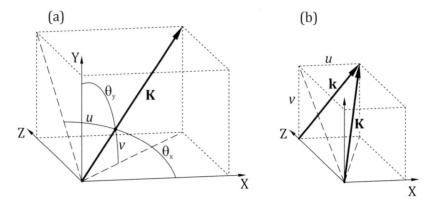

Figure B.2 (a) Unit vector in space showing the angles θ_x, θ_y and their complementaries u and v. (b) For vectors forming a small angle with the Z axis, $\sin u \simeq u$ and $\sin v \simeq v$, and the projection of the vector on the XY plane has approximate coordinates (u, v).

B.6 Interpretation of the Components of a 2D Direction Vector

We are going to discuss how the three components of a unit vector $\mathbf{K} = \begin{bmatrix} k_x & k_y & k_z \end{bmatrix}^1$ can be interpreted in terms of its direction in space, especially when the vector $\mathbf{K}$ is almost perpendicular to the XY plane. Unit vectors are very useful to describe directions in space. For example, the projection of any arbitrary vector $\mathbf{r}$ along direction $\mathbf{K}$ is given by the dot product $r_K = \mathbf{r} \cdot \mathbf{K}$.

Let θ_x, θ_y, and θ_z be the angles formed by $\mathbf{K}$ with the three coordinate axes X, Y, and Z, and let $\mathbf{e}_x$, $\mathbf{e}_y$, and $\mathbf{e}_z$ be the standard basis vectors (unit vectors along the X, Y, and Z axes). The three Cartesian components of $\mathbf{K}$ are given by its dot product with each of the three vectors of the standard basis

$$
\begin{aligned}
k_x &= \mathbf{K} \cdot \mathbf{e}_x = \cos \theta_x, \\
k_y &= \mathbf{K} \cdot \mathbf{e}_y = \cos \theta_y, \\
k_z &= \mathbf{K} \cdot \mathbf{e}_z = \cos \theta_z.
\end{aligned}
\tag{B.22}
$$

The angle θ_x that $\mathbf{K}$ forms with the X axis is complementary to the angle u that $\mathbf{K}$ forms with the plane YZ, that is, $\theta_x + u = \pi/2$ and $\cos \theta_x = \sin u$. In a similar way, the angle θ_y that $\mathbf{K}$ forms with the Y axis is complementary to the angle v that $\mathbf{K}$ forms with the plane XZ, that is, $\theta_y + v = \pi/2$ and $\cos \theta_y = \sin v$ (see Figure B.2(a)). According to this, we can rewrite $\mathbf{K}$ as

$$
\mathbf{K} = \begin{bmatrix} \sin u & \sin v & k_z \end{bmatrix},
$$

which is exact, and valid whenever $\mathbf{K}$ is a unit vector.

---

[1] For commodity, we will use row vectors in this section.

Let us assume now that $\mathbf{K}$ is almost parallel to the Z axis. This means $\theta_z \simeq 0$ and $\cos \theta_z \simeq 1$. Also, both angles u and v will be very small, so $\sin u \simeq u$ and $\sin v \simeq v$. Using these approximations, the vector $\mathbf{K}$ reads

$$\mathbf{K} \simeq \begin{bmatrix} u & v & 1 \end{bmatrix}. \tag{B.23}$$

As the third component is locked to 1, we only need the first two components to describe the direction. The same information is then stored in $\mathbf{K}$ and in its 2-dimensional projection on the XY plane, $\mathbf{k}$

$$\mathbf{k} = \begin{bmatrix} u & v \end{bmatrix}, \tag{B.24}$$

as shown in Figure B.2(b). Of course, $\mathbf{k}$ is not a unit vector and only describes the direction of $\mathbf{K}$ under the above-mentioned approximations.

From a more algebraic point of view, if $\|\mathbf{K}\| = 1$ its three components are not independent but linked by the relation

$$k_z = \sqrt{1 - \left(k_x^2 + k_y^2 \right)},$$

and when the vector is almost parallel to the Z axis, its first two components are very small, so $\left(k_x^2 + k_y^2 \right) \ll 1$ and $k_z \simeq 1$. The approximations and the dual vector description shown here will be profusely used throughout the book, as paraxial rays form small angles with the Z axis, and their directions are well represented by unit vectors and their 2D counterparts.

Appendix C

Introduction to Surface Geometry

C.1 Introduction

The goal of this appendix is to provide an introduction to the mathematical techniques necessary for describing the geometry of a surface. To do so, we will use some of the concepts and methods of differential geometry, but we will always try to keep it at a reasonable level. We will assume that the reader has a background in Cartesian geometry and elementary calculus, so concepts such as coordinates, vectors, derivatives, etc. are not unknown to him/her. Differential geometry is a branch of mathematics focused on the study of curves and surfaces, and it has numerous applications in physics and engineering. In the field of optics, the methods and techniques of differential geometry have been successfully applied in geometrical optics, essentially through the relationship between curvature and optical power. In our study, we will follow the order given by the classical texts on differential geometry, so we will study curves and then surfaces. As previously mentioned, our aim is to introduce the concepts with the lowest mathematical complexity possible, so our presentation will be neither rigorous nor comprehensive. Indeed, at the end of this appendix we will focus on the study of the so-called parabolic approximation due to its relevance in paraxial optics. Therefore, we refer the reader interested in getting a deeper insight on this subject to the numerous and excellent text books on differential geometry available, such as references [412, 413].

C.2 Curves

C.2.1 Curve Definition

Intuitively, one may imagine a curve as a trajectory in Cartesian space $\mathbb{R}^3$. According to this intuitive definition, a curve would be the set of points that a moving object has passed through. For example, the trajectory described by a pendulum is a circumference arc, while a ball describes a parabola after being shot by a cannon. In this sense, our highways and railways are also curves, and our planes describe curves through the air.

In a more operative way, we will define a curve as a vector function $\mathbf{r}(t) = [x(t), y(t), z(t)]$, where $x(t)$, $y(t)$, and $z(t)$ are functions of t. This way of defining curves is known as *parametric form*, and the coordinates of any point of the curve depend on the value

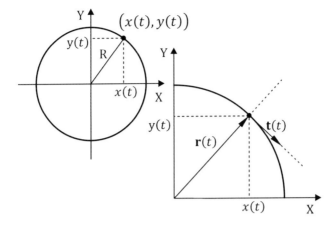

Figure C.1 Example of a plane curve: the circumference. At the right side of the figure, we can see the tangent vector $\mathbf{t}(t)$, which is perpendicular to the position vector $\mathbf{r}(t)$ that defines the curve.

of parameter t. If all the points of a given curve are contained in a plane then we have a *planar curve*, and such curves are defined by a vector function $\mathbf{r}(t) = \left[x(t), y(t)\right]$. In the following, we will focus on planar curves, and, later, we will show how to compute the relevant geometrical properties for 3D curves.

For example, a circumference in $\mathbb{R}^2$ centered at the origin (see Figure C.1) would be described by the following function

$$\mathbf{r}(t) = \left[x(t), y(t)\right] = [R\cos t, R\sin t],\qquad\text{(C.1)}$$

where R is the curvature radius of the circle. If we compute the square of the modulus of the position vector $\mathbf{r}(t)$ of a circle, we have

$$x^2 + y^2 = R^2.\qquad\text{(C.2)}$$

This equation constitutes an alternative definition of a curve, and it is known as the *implicit form* of the curve. In the general case, the implicit form of a plane curve can be formulated as $g(x, y) = 0$, where $g(x, y)$ is a two-dimensional function.

C.2.2 Tangent Vector

We have already seen that a curve is defined by the so-called position vector $\mathbf{r}(t)$. In the following, we will write the position vector $\mathbf{r}(t)$ as a column vector for both the two- and three-dimensional cases. For example, the position vector of a curve in space will be given as

$$\mathbf{r}(t) = \begin{bmatrix} x(t) \\ y(t) \\ z(t) \end{bmatrix}.$$

The successive derivatives of the position vector give important information about the local behavior of the curve. It can be demonstrated [412, 413] that the first derivative $\mathbf{r}'(t) \equiv \frac{d\mathbf{r}}{dt}$ at a point P is a vector tangent to the curve at this point. Indeed, the unitary vector defined as

$$\mathbf{t}(t) = \frac{\mathbf{r}'(t)}{\|\mathbf{r}'(t)\|}, \tag{C.3}$$

is known as *tangent vector* [412, 413]. The tangent vector indicates the local direction of the curve in the same way as the velocity vector indicates the direction of movement of a moving object at a given point. If the tangent vector is constant then the curve is a straight line [412].

Example 50 Calculate the tangent vector of a circle at any point.

To compute the tangent vector of a circle, we will have to calculate the first derivative of the circle's position vector given by equation (C.1). Therefore, we have

$$\mathbf{r}'(t) = \begin{bmatrix} -R\sin t \\ R\cos t \end{bmatrix}. \tag{C.4}$$

If we normalize this vector, the result is the tangent vector of a circle at any point, $\mathbf{t}(t) = [-\sin t, \cos t]^{\mathsf{T}}$, where T stands for the transposition operator. As we can see in Figure C.1, vector $\mathbf{t}(t)$ is tangent to the curve, and, in this particular example, it is also perpendicular to position vector because $\mathbf{r}(t) \cdot \mathbf{t}(t) = 0$.

C.2.3 Normal and Curvature Vectors for a Plane Curve

As we have already seen, the first derivative of a curve gives information about the local direction of the curve through the tangent vector. The second derivative will indicate the speed of change of this local direction and will allow us to introduce the concepts of normal and curvature. For a plane curve, the *normal vector* can be defined as a unitary vector perpendicular to the tangent vector at a given point, see Figure C.2. Mathematically,

$$\mathbf{n}(t) = \frac{\mathbf{t}'(t)}{\|\mathbf{t}'(t)\|}. \tag{C.5}$$

In Figure C.2 we have plotted an arbitrary planar curve that passes through point P, and we have plotted the tangent and normal vectors at this point. Notice that we can define a circle known as an *osculating circle*, which is tangent to the curve at P, so the line that joins the center of the circle and this point coincides with the direction of the normal vector at point P. A second condition for the osculating circle is that the tangent vectors of the curve and the circle have the same derivative at P. In these circumstances, we can define the *curvature* of the osculating circle as the inverse of its radius, $\kappa = R^{-1}$, and the curvature of a curve at a given point P as the curvature of the osculating circle at this point.

Therefore, the osculating circle sets the local behavior of the curve around point P, and we may regard the curvature as a magnitude that indicates the bending ratio of a curve. The more closed a curve, the higher the curvature. This fact can be easily checked as shown

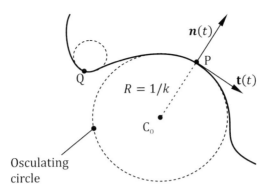

Figure C.2 Plot of a planar curve showing the tangent $\mathbf{t}(t)$ and normal $\mathbf{n}(t)$ vectors at point P and the osculating circle tangent to the curve at P. The osculating circle has the same curvature at point P as the curve. Thus, the osculating circle fixes the local behavior of the curve in the surroundings of point P. Notice that the osculating circle at point Q presents a much lower curvature radius, which indicates a faster bending rate at this point.

in Figure C.2, where we have depicted the osculating circle at a different point, Q. As the curve is more closed at this point, the curvature radius of the osculating circle is lower. Therefore, the curvature at point Q is higher than the curvature at point P. Notice that for a straight line the curvature is zero, and for a circle the curvature is constant.

From equation (C.5), for a general planar curve, the normal vector is defined through the relationship [412]

$$\mathbf{n} = \left(x'^2 + y'^2\right)^{-1/2}\begin{bmatrix} -y' \\ x' \end{bmatrix},$$

and the curvature of a planar curve, given as a function of the Cartesian coordinates, is [412]:

$$\kappa = \frac{\left|x'y'' - x''y'\right|}{\left(x'^2 + y'^2\right)^{3/2}}. \tag{C.6}$$

Therefore, for a curve given by the implicit form $y = f(x)$, we have

$$\kappa = \frac{\left|f''\right|}{\left(1 + f'^2\right)^{3/2}}, \tag{C.7}$$

where $f' \equiv \frac{df}{dx}$. We will make extensive use of this equation throughout Chapter 2.

Finally, we will define the curvature vector as the product of the normal vector by the curvature

$$\mathbf{k} = \pm\kappa \cdot \mathbf{n}. \tag{C.8}$$

Notice the sign uncertainty in this equation. Setting a positive or negative sign in equation (C.8) is equivalent to setting the local orientation of the curve. As a convention, we will take the positive sign, so in the following we will have $\mathbf{k} = \kappa \cdot \mathbf{n}$.

In conclusion, given a plane curve described by a parametric function $\mathbf{r}(t)$, it is possible to know the local behavior at any point of it through its derivatives. This is also true for nonplanar curves, and we will see now how to compute the tangent and normal vectors and the curvature for such curves.

C.2.4 Tangent, Normal and Curvature for Nonplanar Curves

For a 3D curve, the tangent vector is computed through the first derivative of $\mathbf{r}(t)$ using equation (C.3). Regarding the normal vector, it is still perpendicular to the tangent vector, but now all vectors contained in the plane perpendicular to the tangent vector accomplish this condition. Thus, it is necessary to fix an additional requirement, so we will define the normal vector as the unitary vector parallel to the curvature vector $\kappa(t)$. The curvature vector for a 3D curve is defined as [412]

$$\kappa(t) = \frac{d\mathbf{t}}{dt} \bigg/ \left\| \frac{d\mathbf{r}}{dt} \right\|. \tag{C.9}$$

Therefore, the following relationship between the curvature and normal vector (denoted by $\mathbf{N}$ in 3D curves) holds [412]

$$\kappa(t) = \pm \kappa(t) \cdot \mathbf{N}(t). \tag{C.10}$$

Notice the sign uncertainty, similar to what we discussed before for planar curves. Usually, the positive sign is chosen, and the curvature is computed through the following relationship

$$\kappa = \frac{\| \mathbf{r}' \times \mathbf{r}'' \|}{\| \mathbf{r}' \|^3}, \tag{C.11}$$

$\times$ being the cross product.

In addition to tangent and curvature vectors, the local behavior of a 3D curve is guided by the *binormal vector* $\mathbf{B}(t)$, which is a unitary vector perpendicular to both tangent and normal. Thus, the triplet of vectors $\{\mathbf{t}(t), \mathbf{N}(t), \mathbf{B}(t)\}$, together with the curvature $\kappa(t)$ and an additional magnitude called torsion $\tau(t)$, which measures the rate of rotation of the osculating plane, define completely the local behavior of any curve at a given point. For further information, see references [412, 413].

C.3 Surfaces

C.3.1 Surface Definition

We will now proceed to the study of surfaces. As with curves, we are quite familiar with the intuitive notion of surfaces as boundaries of 3D objects. However, the exact mathematical definition of a surface is not trivial, and we will restrict ourselves to an operative

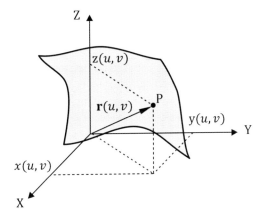

Figure C.3 Definition of surface as a set of points in $\mathbb{R}^3$, where the Cartesian coordinates of any point P located on the surface are $[x(u, v), y(u, v), z(u, v)]$.

definition. Following [413] and [412], we will define a surface as a vector function $\mathbf{r}(u, v) = [x(u, v), y(u, v), z(u, v)]$, which accomplishes some particular mathematical conditions, see details in [412, 413]. In Figure C.3, we have represented an arbitrary surface and one of the points of this surface.

A function $\mathbf{r}(u, v)$, which represents a surface is known as a parametrization of the surface. Usually, we will study the surface under the parametrization known as Monge's chart, so that $\mathbf{r}(x, y) = [x, y, z(x, y)]$, where $z(x, y)$ is a function known as *elevation*. In the following, we will describe surfaces mostly by their Monge's chart. For example, the elevation corresponding to the lower half portion of a sphere would be given as

$$z(x, y) = R\left(1 - \sqrt{1 - \frac{x^2 + y^2}{R^2}}\right), \tag{C.12}$$

where R is the curvature radius of the sphere.

C.3.2 Tangents and Normal Vectors

As with planar curves, the derivatives of a surface provide information about the local behavior of the surface. As surfaces are two-dimensional entities, we will need partial derivatives, and for them we will use the compact notation $f_x \equiv \partial f/\partial x$, $f_y \equiv \partial f/\partial y$, $f_{xy} \equiv \partial^2 f/\partial x \partial y$, and so on. In a first approach, the surface around a given point can be approximated by a plane, known as the *tangent plane* (see Figure C.4). The tangent plane at a given point is defined by two vectors $\mathbf{r}_x$ and $\mathbf{r}_y$, which are tangent to the surface at this point. These vectors can be computed through the first derivatives of the position vector $\mathbf{r}(x, y) = [x, y, z(x, y)]^\mathsf{T}$ and are given by

$$\begin{aligned} \mathbf{r}_x &= (1, 0, z_x), \\ \mathbf{r}_y &= (0, 1, z_y). \end{aligned} \tag{C.13}$$

Introduction to Surface Geometry

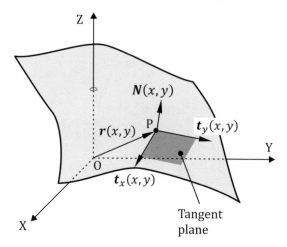

Figure C.4 Definition of the unitary tangents $\mathbf{t}_x$, $\mathbf{t}_y$, and normal, $\mathbf{N}$, of a surface at point P. The two unitary tangent vectors define the tangent plane (darker area).

The existence of the tangent plane is guaranteed by the conditions required for a vector function $\mathbf{r}\,(x, y)$ to define a surface. In particular, tangent vectors $\mathbf{r}_x$ and $\mathbf{r}_y$ should be linearly independent at any point of the surface [412, 413].

From equation (C.13), the unitary tangent vectors are given by

$$\mathbf{t}_x = \frac{\mathbf{r}_x}{\|\mathbf{r}_x\|} = \frac{[1, 0, z_x]}{\sqrt{1 + z_x^2}},$$

$$\mathbf{t}_y = \frac{\mathbf{r}_y}{\|\mathbf{r}_y\|} = \frac{[0, 1, z_y]}{\sqrt{1 + z_y^2}}. \tag{C.14}$$

The unitary vector perpendicular to the tangent plane is the normal vector of the surface, defined by

$$\mathbf{N} = \frac{\mathbf{r}_x \times \mathbf{r}_y}{\|\mathbf{r}_x \times \mathbf{r}_y\|}, \tag{C.15}$$

and, being a function of (x, y), it varies continuously across the surface. The explicit mathematical expression of the normal vector for a surface defined as a Monge's chart is

$$\mathbf{N} = \frac{[-z_x, -z_y, 1]}{\sqrt{1 + z_x^2 + z_y^2}}. \tag{C.16}$$

The first-order derivatives of a surface allow the calculation of the surface area. For a Monge's chart, the area of a surface in a given spatial domain can be computed by

$$A = \iint \sqrt{EG - F^2} dx dy, \tag{C.17}$$

where the integral is extended to the domain of definition of the function. The quantities E, F, and G are the coefficients of the *first fundamental form* of the surface, defined by

$$E = \mathbf{r}_x \cdot \mathbf{r}_x, \ G = \mathbf{r}_y \cdot \mathbf{r}_y, \ F = \mathbf{r}_x \cdot \mathbf{r}_y,$$

which, for the parametrization of a Monge's chart are

$$E = 1 + z_x^2,$$
$$F = z_x z_y,$$
$$G = 1 + z_y^2. \tag{C.18}$$

Those coefficients determine the first-order properties of a surface together with their tangent and normal vectors.

C.3.3 Main Curvatures

In the previous sections, we defined the concept of curvature of a curve. This concept may be extended to surfaces, if we consider the fact that a given surface contains infinite curves. Indeed, we will say that a curve C is contained by a surface S if all the points of the curve are also points of the surface. Therefore, given a point P of a surface S, and a curve C, contained in S, that passes through P, we can always define the curvature vector κ of curve C at P. In these conditions, we will define the *normal curvature* of curve C at P as

$$\kappa_n = \kappa \cdot \mathbf{N}, \tag{C.19}$$

where $\mathbf{N}$ is the normal vector of the surface at P. As there are an infinite number of curves contained in the surface that pass through P, we cannot define a single normal curvature for a surface at a given point, and we have, instead, a set of normal curvatures. It can be demonstrated [413] that there are always two extreme values for this set: a maximum κ_+ and a minimum curvature κ_-, which are known as *main curvatures*. The existence of the main curvatures is one of the key results of differential geometry, and to compute them it is necessary to use the concept of normal section as we will see in the next paragraph.

In Figure C.5, we have represented a normal section at a given point O on the surface as the curve defined by the intersection of the surface with a plane that contains the normal vector of the surface at this point. Considering the reference system depicted in Figure C.5, the equation of the normal section can be written as

$$\mathbf{r}(t) = [t \cos\theta, t \sin\theta, z(t\cos\theta, t\sin\theta)]^\mathsf{T}, \tag{C.20}$$

where θ is the angle formed by the XZ plane and the one generating the normal section. Note that the normal curvature at some direction coincides with the curvature of the normal section in the same direction. Then, we just have to substitute equation (C.20) into (C.11), getting the following expression

$$\kappa_n \equiv \kappa(\theta) = \frac{z_{xx}\cos^2\theta + 2z_{xy}\cos\theta\sin\theta + z_{yy}\sin^2\theta}{\left(1 + z_x^2\cos^2\theta + 2z_xz_y\cos\theta\sin\theta + z_y^2\sin^2\theta\right)^{3/2}}, \tag{C.21}$$

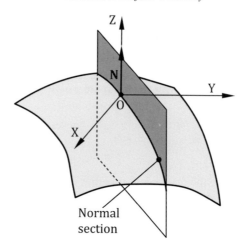

Figure C.5 Normal section at point O, defined as the curve resulting from the intersection of the surface and a plane that contains the normal vector of the surface at O.

where the derivatives are evaluated at point O. The coordinate system can always be chosen [412] in such a way that $z_x = z_y = 0$ when the derivatives are computed at O. In this reference system, equation (C.21) reads

$$\kappa_n = z_{xx} \cos^2 \theta + 2z_{xy} \cos \theta \sin \theta + z_{yy} \sin^2 \theta. \tag{C.22}$$

Although equation (C.22) represents a particular case, we may draw some general conclusions. First, the normal curvature of any curve passing through a point O has the same value as the normal curvature of the normal section that passes through O with the same tangent vector [412]. Second, equation (C.22) can be rearranged as

$$\kappa_n = \begin{bmatrix} \cos \theta & \sin \theta \end{bmatrix} \begin{bmatrix} z_{xx} & z_{xy} \\ z_{xy} & z_{yy} \end{bmatrix} \begin{bmatrix} \cos \theta \\ \sin \theta \end{bmatrix} \equiv \mathbf{u}_\theta^T \mathbb{H} \mathbf{u}_\theta, \tag{C.23}$$

$\mathbf{u}_\theta = \begin{bmatrix} \cos \theta & \sin \theta \end{bmatrix}$ being a unitary vector that points in the direction θ and $\mathbb{H} = \begin{bmatrix} z_{xx} & z_{xy} \\ z_{xy} & z_{yy} \end{bmatrix}$ is the so-called Hessian matrix of the function $z(x, y)$. When the Z axis is perpendicular to the surface at some point, the curvature of a normal section passing through this point is uniquely determined by the coefficients of the Hessian matrix and the direction of the normal section.

The normal curvature is a continuous and periodic function of θ with period π. Then it has to attain maximum and minimum values, κ_+ and κ_-, that can be obtained by solving the equation $d\kappa_n/d\theta = 0$. The directions along which the curvature is maximal or minimal are known as *main or principal directions* and they satisfy

$$\sin 2\theta_\pm = \pm \frac{2z_{xy}}{z_{xx} + z_{yy}},$$

$$\cos 2\theta_\pm = \pm \frac{z_{xx} - z_{yy}}{z_{xx} + z_{yy}}. \tag{C.24}$$

It is not difficult to check that the main directions are always orthogonal, $\theta_+ = \theta_- \pm \pi/2$. If we substitute equations (C.24) into equation (C.22), we get, after some algebra, the values of the main curvatures as

$$\kappa_- = \frac{1}{2}\left(\text{tr}\,(\mathbb{H}) - \sqrt{\text{tr}^2\,(\mathbb{H}) - 4\det(\mathbb{H})}\right),$$

$$\kappa_+ = \frac{1}{2}\left(\text{tr}\,(\mathbb{H}) + \sqrt{\text{tr}^2\,(\mathbb{H}) - 4\det(\mathbb{H})}\right), \tag{C.25}$$

where $\text{tr}\,(\mathbb{H}) = z_{xx} + z_{yy}$ and $\det\,(\mathbb{H}) = z_{xx}z_{yy} - z_{xy}^2$ are the trace and the determinant of the Hessian matrix.

Despite their usefulness, the usage of the equations in (C.25) is limited to those cases in which the reference system can be set with its Z axis perpendicular to the surface. If we want to compute the main curvatures without this limitation, we have to employ the coefficients of the so-called *second fundamental form*, defined by

$$L = -\mathbf{r}_x \cdot \mathbf{N}_x, \; M = -\frac{1}{2}\left(\mathbf{r}_x \cdot \mathbf{N}_y + \mathbf{r}_y \cdot \mathbf{N}_x\right), \; N = -\mathbf{r}_y \cdot \mathbf{N}_x,$$

where $\mathbf{N}_x$ and $\mathbf{N}_y$ are the partial derivatives of the normal vector $\mathbf{N}$ with respect to the x and y directions. Using Monge's chart parametrization, the coefficients of the second fundamental form are

$$L = \frac{z_{xx}}{\sqrt{1 + z_x^2 + z_y^2}}$$

$$M = \frac{z_{xy}}{\sqrt{1 + z_x^2 + z_y^2}} \tag{C.26}$$

$$N = \frac{z_{yy}}{\sqrt{1 + z_x^2 + z_y^2}}.$$

The coefficients of the second fundamental form determine the second-order properties of a surface. To compute the main curvatures from the second fundamental form, it is customary to define the *mean* and *Gaussian* curvatures, respectively H and K, as

$$H = \frac{EN + GL - 2FM}{2\left(EG - F^2\right)}, \tag{C.27}$$

$$K = \frac{LN - M^2}{EG - F^2}, \tag{C.28}$$

and from them

$$\kappa_- = H - \sqrt{H^2 - K},$$

$$\kappa_+ = H + \sqrt{H^2 - K}. \tag{C.29}$$

The inverse relations are straightforward

$$H = \frac{1}{2}(\kappa_- + \kappa_+),$$

$$K = \kappa_- \kappa_+.$$

(C.30)

We will see now two examples that may clarify the procedure that must be followed to compute the main curvatures. Those examples will also be helpful in order to obtain a geometrical interpretation of the mean and Gaussian curvatures.

Example 51 Compute the main curvatures for a sphere defined through equation (C.12) at any point of this surface.

First, we compute the tangent vectors as

$$\mathbf{r}_x = (1, 0, z_x) = \left[1, 0, \frac{x}{\sqrt{R^2 - x^2 - y^2}} \right],$$

$$\mathbf{r}_y = (1, 0, z_y) = \left[0, 1, \frac{y}{\sqrt{R^2 - x^2 - y^2}} \right].$$

Next, we will calculate the normal vector using equation (C.16), so, after some algebra, we have

$$\mathbf{N} = \left[-\frac{x}{R}, -\frac{y}{R}, \sqrt{1 - \frac{x^2 + y^2}{R^2}} \right].$$

From this expression, we can obtain the first derivatives of the normal vector as

$$\mathbf{N}_x = \left[-\frac{1}{R}, 0, \frac{-x}{R\sqrt{R^2 - x^2 - y^2}} \right],$$

$$\mathbf{N}_y = \left[0, -\frac{1}{R}, \frac{-y}{R\sqrt{R^2 - x^2 - y^2}} \right].$$

Thus, we can now compute the coefficients of the first fundamental form

$$E = \mathbf{r}_x \mathbf{r}_x = \frac{R^2 - y^2}{R^2 - x^2 - y^2},$$

$$G = \mathbf{r}_y \mathbf{r}_y = \frac{R^2 - x^2}{R^2 - x^2 - y^2},$$

$$F = \mathbf{r}_x \mathbf{r}_y = \frac{xy}{R^2 - x^2 - y^2},$$

and second fundamental form

$$L = -\mathbf{r}_x \mathbf{N}_x = \frac{1}{R} \frac{R^2 - y^2}{R^2 - x^2 - y^2},$$

$$N = -\mathbf{r}_y \mathbf{N}_y = \frac{1}{R} \frac{R^2 - x^2}{R^2 - x^2 - y^2},$$

$$M = -\frac{1}{2}\left(\mathbf{r}_x \mathbf{N}_y + \mathbf{r}_y \mathbf{N}_x\right) = \frac{1}{R} \frac{xy}{R^2 - x^2 - y^2},$$

so the Gaussian curvature is

$$K = \frac{LN - M^2}{EG - F^2} = \frac{1}{R^2},$$

while the mean curvature would be

$$H = \frac{EL + GN - 2FM}{2\left(EG - F^2\right)} = \frac{1}{R}.$$

Therefore, for any point of a sphere, the Gaussian and mean curvatures are constant. More-over, if we apply equation (C.29), we find that for any point of a sphere $\kappa_- \equiv \kappa_+ = R^{-1}$, so the sphere is the surface that has constant curvature (see Figure C.6), a well-known result of elementary geometry.

The next example deals with an other surface of great interest in ophthalmic optics, the cylinder.

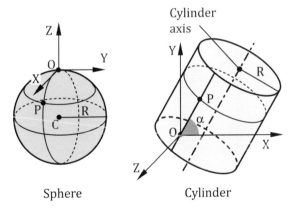

Sphere   Cylinder

Figure C.6 Representation of a sphere (left) and a cylinder (right). All points of the sphere are at the same distance R from the center of the sphere C, so the curvature at any point P is constant. For the cylinder, the two principal curvatures are zero and R^{-1}, where R is the radius of the cylinder. Note that the symmetry axis of the cylinder is tilted, with α being the angle formed by the cylinder axis and the X axis.

Example 52 Following the previous example, compute the main curvatures at any point of a cylinder given by the following elevation $z(x, y) = R - \sqrt{R^2 - (y\sin\alpha - x\cos\alpha)^2}$, where α is the angle formed by the cylinder axis and the X axis.

In this example, we will compute first the partial derivatives of the elevation. Therefore, we have, for the first partial derivatives

$$z_x = \frac{x\cos^2\alpha - y\cos\alpha\sin\alpha}{\sqrt{R^2 - (y\sin\alpha - x\cos\alpha)^2}},$$

$$z_y = \frac{-x\cos\alpha\sin\alpha + y\sin^2\alpha}{\sqrt{R^2 - (y\sin\alpha - x\cos\alpha)^2}},$$

while, the second derivatives are

$$z_{xx} = \frac{R^2\cos^2\alpha}{\sqrt{[R^2 - (y\sin\alpha - x\cos\alpha)^2]^3}},$$

$$z_{xy} = \frac{-R^2\cos\alpha\sin\alpha}{\sqrt{[R^2 - (y\sin\alpha - x\cos\alpha)^2]^3}},$$

$$z_{yy} = \frac{R^2\sin^2\alpha}{\sqrt{[R^2 - (y\sin\alpha - x\cos\alpha)^2]^3}}.$$

Then, we can use equation (C.18) to compute the coefficients of the first fundamental form as

$$E = 1 + z_x^2 = \frac{R^2 - (y\sin\alpha - x\cos\alpha)^2\sin^2\alpha}{R^2 - (y\sin\alpha - x\cos\alpha)^2},$$

$$G = 1 + z_y^2 = \frac{R^2 - (y\sin\alpha - x\cos\alpha)^2\cos^2\alpha}{R^2 - (y\sin\alpha - x\cos\alpha)^2},$$

$$F = z_x z_y = \frac{-(y\sin\alpha - x\cos\alpha)^2\sin\alpha\cos\alpha}{R^2 - (y\sin\alpha - x\cos\alpha)^2}.$$

In a similar way, the coefficients of the second fundamental form can be written as

$$L = \frac{z_{xx}}{\sqrt{1 + z_x^2 + z_y^2}} = \frac{R\cos^2\alpha}{R^2 - (y\sin\alpha - x\cos\alpha)^2},$$

$$N = \frac{z_{yy}}{\sqrt{1 + z_x^2 + z_y^2}} = \frac{R\sin^2\alpha}{R^2 - (y\sin\alpha - x\cos\alpha)^2},$$

$$M = \frac{z_{xy}}{\sqrt{1 + z_x^2 + z_y^2}} = \frac{-R\sin\alpha\cos\alpha}{R^2 - (y\sin\alpha - x\cos\alpha)^2}.$$

Notice that, in this case, $LN - M^2 = 0$, so the Gaussian curvature is also zero, that is, $K = 0$. For the mean curvature, we have

$$H = \frac{EN + GL - 2FM}{2\left(EG - F^2\right)} = \frac{1}{2R}.$$

If we now compute the main curvatures we will obtain $\kappa_- = 0$ and $\kappa_+ = R^{-1}$. Thus, for a cylinder, the main curvatures are constant, regardless of the orientation of the cylinder axis, and the minimum curvature is zero. This is logical if we consider that the cylinder is the surface formed by the rotation of a straight line (thus with no curvature) around a parallel axis, with R being the radius of rotation. Notice that the surface astigmatism is related to the difference of the main curvatures, so, according to equation (C.29), this difference would be $\Delta\kappa = \kappa_+ - \kappa_- = 2\left|\sqrt{H^2 - K}\right|$ so, in this particular case it would be $\Delta\kappa = R^{-1}$.

We have seen two classical examples of the computation of the first and second fundamental form in analytical surfaces and the calculation of the main curvatures. Given the high symmetry of the surfaces studied (the sphere and cylinder), the resulting curvatures are very simple, but the algebra involved, although not formally difficult, is a bit cumbersome. We will see in the next section that in the frame of paraxial optics we can simplify the problem of computing the main curvatures by approximating the surface at a point by a paraboloid using the so-called parabolic approximation.

C.4 Geometry of the Parabolic Approximation

C.4.1 Parabolic Approximation

Within the framework of paraxial geometrical optics we are interested in the properties of surfaces in a region close to the optical axis of the system. We define the *surface vertex* as the point where the surface intercepts the optical axis. If we choose the origin of the coordinate system to coincide with this vertex, the surface can be approximated by its Taylor expansion around the point $(0,0)$

$$z\left(x,y\right) = z_x x + z_y y +$$

$$+ \frac{1}{2}\left(z_{xx}x^2 + 2z_{xy}xy + z_{yy}y^2\right)$$

$$+ \frac{1}{6}\left(z_{xxx}x^3 + 3z_{xxy}x^2y + 3z_{yyx}xy^2 + z_{yyy}y^3\right)$$

$$+ \cdots + \frac{1}{n!}\sum_{m=0}^{n}\binom{n}{m}\frac{\partial^n z}{\partial x^{n-m}\partial y^m}x^{n-m}y^m + \cdots, \tag{C.31}$$

where all the derivatives are evaluated at $(0,0)$, and the first term in the Taylor expansion, $z(0,0)$, is zero because of our previous selection. We can also get rid of the linear terms if

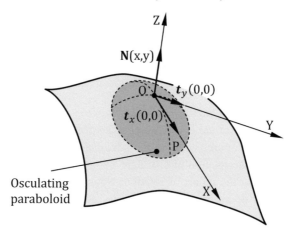

Figure C.7 Parabolic approximation. In the neighborhood of point O, the surface can be approximated by a paraboloid, which is known as an "osculating paraboloid."

we choose the Z axis to be perpendicular to the surface at the origin, as shown in Figure C.7. In this case, the tangent vectors $\mathbf{t}_x$ and $\mathbf{t}_y$ would be parallel to the coordinate axis X and Y, respectively. Mathematically, this latter condition translates into $z_x(0,0) = z_y(0,0) = 0$. Finally, if the surface is not too steep, we can neglect the third- and higher-order terms, with equation (C.31) reducing to

$$z(x,y) \simeq \frac{1}{2}\left(z_{xx}x^2 + 2z_{xy}xy + z_{yy}y^2\right). \tag{C.32}$$

This equation can be arranged as follows

$$z(x,y) = \frac{1}{2}\mathbf{r}^{\mathsf{T}}\mathbb{H}\mathbf{r} \equiv \frac{1}{2}\begin{bmatrix} x & y \end{bmatrix}\begin{bmatrix} h_{11} & h_{12} \\ h_{21} & h_{22} \end{bmatrix}\begin{bmatrix} x \\ y \end{bmatrix}, \tag{C.33}$$

with $\mathbf{r} = \begin{bmatrix} x & y \end{bmatrix}^{\mathsf{T}}$ being the position vector and $h_{11} = z_{xx}(0,0)$, $h_{12} = h_{21} = z_{xy}(0,0)$ and $h_{22} = z_{yy}(0,0)$ the elements of the Hessian matrix at the surface vertex.[1] Equations (C.32) and (C.33) describe a surface as a second degree polynomial. Such surfaces are known as paraboloids [412]. Moreover, equations (C.32) and (C.33) describe the paraboloid that is tangent to the original surface at point O and with the same second derivatives, which is known as the *osculating paraboloid*. The osculating paraboloid is then the second-order approximation to the surface at point O, as can be seen in Figure C.7. We will describe now the geometrical properties of a surface within the framework of this second-order (or parabolic) approximation.

---

[1] In this appendix, we use notation h_{nm} to distinguish the elements of the Hessian matrix evaluated at the origin, as we have been using the general Hessian with elements z_{xx}, z_{xy}, and z_{yy} in previous sections. We will not need the distinction in the rest of the book.

C.4.2 Tangent and Normal Vectors under Parabolic Approximation

First, we will compute the tangent and normal vectors from the first derivatives of the elevation function at a point of coordinates $\mathbf{r} = (x, y)$ located near the surface vertex O. From equation (C.3), we have

$$\mathbf{t}_x = \left[1, 0, h_{11}x + h_{12}y\right]^{\mathsf{T}},$$
$$\mathbf{t}_y = \left[0, 1, h_{12}x + h_{22}y\right]^{\mathsf{T}}. \tag{C.34}$$

Within the framework of the second-order approximation we assume that

$$\|\mathbf{t}_x\|^2 = 1 + (h_{11}x + h_{12}y)^2 \approx 1,$$
$$\|\mathbf{t}_y\|^2 = 1 + (h_{12}x + h_{22}y)^2 \approx 1. \tag{C.35}$$

Therefore, if we consider the tangent vectors as unitary, the normal vector in the parabolic approximation can be computed directly from the cross product of $\mathbf{t}_x$ and $\mathbf{t}_y$ given by

$$\mathbf{N} = \left[-h_{11}x - h_{12}y, -h_{12}x - h_{22}y, 1\right]^{\mathsf{T}}. \tag{C.36}$$

We can derive an important geometrical property from equation (C.36). Let us consider the following vector $\mathbf{n} = [n_1, n_2]^{\mathsf{T}}$, which is the projection of vector $\mathbf{N}$ over the XY plane. According to equation (C.36), the relationship between $\mathbf{n}$ and the position vector r is

$$\mathbf{n} = -\begin{bmatrix} h_{11} & h_{12} \\ h_{12} & h_{22} \end{bmatrix} \begin{bmatrix} x \\ y \end{bmatrix} = -\mathbb{H}\mathbf{r}. \tag{C.37}$$

This equation shows how the geometrical properties of the surface in the proximity of the vertex point are determined by the elements of the Hessian matrix computed at the vertex. This property does not only apply for the direction of the normal vector but also for other important properties of the surface, such as the value and orientation of the main curvatures.

C.4.3 Curvature under Parabolic Approximation

In Figure C.8, we have represented the normal section of a paraboloid passing through point P. We have also drawn the normal vector $\mathbf{N}$ of the paraboloid at P and its projection over the XY plane, vector $\mathbf{n}$. We have also depicted the position vector $\mathbf{r}$ having the x and y coordinates of point P. In general, vectors $\mathbf{n}$ and $\mathbf{r}$ are not parallel, which means the normal vector $\mathbf{N}$ is not contained in the plane that defines the normal section. However, let us see that there always exist two normal sections for which $\mathbf{n}$ and $\mathbf{r}$ are parallel. This condition can also be established as $\mathbf{n}$ and $\mathbf{r}$ being proportional, $\mathbf{n} = -\kappa\mathbf{r}$, where κ is a constant to be determined and the minus sign is added for convinience. After substituting this relationship into the right term of (C.37), we find

$$\left(\mathbb{H} - \kappa\mathbb{I}\right)\mathbf{r} = 0, \tag{C.38}$$

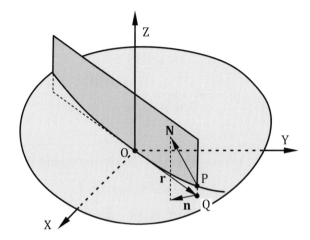

Figure C.8 Orientation of the normal vector **N** of paraboloid at point P belonging to a normal section, defined by the shadowed plane that contains the Z axis, and the paraboloid section. Point Q is the projection of P over the *XY* plane, and **n** is the projection of the normal vector over the *XY* plane. Notice that, in this case, **n** is not parallel to the position vector **r**, indicating that this normal section is not one of the main sections of the surface.

where $\mathbb{I} = \begin{bmatrix} 1 & 0 \\ 0 & 1 \end{bmatrix}$ is the identity matrix. To solve equation (C.38) for κ, we have to impose the following condition

$$\det \begin{bmatrix} h_{11} - \kappa & h_{12} \\ h_{12} & h_{22} - \kappa \end{bmatrix} = 0, \tag{C.39}$$

which leads to the *characteristic equation* of the Hessian,

$$\kappa^2 - (h_{11} + h_{12})\,\kappa + \left(h_{11}h_{22} - h_{12}^2\right) = 0. \tag{C.40}$$

We identify the trace and the determinant of the Hessian matrix in the right side of this expression, so it can be written

$$\kappa^2 - \operatorname{tr}(\mathbb{H})\,\kappa + \det(\mathbb{H}) = 0, \tag{C.41}$$

whose solutions, κ_- and κ_+, are given by exactly the same expressions (C.25) we derived before.

The fundamental forms are particularly simple when evaluated at the vertex. The first one is given by

$$E = G = 1, \ F = 0,$$

while the second one is just the Hessian,

$$L = h_{11}, \ M = h_{12}, \ N = h_{22}.$$

Substituting these into equations (C.30) and (C.28) gives us equally simple expressions for the mean and Gaussian curvatures

$$H = \frac{h_{11} + h_{22}}{2},$$

$$K = h_{11}h_{22} - h_{12}^2, \tag{C.42}$$

which can be substituted into the characteristic equation

$$\kappa^2 + 2H\kappa + K = 0, \tag{C.43}$$

with exactly the same solutions shown in (C.29),

$$\kappa_{\pm} = H \pm \sqrt{H^2 - K}. \tag{C.44}$$

The orientation of the main sections can be obtained by solving $\mathbb{H}\mathbf{r} = \kappa_{\pm}\mathbf{r}$ for the position vector. Any vector $\mathbf{r}$ pointing along the direction that forms an angle θ with the X axis can be written as $(r\cos\theta, r\sin\theta)$. The modulus cancels out in both sides of the matrix equations that transform into

$$\begin{bmatrix} h_{11} & h_{12} \\ h_{12} & h_{22} \end{bmatrix} \begin{bmatrix} \cos\theta_{\pm} \\ \sin\theta_{\pm} \end{bmatrix} = \kappa_{\pm} \begin{bmatrix} \cos\theta_{\pm} \\ \sin\theta_{\pm} \end{bmatrix},$$

that is,

$$(h_{11} - \kappa_{\pm})\cos\theta_{\pm} + h_{12}\sin\theta_{\pm} = 0,$$
$$h_{12}\cos\theta_{\pm} + (h_{22} - \kappa_{\pm})\sin\theta_{\pm} = 0. \tag{C.45}$$

Each of the two equations of the system have just one unknown, so either equation may be used to get the solution

$$\tan\theta_{\pm} = \frac{\kappa_{\pm} - h_{11}}{h_{12}} = \frac{h_{12}}{\kappa_{\pm} - h_{22}}. \tag{C.46}$$

Substitution of (C.42) and (C.44) into the previous equation, and some amount of algebraic manipulation, leads to the fact that the main directions are perpendicular to each other, $\theta_{+} = \theta_{-} \pm \pi/2$.

Example 53 Compute the main curvatures and their orientations for a surface given by the following elevation function (in parabolic approximation) $z(x, y) = 3x^2 - 4xy + 5y^2$.

The coefficients of the second-order polynomial that defines the surface are related to the coefficients of the Hessian matrix. Moreover, if the elevation is measured in meters, we see that the Hessian would be measured in diopters, so, in our case, we have

$$\mathbb{H} = \begin{bmatrix} 6 & -4 \\ -4 & 10 \end{bmatrix} D.$$

486

The trace of the matrix is $\mathrm{tr}\,(\mathbb{H}) = 16\,\mathrm{D}$ and its determinat $\det\,(\mathbb{H}) = 44\,\mathrm{D}^2$. The main curvatures are then

$$\kappa_- = 8 - \sqrt{8^2 - 44} = 8 - \sqrt{20} = 3.53\,\mathrm{D},$$

$$\kappa_+ = 8 + \sqrt{8^2 - 44} = 8 + \sqrt{20} = 12.47\,\mathrm{D}.$$

The orientation of the minor curvature is

$$\theta_- = \arctan\left(\frac{\kappa_- - h_{11}}{h_{12}}\right) = \arctan\left(\frac{3.53 - 6}{-4}\right) = 31.7°,$$

and θ_+ is just $\theta_- \pm 90° = 121.7°$, where we have used the plus sign so that both directions are between 0 and 180°.

To finish this section, we will invert equations (C.44) and (C.46) to obtain the Hessian matrix from the main curvatures and their orientation. We will first demonstrate that if the Hessian is diagonal, the curvatures on its main diagonal are the main curvatures. Let us then assume that $h_{12} = 0$. In that case, $K = h_{11}h_{12}$, and direct substitution into equation (C.44) leads to

$$\kappa_\pm = \frac{h_{11} + h_{22}}{2} \pm \frac{|h_{11} - h_{22}|}{2},$$

that is, $\kappa_+ = h_{11}$ and $\kappa_- = h_{22}$ if $h_{11} > h_{22}$ and the other way around if $h_{22} > h_{11}$. Let us assume for now that $h_{11} < h_{22}$, so the Hessian matrix is

$$\mathbb{H} = \begin{bmatrix} \kappa_- & 0 \\ 0 & \kappa_+ \end{bmatrix}.$$

Similarly, the orientation $\tan\theta_- = h_{12}/(\kappa_- - \kappa_+) = 0$, so the minor curvature lies along the X axis and the maximum curvature along the Y axis. Now, if we rotate the X and Y axis clockwise by an angle $-\alpha$, so that the direction θ_- makes an angle α with the new X axis, the Hessian of the surface in the new reference system is given by the transformation we presented in Appendix B

$$\hat{\mathbb{H}} = \mathbb{R}(-\alpha)\mathbb{H}\mathbb{R}(-\alpha)^{-1} = \mathbb{R}(\alpha)^{-1}\mathbb{H}\mathbb{R}(\alpha),$$

where the rotation matrix is given by

$$\mathbb{R}(\alpha) = \begin{bmatrix} \cos\alpha & \sin\alpha \\ -\sin\alpha & \cos\alpha \end{bmatrix},$$

and its inverse is obtained by changing the sign of θ, as the reader can easily check. After carrying out the matrix multiplications that transform the Hessian we have

$$\hat{\mathbb{H}} = \begin{bmatrix} \kappa_- \cos^2\alpha + \kappa_+ \sin^2\alpha & -(\kappa_+ - \kappa_b)\sin\alpha\cos\alpha \\ -(\kappa_+ - \kappa_b)\sin\alpha\cos\alpha & \kappa_- \cos^2\alpha + \kappa_+ \sin^2\alpha \end{bmatrix}.$$

The diagonal terms can be simplified by adding and substracting $\kappa_- \sin^2\alpha$ to the first one and $\kappa_- \cos^2\alpha$ to the second one, and defining $\Delta\kappa = \kappa_+ - \kappa_-$. After these manipulations, we get to the final expression for the Hessian on the rotated reference system

$$\hat{\mathbb{H}} = \begin{bmatrix} \kappa_- + \Delta\kappa \sin^2\alpha & -\Delta\kappa \sin\alpha \cos\alpha \\ -\Delta\kappa \sin\alpha \cos\alpha & \kappa_- + (\kappa_+ - \kappa_-)\cos^2\alpha \end{bmatrix}.$$

Now, this Hessian corresponds to a surface whose curvature κ_- forms an angle α with the $\hat{X}$ axis of a reference system $\hat{X}\hat{Y}$. In other words, the angle α is just θ_- in the reference system $\hat{X}\hat{Y}$. If we rotate back both surface and reference system, the Hessian will not change. Therefore, the previous expression is fully general, and

$$\mathbb{H} = \begin{bmatrix} \kappa_- + \Delta\kappa \sin^2\theta_- & -\Delta\kappa \sin\theta_- \cos\theta_- \\ -\Delta\kappa \sin\theta_- \cos\theta_- & \kappa_- + \Delta\kappa \cos^2\theta_- \end{bmatrix}, \qquad (C.47)$$

is the Hessian of a surface whose minor curvature section forms an angle θ_- with the X axis of our reference system, whatever it may be.

Appendix D

Local Dioptric Power Matrix

D.1 Introduction

Throughout this book, we have extensively used a matrix formalism for describing the power of ophthalmic lenses (see Chapters 4 and 5). In most cases, we have employed this formalism for studying monofocal lenses, which, within the paraxial approximation, have constant power. However, it is also possible, within some approximations, to use the matrix methods for describing lenses with varying power, such as multifocals. For a lens with variable power, the elements of the dioptric power matrix also present a spatial dependence, and the power is given by a *local* dioptric power matrix. The description of the paraxial properties of a lens with spatially varying power is particularly useful for some of the lens measuring techniques described in Chapter 10. In this appendix, we will derive the mathematical expression that relates the elements of the local dioptric power matrix with the lens surfaces through the derivatives of the so-called thickness function, defined as the difference between the sags of the front and back surfaces. We will present two different derivations of the local dioptric power matrix: First, we will calculate the local power matrix using paraxial ray tracing, while for the second derivation we will make a calculation based on the propagation of a wavefront through the lens.

D.2 Local Dioptric Power Matrix from Paraxial Ray Tracing

Let us suppose the surface of an ophthalmic lens is given by the sag $z(x, y)$. Let us also consider a point P located at this surface whose location is given by $\mathbf{r}_P = \left[x_P, y_P \right]^\mathsf{T}$, and a nearby point Q with position vector $\mathbf{r}_Q = \left[x_Q, y_Q \right]^\mathsf{T}$. Also, we will suppose that the parabolic approximation holds, so that the following expression is valid for describing the surface elevation at point Q

$$z_Q \cong z_P + \mathbf{v}^\mathsf{T} \delta \mathbf{r} + \frac{1}{2} \delta \mathbf{r}^\mathsf{T} \mathbb{H} \delta \mathbf{r}, \qquad (D.1)$$

where $\delta\mathbf{r} = \mathbf{r}_Q - \mathbf{r}_P$, is the difference between the position vectors at points Q and P, respectively, and $\mathbf{v}$ is the vector of first derivatives of the elevation function at point P^1

$$\mathbf{v} = \begin{bmatrix} z_x & z_y \end{bmatrix}^\mathsf{T}.$$

Finally, $\mathbb{H}$ is the Hessian matrix of the surface at point P, given by the following expression:

$$\mathbb{H} = \begin{bmatrix} z_{xx} & z_{xy} \\ z_{xy} & z_{yy} \end{bmatrix}.$$

If the sag derivatives z_x and z_y are small, we can describe the normal vector at point Q as $\mathbf{N} = \begin{bmatrix} n_1 & n_2 & 1 \end{bmatrix}^\mathsf{T}$, where n_1 and n_2 are the elements of a vector $\mathbf{n} = \begin{bmatrix} n_1 & n_2 \end{bmatrix}^\mathsf{T}$ given by [362]

$$\mathbf{n} = -\mathbf{v} - \mathbb{H}\delta\mathbf{r}. \tag{D.2}$$

Let us consider now an ophthalmic lens formed by two surfaces with elevations $z_1(x, y)$ and $z_2(x, y)$, for the front and back surface, respectively. Within the parabolic approximation, those elevations can be described by equation (D.1). A paraxial ray, whose direction vector is $\mathbf{K}_1 = \begin{bmatrix} k_{1x} & k_{1y} & 1 \end{bmatrix}^\mathsf{T}$, impinges on the front surface of the lens at point Q, located close to P. Let us give the notation $\mathbf{k}_1 = \begin{bmatrix} k_{1x} & k_{1y} \end{bmatrix}^\mathsf{T}$ to the vector formed by the x and y components of the incident vector. If n is the refractive index of the lens, the paraxial form of the three-dimensional Snell's law (see equation (3.18) in Chapter 3) states that the direction vector of the ray after refraction at the first surface of the lens would be given as

$$n\mathbf{K}_1' = \mathbf{K}_1 + (n - 1)\,\mathbf{N}_1, \tag{D.3}$$

which can be written as

$$n\mathbf{k}_1' = \mathbf{k}_1 - (n - 1)\,\mathbf{v}_1 - (n - 1)\,\mathbb{H}_1\delta\mathbf{r},$$

where $\mathbb{H}_1$ is the Hessian of the front surface at point P and $\mathbf{v}_1$ the corresponding first derivatives vector. To compute the refraction of the emergent ray at the second surface of the lens, it would be necessary to locate the intersection point Q' of this ray with the back surface. If we consider the lens as a thin system, then we find that $\mathbf{r}_{Q'} \approx \mathbf{r}_Q$, so the equation for the ray refraction at the second surface of the lens becomes

$$\mathbf{k}_2' = n\mathbf{k}_2 - (1 - n)\,\mathbf{v}_2 - (1 - n)\mathbb{H}_2\delta\mathbf{r}. \tag{D.4}$$

Equations (D.3) and (D.4) can be combined, taking into account that $\mathbf{k}_2 = \mathbf{k}_1'$, obtaining

$$\mathbf{k}_2' = \mathbf{k}_1 - (n - 1)\,(\mathbf{v}_1 - \mathbf{v}_2) - (n - 1)\,(\mathbb{H}_1 - \mathbb{H}_2)\,\delta\mathbf{r}.$$

---

[1] Notice that, in this appendix, we will denote the first and second partial derivatives of a given function f as f_x, f_y, f_{xx}, f_{xy}, and f_{yy}.

By definition, the prismatic power at point Q is $\mathbf{p_Q} = \mathbf{k_2'} - \mathbf{k_1}$, so we have

$$\mathbf{p_Q} = \mathbf{p_P} - \mathbb{P}\delta\mathbf{r}, \tag{D.5}$$

where $\mathbf{p_P} = (1-n)(\mathbf{v_1} - \mathbf{v_2})$ is the prismatic power at point P, which is known as the *local ground prism* [414], and $\mathbb{P}$ is the local dioptric power matrix, defined as

$$\mathbb{P} = (n-1)(\mathbb{H}_1 - \mathbb{H}_2). \tag{D.6}$$

Therefore, equation (D.5) is analogous to Prentice's law defined in Chapter 5, but in equation (D.5) the ground prism and the dioptric power matrix change along the lens surface and they are no longer constant. This is why we will refer to equation (D.5) as the *local Prentice's law*.

From equation (D.6), the local dioptric power matrix can be written as

$$\mathbb{P}(x, y) = (n-1)\left(\begin{bmatrix} z_{1xx} & z_{1xy} \\ z_{1xy} & z_{1yy} \end{bmatrix} - \begin{bmatrix} z_{2xx} & z_{2xy} \\ z_{2xy} & z_{2yy} \end{bmatrix}\right) \tag{D.7}$$

$$= (n-1)\begin{bmatrix} t_{xx} & t_{xy} \\ t_{xy} & t_{xx} \end{bmatrix}$$

where $t(x, y) = z_1(x, y) - z_2(x, y)$ the *thickness function* [362]. Therefore, as shown by equation (D.13), the elements of the local power of a lens are given by the Hessian of the second derivatives of the thickness function. The role of the thickness function will be clarified in the next section, where we derive the local power matrix from the propagation of a wavefront through a lens.

D.3 Local Dioptric Power Matrix from Wavefront

In Figure D.1(a), we have represented an incident planar wavefront W over an ophthalmic lens and the corresponding exit wavefront W'. We will compute now the optical path between a point A of W and point D of W' (see Figure D.1(a)). The optical path difference between A and D is the sum of the optical paths at the different media between the two extreme points, so

$$[AD] = [AB] + [BC] + [CD],$$

where $[PQ] = n\overline{PQ}$ is the optical path between any pair of points P and Q. From this definition, the optical path [AD] is given by

$$[AD] = z_1(x, y) + nt_B(x, y) + t_2(x, y). \tag{D.8}$$

Where $z_1(x, y) = \overline{AB}$, $t_B(x, y) = \overline{BC}$ and $t_2(x, y) = \overline{CD}$, respectively. Notice that these distances are measured along the Z-axis, so equation (D.8) is a paraxial approximation of the real optical path. Without the paraxial approximation, the optical path must be measured along the refracted rays.

From Fermat's principle, the optical path difference along any ray between two wavefronts is a constant. Therefore, from the optical path difference at the lens center

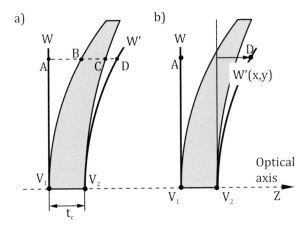

Figure D.1 a) Optical path between the points A of a planar incident wavefront W and D of the exit wavefront W' of an ophthalmic lens. b) Elevation function $W'(x, y)$ of the emergent wavefront for the same lens at point D.

$[V_1 V_2] = nt_c$, where t_c is the center thickness, we have

$$[V_1 V_2] = [AB] \Rightarrow nt_c = z_1 (x, y) + nt_B (x, y) + t_2 (x, y) . \tag{D.9}$$

As can be seen in Figure D.1(b), the elevation of the refracted wavefront W' at point D is

$$W' (x, y) = z_2 (x, y) + t_2 (x, y) . \tag{D.10}$$

From equation (D.9) we have

$$t_2 (x, y) = nt_c - z_1 (x, y) - nt (x, y) .$$

Substituting this expression in equation (D.10) we get

$$W' (x, y) = z_2 (x, y) + nt_c - z_1 (x, y) - nt_B (x, t) . \tag{D.11}$$

Finally, if we take into account the expression for the local thickness, $t_B (x, y) = t_c + z_2 (x, y) - z_1 (x, y)$, from equation (D.11) the output wavefront is given by

$$W'(x, y) = (n - 1) \left[z_1 (x, y) - z_2 (x, y) \right] = (n - 1) t (x, y) . \tag{D.12}$$

Equation (D.12) states that for a planar incident wavefront, the output wavefront is proportional to the thickness function. As the input vergence is zero, the power matrix of the lens is the Hessian of the output wavefront, $\mathbb{P} = \mathbb{H} (W')$; therefore, the dioptric power matrix is given by

$$\mathbb{P} (x, y) = (n - 1) \begin{bmatrix} t_{xx} & t_{xy} \\ t_{xy} & t_{yy} \end{bmatrix}, \tag{D.13}$$

which is the same result we obtained previously.

Appendix E

Seidel Aberrations and Zernike Polynomials

E.1 Sign Conventions and Coordinate Systems

In this appendix we present a brief summary of Seidel aberrations and their relation with Zernike polynomials, which were already introduced in Chapter 2. We will adhere to most of the notation and structure used by Wyant and Creath, [77]. The reader may also consult the books by Welford [78], Malacara [376], or Mahahan [415] to deepen their understanding of the theory. In this section, we are going to provide the definitions, the sign convention, and the coordinate systems that will be used for the description of Seidel aberrations and Zernike polynomials. As shown in Figure E.1(a), when we image an off-axis point through an optical system, the *principal ray* is the one going from the object to the image through the center of the entrance pupil. The *tangential plane* contains the principal ray and the optical axis. The rays coming from the object point and contained in this plane are called tangential rays. In general, the planes that contain the optical axis are called meridional planes, so the tangential plane is a meridional one. We will use a standard reference system with the Z axis matching the optical axis and pointing from O to O', the Y axis pointing to the right, and the X axis pointing upward. The *sagittal plane* also contains the principal ray and is orthogonal to the tangential plane. The rays coming from the object point that propagate in this plane are called sagittal rays, and they intersect the entrance pupil along a line that is perpendicular to the tangential plane and passes through the optical axis (our Y axis). With this arrangement, sagittal rays intersect the pupil at height $x = 0$, while tangential rays intersect at $y = 0$. The principal ray is both sagittal and meridional and intersects the pupil at $(0, 0)$. Any other sagittal ray is a *skew ray*, that is, it does not intersect the optical axis and neither is it parallel to it. Figure E.1(b) summarizes these axis conventions, showing the Cartesian coordinate system we are going to use in this section.

Angles are measured relative to the reference axis or plane using the right hand rule, following the standard procedure in geometrical optics [47]. The angle between ray and axis is positive if we must rotate the axis toward the ray counterclockwise, as shown in Figure E.2(a). At the pupil plane, the relation between polar and Cartesian coordinates is defined as shown in Figure E.2(b). The polar angle θ increases counterclockwise and $\theta = 0$ is along the +X axis when looking at the exit pupil from the image plane.

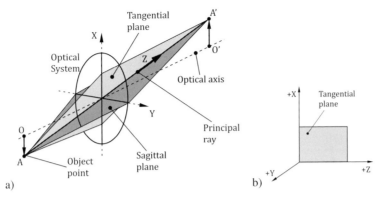

Figure E.1 Tangential and sagittal planes. The plane formed by the object point and the optical axis is the tangential plane. The sagittal plane is perpendicular to it and contains the principal ray.

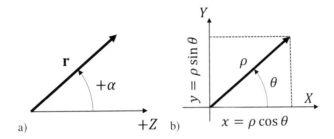

Figure E.2 a) Angle sign convention and b) relation between polar and Cartesian coordinates at the exit pupil as seen from the image plane.

One way to describe the aberrations of optical systems is through the wavefront concept. As discussed in Chapter 3, for any given object point, the wavefront can be defined as the surface formed by the points that are located at the same optical path from the object point [47]. This means that the light coming from the object point reaches any position of the wavefront at the same time because it has traveled the same optical path. For a perfect system, the wavefront passing through the center of the exit pupil is a sphere known as the *Gaussian sphere*, labeled W_{GS} in Figure E.3(a). Its center of curvature is located at the paraxial image point P′. This ideal wavefront is spherical because its associated rays, perpendicular to its surface, converge at the paraxial image point. For a real system, the exit wavefront in no longer a sphere. The difference between the real wavefront, W, and the Gaussian sphere, measured at the exit pupil, is the *wave aberration*. As shown in Figure E.3, any aberrated (real) ray perpendicular to W at point Q (x, y) on the exit pupil plane XY and with direction vector **k** intersects the Gaussian plane $X_i Y_i$ at point P″. If the aberrated ray intersects the Gaussian sphere at point Q_{GS} (x, y), the wave error is defined as the optical path difference between points Q and Q_{GS} along the ray **k**,

$$\Delta W(x, y) = W(x, y) - W_{GS}(x, y) = n\overline{QQ_{GS}}. \tag{E.1}$$

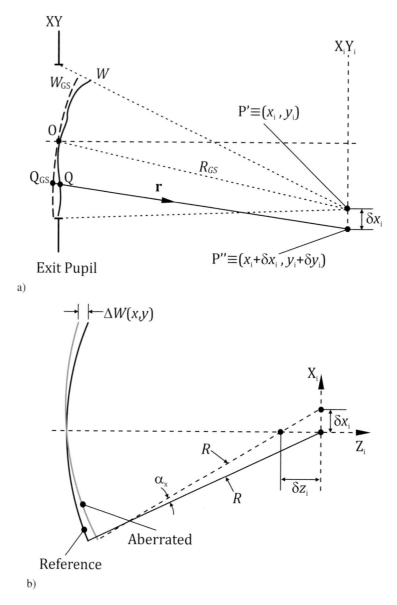

a)

b)

Figure E.3 Wavefront error definition. a) For a perfect system, the output wavefront at the exit pupil is the Gaussian sphere W_{GS} with radius R_{GS} and curvature center at the paraxial image point P'. A real system will have an aberrated wavefront, W. A ray passing through point (x, y) will intersect the Gaussian plane at point P''. The wave aberration for ray $\mathbf{k}$ is $\Delta W(x, y) = n\overline{QQ_{GS}}$ b) α_x, δx_i, and δz_i are the angular, transverse, and longitudinal aberrations.

From its definition, the wave error is positive if the aberrated wavefront is ahead of the Gaussian sphere along the direction of propagation. If $\Delta W(x, y)$ is small, the angular deviation between the normals of W and W_{GS} at point (x, y) denoted by α_x and α_y (contained in the planes XZ and YZ, respectively) will be

$$(\alpha_x, \alpha_y) = -\frac{1}{n} (\partial_x \Delta W, \partial_y \Delta W), \qquad (E.2)$$

where the notation ∂_j denotes partial derivative with respect to j. These two angles are named the *angular aberrations* and determine the deviation of the aberrated ray as shown in Figure E.3(b). In air, with refractive index $n = 1$, the *transverse ray aberrations* are defined as the lateral displacement δx_i and δy_i of the aberrated point P″ with respect to the Gaussian image P′. For small wave errors they can be calculated as

$$(\delta x_i, \delta y_i) = R_{GS} (\alpha_x, \alpha_y) = -\frac{R_{GS}}{n} (\partial_x \Delta W, \partial_y \Delta W), \qquad (E.3)$$

where R_{GS} is the radius of the Gaussian sphere. It is interesting to note that the transverse ray aberrations depend on the direction of the incident ray. Equation (E.3) shows how the (nonparaxial) rays of the beam do not converge on a single image point, but they would rather form a cloud of points (known as blurred or aberrated spot) on the image plane. This aberrated spot causes the loss of image quality associated with the presence of aberrations in an optical system. On the other hand, equation (E.3) is the practical way to numerically compute the exit wave error. By using exact ray tracing, we can obtain the transverse aberration for the rays crossing the exit pupil at a plurality of points. If the number of rays is large enough, we can numerically integrate equation (E.3) from the transverse aberration data $[\delta x_i (x, y), \delta y_i (x, y)]$, thus obtaining the wavefront error at the exit pupil, $\Delta W (x, y)$.

Aberrations can also be defined axially, as the distance between the point a given ray crosses the optical axis and the paraxial image plane, δz_i. This is named *longitudinal ray aberration*, illustrated in Figure E.3(b). If the wave error is small, we can approximate $\delta z_i / \delta x_i \approx R_{GS}/(x_i - \delta x_i)$, and taking into account that $\delta x_i \ll x_i$, we obtain

$$\delta z_i \approx \frac{R_{GS}}{x_i} \delta x_i = -\frac{R_{GS}^2}{x_i} \partial_x \Delta W. \qquad (E.4)$$

An important final consideration is that the wavefront error may depend on the position of the paraxial image P′. That is to say that the aberration will in general depend on the transverse location of the object point, or if we consider the image formation of an extended object, the aberration will act in a different way to object points located at different distances from the optical axis. For this reason and to be more precise, we must write the wave error function as

$$\Delta W (x, y, x_i, y_i) = W (x, y, x_i, y_i) - W_{GS} (x, y, x_i, y_i),$$

where we have included the dependence of ΔW with the position at the exit pupil, (x, y), and the position of the paraxial image, (x_i, y_i).

E.2 Seidel Aberrations

E.2.1 Introduction

In this section we are going to review the monochromatic aberrations of a rotationally symmetrical optical system. In this case, for every object point, the output wavefront at the exit pupil will be a quasi-sphere centered at the paraxial image point. Due to this rotational symmetry, the exit wavefront will not change in the case of a rigid rotation of the XY and $X_i Y_i$ Cartesian axes about the Z axis, and this implies the wavefront aberration must be a function of the combinations $x^2 + y^2$, $xx_i + yy_i$, and $x_i^2 + y_i^2$ (see [78], pp. 105–107). Rotational symmetry also allows us to pick the XZ plane as the meridional or tangential one. The paraxial image point will then lie on the X_i axis of the paraxial image plane, and we can set its coordinates as $y_i = 0$ and $x_i = h_i$. The wavefront error will then be a function of the three variables x, y, and h_i, and can be expanded as a polynomial on the combinations $x^2 + y^2$, xh_i, and h_i^2. If we call these combinations A, B, and C, respectively, the expansion of the wavefront up to second order on the three quantities would contain the terms A, B, C, A^2, B^2, C^2, AB, AC, and BC, that is, [77],

$$\Delta W(x, y, h_i) = a_d \left(x^2 + y^2 \right) + a_t x h_i + a_p h_i^2$$
$$+ b_s \left(x^2 + y^2 \right)^2 + b_c x h_i \left(x^2 + y^2 \right) + b_a x^2 h_i^2 \qquad (E.5)$$
$$+ b_f h_i^2 \left(x^2 + y^2 \right) + b_d x h_i^3 + b_p h_i^4$$
$$+ \text{sixth and higher order terms.}$$

Each term in this expansion has a precise optical meaning, that is, each term is related with a specific pattern of ray aberrations. The first three terms have order two in the rectangular coordinates and are known as defocus (longitudinal shift of the center of the Gaussian wavefront), tilt (lateral displacement of the center of the Gaussian wavefront), and piston. The respective coefficients a_d, a_t, and a_p determine the amount of these aberrations. Defocus is quadratic in x and y, tilt is linear in x, and the piston term does not change across the exit pupil. The terms in the second group have order four in the rectangular coordinates and are known as the *Seidel aberrations*. Each term receives the name of a particular aberration: spherical b_s, coma b_c, astigmatism b_a, field curvature b_f, and distortion b_d. The last term in h_i^4 is a constant shift with no consequences on the final image quality and will be dropped from the discussion. As ray aberrations are obtained as derivatives of the wavefront error, second- and fourth-order wavefront aberrations have one degree less in terms of ray aberrations, and for this reason they are usually named first- and third-order aberrations.

The analysis of the Seidel aberrations is usually conducted in polar coordinates at the exit pupil. By using the relations shown in Figure E.2(b), equation (E.5) can be rewritten as

$$\Delta W(dx, dy, h_i) = a_d \rho^2 + a_t h_i \rho \cos\theta + a_p h_i^2$$
$$+ b_s \rho^4 + b_c \rho^3 h_i \cos\theta + b_a h_i^2 \rho^2 \cos^2\theta \qquad (E.6)$$
$$+ b_f h_i^2 \rho^2 + b_d h_i^3 \rho \cos\theta,$$

Table E.1 *First-order and Seidel aberrations. The column "field" indicates the field dependence and the column "pupil size" the dependence with the exit pupil radius.*

Coefficient	Order	Field	Pupil Size	Form	Name
a_p	first	h^2	no	1	piston
a_t	first	h	d	$\rho \cos\theta$	tilt
a_d	first	no	d^2	ρ^2	defocus
b_s	third	no	d^4	ρ^4	spherical
b_c	third	h	d^3	$\rho^3 \cos\theta$	coma
b_a	third	h^2	d^2	$\rho^2 \cos^2\theta$	astigmatism
b_f	third	h^2	d^2	ρ^2	field curvature
b_d	third	h^3	d	$\rho \cos\theta$	distortion

where d is the radius of the pupil and $\rho = r/d$ is the normalized radial coordinate. Note that piston, tilt, and defocus are still included in equation (E.6), although they are not Seidel aberrations. The eight terms are summarized in Table E.1. The columns named "field" and "pupil size" describe the dependence of each aberration with the lateral field and the size of the exit pupil.

With the exception of the piston term, all the aberrations depend on the pupil size d, as it determines the maximum value the radial coordinate r can have. However, the pupil size does not affect all the aberrations equally, due to the different powers of the radial pupil coordinate in equation (E.6). Assuming similar values of the coefficients, low-order aberrations such as tilt and defocus would dominate for small pupil sizes, while higher-order aberrations such as coma or spherical would dominate for larger pupils. Equally important is the field dependence of the wavefront aberration, given by the paraxial image height h_i. The dependence is different to that with the pupil size: defocus and spherical do not have field dependence at all (so these aberrations are defined for objects on the optical axis), while astigmatism, field curvature, and distortion have the strongest dependence with the field, which is the reason they are called *field aberrations*. Tilt and coma show an intermediate linear relation with the field coordinate. For any aberration other than defocus and spherical, the farther the object from the optical axis, the larger the aberration term. Another interesting aspect of the Seidel aberrations is that the functional form of defocus and field curvature with respect to the coordinates at the exit pupil plane is the same. They, however, have a different dependency with the lateral field. The same can be said with respect to tilt and distortion. Because of this, several object points at different distances from the optical axis are needed to differentiate focus from field curvature and distortion from tilt.

The coefficients of the first- and third-order aberrations in (E.6) may have negative or positive signs, so the combination of two or more optical systems may result in a reduction of the total wavefront error by balancing the aberrations between them. Indeed, the balance of aberrations constitutes one of the fundamental goals in almost any optical design task, and it explains the need for complex systems with numerous lenses or mirrors that characterize many optical systems. In the simpler case of an ophthalmic lens, its ideal shape is determined by the condition that the astigmatism and field curvature produced by the first surface should be canceled by the same aberrations at the second surface.

We will describe next each one of the Seidel aberrations, their effects on the image quality, and their relative importance in the field of ophthalmic optics. For a deeper description of the subject the reader is directed to [376, 77, 415, or 78].

E.2.2 Spherical Aberration

The spherical aberration is given by the fourth term in equation (E.6), and the corresponding wavefront aberration is

$$\Delta W\left(\rho, \theta, h_i\right) = b_s\left(x^2 + y^2\right)^2 = b_s\rho^4. \tag{E.7}$$

As the spherical aberration does not have field dependence, it is usual to choose an axial point to study its effects. From equation (E.3), the spherical transverse ray aberration is (note that we are not using normalized coordinates)

$$\delta x_i = -4R_{GS}b_s\rho^2 x, \\ \delta y_i = -4R_{GS}b_s\rho^2 y. \tag{E.8}$$

From this equation, it follows that the distribution of impact points on the image plane (the point spread function, see Section E.4.1) corresponding to the spherical aberration has rotational symmetry. We can also see that the rays that pass through the border of the pupil present the greater amount of spherical aberration. The transverse distance from the paraxial image point to the impact point of a border ray is

$$TSA = 4R_{GS}b_s d^3, \tag{E.9}$$

where d is the radius of the exit pupil of the system. This distance is known as the *transverse spherical aberration* and quantifies the spherical aberration present in an optical system.

Another measure of spherical aberration is shown in Figure E.4, where we have depicted the refraction of a beam of rays coming from an axial object point O. Because of the spherical aberration, the rays of the beam intercept the optical axis at different points, forming the so-called *caustic*. As we can see in Figure E.4, those rays passing through the border of the exit pupil experience the greater deviation. If the pupil is circular, all the border rays intersect the optical axis at point B, which is located farther away from the paraxial image O' than the remaining intercept points. The distance between O' and B is therefore a measure of the amount of spherical aberration. This distance is known as *longitudinal spherical aberration* [415].

Due to the symmetry of the spherical aberration, its effect on the image quality is quite similar to the effect of defocus. Both produce blurring and loss of image resolution, but, if their coefficients are not too large, less noticeable than that produced by other aberrations. Indeed, a small amount of spherical aberration has even been considered beneficial in

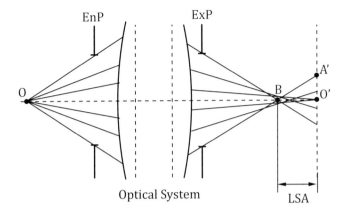

Figure E.4 Ray tracing of a system presenting spherical aberration. The distance between the points O' and A' is the transverse spherical aberration, while the distance between O and B is the longitudinal spherical aberration.

some applications. For example, some old portrait lenses had some amount of spherical aberration that would slightly blur the image, smoothing out some picture imperfections.

E.2.3 Coma

Coma is similar to spherical aberration, but it has a field dependence that destroys the revolution symmetry. When an optical system presents coma, the image of an off-axis object point is a spot with a tail (coma) like that of a comet, hence the name. The formation of this spot is illustrated in Figure E.5. The asymmetrical spot associated with coma distorts the imaging process in a quite unpleasant way, so coma is one of the most noticeable and disturbing aberrations.

Mathematically, if we consider the wavefront aberration associated with coma

$$\Delta W(\rho, \theta, h_i) = b_c x h_i (x^2 + y^2) = b_c h_i \rho^3 \cos \theta,$$

we can write the transverse ray aberration as

$$\delta x_i = -R_{GS} b_c h_i (3x^2 + y^2) = -R_{GS} b_c h_i \rho^2 (2 + \cos 2\theta),$$
$$\delta y_i = -2R_{GS} b_c h_i x y = -R_{GS} b_c h_i \rho^2 \sin 2\theta, \tag{E.10}$$

which corresponds to the parametric equation of a circumference with radius $r_c(\rho) = R_{GS} b_c h_i \rho^2$ centered at the point of coordinates $(0, -2R_{GS} b_c h_i \rho^2)$ at the paraxial image plane. In other words, all the rays hitting the exit pupil at radial distance ρ draw a displaced circle on the Gaussian plane. Two of these circumferences, corresponding to different values of ρ, are depicted in Figure E.5. In a similar way as we defined the transverse spherical aberration, it is possible to characterize the magnitude of coma from the points that present the greater distance with the paraxial image point. Due to the comma asymmetry,

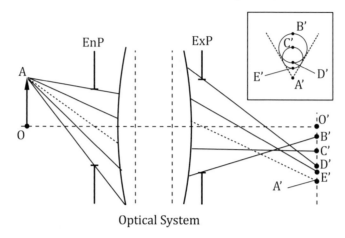

Optical System

Figure E.5 Optical scheme of a system with coma. A′ is the paraxial image of A. The rays that pass through the border of the pupil (points B′ and D′) experience greater deviation than the rays closer to the pupil center (points C′ and E′). When considering the rays outside of the tangential plane, the resulting image of an object point has a comet tail-like shape that gives the name of the aberration (see insert at the upper right corner).

we define those points along the tangential plane and the sagittal plane (see Figure E.1). Therefore, we can define the tangential coma (TCA), and the sagittal coma (SCA) as the maximum lateral distances in the image plane measured along the tangential and sagittal directions. These transverse aberrations are given by

$$TCA = 3R_{GS}b_ch_id^2,$$
$$SCA = TCA/3. \tag{E.11}$$

As they are not independent, it is only necessary to consider one of them.

E.2.4 Oblique Astigmatism

Spherical and coma aberrations show a strong dependence on the pupil size. In ophthalmic lenses, the eye pupil acts as the aperture stop for the lens-eye system, and, as the pupil is much smaller than the lens itself and its curvature radii, the role of both spherical aberration and coma is quite small. However, this is not the case for field aberrations such as oblique astigmatism, which depend more on the oblique incidence of rays upon the lens surfaces and less on pupil size. In Figure E.6, we show the refraction of a light beam coming from an off-axis object point in a system that presents oblique astigmatism. As we can see in this figure, after refraction, an astigmatic beam forms a Sturm's conoid with its two characteristic focal lines, tangential and sagittal.

From Table E.1, the wavefront error of a system having oblique astigmatism is given by:

$$\Delta W(\rho, \theta, h_i) = b_a h_i^2 x^2 = b_a h_i^2 \rho^2 \cos^2 \theta, \tag{E.12}$$

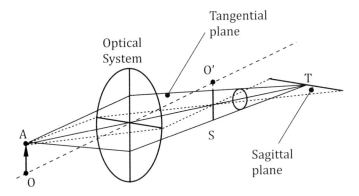

Figure E.6 Formation of an astigmatic beam (Sturm's conoid) due to the presence of oblique astigmatism.

so the transverse ray aberration will be

$$\delta x_i = 2R_{GS}b_ah_i^2\rho\cos\theta,$$

$$\delta y_i = 0.$$

This is the equation of a vertical line along the tangential direction. Thus a focal line (labeled S in Figure E.6) is formed in the tangential direction of the Gaussian plane. If we defocus the system by displacing the image plane by an amount Δz along the Z axis, the wave error will be [77]

$$\Delta W(\rho,\theta,h_i) = b_ah_i^2x^2 + \frac{\Delta z}{2R_{GS}}(x^2+y^2), \tag{E.13}$$

and the transverse ray aberration becomes

$$\delta x_i = \left(2R_{GS}b_ah_i^2 + \frac{\Delta z}{2R_{GS}}\right)x,$$

$$\delta y_i = \frac{\Delta z}{R_{GS}}y.$$

If the focal shift is selected so that

$$2R_{GS}b_ah_i^2 + \frac{\Delta z}{2R_{GS}} = 0,$$

and

$$\Delta z = -4b_aR_{GS}^2h_i^2,$$

then the horizontal transverse ray aberration becomes zero and a horizontal focal line (labeled T in Figure E.6) is formed along the sagittal direction at a distance Δz of the focal line S. The ray caustic between the two focal lines is Sturm's conoid. In ophthalmic optics this effect is associated with the presence of unwanted cylindrical power at oblique sight directions.

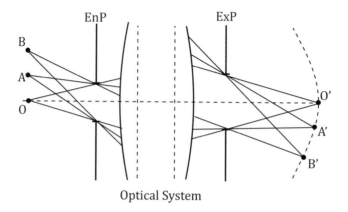

Figure E.7 Image formation of a system presenting field curvature.

E.2.5 Field Curvature

The effect of field curvature is the breach of the second Maxwell condition for an ideal image-forming system. This condition states that the image of a plane is another plane. As can be seen in Figure E.7, when a system presents field curvature, the image points corresponding to a set of object points {O, A, B, ...} contained in the same plane no longer form a plane in the image space, but a spherical surface (known as Petzval's sphere or Peztval's surface).

The wavefront aberration that describes the field curvature is

$$\Delta W(\rho, \theta, h_i) = b_f h_i (x^2 + y^2) = b_f h_i^2 \rho^2, \tag{E.14}$$

and, as expected, this aberration can be understood as a field-dependent defocus. The effects of both astigmatism and field curvature are usually combined, and both aberrations have the same order of magnitude. Coddington's equations [56] are used to compute the coefficients for these aberrations for a spherical surface separating two media with different refractive indexes. These equations relate the curvature radius of the spherical surface and the incident and exit refractive indexes to the location of the tangential and sagittal focal lines, as a function of the object height h_i.

E.2.6 Distortion

Finally, the last of the five classical Seidel aberrations is distortion. This aberration changes the appearance rather than the sharpness of the image. Distortion can be explained as an offense against the third Maxwell condition, which establishes that the object and image must be similar, that is, they must have equal shape, except for a scale factor. For example, if the object is an isosceles triangle, the image should also be another isosceles triangle with the same proportion among their sides.

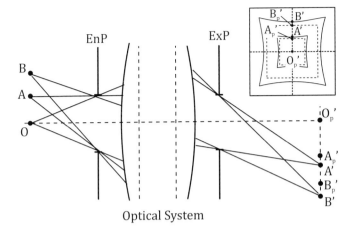

Optical System

Figure E.8 Image distortion. The object and its image are not similar. In this example, the distortion coefficient is positive, which produces the so-called pincushion distortion, so the image of a square grid is distorted, as depicted in the inset at the upper right corner of the figure.

Distortion may be interpreted as a variable magnification that depends on the object height. Indeed, given the wavefront aberration associated with distortion

$$\Delta W(\rho, \theta, h_i) = b_d h_i^3 x = b_d h_i^3 \rho \cos \theta, \qquad (E.15)$$

the corresponding ray aberrations will be

$$\delta x_i = b_d h_i^3,$$
$$\delta y_i = 0,$$

which does not depend on the exit pupil location, so all the rays coming from an object point will intercept the paraxial image plane at the same image point (this is why the sharpness of the image is not affected), but this point is displaced a distance δx_i in the direction of the object, the tangential plane. Depending on the sign of the coefficient b_d, we have positive distortion (or "pincushion" distortion as in Figure E.8) or negative distortion (also known as "barrel" distortion) [416]. In ophthalmic optics so-called dynamic distortion is also important, and this is discussed in Section 6.3.1.

E.3 Zernike Polynomials

Despite their usefulness for explaining the behavior of aberrated optical systems, we must keep in mind that Seidel aberrations constitute just a third-order approximation to the actual wavefront. Moreover, written as in equation (E.6), they can only describe rotationally symmetrical wave errors. Nowadays, it is possible to calculate and measure complex wavefronts with no symmetries, as, for example, those of the human eye [74]. It is therefore necessary to describe and handle general aberrated wavefronts with high accuracy, and the most common technique is the expansion of wavefront in a series of polynomials defined over a given domain. Optical systems usually have circular apertures, so both the entrance and exit pupils are circular, and consequently, the preferred polynomial expansion are the Zernike polynomials [77, 80, 416, 417], as they are orthonormal in the unit circle (all the points inside a centered circumference or radius 1). Additionally, Zernike polynomials are made of terms that can be interpreted as Seidel aberrations.

Zernike polynomials are usually defined in normalized polar coordinates at the exit pupil of the system, see Figure E.2. Each Zernike polynomial, also known as *Zernike mode*, consists of two main components, one radial and the other azimuthal. The radial components are polynomials in r, and the azimuthal components are harmonic sinusoidal functions of the azimuthal angle θ. To unambiguously describe Zernike polynomials, a double indexing scheme with two integers n, m is used. Index n is the highest order of the radial polynomial, and the index m describes the azimuthal harmonics of the sinusoidal components. By this scheme, Zernike polynomials are defined as [80]

$$Z_n^m(r,\theta) = \begin{cases} N_n^m R_n^m(r)\cos m\theta, & m \geq 0 \\ -N_n^m R_n^m(r)\sin m\theta, & m < 0 \end{cases},$$

where $r \leq 1$ and the radial polynomials are

$$R_n^m(r) = \sum_{s=0}^{(n-m)/2} \frac{(-1)^s (n-s)!}{s!\left(\frac{n+|m|}{2}-s\right)!\left(\frac{n-|m|}{2}-s\right)!} r^{n-2s},$$

and N_n^m is a normalization factor given by

$$N_n^m = \sqrt{\frac{2(n+1)}{1+\delta_{m0}}},$$

where δ_{m0} is the Kronecker delta function for index m, i.e. $\delta_{m0} = 1$ for $m = 0$ and $\delta_{m0} = 0$ for $m \neq 0$. In this definition of Zernike polynomials, $n-m \geq 0$ and $n = 2p+m$ where p is a positive integer. This means that for a given n, index m can only take the values $-n, -n+2, -n+4, \ldots n$. The radial polynomials are defined so that $R_n^m(1) = 1$. Also, Zernike polynomials are limited to the range $[-1, 1]$ inside the unit circle, that is, $|Z_n^m(r \leq 1, \theta)| \leq 1$. Finally, Zernike polynomials are orthogonal in the unit circle,

$$\frac{\int_0^1 \int_0^{2\pi} Z_n^m(r,\theta) Z_{n'}^{m'}(r,\theta)\, r dr d\theta}{\int_0^1 \int_0^{2\pi} r dr d\theta} = \delta_{nn'}\delta_{mm'}.$$

Given an arbitrary optical system with or without rotational symmetry and with pupil size d, we can define the normalized radial coordinate $\rho = r/d$, and we can express its wavefront error as an expansion on Zernike polynomials

$$\Delta W(r,\theta) = \sum_{n=0}^{N} \sum_{m=-n}^{n} c_{nm} Z_n^m(\rho,\theta).$$ (E.16)

As ρ is a normalized coordinate, Zernike polynomials are dimensionless, and the coefficients of the expansion, c_{nm}, have the same dimension as ΔW, that is, length. The previous expression can be used to construct the wavefront if we have previous knowledge of the coefficients. If we know the wavefront, either from theoretical derivations or from measurement, we can compute the coefficients with the expression

$$c_{nm} = \frac{1}{\pi} \int_0^1 \int_0^{2\pi} W(d\rho,\theta) Z_n^m(\rho,\theta) \rho d\rho d\theta,$$

which is a consequence of the orthogonality of Zernike polynomials. Another important consequence is that the expansion (E.16) is the best fit for the function ΔW, for the given number of polynomials used in the expansion. We could even use a set of nonconsecutive polynomials, even only one, and the expansion would still be the best possible fit to ΔW with *that set* of Zernike polynomials. From the practical point of view, the wavefront is usually known from measurement of a finite set of M points with polar coordinates (r_k, θ_k). In this case, the computation of the coefficients is obtained approximating the integral by a sum

$$c_{nm} = \frac{1}{\pi} \sum_{k=1}^{M} \Delta W(r_k, \theta_k) Z_n^m(\rho_k, \theta_k),$$ (E.17)

where $\rho_k = r_k/d$. This expression is only approximate, as the orthogonality condition breaks down when the integrals are substituted by sums. There are mathematical techniques that tackle this issue, such as computing the coefficients by optimization techniques, or slightly modifying the Zernike polynomials so that the resulting ones are still orthogonal when integrals are substituted by sums over the set (ρ_k, θ_k), but we will not discuss them here. See for example [34].

Table E.2 contains the first twelve Zernike modes, also called Zernike aberrations, together with their usual names. These aberrations should not be confused with Seidel aberrations. For example, Zernike coma includes Seidel coma and Seidel field curvature. This is because of the *minimum fitting variance* of orthogonal polynomials described in Chapter 2. Seidel aberrations are terms of a power expansion, and only describe correctly ΔW if all of them are included in the expansion. Each Zernike mode fits the error function; the bigger the number of modes included in the sum, the better the fit, but just a single term with the coefficient computed with (E.17) will provide the best possible fit (with this single mode).

Applications of Zernike polynomials involve the use of mathematical expressions involving indexed quantities. If a single index could be used to address the polynomials, the handling of the equations would be a lot easier. To enumerate the two-index Zernike modes with a single index, we must choose a mapping that associates a single integer j

Table E.2 *List of the first Zernike aberrations. The polynomials are sorted using the index mapping of Wyant [77]. The OSA index [80] and the Noll index [417] are included for reference.*

Mode	j (Wyant)	j (OSA)	j (Noll)	n	m	Z_j	Name
Z_0^0	0	0	1	0	0	1	piston
Z_1^1	1	2	2	1	1	$\rho\cos\theta$	x-tilt
Z_1^{-1}	2	1	3	1	-1	$\rho\sin\theta$	y-tilt
Z_2^0	3	4	4	2	0	$2\rho^2-1$	focus
Z_2^2	4	5	6	2	2	$\rho^2\cos 2\theta$	primary astigmatism @0°
Z_2^{-2}	5	3	5	2	-2	$\rho^2\sin 2\theta$	primary astigmatism @45°
Z_3^1	6	8	8	3	1	$(3\rho^3-2\rho)\cos\theta$	primary x coma
Z_3^{-1}	7	7	7	3	-1	$(3\rho^3-2\rho)\sin\theta$	primary y coma
Z_4^0	8	12	11	4	0	$6\rho^4-6\rho^2+1$	primary sphere
Z_3^3	9	9	10	3	3	$\rho^3\cos 3\theta$	trefoil @0°
Z_3^{-3}	10	6	9	3	-3	$\rho^3\sin 3\theta$	trefoil @45°
Z_4^2	11	13	12	4	2	$(4\rho^4-3\rho^2)\cos 2\theta$	secondary astigmatism @0°
Z_4^{-4}	12	11	13	4	-4	$(4\rho^4-3\rho^2)\sin 2\theta$	secondary astigmatism @45°

to the pair (n,m), $Z_n^m \rightarrow Z_j$. Using this single index, a wavefront error function can be written as

$$\Delta W(r,\theta) = \sum_{j=0}^{\infty} c_j Z_j(\rho,\theta).$$

The double index scheme used to define the Zernike polynomials is unambiguous, but there is no universal mapping to a single index. There are three recognized indexing schemes in the literature: Noll [417], Wyant [77], and the OSA standard for reporting eye aberrations [80]. In Table E.2 we have ordered the Zernike aberrations using the Wyant indexing [77], but we have also included the two other index schemes. In any case, as the OSA white paper recommends, when talking about individual Zernike aberrations, the two index scheme should always be used to avoid confusion [80]. As we can see in Table E.2, each Zernike aberration contains the appropriate amount of each lower term so that the set is kept orthogonal and the minimum variance principle is satisfied.

E.4 Relation between Zernike and Seidel Aberrations

In many optical applications the quantities of interest are not Zernike aberrations, but Seidel aberrations. Inspecting Tables E.1 and E.2, it is evident that there is a relationship between them. In fact, third-order aberrations can be obtained from the first nine Zernike modes as listed in Table E.2. This approximation assumes that there are no significant Zernike aberrations above Z_4^0 as ordered in Table E.2. This is evident in the case of Zernike secondary astigmatism that includes Seidel astigmatism and Seidel sphere. The disregard of this approximation may introduce big interpretation errors if we try to calculate Seidel aberrations from a Zernike expansion. Using the first nine Zernike aberrations of Table E.2, we can write the wavefront error as

$$
\begin{aligned}
\Delta W (r, \theta) &= \sum_{j=0}^{8} c_j Z_j (\rho, \theta) \\
&= c_0 + c_1 \rho \cos \theta + c_2 \rho \sin \theta \\
&\quad + c_4 \rho^2 \cos 2\theta + c_5 \rho^2 \sin 2\theta \\
&\quad + c_6 (3\rho^2 - 2) \rho \cos \theta + c_7 (3\rho^2 - 2) \rho \sin \theta \\
&\quad + c_8 (6\rho^4 - 6\rho^2 + 1).
\end{aligned}
\tag{E.18}
$$

By using the identity

$$
a \cos \alpha + b \sin \beta = (a^2 + b^2)^{1/2} \cos \left[\alpha - \arctan (b/a) \right],
$$

equation (E.18) can be rearranged as

$$
\begin{aligned}
\Delta W (r, \theta) &= a_p + a_t \rho \cos(\theta - \phi_t) + a_d \rho^2 \\
&\quad + b_a \rho^2 \cos^2(\theta - \phi_a) \\
&\quad + b_c \rho^3 \cos(\theta - \phi_c) \\
&\quad + b_s \rho^4,
\end{aligned}
\tag{E.19}
$$

where the coefficients and the phases ϕ_t, ϕ_a, and ϕ_c are related to the Zernike coefficients in (E.18) according to the expression in Table E.3. The appearing of rotation phases in the arguments of the cosine functions in the last equation indicates that a general Zernike expansion can represent wavefront errors in systems without rotational symmetry. We must also keep in mind that without field dependence the terms of equation (E.19) are not true Seidel aberrations. Some optical methods used to measure wavefronts, such as those described in Chapter 10, obtain wavefront data from a single object point. For this reason, field curvature looks like defocus and distortion like tilt. Therefore, as we have already discussed, we should obtain a set of wavefront measurements from different object points to determine the Seidel aberrations unambiguously from a Zernike expansion.

E.4.1 Measuring Image Quality: PSF and MTF

An ideal optical system must comply with Maxwell's stigmatic condition, which states that the image of a point is another point. In other words, a system is stigmatic when all the rays that come from a given object point converge on the same image point after

Table E.3 *Relation between Zernike coefficients and Seidel coefficients and rotation phases. The sign in the defocus coefficient has to be chosen to minimize the magnitude of the coefficient. Once selected, the opposite sign has to be chosen for the astigmatism coefficient.*

Name	Coefficient	Phase
piston	$a_t = c_0 - c_3 + c_8$	
tilt	$a_t = \sqrt{(c_1 - 2c_6)^2 + (c_2 - 2c_7)^2}$	$\arctan\left(\frac{c_2 - 2c_7}{c_1 - 2c_6}\right)$
defocus	$a_d = \left(2c_3 - 6c_8 \pm \sqrt{c_4^2 + c_5^2}\right)$	
astigmatism	$b_a = \mp 2\sqrt{c_4^2 + c_5^2}$	$\frac{1}{2}\arctan\left(\frac{c_5}{c_4}\right)$
coma	$b_c = 3\sqrt{c_6^2 + c_7^2}$	$\arctan\left(\frac{c_7}{c_6}\right)$
spherical	$b_s = 6c_8$	

passing through the system. As we have seen, aberrations prevent the accomplishment of the stigmatic condition. In aberrated systems the impact points of the rays coming from the same object point are spread over the paraxial image plane. Therefore, the image of a point is no longer a point but a blurred spot of light, which can be characterized by a function $PSF(x_iy_i)$ that gives the irradiance in the Gaussian plane X_iY_i. This function is known as the *point spread function* (PSF). For monochromatic illumination, the PSF of an optical system can be calculated from the pupil function. The pupil function or aperture function describes how an incident wavefront is affected when it is transmitted through an optical system. It is a complex function of the position in the exit pupil that indicates the relative change in amplitude and phase of the wavefront. For a circular pupil, the pupil function is given by [71]

$$p(x, y) = \text{cyl}\left(\frac{\rho}{2}\right) \exp\left(i2\pi \, \Delta W(x, y) / \lambda\right),$$

where ρ is the normalized radial coordinate at the exit pupil, λ is the wavelength, and $\Delta W(x, y)$ the wave error. The cylinder function, $\text{cyl}(\rho/2)$, is a function that is 1 for $\rho < 1$, 0 otherwise. The PSF is calculated from the Fourier transform of the pupil function, scaled to the Gaussian image plane [71]

$$PSF(x_i, y_i) = |P(x_i/\lambda R_{GS}, y_i/\lambda R_{GS})|^2,$$

where $P(\xi, \eta) = \mathcal{F}[p(x, y)]$ is the Fourier transform of the pupil function, R_{GS} is the radius of the Gaussian sphere, and (x_i, y_i) are the coordinates at the paraxial image plane, see Figure E.3. If $\Delta W = 0$, the PSF is the Airy function and the system is said to be diffraction limited.

The point spread function is one of the tools employed to determine the image quality of an optical system because, in application of the linear systems theory, the PSF is related to the optical transfer function (OTF) of the system. For this reason, the knowledge of the PSF leads to a quantitative determination of the image quality of an optical system. The size of the blurred spot is also related to the optical resolution of a system, so the PSF constitutes a useful tool to evaluate image degradation, particularly for points close to the limit of the field of view. A useful form of representing the PSF is the ray spot diagram, which is formed by the interception points of a number of rays coming from the object point (usually arranged in such a way that the pupil is equally sampled) with a plane perpendicular to the optical axis. On many occasions, the plane of interception is the paraxial image plane. In order to relate the PSF to the image quality, it is customary to compute the size of the PSF as

$$R_{PSF}^2 = \iint \left[(\xi - x_{i0})^2 + (\eta - y_{i0})^2 \right] PSF(\xi, \eta) \, d\xi \, d\eta \,, \tag{E.20}$$

where (x_{i0}, y_{i0}) are the coordinates of the centroid of the PSF spot. If we do not take into account the effects of diffraction, the value of R_{PSF} is directly related to the limit of resolution of the optical system [416]. If diffraction is taken into account, a good indicator of the quality of an optical system is the comparison of R_{PSF} with the radius of the first ring of the Airy pattern produced by the system aperture, R_{Airy}. If $R_{PSF} \cong R_{Airy}$, the system is very close to the diffraction limit, so the effects of aberrations are negligible. To reach the so-called diffraction-limited systems is the ultimate goal of any optical system designer [110].

Closely related with the OTF is the modulation transfer function (MTF). A rigorous definition of the MTF requires the use of the methods of Fourier optics [71], which is completely outside of the scope of this book. However, an operative definition can be developed using an analogy with the contrast transfer function (CTF), widely employed in visual optics. If we consider a periodic object composed of a pattern of bright and dark lines of the same width, we can define two important magnitudes: the spatial frequency as the inverse of the distance between two lines, f_s (in mm^{-1}); and the contrast (or modulation), which is defined as $C = (L_{max} - L_{min}) / (L_{max} + L_{min})$, where L_{max} and L_{min} are the luminances corresponding to the bright and dark lines, respectively. The image of this periodic object will also be periodic but with some degradation, including a loss of contrast that depends on the spatial frequency of the object. The CTF is defined as

$$CTF(f_s) = \frac{C_{image}(f_s)}{C_{object}(f_s)}, \tag{E.21}$$

where C_{image} and C_{object} are the image and object contrast, respectively. The MTF is defined in the same way as the CTF but considering a pattern with a sinusoidal profile, instead of the square profile that would correspond to the purely black and white fringes. Indeed, there is a relationship between the CTF and the MTF that can be derived from the properties of Fourier series [71]. The MTF of a diffraction-limited system is almost a straight line descending from its maximum value of 1 obtained for frequency $f_s = 0$ lines/mm to the minimum value of 0 for the so-called cut-off frequency f_c. For a diffraction-limited system, the inverse of the cut-off frequency is proportional to the inverse of the radius of the Airy pattern and, thus, it corresponds to the resolution limit of the optical system.

Compared to a diffraction-limited system, the MTF of a system presenting aberrations always fulfills the relationship $MTF (f_s) \leq MTF_d (f_s)$ $\forall f_s$, where $MTF_d (f_s)$ is the MTF of the diffraction-limited system. Therefore, the area under the MTF curve for a real optical system will always be lower than the area under the MTF curve for the equivalent (same aperture and or power) diffraction-limited system. This property provides a quantitative measure of comparing the quality of two imaging forming instruments by computing the area under the MTF curve, as the system presenting the greater area under the MTF curve will have the greater image quality.

Appendix F

Abelès Theory of Multilayer Films

F.1 Introduction

In this appendix, we will present the basic theory of transfer matrix for the computation of the optical properties of a multilayer film. The reader must be aware that there are two approaches for the computation of a multilayer film. One is based on the application of the principle of interference of multiple sources (see [216] for details), while the other is the so-called Abelès transfer matrix theory. We will adopt here the latter approach, and we will follow the method employed by Pedrotti et al. [389] because of its simplicity. In order to follow this appendix properly, the reader must have some background in physical optics and knowledge of some mathematical tools, basically the algebra of complex numbers.

To explain the theory of multilayer films, we will follow the classical approach of first computing the transfer matrix for a single layer film, and afterward we will derive the equations for computing the coefficients of reflection and transmission for single and multilayer films. These results have been used in Chapter 11 for explaining the way in which antireflective coatings for ophthalmic lenses work.

The Abelès theory of multilayers is based on the propagation of planar electromagnetic waves. These waves are formed by the combination of an electric $\mathbf{E}\left(\mathbf{r},t\right)$ and a magnetic $\mathbf{B}\left(\mathbf{r},t\right)$ field given by

$$\mathbf{E}\left(\mathbf{r},t\right) = \mathbf{E}_0 e^{i(\mathbf{kr}-\omega t)},$$
$$\mathbf{B}\left(\mathbf{r},t\right) = \mathbf{B}_0 e^{i(\mathbf{kr}-\omega t)}, \tag{F.1}$$

where i is the imaginary unit, $\mathbf{r}$ the position vector, t the time, $\mathbf{E}_0$ and $\mathbf{B}_0$ are amplitudes of the fields, $\mathbf{k}$ is the propagation vector, and ω the angular frequency of the wave. The propagation vector and the frequency are related through the speed of light in the medium as $\|\mathbf{k}\| = \omega/v$, where $v = c/n$ is the speed of light in the medium, with n the refractive index of the medium and c the speed of light in vacuum. Alternatively, the modulus of the propagation vector can be written as $\|\mathbf{k}\| = 2\pi n/\lambda$, where λ is the wavelength (or spatial period) of the light wave.

Vectors $\mathbf{E}(\mathbf{r}, t)$ and $\mathbf{B}(\mathbf{r}, t)$ are not independent, as they are related through Maxwell equations. If the wave is propagating in a homogeneous and isotropic medium, they are related by [389]

$$\mathbf{E}(\mathbf{r}, t) = \frac{c}{n}\mathbf{B}(\mathbf{r}, t). \tag{F.2}$$

Finally, the energy carried by a planar electromagnetic wave is given by the modulus of its Poynting vector

$$\mathbf{S} = \epsilon_0 c^2 \mathbf{E} \times \mathbf{B}, \tag{F.3}$$

with ϵ_0 being the dielectric constant and $\times$ representing the vector cross product. From the Poynting vector, it is possible to compute the irradiance as

$$I = \langle \|\mathbf{S}\| \rangle = \frac{1}{T} \int_0^T \|\mathbf{S}\| dt, \tag{F.4}$$

with $T = 2\pi/\omega$ being the temporal period of the wave that is related to the wavelength through the relationship $c = \lambda/T = \lambda\nu$, where $\nu = 1/T$ is the frequency of the light wave. Further details on the basic properties of planar electromagnetic waves are given in [15], [216], and [389].

F.2 Transfer Matrix

Let us consider now a single-layer film made of a material with refractive index n_1 deposited over a substrate of refractive index n_s, as shown in Figure F.1. As depicted in this figure, a ray impinges on the surface that separates the external medium and the film, forming an incident angle θ_0 with the normal. Thereafter, part of the light is reflected and part is transmitted. After propagating into the film, the transmitted ray arrives at the boundary between the film and the substrate, and it is again partially reflected and partially transmitted. The reflected ray is back propagated to the surface between the film and the layer, and part of it passes through the external medium. Therefore, we have a phenomenon of multiple reflections in a single-layer film. Indeed, the physical principle under which a multilayer film operates is the interference between those multiple reflections.

In terms of geometrical optics, if the angle of incidence of the first ray over the external surface of the film is θ_0, then the angle of transmission is θ_{t1} and both angles are related through Snell's law $n_0 \sin(\theta_0) = n_1 \sin(\theta_{t1})$. Notice that θ_{t1} is also the angle of incidence of the reflected ray when it impinges on the surface between the film and the external medium (see Figure F.1).

In order to compute the multiple wave interference produced by a single-layer film, we must adopt the paradigm of electromagnetic optics and, thus, it is necessary to use the electric and magnetic fields. In Figure F.2 we have represented a particular case when a planar wave of linearly polarized light impinges on the single-layer film with the electric vector $\mathbf{E}_0$ parallel to the z-axis and, thus, pointing out of the plane of the figure. The direction of

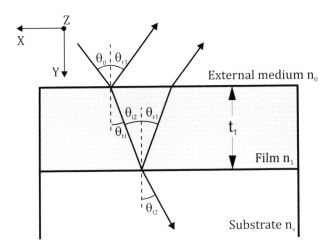

Figure F.1 Representation of a single-layer film with refractive index n_1 and thickness t_1, deposited over a substrate of index n_s. An incident ray coming from the external medium of index n_0 impinges over the external surface of the film with an incidence angle θ_0. As a result, the light propagates through the structure with multiple reflexions at the surfaces of the film, the external medium, and the substrate.

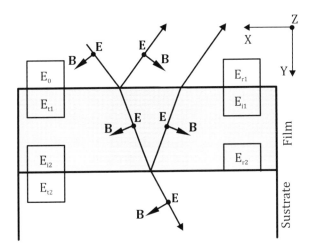

Figure F.2 Electric and magnetic fields for a linearly polarized planar wave it propagates through a single-layer film. Notice the disposition of the electric field that is parallel to the Z direction so it points perpendicular to the plane of the figure. According to the laws of electromagnetic optics, the magnetic field vector is perpendicular to both the electric field and the ray propagation direction.

propagation of the incident ray is given by the vector $\mathbf{k}_0 = \frac{2\pi n_0}{\lambda}\left(-\sin\theta_0, \cos\theta_0\right)$, and the vector that represents the magnetic field is perpendicular to both the electric field vector and the propagation vector, so $\mathbf{B}_0 = B_0\left(\cos\theta_0, \sin\theta_0\right)$. We are dropping the third component for simplicity, as we will only consider the propagation in a plane.

After refraction in the boundary between the external medium and the film, the magnitude of the transmitted electric field is E_{t1} and the propagation vector is now $\mathbf{k}_{t1} = \frac{2\pi n_1}{\lambda}(-\sin\theta_{t1}, \cos\theta_{t1})$. Then the wave propagates through the film until it arrives at the boundary between the film layer and the substrate. At this point, the magnitude of the electric field vector is now E_{i2}. This field is then split into a transmitted field E_{t2} and a reflected field, which is back propagated until it arrives at the boundary between the film and the external medium as a vector with magnitude E_{i1}, see Figure F.2 for details. As the field E_{i1} has been propagated through the film, there is a phase difference $2\delta_1$ between E_{i1} and E_{t1}. It can be demonstrated [389] that

$$\delta_1 = \frac{2\pi}{\lambda}n_1 t_1 \cos\theta_{t1}, \tag{F.5}$$

where t_1 is the layer thickness and n_1 its refractive index.

In order to get the relationship between the fields at the film-substrate and the film-external medium boundaries, we have to apply the boundary conditions for the electric and magnetic fields. According to the classical electromagnetic theory, given a planar surface that separates two mediums with different electric permittivity ϵ, the following boundary conditions apply for both the electric and magnetic fields

$$\mathbf{E}_{2T} = \mathbf{E}_{1T},$$
$$\mathbf{B}_{2T} = \mathbf{B}_{1T}, \tag{F.6}$$

where $\mathbf{E}_{1T}$ and $\mathbf{E}_{2T}$ are the tangential components (with respect to the boundary surface) of the electric fields and $\mathbf{B}_{1T}$ and $\mathbf{B}_{2T}$ are the corresponding tangential components of the magnetic fields.

We will now apply equations (F.6) for both the electric and magnetic fields at the two interfaces shown in Figures F.1 and F.2. As the electric field has been already chosen parallel to the planar boundaries, we have

$$E_0 + E_{r1} = E_{t1} + E_{i1},$$
$$E_{i2} + E_{r2} = E_{t2}. \tag{F.7}$$

Regarding the magnetic fields, as can be seen in Figure F.2, the component of the magnetic field vector that is tangent to the surface is the horizontal component, so the boundary conditions for the magnetic field are

$$B_0 \cos\theta_0 - B_{r1}\cos\theta_0 = B_{t1}\cos\theta_{t1} - B_{i1}\cos\theta_{t1},$$
$$B_{i2}\cos\theta_{t1} - B_{r2}\cos\theta_{t1} = B_{t2}\cos\theta_{t2}, \tag{F.8}$$

where B_0 is the incident field that is split into a reflected B_{r1} and transmitted B_{t1} fields at the external surface of the film. The transmitted field is propagated into the film and impinges on the surface that separates the film layer and the substrate with a magnitude B_{i2}. This field is then divided into a transmitted field B_{t2} and a reflected B_{r2} one, which is back propagated to the interface between the film and the external medium, where it arrives with magnitude B_{i1}. Finally, this field is transmitted to the external medium, where it contributes to the reflected field B_{r2}. The cosine factors that appear in equation (F.8) are due to the fact that the magnetic field vector is not parallel to the film surface. Finally, as can be seen in Figure F.2, the orientations of the magnetic fields corresponding to the incident and transmitted waves are opposed to the orientation of the magnetic fields corresponding to the reflected waves, and thus the fields corresponding to the incident and transmitted wavefronts appear with a minus sign in equation (F.8).

We will now use the relationship between electric and magnetic fields stated in (F.2) to write equation (F.8) in terms of the electric fields, so we arrive to the following equation

$$\gamma_0 E_0 - \gamma_0 E_{r1} = \gamma_1 E_{t1} - \gamma_1 E_{i1},$$
$$\gamma_1 E_{i2} - \gamma_1 E_{r2} = \gamma_s E_{t2},$$
(F.9)

with $\gamma_0 = \frac{n_0}{c} \cos\theta_0$, $\gamma_1 = \frac{n_1}{c} \cos\theta_{t1}$, and $\gamma_s = \frac{n_s}{c} \cos\theta_{t2}$. Equations (F.7) and (F.9) form a system of equations from which it is possible to determine the values of the reflected field E_{r1} and the transmitted field E_{t2} of the layer as a function of the incident field E_0. From these fields, the transmission and reflection coefficients of the layer can be obtained.

To do so, we will define the resultant electric and magnetic fields at the external medium and at the substrate. For the first ones, we have

$$E_a = E_0 + E_{r1} = E_{t1} + E_{i1},$$
$$B_a = \gamma_0 E_0 - \gamma_0 E_{r1} = \gamma_1 E_{t1} - \gamma_1 E_{i1}.$$
(F.10)

with E_a and B_a as the resultant fields at the external medium. For the fields at the substrate we have

$$E_b = E_{t2} = E_{i2} + E_{r2},$$
$$B_b = \gamma_s E_{t2} = \gamma_1 E_{i2} - \gamma_1 E_{r2}.$$
(F.11)

Finally, in order to solve the equation system defined by (F.7) and (F.9), we will take into account that

$$E_{i2} = E_{t1} e^{-i\delta_1},$$
$$E_{i1} = E_{r2} e^{-i\delta_1}.$$
(F.12)

The origin of these relationships is that E_{i2} corresponds to the field E_{t1} propagated from the external to the inner surfaces of the film layer and the same applies for E_{i1} and E_{r2}. At the begining of this section, we saw that the phase difference $2\delta_1$ corresponds to a double pass though the film layer, so that $E_{i1} = E_{t1} e^{-2i\delta_1}$. Thus, the phase difference that corresponds to a single pass (or transverse according to [389]) of the film layer, from E_{t1} to E_{i2} and from E_{r2} to E_{i1} is then δ_1. After substituting the fields defined in equation (F.12), in (F.10), and using some algebra, we get the following matrix equation

$$\begin{bmatrix} E_a \\ B_a \end{bmatrix} = \begin{bmatrix} 1 & e^{-i\delta_1} \\ \gamma_1 & -\gamma_1 e^{-i\delta_1} \end{bmatrix} \begin{bmatrix} E_{t1} \\ E_{r2} \end{bmatrix}.$$
(F.13)

In a similar way, for the fields at the substrate, we get for E_{t1} and E_{r2}

$$\left[\begin{array}{c} E_b \\ B_b \end{array}\right] = \left[\begin{array}{cc} e^{-i\delta_1} & 1 \\ \gamma_1 e^{-i\delta_1} & -\gamma_1 \end{array}\right]\left[\begin{array}{c} E_{t1} \\ E_{r2} \end{array}\right]. \tag{F.14}$$

This equation can be inverted so that

$$\left[\begin{array}{c} E_{t1} \\ E_{r2} \end{array}\right] = \frac{1}{2\gamma_1}\left[\begin{array}{cc} \gamma_1 e^{i\delta_1} & e^{i\delta_1} \\ \gamma_1 & -1 \end{array}\right]\left[\begin{array}{c} E_b \\ B_b \end{array}\right]. \tag{F.15}$$

Therefore, if we substitute the inner fields E_{t1} and E_{r2} from equation (F.15) in equation (F.13) we find that

$$\left[\begin{array}{c} E_a \\ B_a \end{array}\right] = \frac{1}{2\gamma_1}\left[\begin{array}{cc} 1 & e^{-i\delta_1} \\ \gamma_1 & -\gamma_1 e^{-i\delta_1} \end{array}\right]\left[\begin{array}{cc} \gamma_1 e^{i\delta_1} & e^{i\delta_1} \\ \gamma_1 & -1 \end{array}\right]\left[\begin{array}{c} E_b \\ B_b \end{array}\right], \tag{F.16}$$

and after computing the matrix product, we arrive at the following equation

$$\left[\begin{array}{c} E_a \\ B_a \end{array}\right] = \frac{1}{2\gamma_1}\left[\begin{array}{cc} \gamma_1\left(e^{i\delta_1}+e^{-i\delta_1}\right) & \left(e^{i\delta_1}-e^{-i\delta_1}\right) \\ \gamma_1^2\left(e^{i\delta_1}-e^{-i\delta_1}\right) & \gamma_1\left(e^{i\delta_1}+e^{-i\delta_1}\right) \end{array}\right]\left[\begin{array}{c} E_b \\ B_b \end{array}\right]. \tag{F.17}$$

We will apply now Euler's relationships $e^{i\delta_1} = \cos\delta_1 + i\sin\delta_1$ and $e^{-i\delta_1} = \cos\delta_1 - i\sin\delta_1$ to rewrite equation (F.17) in its final form:.

$$\left[\begin{array}{c} E_a \\ B_a \end{array}\right] = \left[\begin{array}{cc} \cos\delta_1 & \frac{i}{\gamma_1}\sin\delta_1 \\ i\gamma_1\sin\delta_1 & \cos\delta_1 \end{array}\right]\left[\begin{array}{c} E_b \\ B_b \end{array}\right], \tag{F.18}$$

where the matrix

$$M_1 = \left[\begin{array}{cc} \cos\delta_1 & \frac{i}{\gamma_1}\sin\delta_1 \\ i\gamma_1\sin\delta_1 & \cos\delta_1 \end{array}\right] \tag{F.19}$$

is known as the *transfer matrix* of a single-layer film [389]. The concept of transfer matrix is not limited to single-layer films, so any multilayer film will be characterized by a transfer matrix [389]

$$M = \left[\begin{array}{cc} m_{11} & m_{12} \\ m_{21} & m_{22} \end{array}\right], \tag{F.20}$$

which is related to the transfer matrices of the individual layers that form the multilayer structure by

$$M = M_1 M_2 M_3 \ldots M_N, \tag{F.21}$$

where N is the number of layers that form the multilayer film and

$$M_j = \left[\begin{array}{cc} \cos\delta_j & \frac{i}{\gamma_j}\sin\delta_j \\ i\gamma_j\sin\delta_j & \cos\delta_j \end{array}\right], \tag{F.22}$$

is the transfer matrix of the jth layer, with $\delta_j = \frac{2\pi}{\lambda}t_j n_j\cos\theta_{tj}$ being the phase difference introduced by this layer, and $\gamma_j = \frac{n_j}{c}\cos\theta_{tj}$ with t_j the thickness, n_j the refractive index,

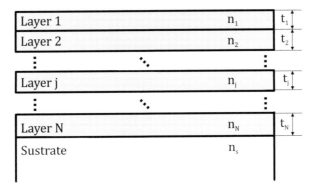

Figure F.3 Structure of a multilayer film. Notice the order of deposition being Layer 1, the most external, and Layer N, the most internal, layers of the film.

and θ_{tj} the direction of propagation of light within the jth layer. In Figure F.3 the typical geometry of a multilayer film is shown with the deposition order indicated, so that the layer with index n_1 (labeled as Layer 1 in the figure) is the most external and the layer with index n_N (Layer N in figure) the most internal.

Once the transfer matrix has been defined for both a single- and a multilayer film, we will calculate the reflectance of a multilayer film, which is the relevant magnitude when studying anti-reflective coatings such as those employed by ophthalmic lenses.

F.3 Reflectance of a Multilayer Film

In order to compute the reflectance of a multilayer film, we will start from the general relationship between the incident (E_a, B_a) and transmitted fields (E_b, B_b) of the multilayer. According to the definition of the transfer matrix, we have

$$\begin{bmatrix} E_a \\ B_a \end{bmatrix} = \begin{bmatrix} m_{11} & m_{12} \\ m_{21} & m_{22} \end{bmatrix} \begin{bmatrix} E_b \\ B_b \end{bmatrix}, \tag{F.23}$$

which can be written as

$$E_a = m_{11}E_b + m_{12}B_b,$$
$$B_a = m_{21}E_b + m_{22}B_b. \tag{F.24}$$

From equation (F.11) we know that the transmitted fields (E_b, B_b) can be written as

$$E_b = E_{t2},$$
$$B_b = \gamma_s E_{t2}, \tag{F.25}$$

while for the incident (E_a, B_a) fields we have a similar equation

$$E_a = E_0 + E_{r1},$$
$$B_a = \gamma_0 E_0 - \gamma_0 E_{r1}. \tag{F.26}$$

After substituting equations (F.25) and (F.26) in equation (F.24) we have

$$E_0 + E_{r1} = m_{11}E_{t2} + \gamma_s m_{12}E_{t2},$$
$$\gamma_0 E_0 - \gamma_0 E_{r1} = m_{21}E_{t2} + \gamma_s m_{22}E_{t2}. \tag{F.27}$$

This equation can be rearranged as follows

$$E_{t2} = \frac{E_0 + E_{r1}}{m_{11} + \gamma_s m_{12}},$$
$$E_{t2} = \frac{\gamma_0 E_0 - \gamma_0 E_{r1}}{m_{21} + \gamma_s m_{22}}. \tag{F.28}$$

Therefore, we can remove E_{t2} from the two terms of equation (F.28) and get

$$(m_{21} + \gamma_s m_{22})(E_0 + E_{r1}) = (m_{11} + \gamma_s m_{22})(\gamma_0 E_0 - \gamma_1 E_{r1}). \tag{F.29}$$

After some algebra, the relation between the incident E_0 and the reflected E_{r1} fields at the external multilayer boundary is

$$(\gamma_0 m_{11} + \gamma_0 \gamma_s m_{12} + m_{21} + \gamma_s m_{22}) E_{r1} =$$
$$= (\gamma_0 m_{11} + \gamma_0 \gamma_s m_{12} - m_{21} - \gamma_s m_{22}) E_0. \tag{F.30}$$

Finally, if we define the *reflection coefficient* of the multilayer as the quotient between the reflected and the incident field, we have

$$r = \frac{E_{r1}}{E_0} = \frac{\gamma_0 m_{11} + \gamma_0 \gamma_s m_{12} - m_{21} - \gamma_s m_{22}}{\gamma_0 m_{11} + \gamma_s \gamma_0 m_{12} + m_{21} + \gamma_s m_{22}}. \tag{F.31}$$

Notice that the reflection coefficient is, in general, a complex number, as it is the relationship between two complex fields. In practice, we are interested in an other magnitude, the *reflectance*, which gives us the ratio between the incident and reflected energy at the multilayer. As the energy transported by a electromagnetic wave is given by the modulus of the Poynting vector, then it is proportional to the squared modulus of the electric field (see equations (F.2), (F.3), and (F.4)). Taking this into account, the reflectance can be defined as the squared modulus of the reflection coefficient

$$R = \|r\|^2. \tag{F.32}$$

Let us now consider the problem of polarization of the incident light. To do so, we must recall that equation (F.31) has been obtained for a linearly polarized incident wave with the electrical field perpendicular to the plane of incidence. Thus, the reflectance computed by equation (F.31) is the perpendicular reflectance or $r_\perp$. In the general case, the incident electric field vector $\mathbf{E}$ will have a component perpendicular to the plane of incidence, $\mathbf{E}_\perp$, and a component parallel to this plane, $\mathbf{E}_\parallel$. Therefore, for a general incident filed, we must repeat the procedure described above to compute the parallel coefficient of reflection $r_\parallel$. The only change would be that instead of defining $\gamma_j = \frac{n}{c}\cos\theta_{tj}$, we should define

$\gamma_j = \frac{n}{c \cos \theta_{tj}}$ for all the relevant equations (see [389] for details). Therefore, in the general case, if the incident field vector is given as

$$\mathbf{E}_0 = \alpha_\perp \mathbf{E}_{0\perp} + \alpha_\| \mathbf{E}_{0\|},$$ (F.33)

then, the reflected field would be

$$\mathbf{E}_r = \alpha_\perp r_\perp \mathbf{E}_{0\perp} + \alpha_\| r_\| \mathbf{E}_{0\|},$$ (F.34)

and the reflectance will be given by

$$R = \frac{\|\mathbf{E}_r\|^2}{\|\mathbf{E}_0\|^2} = \frac{\alpha_\perp^2 R_\perp + \alpha_\|^2 R_\|}{\alpha_\perp^2 + \alpha_\|^2}.$$ (F.35)

The multilayer films employed in ophthalmic optics will operate in a natural environment, and they will be used with natural (or depolarized) light. In this case, we can consider that $\alpha_\perp = \alpha_\| = 1/2$, and the reflectance becomes $R = \frac{1}{2} R_\perp + \frac{1}{2} R_\|$, which is the usual equation used when designing anti-reflective coatings for ophthalmic lenses.

References

[1] S. Musikant, *Optical Materials*, 1st ed. New York: Marcel Dekker, Inc., 1985.

[2] (2014, November) Spanish Ministry of Education, America's Museum web page. Museo de America, Spanish Ministry of Education. [Online]. Available: http://www.mecd.gob.es/museodeamerica/coleccion2/seleccion-de-piezas2/Arqueolog-a/espejo-de-obsidiana.html.

[3] P. the Elder, *The Natural History*, H. R. John Bostock, Ed. Taylor and Francis, 1855.

[4] S. Rasmussen, *How Glass Changed the World*. Springer Verlag, 2012.

[5] H. Tait, Ed., *Five Thousand Years of Glass*, revised edition ed. University of Pennsylvania Press, 2004.

[6] C. Braghin, Ed., *Archaeological Studies on the Uses and Social Context of Glass Artefacts from the Warring States to the Northern Song Period*. Leo S. Olschki Editore (Orientalia Venetiana XIV), 2002.

[7] C. F. Kim. (2012) Early chinese lead-barium glass its production and use from the warring states to the han period (475 BCE - 220 CE). John Brown University.

[8] C. M. Jackson, "Making colourless glass in the roman period," *Archaeometry*, vol. 47, no. 4, pp. 763–780, 2005.

[9] V. I. Evison, *Glass in Britain and Ireland AD 350-1100. British Museum Occasional paper 127*. British Museum Press, 2000, ch. Glass vessels in England, 400-1100 CE, pp. 47–104.

[10] D. L. Hamilton. (1998) Methods of conserving underwater archaeological material culture. conservation files: Anth 605, conservation of cultural resources i. nautical archaeology program. Texas A&M University. http://nautarch.tamu.edu/CRL/conservationmanual/ConservationManual.pdf.

[11] F. Twyman, *Prism and Lens Making*, 2nd ed. New York: Adam Hilger, IOP Publishing Ltd., 1988, second edition published under Adam Hilger imprint by IOP Publishing Ltd.

[12] I. Newton, *Opticks: Or, a Treatise of the Reflections, Refractions, Inflections and Colours of Light*, fourth edition ed. Willyam Innys, 1730.

[13] J. Kirmeier, *Fraunhofer in Benediktbeuern Glassworks and Workshop*. Fraunhofer Gesellschaft, 2008, ch. Research and production. Glassworks and Optical Institute in Benediktbeuern, pp. 18–31.

[14] E. Preston, "Platinum in the glass industry," *Platinum Metals Rev*, vol. 4, no. 1, pp. 2–9, 1960.

[15] E. Born, M. Wolf, *Principles of Optics*, 6th ed. Pergamon Press, 1980.

[16] (2015, February) Interactive abbe diagram. Schott America. [Online]. Available: http://www.us.schott.com/.

[17] *Use of impact-resistant lenses in eyeglasses and sunglasses*, Code of federal regulations. title 21. chapter i, subchapter h, part 801, sec 801.410 ed., Food and Drug Administration, 2014. [Online]. Available: http://www.accessdata.fda.gov/scripts/cdrh/cfdocs/cfCFR/CFRSearch.cfm?fr=801.410

[18] *EN165:1995 Personal eye-protection. Vocabulary*, European Std., 1995.

[19] *EN166:1995 Personal eye-protection. Specifications*, European Std., 1995.

[20] *ISO14889:1997 Ophthalmic Optics. Spectacle Lenses. Fundamental requirements for uncut finished spectale lenses. (ISO 14889:1997)*, International Standard Organization Std., 1997.

[21] (2015, March) Plexiglas(r). Evonik industries. http://geschichte.evonik.de/sites/geschichte/en/inventions/plexiglas/pages/default.aspx.

[22] J. L. Bruneni, *More Than Meets the Eye: The Stories behind the Development of Plastic Lenses*. PPG Industries, Inc, 1997.

[23] *Watching the World: The MR Series*, Mitsui Chemicals, 2012. [Online]. Available: www.mitsuichem.com/special/mr.

[24] D. V. Krevellen, *Properties of Polymers*, 3rd ed. Elsevier, 1990.

[25] C. Fowler and K. Latham, *Spectacle Lenses: Theory and Practice*. Elsevier Science & Technology Books, 1995. [Online]. Available: https://books.google.es/books?id=6oQ3kgEACAAJ.

[26] J. Alonso, A. Peral, J. C. Sanz, and E. Bernabeu, "Measurement of mechanical warpage in CR-39 lenses," *Ophthalmic Physiological Optics*, vol. 17, pp. 81–87, 1997.

[27] A. Peral, J. Alonso, J. C. Sanz, and E. Bernabeu, "Deflectometric measurement of mechanical spectacle lens deformation," *Ophthamic and Physiological Optics*, vol. 20, no. 6, pp. 473–479, 2000.

[28] R. Bellucci, *Cataract. ESASO Course Series*. Karger, 2013, ch. An Introduction to Intraocular Lenses: Material, Optics, Haptics, Design and Aberration, pp. 38–55.

[29] M. P. Keating, "A system matrix for astigmatic optical systems: I. introduction and dioptric power relations," *American Journal of Optometry and Physiological Optics*, vol. 58, no. 10, pp. 810–819, 1981.

[30] M. Audin, *Geometry*. Springer Verlag, 2003.

[31] W. W. Merte, "Anastigmatic lens," US Patent 2 530 397, 1950.

[32] M. Jalie, "Ophthalmic spectacle lenses having a hyperbolic surface," US Patent 4 289 387A, 1981.

[33] B. G. S. Doman, *The Classical Orthogonal Polynomials*. World Scientific Publishing, 2014.

[34] G. M. Dai, *Wavefront Optics for Vision Correction*. SPIE Press, 2008.

[35] J. Wang and D. Silva, "Wave-front interpretation with Zernike polynomials," *Applied Optics*, vol. 19, no. 9, pp. 1510–1518, May 1980.

[36] R. Luneburg, *Mathematical Theory of Optics*. Unversity of California Press, 1964.

[37] G. A. Deschamps, "Ray techniques in electromagnetics," *Proceedings of the IEEE*, vol. 60, no. 9, pp. 1022–1035, Sept 1972.

[38] J. Arnaud, *Progress in Optics, XI*. North Holland, 1973, vol. XI, ch. Hamiltonian theory of beam mode propagation, pp. 248–303.

[39] ——, *Beam and Fiber Optics*. Academic Press, January 1976.

[40] M. Bastiaans, "Wigner distribution function and its application to first-order optics," *Journal of the Optical Society of America*, vol. 69, no. 12, pp. 1710–1716, Dec 1979. [Online]. Available: http://www.osapublishing.org/abstract.cfm?URI=josa-69-12-1710.

[41] H. H. Arsenault, "A matrix representation for non-symmetrical optical systems," *Journal of Optics*, vol. 11, no. 2, p. 87, 1980. [Online]. Available: http://stacks.iop.org/0150-536X/11/i=2/a=002.

[42] B. Macukow and H. Arsenault, "Extension of the matrix theory for nonsymmetrical optical systems," *Journal of Optics*, vol. 15, no. 3, p. 145, 1984. [Online]. Available: http://stacks.iop.org/0150-536X/15/i=3/a=003

[43] R. Blendowske, "Hans-Heinrich Fick: Early contributions to the theory of astigmatic systems," *South African Optometrist*, vol. 62, no. September, pp. 105–110, 2003.

[44] W. F. Long, "A matrix formalism for decentration problems," *American Journal of Optometry and Physiological Optics*, vol. 57, pp. 27–33, 1976.

[45] M. P. Keating, "A system matrix for astigmatic systems: Ii. corrected systems including an astigmatic eye," *American Journal of Optometry and Physiological Optics*, vol. 58, no. 11, pp. 919–929, 1981.

[46] W. F. Harris, "Astigmatism," *Ophthalmic and Physiological Optics*, vol. 20, no. 1, pp. 11–30, 2000b.

[47] J. Greivenkamp, *Field Guide to Geometrical Optics*, ser. Field Guide Series. SPE International Society for Optics and Photonics, 2004. [Online]. Available: https://books.google.com/books?id=1YfZNWZAwCAC.

[48] W. Gartner, "Astigmatism and optometric vectors," *American Journal of Optometry and Archives of American Academy of Optometry*, vol. 42, no. 8, p. 4, 1965.

[49] F. C. Deal and J. Toop, "Recommended coordinate systems for thin spherocylindrical lenses," *Optometry and Vision Science*, vol. 70, no. 5, pp. 409–413, 1993.

[50] L. Levin, S. Nilsson, J. Ver Hoeve, W. Samuel, P. Kaufman, and A. Alm, *Adler's Physiology of the Eye*, 11th ed. Saunders, 2011.

[51] H. Davson, *Physiology of the Eye*, 5th ed. Macmillan Press, 1990.

[52] Y. LeGrand and S. ElHage, *Physiological Optics*, 1st ed., ser. Springer Series in Optical Sciences. Springer-Verlag Berlin Heidelberg, 1980, vol. 13.

[53] D. Atchinson and G. Smith, *Optics of the Human Eye*. Butterworth-Heinemann, February 2000.

[54] P. Artal, Ed., *Handbook of Visual Optics, Volume One: Fundamentals and Eye Optics*. CRC Press, 2017, vol. 1.

[55] D. A. Atchison, "Spectacle lens design: A review," *Applied Optics*, vol. 31, no. 19, pp. 3579–3585, 1992.

[56] M. Jalie, *The Principles of Ophthalmic Lenses*, 4th ed. The Association of British Dispensing Opticians, 1984.

[57] R. S. Park and G. E. Park, "The center of ocular rotation in the horizontal plane," *American Journal of Physiology-Legacy Content*, vol. 104, no. 3, pp. 545–552, 1933. [Online]. Available: https://doi.org/10.1152/ajplegacy.1933.104.3.545.

[58] G. Fry and W. Hill, "The center of rotation of the eye," *American Journal of Optometry and Archives of American Academy of Optometry*, vol. 39, no. 11, pp. 581–595, 1962.

[59] E. Perkins, B. Hammond, and A. Milliken, "Simple method of determining the axial length of the eye," *British Journal of Ophthalmology*, vol. 60, no. 4, pp. 266–270, 1967. [Online]. Available: www.ncbi.nlm.nih.gov/pmc/articles/PMC1017489/.

[60] D. A. Atchinson and L. N. Thibos, "Optical models of the human eye," *Clinical and Experimental Optometry*, vol. 99, no. 2, pp. 99–106, March 2016.

[61] R. S. Spratt, D. Meister, and T. Kratzer, "Wavefront optimized progressive lens," US Patent 8 985 767, 2015.

[62] R. Li, Z. Wang, Y. Liu, and G. Mu, "A method to design aspheric spectacles for correction of high-order aberrations of human eye," *Science China Technological Sciences*, vol. 55, no. 5, pp. 1391–1401, May 2012.

[63] T. Yamakaji and T. Hatanaka, "Spectacle lens and manufacturing method therefor," US Patent 6 637 880, 2003.

[64] M. A. Berthezene, C. Carimalo, D. de Gaudemaris, and C. Guillous, "Method for the determination of a progressive ophthalmic lens," EP Patent 1 837 699, 2007.

[65] R. I. Adygamovich and A. S. Kovychev, "Method for prevention and treatment of refractive disorders vision and device for carrying out," RU Patent 2 501 538, 2012.

[66] S. R. Varnas, "Ophthalmic lens element," US Patent 8 540 365B2, 2013.

[67] E. L. Smith, N. Greeman, P. Greeman, A. Ho, and B. A. Holden, "Methods and apparatuses for altering relative curvature of field and positions of peripheral, off-axis focal positions," US Patent 0 161 065A1, 2009.

[68] P. Rojo, S. Royo, J. Caum, J. Ramirez, and I. Madariaga, "Generalized ray tracing method for the calculation of the peripheral refraction induced by an ophthalmic lens," *Optical Engineering*, vol. 54, no. 2, p. 025106, February 2015.

[69] J. Wallman and J. Winawer, "Homeostasis of the eye growth and the question of myopia," *Neuron*, vol. 43, pp. 447–468, 2004.

[70] W. N. Charman and H. Radhakrishnan, "Peripheral refraction and the development of refractive error, a review," *Ophthalmic and Physiological Optics*, vol. 30, pp. 321–338, 2010.

[71] J. Goodman, *Introduction to Fourier Optics*, 3rd ed. W. H. Freeman, 2004.

[72] M. Smirnov, "Measurements of the wave aberration of the human eye," *Biofizika*, vol. 6, pp. 776–795, February 1961.

[73] R. H. Webb, C. M. Penney, and K. P. Thompson, "Measurement of ocular local wavefront distortion with a spatially resolved refractometer," *Applied Optics*, vol. 31, no. 19, pp. 3678–3686, Jul 1992. [Online]. Available: http://ao.osa.org/abstract.cfm?URI=ao-31-19-3678.

[74] J. Liang, B. Grimm, S. Goelz, and J. F. Bille, "Objective measurement of wave aberrations of the human eye with the use of a Hartmann-Shack wave-front sensor," *Journal of the Optical Society of America A*, vol. 11, no. 7, pp. 1949–1957, 1994.

[75] P. Artal, I.Iglesias, N. López-Gil, and D. Green, "Doublepass measurements of the retinal-image quality with unequal entrance and exit pupil sizes and the reversibility of the eye's optical system," *Journal of the Optical Society of America A*, vol. 12, pp. 2358–2366, 1995.

[76] E. Moreno-Barriuso and R. Navarro, "Laser ray tracing versus hartmann-shack sensor for measuring optical aberrations in the human eye," *Journal of the Optical Society of America A*, vol. 17, no. 6, pp. 974–985, Jun 2000. [Online]. Available: http://josaa.osa.org/abstract.cfm?URI=josaa-17-6-974.

[77] J. Wyant and K. Creath, *Basic Wavefront Aberration Theory for Optical Metrology*. Academic Press, 1992, vol. XI, ch. Basic Wavefront Aberration Theory for Optical Metrology, pp. 27–39.

[78] W. Welford, *Aberrations of optical systems*. Adam Hilger, 1986.

[79] J. Schwiegerling, *Field Guide to Visual and Ophthalmic Optics*. SPIE Optical Engineering Press, 2004.

[80] L. Thibos, R. A. Applegate, J. T. Schwiegerling, and R. Webb, "Standards for reporting the optical aberrations of eyes," in *Vision Science and its Applications*. Optical Society of America, 2000, p. SuC1. [Online]. Available: www.osapublishing.org/abstract.cfm?URI=VSIA-2000-SuC1.

[81] H. Saunders, "The algebra of sphero-cylinders," *Ophthalmic and Physiological Optics*, vol. 5, no. 2, pp. 157–163, 1985.

[82] M. P. Keating, *Geometric, Physical and Visual Optics*. Butterworth-Heinemann, 2001.

[83] L. N. Thibos, W. Wheeler, and D. Horner, "Power vectors: an application of fourier analysis to the description of statistical analysis of refractive error," *Optometry and Vision Science*, vol. 74, no. 6, pp. 367–375, June 1997.

[84] W. F. Harris, "Power vectors versus power matrices, and the mathematical nature of dioptric power," *Optometry and Vision Science*, vol. 84, no. 6, pp. 1060–1063, 2007.

[85] ——, "Algebra of sphero-cylinders and refractive errors, and their means, variance and standard deviation," *Optometry and Physiological Optics*, vol. 65, no. 10, pp. 794–802, 1988.

[86] N. J. Higham, *Functions of Matrices: Theory and Computation*, ser. Other Titles in Applied Mathematics. Society of Industrial and Applied Mathematics, 2008, ch. 6. Matrix square root, pp. 133–171. [Online]. Available: https://doi.org/10.1137/1.9780898717778.

[87] W. F. Harris, D. J. Malan, and A. Rubin, "The distribution of dioptric power: ellipsoids of constant probability density," *Ophthalmic and Physiological Optics*, vol. 11, no. 4, pp. 381–384, 1991.

[88] W. F. Harris, D. Malan, and A. Rubin, "Ellipsoidal confidence regions for mean refractive status," *Optometry and Vision Science*, vol. 68, no. 12, pp. 950–953, 1991.

[89] W. F. Harris, "Clinical measurement, artifact, and data analysis in dioptric power space," *Optometry and Vision Science*, 2001.

[90] ——, "Wavefronts and their propagation in astigmatic optical systems," *Optometry and Vision Science*, vol. 73, no. 9, pp. 606–612, 1996.

[91] ——, "A unified paraxial approach to astigmatic optics," *Optometry and Vision Science*, vol. 76, no. 7, pp. 480–499, 1999.

[92] J. Liang, D. R. Williams, and D. T. Miller, "Supernormal vision and high-resolution retinal imaging through adaptive optics," *Journal of the Optical Society of America A*, vol. 14, no. 11, pp. 2884–2892, 1997.

[93] R. Navarro, E. Moreno-Barriuso, S. Bara, and T. Mancebo, "Phase plates for wave-aberration compensation in the human eye," *Optics Letters*, vol. 25, no. 4, pp. 236–238, 2000.

[94] A. Y. Yi and T. W. Raasch, "Design and fabrication of freeform phase palate for high-order ocular aberration correction," *Applied Optics*, vol. 44, no. 32, pp. 6869–6876, November 2005.

[95] S. Marcos, P. Pérez-Merino, and C. Dorronsoro, *Handbook of Visual Optics. Volume I*. CRC Press, 2017, ch. Monochromatic aberrations, pp. 293–311.

[96] J. F. Castejon-Mochon, N. Lopez-Gil, A. Benito, and P. Artal, "Ocular wave-front aberration statistics in a normal young population," *Vision Research*, vol. 42, pp. 1611–1617, 2002.

[97] S. Bara, T. Mancebo, and E. Moreno-Barriuso, "Positioning tolerances for phase plates compensating aberrations of the human eye," *Applied Optics*, vol. 39, no. 19, pp. 3413–3420, July 2000.

[98] A. Guirao, D. R. Williams, and I. Cox, "Effect of rotation and translation on the expected benefit of an ideal method to ceorrect the eye's higher order aberrations," *Journal of the Optical Society of America A*, vol. 18, no. 5, pp. 1003–1015, May 2001.

[99] I. Cox, *Handbook of Visual Optics, Volume II*. CRC Press, 2017, ch. Contact Lenses, pp. 187–202.

[100] N. Lopez-Gil, J. F. Castejon-Mochon, and V. Fernandez-Sanchez, "Limitations of the ocular wavefront correction with contact lenses," *Vision Research*, vol. 49, pp. 1729–1737, 2008.

[101] G. Kleinmann, E. I. Assia, and D. J. Apple, Eds., *Premium and Specialized Intraocular Lenses*. Bentham Science, 2014.

[102] S. Strenstrom, "Investigation of the variation and the correlation of the optical eelement of human eyes," *American Journal of Optometry and Archives of American Academy of Optometry*, vol. 25, pp. 496–504, 1948.

[103] C. S. Kee, "Astigmatism and its role in emmetropization," *Experimental Eye Research*, vol. 114, pp. 89–95, 2013.

[104] F. J. Rucker, "The role of luminance and chromatic cues in emmetropization," *Ophthalmic Physiological Optics*, vol. 33, pp. 196–214, 2013.

[105] D. I. Flitcroft, "Emmetropisation and the aetiology of refractive errors," *Eye*, vol. 28, no. 2, pp. 169–179, 2014. [Online]. Available: http://dx.doi.org/10.1038/eye.2013.276.

[106] D. O. Mutti, G. L. Mitchell, L. A. Jones, N. E. Friedman, S. L. Frane, W. K. Lin, M. L. Moeschberger, and K. Zadnik, "Axial growth and changes in lenticular and corneal power during emmetropization in infants," *Investigative Ophthalmology and Visual Science*, vol. 46, pp. 3074–3080, 2005.

[107] P. Chen-Wei, D. Mohamed, C. Ching-Yu, W. Tien-Yin, and S. Seang-Mei, "The age-specific prevalence of myopia in asia: A meta-analysis," *Optometry and Vision Science*, vol. 92, no. 3, pp. 258–266, March 2015.

[108] M. Day and L. A. Duffy, "Myopia and defocus: the current understanding," *Scandinavian Journal of Optometry and Visual Science*, vol. 4, no. 1, pp. 1–14, June 2011.

[109] J. Tabernero, D. Vazquez, A. Seidemann, D. Uttenweiler, and F. Schaeffel, "Effects of myopic spectacle correction and radial refractive gradient spectacles on peripheral refraction," *Vision Research*, vol. 49, pp. 2176–2186, 2009.

[110] E. L. Smith, A. Ho, B. A. Holden, and P. Greeman, "Methods and apparatuses for altering relative curvature of field and positions of peripheral, off-axis focal positions," WO Patent 2007/092 853A3, 2007.

[111] P. Sankaridurg, L. Donovan, S. Varnas, A. Ho, X. Chen, A. Martinez, S. Fisher, Z. Lin, E. L. I. Smith, J. Ge, and B. Holden, "Spectacle lenses designed to reduce progression of myopia: 12-month results," *Optometry and Vision Science*, vol. 87, no. 9, pp. 631–641, September 2010.

[112] J. Huang, D. Wen, Q. Wang, C. McAlinden, I. Flitcroft, H. Chen, S. M. Saw, H. Chen, F. Bao, Y. Zhao, L. Hu, X. Li, R. Gao, W. Lu, Y. Du, Z. Jinag, A. Yu, H. Lian, Q. Jiang, Y. Yu, and J. Qu, "Efficacy comparison of 16 interventions for myopia control in children," *Ophthalmology*, vol. 123, no. 4, pp. 697–708, April 2016.

[113] I. L. Bailey and J. E. Lovie, "New design principles for visual acuity letter charts," *American Journal of Optometry and Physiological Optics*, vol. 53, no. 11, pp. 740–745, 1976.

[114] P. G. Barten, *Contrast Sensitivity of the Human Eye and Its Effects on Image Quality*. SPIE Optical Engineering Press, 1999.

[115] L. Chen, B. Singer, A. Guirao, J. Porter, and D. R. Williams, "Image metrics for predicting subjective image quality," *Optometry and Vision Science*, vol. 82, no. 5, pp. 358–369, 2005.

[116] R. A. Applegate, J. Marsack, and L. N. Thibos, "Metrics of retinal image quality predict visual performance in eyes with 20/17 or better visual acuity," *Optometry and Vision Science*, vol. 83, no. 9, pp. 635–640, September 2006.

[117] E. Villegas, C. Gonzalez, B. Bourdoncle, T. Bonnin, and P. Artal, "Correlation between optical and psychophysical parameters as a function of defocus," *Optometry and Vision Science*, vol. 79, no. 1, pp. 60–67, January 2002.

[118] J. Bühren, K. Pesudovs, T. Martin, A. Strenger, G. Yoon, and T. Kohen, "Comparison of optical quality metrics to predict subjective quality of vision after laser in situ keratomileusis," *Journal of Cataract and Refractive Surgery*, vol. 35, pp. 846–855, 2009.

[119] Y. Nochez, G. VanderMeer, Y. Benard, P.-J. Pisella, and R. Legras, "Comparison of optical quality metrics to predict subjective quality of vision in multifocal contact lenses wearers," *Investigative Ophthalmology and Visual Science*, vol. 53, p. 1393, 2012.

[120] J. T. Winthrop, "Progressive power ophthalmic lenses," US Patent 4 514 061, 1985.

[121] B. C. Wooley, "Method for designing progressive addition lenses," US Patent 6 956 682B2, 2005.

[122] J. M. Lindacher, "Premium vision ophthalmic lenses," US Patent 8 152 300, 2012.

[123] C. Pedrono, "Method for determination of a pair of progressive ophthalmic lenses," US Patent 20 080 106 697, 2008.

[124] O. Nestares, R. Navarro, and B. Antona, "Bayesian model of snellen visual acuity," *Journal of the Optical Society of America A*, vol. 20, no. 7, pp. 1371–1381, 2003.

[125] E. Dalimier, E. Pailos, R. Rivera, and R. Navarro, "Experimental validation of a bayesian model of visual acuity," *Journal of Vision*, vol. 9, no. 2009, pp. 1–16, 2009. [Online]. Available: http://jov.highwire.org/content/9/7/12.short.

[126] T. W. Raasch, "Spherocylindrical refractive errors and visual acuity," *Optometry and Vision Science*, vol. 72, no. 4, pp. 272–275, 1995.

[127] M. Pincus, "Unaided visual acuities correlated with refractive errors," *American Journal of Ophthalmology*, vol. 29, pp. 853–858, 1946.

[128] R. Blendowske, "Unaided visual acuity and blur: A simple model," *Optometry and Vision Science*, vol. 92, no. 6, pp. 121–125, 2015.

[129] G. Smith, "Relation between spherical refractive error and visual-acuity," *Optometry and Vision Science*, pp. 591–598, 1991.

[130] J. Gomez-Pedrero and J. Alonso, "Phenomenological model of visual acuity," *Journal of Biomedical Optics*, vol. 21, no. 12, pp. 125 005, December 2016.

[131] M. P. Keating, "A system matrix for astigmatic optical systems: II corrected systems including an astigmatic eye," *American Journal of Optometry and Physiological Optics*, vol. 58, no. 10, pp. 810–819, 1981.

[132] T. Fannin and T. Grosvenor, *Clinical Optics*, 2nd ed. Butterworth-Heinemann, 1996.

[133] J. N. Trachman and R. F. Dippner, "Psychophysical scaling of the prism diopter unit," *Bulletin of the Psychonomic Society*, vol. 6, no. 2, pp. 140–142, 1975.

[134] W. F. Harris, "Ocular rotation factor and orientational demand with astigmatic lenses," *Opthalmic and Physiological Optics*, vol. 13, no. 3, pp. 309–312, 1993.

[135] A. Remole, "Determining exact prismatic deviations in spectacle corrections," *Optometry and Vision Science*, vol. 76, no. 11, 1999.

[136] A. Remole, "New equations for determining ocular deviations produced by spectacle corrections," *Optometry and Vision Science*, vol. 77, no. 10, pp. 555–563, 2000.

[137] J. R. Flores, "Paraxial ocular rotation with astigmatic lenses," *Optometry and Vision Science*, vol. 87, no. 12, pp. E1044–E1052, 2010.

[138] W. Becken, H. Altheimer, G. Esser, W. Mueller, and D. Uttenweiler, "Optical magnification matrix: near objects in the paraxial case," *Optometry and Vision Science*, vol. 85, no. 7, pp. 581–592, 2008. [Online]. Available: http://www.ncbi.nlm.nih.gov/pubmed/18594337.

[139] W. F. Harris and M. P. Keating, "Proof that the prismatic effect is perpendicular to the lens thickness contour," *Optometry and Vision Science*, vol. 68, no. 6, pp. 459–460, 1991.

[140] I. P. Howard and B. J. Rogers, *Binocular Vision and Stereopsis*. Oxford University Press, 1995.

[141] M. Scheiman and B. Wick, *Clinical Management of Binocular Vision*, 4th ed. Lippincott, Williams & Wilkins, 2014.

[142] R. Penrose, "A generalized inverse for matrices," *Proceedings of the Cambridge Philosophical Society*, vol. 51, pp. 406–413, 1955.

[143] W. H. Press, S. A. Teukolsky, W. T. Vetterling, and B. P. Flannery, *Numerical Recipes. The Art of Scientific Computing*. Cambridge University Press, 2007.

[144] W. F. Harris, "Solving the matrix form of Prentice's equation for dioptric power," *Optometry and Vision Science*, vol. 68, no. 3, pp. 178–182, 1991.

[145] H. Buchdahl, *An Introduction to Hamiltonian Optics*. Cambridge University Press, 1970.

[146] D. Onciul, "ABCD propagation law for misaligned general astigmatic gaussian beams," *Journal of Optics*, vol. 23, no. 4, pp. 163–165, 1992.

[147] W. F. Harris, "Paraxial ray tracing through noncoaxial astigmatic optical systems, and 5x5 augmented system matrix," *Optometry and Vision Science*, vol. 71, no. 4, pp. 282–285, 1994.

[148] W. F. Harris, "Astigmatic optical systems with separated and prismatic and noncoaxial elements: system matrices and system vectors," *Optometry and Vision Science*, vol. 70, no. 6, pp. 545–551, 1993.

[149] A. Keirl and C. Christie, Eds., *Clinical Optics and Refraction. A Guide for Optometrists, Contact Lens Opticians and Dispensing Opticians*. Butterworth-Heinemann, 2007.

[150] A. Remole, "A new method for determining prismatic effects in cylindrical spectacle corrections," *Optometry and Vision Science*, vol. 77, no. 4, pp. 211–220, 2000.

[151] S. Barbero and J. Portilla, "Geometrical interpretation of dioptric blurring and magnification in ophthalmic lenses," *Optics Express*, vol. 23, no. 10, pp. 13 185–13 199, 2015. [Online]. Available: https://www.osapublishing.org/abstract.cfm?URI=oe-23-10-13185.

[152] S. Barbero and J. Portilla, "The relationship between dioptric power and magnification in progressive addition lenses," *Ophthalmic and Physiological Optics*, vol. 36, no. 4, pp. 421–427, 2016.

[153] M. P. Keating, "A matrix formulation of spectacle magnification," *Ophthalmic and Physiological Optics*, vol. 2, no. 2, pp. 145–158, 1982.

[154] M. P. Keating, "The aniseikonic matrix," *Ophthalmic and Physiological Optics*, vol. 2, no. 3, pp. 193–204, 1982.

[155] W. F. Harris, "Magnification, blur, and ray state at the retina for the general eye with and without a general optical instrument in front of it: 1. distant objects," *Optometry and Vision Science*, vol. 78, no. 78, pp. 888–900, December 2001.

[156] W. F. Harris, "Magnification, blur, and ray state at the retina for the general eye with and without a general optical instrument in front of it: 2. near objects," *Optometry and Vision Science*, vol. 78, no. 12, pp. 901–905, 2001.

[157] E. Acosta and R. Blendowske, "Paraxial propagation of astigmatic wavefronts in optical systems by an augmented stepalong method for vergences," *Optometry and Vision Science*, vol. 82, no. 10, pp. 923–932, 2005. [Online]. Available: www.ncbi.nlm.nih.gov/pubmed/16276326.

[158] W. Becken, H. Altheimer, G. Esser, W. Mueller, and D. Uttenweiler, "Optical magnification matrix: near objects and strongly oblique incidence," *Optometry and Vision Science*, vol. 85, no. 85, pp. E593–E604, July 2008.

[159] J. Rubinstein and G. Wolansky, "Differential relations for the imaging coefficients of asymmetric systems," *Journal of the Optical Society of America A*, vol. 20, no. 12, pp. 2365–2369, December 2003.

[160] W. J. Smith, *Modern Optical Engineering*, 4th ed. McGraw-Hill Education, 2007.

[161] S. Barbero and M. Faria-Ribeiro, "Foveal vision power errors induced by spectacle lenses designed to correct peripheral refractive errors," *Ophthalmic and Physiological Optics*, vol. 38, pp. 317–325, 2018.

[162] B. Howland and H. Howland, "Subjective measurement of high-order aberrations of the eye," *Science*, vol. 193, pp. 580–582, 1976.

[163] J. Porter, A. Guirao, I. Cox, and D. Williams, "Monochromatic aberrrations of the hyman eye in large populations," *Journal of the Optical Society of America A*, vol. 18, no. 8, pp. 1793–1803, 2001.

[164] S. Plainis and I. Pallikaris, "Ocular monochromatic aberration statistics in a large emmetropic population," *Journal of Modern Optics*, vol. 55, no. 4-5, pp. 759–772, 2008.

[165] R. Calver, M. Cox, and D. Elliot, "Effect of aging on the monochromatic aberrations of the hyman eye," *Journal of the Optical Society of America A*, vol. 19, no. 9, pp. 2069–2078, 1999.

[166] J. He, S. Burns, and S. Marcos, "Monochromatic aberrations in the accommodated human eye," *Vision Research*, vol. 40, pp. 41–48, 2000.

[167] X. Cheng, A. Bradley, X. Hong, and L. Thibos, "Relationship between refractive error and monochromatic aberrations of the eye," *Optometry and Vision Science*, vol. 80, no. 1, pp. 43–49, 2003.

[168] P. de Gracia, C. Dorronsoro, E. Gambra, G. Marin, M. Hernandez, and S. Marcos, "Combining coma with astigmatism can improve retinal image over astigmatism alone," *Vision Research*, vol. 50, no. 19, pp. 2008–2014, 2010. [Online]. Available: www.sciencedirect.com/science/article/pii/S0042698910003615.

[169] P. de Gracia, C. Dorronsoro, G. Marin, M. Hernández, and S. Marcos, "Visual acuity under combined astigmatism and coma: Optical and neural adaptation effects," *Journal of Vision*, vol. 11, no. 2, p. 5, 2011. [Online]. Available: + http://dx.doi.org/10.1167/11.2.5.

[170] W. Wollaston, "Lxi. on an improvement in the form of spectacle glasses," *Philosophycal Magazine (Series 1)*, vol. 17, pp. 327–329, 1804.

[171] R. Kingslake, "Who discovered Coddington's equations?" *Optics & Photonics News*, vol. 5, no. 8, pp. 20–23, 1994.

[172] D. A. Atchison and W. N. Charman, "Thomas young's contributions to geometrical optics," *Clinical and Experimental Optometry*, vol. 94, no. 4, pp. 333–340, July 2011.

[173] F. Ostwalt, "Ueber periscopic glaser," *Arch. f. Ophth.*, vol. 46, p. 475, 1898.

[174] M. Tscherning, "Dioptrique oculaire," *Encyclopedie francaise d'ophthalmologie*, vol. 3, p. 105, 1904.

[175] A. S. Percival, "Periscopic lenses," *The British Journal of Opthalmology*, vol. 10, pp. 369–379, July 1926.

[176] M. von Rohr, *Die Brille als optisches Intrument*. Springer-Verlag Berlin Heidelberg. 1921.

[177] B. A.G., "An historical review of optometric principles and techniques," *Opthalmic and Physiological Optics*, vol. 6, no. 1, pp. 3–21, 1986.

[178] E. Tillyer, "Lens," US Patent 1 356 670, 1920.

[179] J. K. Davis, H. G. Fernald, and A. W. Rayner, "An analysis of opthalmic lens design," *American Journal of Optometry and Archives of American Academy of Optometry*, vol. 41, no. 7, pp. 400–421, July 1964.

[180] B. Bourdoncle, J. P. Chauveau, and J. L. Mercier, "Traps in displaying optical performances of a progressive-addition lens," *Applied Optics*, vol. 31, no. 19, pp. 3586–3593, 1992. [Online]. Available: http://ao.osa.org/abstract.cfm?URI=ao-31-19-3586.

[181] E. A. Villegas and P. Artal, "Spatially resolved wavefront aberrations of ophthalmic progressive-power lenses in normal viewing conditions," *Optometry and Vision Science*, vol. 80, no. 2, pp. 106–114, 2003.

[182] D. A. Atchison, "Third-order theory and aspheric spectacle lens design," *Ophthalmic and Physiological Optics*, vol. 4, no. 2, pp. 179–186, 1984.

[183] A. Whitwell, "On the best form of spectacle lenses-xix," *The Optician*, vol. 61, pp. 241–246, 1921.

[184] G. Smith and D. A. Atchison, "Effect of conicoid asphericity on the tscherning ellipses of ophthalmic spectacle lenses," *Journal of the Optical Society of America*, vol. 73, no. 4, pp. 441–445, 1983. [Online]. Available: www.ncbi.nlm.nih.gov/pubmed/6864357.

[185] D. A. Atchison, "Modern optical design assessment and spectacle lenses," *Optica Acta: International Journal of Optics*, vol. 32, no. 5, pp. 607–634, 2010.

[186] H. Hopkins, *Wave Theory of Aberrations*. London: Oxford University Press, 1950.

[187] R. Kingslake, "Who discovered coddington's equations?" *Optics and Photonics News*, vol. 5, no. 8, pp. 20–23, Aug 1994. [Online]. Available: www.osa-opn.org/abstract.cfm?URI=opn-5-8-20.

[188] M. Katz, "Aspherical surfaces used to minimize oblique astigmatic error, power error, and distortion of some high positive and negative power ophthalmic lenses," *Applied Optics*, vol. 21, no. 16, pp. 2982–2991, 1982.

[189] Z. Malacara and D. Malacara, "Tscherning ellipses and ray tracing in aspheric opthalmic lenses," *American Journal of Optometry and Physiological Optics*, vol. 62, no. 7, pp. 456–462, 1985.

[190] Z. Malacara and D. Malacara, "Aberrations of sphero-cylindrical opthalmic lenses," *Optometry and Vision Science*, vol. 67, no. 4, pp. 268–276, 1990.

[191] A. Miks, J. Novak, and P. Novak, "Third-order design of aspheric spectacle lenses," *Optik*, vol. 121, pp. 2097–2104, 2010.

[192] D. A. Atchison and G. Smith, "Spectacle lenses and third-order distortion," *Ophthalmic and Physiological Optics*, vol. 7, no. 3, pp. 303–308, 1987.

[193] F. LeTexier, W. Lenne, and J.-L. Mercier, "Generalization of the tscherning theory: optimization of aspheric opthalmic lenses," *Ophthalmic and Physiological Optics*, vol. 7, no. 1, pp. 63–72, 1987.

[194] J. E. A. Landgrave and J. R. Moya-Cessa, "Generalized coddington equations in ophthalmic lens design," *Journal of the Optical Society of america A*, vol. 13, no. 8, pp. 1637–1644, 1996.

[195] C. E. Campbell, "Generalized coddington equations found via an operator method," *Journal of the Optical Society of America A*, vol. 23, no. 7, pp. 1691–1698, 2006.

[196] S. Barbero, "Minimum tangential error ophthalmic lens design without multi-parametric optimization," *Optics Communications*, vol. 285, pp. 2769–2773, 2012.

[197] P. Rojo, S. Royo, J. Ramírez, and I. Madariaga, "Numerical implentation of generalized coddington equations for ophthalmic lens design," *Journal of Modern Optics*, vol. 61, no. 3, pp. 204–214, 2014, DOI:10.1080/09500340.2013.878964

[198] D. Crespo, J. Alonso, and J. M. Cleva, "Lentes oftÃ¡lmicas monofocales," WO Patent 040 452A1, 2009.

[199] O. Stavroudis, *The Optics of Rays, Wavefronts, and Caustics*. Academic Press, 2012.

[200] P. Rojo, "Generalized ray tracing method for the calculation of the indiced peripheral refraction by an ophthalmic lens," Ph.D. dissertation, Universitat Politecnica de Catalunya, BarcelonaTech, 2015.

[201] B. Holden, T. Fricke, D. Wilson, M. Jong, K. Naidoo, P. Sankaridurg, T. Wong, T. Naduvilath, and S. Resnikoff, "Global prevalence of myopia and high myopia and temporal trends from 2000 through 2050," *Ophthalmology*, vol. 123, no. 5, pp. 1036–1042, May 2016.

[202] E. L. Smith III, "Optical treatment strategies to slow myopia progression: Effects of the visual extent of the optical treatment zone," *Experimental Eye Research*, vol. 114, pp. 77–88, 2013.

[203] W. N. Charman, "Developments in the correction of presbyopia I: Spectacle and contact lenses," Ophthalmic & Physiological Optics, vol. 34, pp. 8–29, 2014.

[204] D. A. Atchison, "Third-order theory of spectacle lenses applied to correction of peripheral refractive errors," *Optometry and Vision Science*, vol. 88, no. 2, pp. 227–233, February 2011.

[205] R. B. Mandell, *Contact Lens Practice*, 4th ed. Charles C Thomas Pub Ltd, 1988.

[206] W. A. Douthwaite, *Contact Lens Optics & Lens Design*, 2nd ed. Butterworth-Heinemann, 1995.

[207] L. Sorbara and C. Woods, *Correction of Presbyopia with GP Contact Lenses*. The Centre for Contact Lens Research, School of Optometry, University of Waterloo, 2008.

[208] A. Goncharov and C. Dainty, "Wide-field schematic eye models with gradient-index lens," *Journal of the Optical Society of America, A*, vol. 24, no. 8, pp. 2157–2174, 2007.

[209] J. Ares, R. Flores, S. Bara, and Z. Jaroszewick, "Presbyopia compensation with a quartic axicon," *Optometry and Vision Science*, vol. 82, no. 12, pp. 1071–1078, 2005.

[210] D. Madrid-Costa, J. Ruiz-Alcocer, S. García-Lázaro, T. Ferrer-Blasco, and R. Montés-Micó, "Optical power distribution of refractive and aspheric multifocal contact lenses: Effect of pupil size," *Contact Lens & Anterior Eye*, vol. 38, pp. 317–321, 2015.

[211] K. J. Hoffer, *IOL Power*. SLACK Incorporated, 2011.

[212] T. Olsen, "Calculation of intraocular lens power: a review," *Acta Ophthalmologica Scandinavica*, vol. 85, no. 5, pp. 472–485, 2007. [Online]. Available: http://dx.doi.org/10.1111/j.1600-0420.2007.00879.x.

[213] W. Drexler, O. Findl, R. Menapace, G. Rainer, C. Vass, C. K. Hitzenberger, and A. F. Fercher, "Partial coherence interferometry: a novel approach to biometry in cataract surgery," *American Journal of Ophthalmology*, vol. 126, no. 4, pp. 524–534, 1998.

[214] T. Olsen, "Prediction of the effective postoperative (intraocular lens) anterior chamber depth," *Journal of Cataract and Refractive Surgery*, vol. 32, no. 3, pp. 419–424, 2006.

[215] A. Maxwell and L. T. Nordan, Eds., *Current Concepts of Multifocal Intraocular Lenses*. SLACK Incorporated, 1991.

[216] E. Hetch and A. Zajac, *Optics*. Addison-Wesley, 1974.

[217] J. A. Davison and M. J. Simpson, "History and development of the apodized diffractive intraocular lens," *Journal Cataract Refractive Surgery*, vol. 32, pp. 849–858, 2006.

[218] R. T. Ang, "Are extended-depth-of-focus IOLs hitting the visual sweet spot?" *Ophthalmology Times*, vol. 12, no. 2, pp. 24–27, March 2016.

[219] P. Artal, S. Manzanera, P. Piers, and H. Weeber, "Visual effect of the combined correction of spherical and longitudinal chromatic aberrations," *Optics Express*, vol. 18, pp. 1637–1648, 2010.

[220] H. Weeber, "Multi-ring lens, system and methods for extended depth of focus," US Patent US 2014/0 168 602 A1, 2014.

[221] M. S. Millan and F. Vega, "Extended depth of focus intraocular lens: chromatic performance," *Biomedical Optics Express*, vol. 8, no. 9, pp. 4294–4309, 2017.

[222] M. Choi, X. Hong, and Y. Liu, "Multifocal diffractive ophthalmic lens using supressed diffractive order," US Patent US 9335564 B2, 2016.

[223] Y. Houbrechts, C. Pagnouille, and D. Gatinel, 'Intraocular lens," European Patent EP 2 503 962 B, 2013.

[224] (2016, September) Crystalens and Trulign toric: Accommodating IOLs for cataract surgery. All About Vision. [Online]. Available: www.allaboutvision.com/conditions/accommodating-iols.htm.

[225] L. F. Garner and M. K. H. Yap, "Changes in ocular dimensions and refraction with accommodation," *Ophthalmic and Physiological Optics*, vol. 17, no. 1, pp. 12–17, 1997. [Online]. Available: http://dx.doi.org/10.1046/j.1475-1313.1997.96000506.x.

[226] W. N. Charman, "The eye in focus: Accommodation and presbyopia," *Clinical and Experimental Optometry*, vol. 91, no. 3, pp. 207–225, 2008.

[227] P. Rosales, M. Dubbelman, S. Marcos, and R. van der Heijde, "Crystalline lens radii of curvature from Purkinje and Scheimpflug imaging," *Journal of Vision*, vol. 6, no. 2006, pp. 1057–1067, 2006.

[228] G. G. Heath, "Components of accommodation," *Optometry and Vision Science*, vol. 33, no. 11, pp. 569–567, 1959.

[229] D. Hamasaki, J. Ong, and E. Marg, "The amplitude of accommodation in presbyopia," *American Journal of Optometry and Archives*, vol. 33, no. 1, pp. 3–14, 1956.

[230] F. Sun, L. Stark, A. Nguye, J. Wong, V. Lakshminaryanan, and E. Mueller, "Changes in accommodation with age: Static and dynamic," *American Journal of Optometry and Physiological Optics*, vol. 65, no. 6, pp. 492–498, 1988.

[231] H. Hofstetter, "A longitudinal study of amplitude changes in presbyopia," *American Journal of Optometry and Archives of American Academy of Optometry*, vol. 42, no. 1, pp. 3–8, January 1965.

[232] W. Charman, "The path to presbyopia: straight or crooked?" *Ophthalmic and Physiological Optics*, vol. 9, no. 4, pp. 424–430, 1989.

[233] M. Gogging, Ed., *Astigmatism - Optics, Physiology and Management*. InTech, 2012.

[234] M. A. Brzezinski, "Review. astigmatic accommodation (sectional accommodation). a form of dynamic astigmatism," *Clinical and Experimental Optometry*, vol. 65, no. 1, pp. 5–11, 1982.

[235] K. Ukai and Y. Ichihashi, "Changes in ocular astigmatism over the whole range of accommodation," *Optometry and Vision Science*, vol. 68, pp. 813–818, 1991.

[236] L. R. Stark, N. C. Strang, and D. A. Atchison, "Dynamic accommodation response in the presence of astigmatism," *Journal of the Optical Society of America A*, vol. 20, pp. 2228–2236, 2003.

[237] H. Radhakrishnan and W. N. Charman, "Changes in astigmatism with accommodation," *Ophthalmic and Physiological Optics*, vol. 27, no. 3, pp. 275–280, 2007.

[238] H. Hofstetter, "The correction of astigmatism for near work," *Optometry and Vision Science*, vol. 22, no. 3, pp. 121–134, 1945.

[239] The Politician. The College of Optometry. 42 Craven Street, London, WC2N 5NG. [Online]. Available: www.college-optometrists.org/the-college/museum/online-exhibitions/virtual-art-gallery/the-politician.html.

[240] C. E. Letocha, "The invention and early manufacture of bifocals," *Survey of Opthalmology*, vol. 35, pp. 226–235, 1990.

[241] W. J. Rosenthal, *Spectacles and Other Vision Aids: A History and Guide to Collecting*. Norman Publishing, 1996.

[242] W. F. Harris, "Prismatic effect in bifocal lenses and the location of the near optical centre," *Ophthalmic and Physiological Optics*, vol. 14, no. 2, pp. 203–209, April 1994.

[243] M. Freeman, "A compact guide to understand the executive bifocal lens," American Optical, Tech. Rep., 1975.

[244] O. Aves, "Improvements in and relating to multifocal lenses and the like and the method of grinding same," UK Patent 15 735, 1908.

[245] D. Volk and J. W. Weinberg, "The omnifocal lens for presbyopia," *Archives of Ophthalmology*, vol. 68, pp. 776–784, December 1962.

[246] G. Minkwitz, "Über den flächenastigmatismus Bei Gewissen Symmetrischen Asphären," *Optica Acta: International Journal of Optics*, vol. 10, no. 3, pp. 223–227, 1963.

[247] J. E. Sheedy, "Correlation analysis of the optics of progressive addition lenses," *Optometry and Vision Science*, vol. 81, no. 5, pp. 350–361, 2004.

[248] J. E. Sheedy, "Progressive addition lenses - matching the specific lens to patient needs," *Optometry*, vol. 75, no. 2, pp. 83–102, 2004.

[249] G. Minkwitz, "Bemerkungen ÃŒber Nabelpunktslinien auf FlÃ chenstÃŒcken," *Monatsberichte der deutschen Akademie der Wissenschaften*, vol. 7, pp. 608–610, 1965.

[250] A. Schönhofer, "Bemerkungen zu Einem Satz von G. Minkwitz über den Astigmatismus Asphärischer Flächen," *Optica Acta*, vol. 23, pp. 153–159, 1976.

[251] G. Esser, W. Becken, H. Altheimer, and W. Müller, "Generalization of the Minkwitz theorem to nonumbilical lines of symmetrical surfaces," *Journal of the Optical Society of America A*, vol. 34, no. 3, pp. 441–448, 2017.

[252] R. Blendowske, "Simple approach to the generalized Minkwitz theorem," *Journal of the Optical Society of America A*, vol. 34, no. 9, pp. 1481–1483, 2017.

[253] H. Mukaiyama and K. Kato, "Progressive multifocal lens and manufacturing method of eyeglass lens and progressive multifocal lens," US Patent 6 019 470, 2000.

[254] C. Yamamoto, "Progressive power spectacle lens," US Patent 6 354 704B2, 2002.

[255] P. R. Wilkinson, "Progressive power ophthalmic lenses," EP Patent 0 027 339A2, 1981.

[256] H. Altheimer, G. Esser, H. Pfeiffer, R. Barth, M. Fuess, and W. Haimerl, "Progressive ophthalmic lens," US Patent 6 213 603B1, 2001.

[257] A. G. Poullain and D. H. J. Cornet, "Optical lens," US Patent 1 143 316, 1915.

[258] H. O. Gowlland, "Muti-focal lens," CA Patent 159 395, 1914.

[259] H. O. Gowlland, "Improvements in or relating to lenses for spectacles and eye-glasses," GB Patent 191 505 583, 1915.

[260] B. Cretin-Maitenaz, "Multifocal lens having a locally variable power," US Patent 2 869 422, 1959.

[261] B. Maitenaz, "Four steps that led to Varilux," *Optometry and Vision Science*, vol. 43, no. 7, pp. 441–450, July 1966.

[262] D. Volk and J. W. Weinberg, "Brillenlinse und verfahren zum herstellen derselben," DE Patent 1 422 558, 1969.

[263] D. A. Atchison, "New thinking about presbyopia," *Clinical and Experimental Optometry Clinical*, vol. 91, no. May, pp. 205–206, 2008.

[264] D. J. Meister and S. W. Fisher, "Progress in the spectacle correction of presbyopia. part 1: Design and development of progressive lenses," *Clinical and Experimental Optometry*, vol. 91, no. 3, pp. 240–250, 2008.

[265] D. J. Meister and S. W. Fisher, "Progress in the spectacle correction of presbyopia. part 2: Modern progressive lens technologies," *Clinical and Experimental Optometry*, vol. 91, no. 3, pp. 251–264, 2008.

[266] D. R. Pope, "Progressive addition lenses : History , design , wearer satisfaction and trends," *Vision Science and its Applications*, vol. 35, p. NW9, 2000.

[267] W. Köppen. (2018, April) Progressive memories and calculus. [Online]. Available: www.wernerkoeppen.com/home/.

[268] C. Fowler, "Recent trends in progressive power lenses," in *Ophthalmic and Physiological Optics*, vol. 18, no. 2, 1998, pp. 234–237.

[269] G. Kelch, H. Lahres, and H. Wietschorke, "Spectacle lens," US Patent 5 444 503, 1995.

[270] R. A. Chipman, P. Reardon, and A. Gupta, "Ophthalmic optic devices," US Patent 6 000 798, 1999.

[271] E. V. Menezes, "Progressive addition lenses," US Patent 2004/0 080 711A1, 2004.

[272] G. Esser, H. Altheimer, W. Becken, A. Welk, and D. Uttenweiler, "Method for calculating a spectacle lens using viewing angle-dependent prescription data," US Patent 8 915 589B2, 2014.

[273] M. Menozzi, A. V. Buol, H. Krueger, and C. Miege, "Direction of gaze and comfort: discovering the relation for the ergonomic optimization of visual tasks," *Ophthalmic and Physiological Optics*, vol. 14, no. 4, pp. 393–399, 1994.

[274] H. Heuer and D. A. Owens, "Vertical gaze direction and the resting posture of the eyes," *Perception*, vol. 18, no. 3, pp. 363–377, 1989.

[275] H. Heuer, M. Brüwer, T. Römer, H. Kröger, and H. Knapp, "Preferred vertical gaze direction and observation distance," *Ergonomics*, vol. 34, no. 3, pp. 379–392, 1991.

[276] G. Guilino and R. Barth, "Progressive power opthalmic lens," US Patent 4 315 673, 1982.

[277] J. E. Sheedy, C. Campbell, E. King-Smith, and J. R. Hayes, "Progressive powered lenses: the Minkwitz theorem," *Optometry and Vision Science*, vol. 82, no. 10, pp. 916–922, 2005. [Online]. Available: http://content.wkhealth.com/linkback/openurl? sid=WKPTLP:landingpage&an=00006324-200510000-00014.

[278] D. Rodriguez, J. Alonso, and J. A. Quiroga, "Squareness error calibration of a CMM for quality control of ophthalmic lenses," *International Journal of Advanced Manufacturing*, vol. 68, no. 1-4, pp. 487–493, 2013.

[279] R. Arroyo, D. Crespo, and J. Alonso, "Scoring of progressive power lenses by means of user power maps," *Optometry and Vision Science*, vol. 89, no. 4, pp. E489–501, 2012.

[280] T. Shinohara and S. Okazaki, "Progressive multifocal ophthalmic lens," US Patent 4 537 479, 1985.

[281] G. Savio, G. Concheri, and R. Meneghello, "Parametric modelling of free-form surfaces for progressive addition lens," *Proceedings of the IMProVe 2011*, pp. 167–176, 2011.

[282] J. Loos, G. Greiner, and H. Seidel, "A variational approach to progressive lens design," *Computer-Aided Design*, vol. 30, no. 8, pp. 595–602, 1998. [Online]. Available: www.sciencedirect.com/science/article/pii/S0010448597001024.

[283] D. M. Hasenauer, "Design and analysis of opthalmic progressive addition lenses," in *International Optical Design Conference*, ser. Proceedings SPIE, vol. 3482, SPIE. Kona, HI: SPIE, September 1998, pp. 647–655.

[284] X. Jonsson, "Méthodes de points intérieurs et de régions de confiance en optimisation non linéare et application Ã la conception de verres ophthalmiques progressifs," Ph.D. dissertation, UniversitÃ© Paris-VI, 2002.

[285] J. Wang, R. Gulliver, and F. Santosa, "Analysis of a variational approach to progressive lens design," *SIAM Journal on Applied Mathematics*, vol. 64, no. 1, pp. 277–296, 2003. [Online]. Available: https://doi.org/10.1137/S0036139902408941.

[286] W. Jiang, W. Bao, Q. Tang, and H. Wang, "A variational-difference numerical method for designing progressive-addition lenses," *Computer-Aided Design*, vol. 48, pp. 17–27, 2014. [Online]. Available: http://dx.doi.org/10.1016/j.cad.2013.10.011.

[287] J. F. Isenberg, "Formulation of an optical surface for a progressive addition ophthalmic lens," in *International Optical Design Conference*, ser. Proceedings SPIE, vol. 3482, SPIE. Kona, HI: SPIE, September 1998, pp. 627–633.

[288] C. Dürsteler, "Sistema de diseño de lentes progresivas asistido por ordenador," Ph.D. dissertation, Universitat Politecnica de Catalunya, Barcelona, 1991.

[289] M. Tazeroualti, *Curves and Surfaces in Geometric Design*. A. K. Peters, 1994, ch. Designing a progressive lens, pp. 467–474.

[290] G. Mendiola-Anda, "Design of surfaces under physical constraints and its application to the design of ophthalmic lenses," Ph.D. dissertation, School of Computing Sciences, University of East Anglia, Norwich, England, April 2006.

[291] J. T. Winthrop, "Progressive addition spectacle lens," US Patent 4 861 153, 1989.

[292] Y. Tang, Q. Wu, X. Chen, and H. Zhang, "A personalized design for progressive addition lenses," *Optics Express*, vol. 25, pp. 28 100–28 111, 2017.

[293] J. L. Preston, "Progressive addition spectacle lenses: design preferences and head movements while reading," Ph.D. dissertation, Ohio State University, 1998.

[294] S. Wittenberg, "Field study of a new progressive addition lens," *Journal of the Americal Optometric Association*, vol. 49, pp. 1013–1021, 1978.

[295] D. Chapman, "One clinic's experience with varilux 2-the first 400 patients," *Optometry Monthly*, vol. 10, pp. 43–46, 1978.

[296] A. Augsburger, S. Cook, R. Deutch, S. Shackleton-Hartenstein, and R. Wheeler, "Patient satisfaction with progressive addition lenses in a teaching clinic," *Optometry Monthly*, vol. 2, pp. 67–72, 1984.

[297] C. M. Sullivan and C. W. Fowler, "Analysis of a progressive addition lens population," *Ophthalmic and Physiological Optics*, vol. 9, no. 2, pp. 163–170, 1989.

[298] M. H. Cho, C. B. Barnette, B. Aiken, and M. Shipp, "A clinical study of patient acceptance and satisfaction of varilux plus and varilux infinity lenses," *Journal of the American Optometric Association*, vol. 62, no. 6, pp. 449–453, 1991.

[299] J. Gresset, " Subjective evaluation of a new multi-design progressive lens," *Journal of the American Optometric Association*, vol. 62, no. 9, pp. 691–698, 1991.

[300] J. M. Young and I. M. Borish, " Adaptability of a broad spectrum of randomly selected patients to a variable design progressive lens: report of a nationwide clinical trial," *Journal of the American Optometric Association*, vol. 65, no. 6, pp. 445–450, 1994.

[301] M. J. Kris, " Practitioner trial of sola percepta progressive lenses," *Clinical and Experimental Optometry*, vol. 82, no. 5, pp. 187–190, 1999.

[302] D. H. Spaulding, " Patient preference for a progressive addition multifocal lens (varilux2) vs a standard multifocal lens design (st-25)," *Journal of the American Optometric Association*, vol. 52, no. 10, pp. 789–794, 1981.

[303] D. N. Schultz, " Factors influencing patient acceptance of varilux 2 lenses," *Journal of the American Optometric Association*, vol. 54, no. 6, pp. 513–520, 1983.

[304] S. A. Hitzeman and C. O. Myers, " Comparison of the acceptance of progressive addition multifocal vs. a standard multifocal lens design," *Journal of the American Optometric Association*, vol. 56, no. 9, pp. 706–710, 1985.

[305] H. J. Boroyan, M. H. Cho, B. C. Fuller, R. A. Krefman, J. H. McDougall, J. L. Schaeffer, and R. L. Tahran, " Lined multifocal wearers prefer progressive addition lenses," *Journal of the American Optometric Association*, vol. 66, no. 5, pp. 296–300, 1995.

[306] K. Krause, "Acceptance of progressive lenses," *Klinische Monatsblatter fur Augenheilkunde*, vol. 209, no. 2-3, pp. 94–99, 1996.

[307] S. Wittenberg, P. N. Richmond, J. Cohen-Setton, and R. R. Winter, " Clinical comparison of the truvision omni and four progressive addition lenses," *Journal of the American Optometric Association*, vol. 60, no. 2, pp. 114–121, 1989.

[308] K. E. Brookman, E. A. Hall, and M. J. Jensen, " A comparative study of the seiko p-3 and varilux 2 progressive addition lenses," *Journal of the American Optometric Association*, vol. 59, no. 5, pp. 406–410, 1988.

[309] R. Krefman, "Comparison of three progressive addition lens designs: a clinical trial," *Southern Journal of Optometry*, vol. 9, pp. 8–14, 1991.

[310] I. M. Borish, S. A. Hitzeman, and K. E. Brookman, " Double masked study of progressive addition lenses," *Journal of the American Optometric Association*, vol. 51, no. 10, pp. 933–943, 1980.

[311] J. Rozas, "Rodenstock impression ilt versus essilor varilux panamic," Tech. Rep., 2002.

[312] S. Fisher, "Relationship between contour plots and the limits of "clear and comfortable" vision in the near zone of progressive addition lenses," *Optometry and Vision Science*, vol. 74, no. 7, pp. 527–531, 1997. [Online]. Available: http://eutils.ncbi.nlm.nih.gov/entrez/eutils/elink.fcgi?dbfrom=pubmed& id=9293521&retmode\penalty-\@M=ref&cmd=prlinks%5Cnpapers2://publication/ uuid/03420C3A-B3F0-4B09-A289-D587E43DA4EB.

[313] D. R. Pope, S. W. Fisher, and A. M. Nolan, "Visual ergonomics, blur tolerance and progressive lens design," in *Vision Science and its Applications*, ser. OSA Trends in Optics and Photonics, A. Sawchuk, Ed., vol. 53. Monterey, California: Optical Society of America, 2001, p. MA1. [Online]. Available: www.osapublishing.org/ abstract.cfm?URI=VSIA-2001-MA1.

[314] J. S. Solaz, R. Porcar-Seder, B. Mateo, M. J. Such, J. C. Dürsteler, A. Giménez, and C. Prieto, "Influence of vision distance and lens design in presbyopic user preferences," *International Journal of Industrial Ergonomics*, vol. 38, no. 1, pp. 1–8, 2008.

[315] W. Jaschinski, M. Konig, T. M. Mekontso, A. Ohlendorf, and M. Welscher, "Comparison of progressive addition lenses for general purpose and for computer vision: an office field study," *Clinical and Experimental Optometry*, vol. 98, no. 3, pp. 234–243, may 2015.

[316] A. Selenow, E. Bauer, S. R. Ali, and W. Spencer, "Progressive lenses: new techniques for assessing visual performance," in *Vision Science and its Applications*, ser. OSA Technical Digest Series. Santa Fe, New Mexico: Optical Society of America, 2000, p. MD4. [Online]. Available: www.osapublishing.org/abstract.cfm? URI=VSIA-2000-MD4.

[317] A. Selenow, E. A. Bauer, S. R. Ali, L. W. Spencer, and K. J. Ciuffreda, "Assessing visual performance with progressive addition lenses," *Optometry and Vision Science*, vol. 79, no. 8, pp. 502–505, 2002. [Online]. Available: www.ncbi.nlm.nih.gov/ pubmed/12199542.

[318] A. Selenow, E. Bauer, L. W. Spencer, and S. R. Ali, "Can contrast sensitivity be used to the performance of pals ?" pp. 136–139, 2001.

[319] Y. H. Ciuffreda, K. J. Ciuffreda, A. Selenow, E. Bauer, S. R. Ali, and L. W. Spencer, "Eye and head movements during low contrast reading with single vision and prgressive lenses," in *Vision Science and its Applications*, ser. OSA Trends in Optics and Photonics, A. Sawchuk, Ed., vol. 53. Monterey, California: Optical Society of America, 2001, p. MA4.

[320] Y. Han, K. J. Cuffreda, A. Selenow, and S. R. Ali, "Dynamic interactions of eye and head movements when reading with single-vision and progressive lenses in a simulated computer-based environment," *Investigative Ophthalmology & Visual Science*, vol. 44, no. 4, pp. 1534–1545, 2003.

[321] Y. Han, K. J. Cuffreda, A. Selenow, E. Bauer, S. R. Ali, and W. Spencer, "Static aspects of eye and head movements during reading in a simulated computer-based environment with single-vision and progressive lenses," *Investigative Ophthalmology & Visual Science*, vol. 44, no. 1, pp. 145–153, 2003.

[322] P. L. Hendicott, "Spatial perception and progressive addition lenses," Ph.D. dissertation, School of Optometry, Queensland University of Technology, Brisbane, Australia, 2007.

[323] N. Hutchings, E. L. Irving, N. Jung, L. M. Dowling, and K. A. Wells, "Eye and head movement alterations in naïve progressive addition lens wearers," *Ophthalmic and Physiological Optics*, vol. 27, no. 2, pp. 142–153, 2007.

[324] B. Mateo, R. Porcar-Seder, J. S. Solaz, and J. C. Dursteler, "Experimental procedure for measuring and comparing head-neck-trunk posture and movements caused by different progressive addition lens designs," *Ergonomics*, vol. 53, no. 7, pp. 904–913, 2010.

[325] E. A. Villegas and P. Artal, "Visual acuity and optical parameters in progressive-power lenses," *Optometry and Vision Science*, vol. 83, no. 9, pp. 672–681, 2006.

[326] B. Bourdoncle, "Varilux panamic: the design process," *Points de Vue*, no. 42, pp. 1–8, 2000.

[327] H. Minkowski, "Volumen und oberfläche," *Mathematische Annalen*, vol. 57, no. 4, pp. 447–495, 1903.

[328] J. Sheedy, R. F. Hardy, and J. R. Hayes, "Progressive addition lenses - measurements and ratings," *Optometry*, vol. 77, no. 1, pp. 23–39, 2006.

[329] R. Arroyo, D. Crespo, and J. Alonso, " Influence of the base curve in the performance of customized and classical progressive lenses," *Optometry and Vision Science*, vol. 90, no. 3, pp. 282–292, 2013.

[330] E. Chamorro, J. M. Cleva, P. Concepcion, M. Subero, and J. Alonso, "Lens design techniques to improve satisfaction in free-form progressive addition lens users," *JOJ Opthalmology*, vol. 6, no. 3, p. 555688, 2018.

[331] W. Haimerl and P. Baumbach, "State of the art optimization of spectacle lenses (pal)," *Bulgarian Journal of Physics*, vol. 27, no. 1, pp. 7–12, 2000.

[332] J. Loos, P. Slusallek, and H. P. Seidel, "Using wavefront tracing for the visualization and optimization of progressive lenses," *Computer Graphics Forum*, vol. 17, no. 3, pp. 255–265, 1998. [Online]. Available: http://dx.doi.org/10.1111/1467-8659.00272 www.blackwell-synergy.com/links/doi/10.1111/1467-8659.00272.

[333] C. Dickinson, *Low Vision: Principles and Practice*. Butterworth-Heinemann, 1998.

[334] R. L. Brilliant, Ed., *Essentials of Low Vision Practice*. Butterworth-Heinemann, 1999.

[335] K. F. Freeman, *Care of the Patient with Visual Impairement*. www.aoa.org/documents/optometrists/CPG-14.pdf: American Optometric Association, 2007.

[336] (2015, December) International Disease Classification IDC-10, category H54 Visual Impairment Including Blindness. World Health Organization. [Online]. Available: http://apps.who.int/classifications/icd10/browse/2016/en.

[337] (2015, December) Membership requirements. Organización Nacional de Ciegos Españoles (ONCE). [Online]. Available: www.once.es/new/otras-webs/english/Member.

[338] (2014, August) Visual impairment and blindness, fact sheet n°282. World Health Organization (WHO). [Online]. Available: www.who.int/mediacentre/factsheets/fs282/en/.

[339] (2016, January) Microscopyu, the source for microscopic education. Nikon, Inc.; Florida State University; Molecular Expressions. [Online]. Available: www.microscopyu.com/.

[340] D. R. Sanders, J. Retzlaff, and M. C. Kraff, "Comparison of empirically derived and theoretical aphakic refraction formulas," *Archives Ophthalmology*, vol. 101, no. 6, pp. 965–967, 1983.

[341] L. L. Lin, Y. F. Shih, C. B. Tsai, C. J. Cheng, L. A. Lee, P. T. Hung, and P. K. Hou, "Epidemiologic study of ocular refraction among schoolchildren in Taiwan in 1995," *Optometry & Vision Science*, vol. 76, no. 5, pp. 275–281, 1999.

[342] L. Lin, Y. F. Shih, C. K. Hsiao, C. J. Chen, L. A. Lee, and P. T. Hung, "Epidemiologic study of the prevalence and severity of myopia among schoolchildren in Taiwan in 2000," *Journal of the Formosan Medical Association*, vol. 100, pp. 684–691, 2001.

[343] J. Schwiegerling, *Optical Specification, Fabrication, and Testing*, ser. SPIE Digital Library. Society of Photo Optical Engineers, 2014. [Online]. Available: https:// books.google.com/books?id=5QuuoQEACAAJ.

[344] C. Brooks, *Understanding Lens Surfacing*. Butterworth-Heinemann, 1992. [Online]. Available: https://books.google.com/books?id=kpdlQgAACAAJ.

[345] *ISO8980-2:2003 Ophthalmic Optics, Uncut Finished Spectacle Lens, Part 2: Specifications for Progressive Power Lenses*, International Standard Organization Std., 2004.

[346] D. Meister and J. E. Sheedy, "Introduction to opthalmic optics," Carl Zeiss Vision, Tech. Rep., 2008.

[347] R. Williamson, *Field Guide to Optical Fabrication*, ser. Field Guide Series. SPIE, 2011. [Online]. Available: https://books.google.com/books?id=NH5LXwAACAAJ.

[348] M. Jalie, "Modern lens manufacturing techniques. part VIII," *MAFO*, no. 6, pp. 20–22, 2015.

[349] D. Horne, *Spectacle Lens Technology*. Crane, Russak, 1978. [Online]. Available: https://books.google.es/books?id=aZ0fMQAACAAJ.

[350] Zeiss. (2018, April) How is a lens produced? | zeiss international. [Online]. Available: https://goo.gl/UabR5E.

[351] *Coatings*, Ophthalmic optics files ed., Essilor, 2016. [Online]. Available: www.essiloracademy.eu/sites/default/files/10.Coatings.pdf.

[352] DVI. (2017, April) DVI LMS. [Online]. Available: www.dvirx.com/

[353] Ocuco. (2018, April) Ocuco LMS. [Online]. Available: https://www.ocuco.com/ innovations-features/.

[354] CCSystems. (2018, April) CCSystems LMS. [Online]. Available: http:// opticalonline.com/index.php/products/.

[355] DCS, *Data Communications Standard*, Data Communication standard committee Std., 2017. [Online]. Available: https://www.thevisioncouncil.org/sites/default/files/ DCS-v3-11-FINAL1.pdf.

[356] DCS, *Lens Description Standard*, Data Communication standard committee Std., 2014. [Online]. Available: https://www.thevisioncouncil.org/sites/default/files/DCS-v3-11-FINAL1.pdf.

[357] Essilor. (2018, April) Lens design software. [Online]. Available: http://essiloriig.com/for-your-lab/digital-surfacing/lens-design-software/.

[358] Zeiss. (2018, April) Zeiss lds. [Online]. Available: www.zeiss.com.

[359] IOT. (2018, April) IOT Futura. [Online]. Available: www.iot.es/lensdesignsoftware .html.

[360] Automation and Robotics. (2018, April) Dual lens mapper. [Online]. Available: www.ar.be/en/products/dual-lensmapper.

[361] W. E. Humphrey, "Lens meter with automated readout," U.S. Patent 4 180 325, 1979. [Online]. Available: www.google.ch/patents/US4180325.

[362] J. A. Gomez-Pedrero, J. Alonso, H. Canabal, and E. Bernabeu, "A generalization of prentice's law for lenses with arbitrary refracting surfaces," *Ophthalmic and Physiological Optics*, vol. 18, no. 6, pp. 514–520, 1998.

[363] C. E. Campbell, "Lensmeter with correction for refractive index and spherical aberration," U.S. Patent 5 175 594, 1992. [Online]. Available: www.google.com/ patents/US5175594.

[364] (2018, April) Refractive index info. [Online]. Available: https://refractiveindex.info.

[365] C. Huang, "Measurement and comparison of progressive addition lenses by three techniques," Ph.D. dissertation, Ohio State University, 2011.

[366] Visionix. (2018, April) Dual lens mapper. [Online]. Available: http://visionixusa .com/ecp-products/refraction-instruments/lensometers.

[367] K. R. Freischlad and C. L. Koliopoulos, "Modal estimation of a wave front from difference measurements using the discrete fourier transform," *J. Opt. Soc. Am. A*, vol. 3, no. 11, pp. 1852–1861, 1986. [Online]. Available: http://josaa.osa.org/ abstract.cfm?URI=josaa-3-11-1852.

[368] R. I. Hartley and A. Zisserman, *Multiple View Geometry in Computer Vision*, 2nd ed. Cambridge University Press, ISBN: 0521540518, 2004.

[369] J. A. Quiroga, D. Crespo, and E. Bernabeu, "Fourier transform method for automatic processing of moire deflectograms," *Optical Engineering*, vol. 38, no. 6, pp. 974–982, 1999.

[370] J. Vargas, J. A. Gomez-Pedrero, J. Alonso, and J. A. Quiroga, "Deflectometric method for the measurement of user power for ophthalmic lenses," *Applied Optics*, vol. 49, no. 27, pp. 5125–5132, 2010.

[371] Lambda-X. (2018, April) Lens mappers. [Online]. Available: www.lambda-x.com/ ophthalmics.

[372] L. Joannes, F. Dubois, and J.-C. Legros, "Phase-shifting Schlieren: high-resolution quantitative Schlieren that uses the phase-shifting technique principle," *Appl. Opt.*, vol. 42, no. 25, pp. 5046–5053, 2003. [Online]. Available: http://ao.osa.org/abstract .cfm?URI=ao-42-25-5046.

[373] G. H. Markus C. Knauer, Jurgen Kaminski, "Phase measuring deflectometry: a new approach to measure specular free-form surfaces," *Proc.SPIE*, vol. 5457, pp. 5457 – 5457 – 11, 2004. [Online]. Available: https://doi.org/10.1117/12.545704.

[374] H. Rapp, *Reconstruction of Specular Reflective Surfaces using Auto-Calibrating Deflectometry*, ser. Schriftenreihe / Institut für Mess- und Regelungstechnik, Karlsruher Institut für Technologie. KIT Scientific Publishing, 2012. [Online]. Available: https://books.google.es/books?id=0NyU9k8zcwoC.

[375] 3DShape. (2018, April) Specgage3d. [Online]. Available: www.3d-shape.com.

[376] D. Malacara, *Optical Shop Testing*, ser. Wiley Series in Pure and Applied Optics. Wiley, 2007.

[377] O. Kafri and I. Glatt, *The physics of moiré metrology*, ser. Wiley series in pure and applied optics. Wiley, 1990. [Online]. Available: https://books.google.es/books? id=ci5RAAAAMAAJ.

[378] Rotlex. (2018, April) Lens mappers. [Online]. Available: www.rotlex.com/products/ spectacle-lenses.

[379] R. Sudol and B. J. Thompson, "Lau effect: theory and experiment," *Appl. Opt.*, vol. 20, no. 6, pp. 1107–1116, 1981. [Online]. Available: http://ao.osa.org/abstract .cfm?URI=ao-20-6-1107.

[380] D. Crespo, J. Alonso, and E. Bernabeu, "Generalized grating imaging using an extended monochromatic light source," *J. Opt. Soc. Am. A*, vol. 17, no. 7, pp. 1231–1240, 2000. [Online]. Available: http://josaa.osa.org/abstract.cfm?URI=josaa-17-7-1231.

[381] Renishaw. (2018, April) Styli and accessories. [Online]. Available: www.renishaw .com/en/styli-for-touch-probes–6333.

[382] D. Rodríguez-Ibáñez, J. Alonso, and J. A. Quiroga, "Squareness error calibration of a CMM for quality control of ophthalmic lenses," *The International Journal of Advanced Manufacturing Technology*, vol. 68, no. 1, pp. 487–493, 2013. [Online]. Available: https://doi.org/10.1007/s00170-013-4746-y.

[383] C.-Y. Huang, T. W. Raasch, A. Y. Yi, J. E. Sheedy, B. Andre, and M. A. Bullimore, "Comparison of three techniques in measuring progressive addition lenses," *Optometry and Vision Science*, vol. 89, no. 11, pp. 1564–1573, 2012.

[384] D. Larrue and M. Legeard, "A comparison of four different lens mappers," *Optometry and Vision Science*, p. 1, 2014. [Online]. Available: http://dx.doi.org/10.1097/OPX.0000000000000396.

[385] Automation and Robotics, "Process control and lens inspection of freeform lenses," 2017. [Online]. Available: https://www.ar.be/sites/default/files/DLM%20Whitepapers/Process%20control%20&%20Lens%20inspection.pdf

[386] Automation and Robotics, "Freeform lens mapping error patterns," 2017. [Online]. Available: www.ar.be/sites/default/files/DLMWhitepapers/FreeformLensMapping.pdf.

[387] R. Henderson and K. Schulmeister, *Laser Safety*. Taylor and Francis, 2004.

[388] D. van Norren and T. Gorgels, "The action spectrum of photochemical damage to the retina: A review of monochromatic threshold data," *Photochemistry and Photobiology*, vol. 87, pp. 747–753, 2011.

[389] F. L. Pedrotti, L. M. Pedrotti, and L. S. Pedrotti, *Introduction to Optics*, 3rd ed. Addison-Wesley, 2006. [Online]. Available: www.amazon.com/exec/obidos/redirect?tag=citeulike07-20&path=ASIN/0131499335.

[390] *ISO8980-3:2003 Ophthalmic Optics, Uncut Finished Spectacle Lens, Part 3: Transmittance Specification and Test Methods*, International Standard Organization Std., 2003.

[391] K. Fujita, S. Hatano, D. Kato, and A. J., "Photochromism of a radical diffusion-inhibited hexaarylbiimidazole derivative with intense coloration and fast decoloration performance," *Optics Letters*, vol. 14, pp. 3105–3108, 2008.

[392] J. C. Crano and R. J. Guglielmeti, Eds., *Organic Photochromic and Thermochromic Compounds. Volume 1: Main Photochromic Families*. Springer Verlag, 2002.

[393] (2018, April) History of Transitions Optical Company. Transitions Optical Inc. [Online]. Available: www.transitions.com/en-us/our-company/history/.

[394] F. Eperjesi, C. W. Fowler, and B. J. W. Evans, "Do tinted lenses or filters improve visual performance in low vision? a review of the literature," *Ophthalmic and Physiological Optics*, vol. 22, pp. 68–77, 2002.

[395] T. W. Leung, R. W. Hong Li, and C. Su Kee, "Blue-light filtering spectacle lenses: Optical and clinical performances," *PLos ONE*, vol. 12, no. 1, p. e0169114, 2017.

[396] Essilor, "Cahiers d'optique oculaire. 9. les traitements," 1997.

[397] B. Kooi and C. Bergman, "An approach to the study of ancient archery using mathematical modelling," *Antiquity*, vol. 71, no. 271, pp. 124–134, 1997.

[398] E. Rosen, "The invention of eyeglasses, part i," *Journal of the History of Medicine and Allied Sciences*, vol. XI, no. 1, pp. 13–46, 1956.

[399] E. Rosen, "The invention of eyeglasses, part ii," *Journal of the History of Medicine and Allied Sciences*, vol. XI, pp. 183–218, 1956.

[400] E. C. Watson, "The invention of spectacles," *Science in Art. Engineering and Science*, vol. 17, no. 6, p. 21, 1954.

[401] (2016, August) The invention of spectacles. The College of Optometrists. 42 Craven Street, London, WC2N 5NG. [Online]. Available: https://www.college-optometrists.org/the-college/museum/online-exhibitions/virtual-spectacles-gallery/the-invention-of-spectacles.html.

[402] A. Gonzalez-Cano, "Eye gimnastics and a negative opinion on eyeglasses by the spanish renaissance physician Cristóbal Méndez," *Atti della Fondazione Giorgio Ronchi*, vol. LIX, no. 4, pp. 559–565, 2004.

[403] (2018, April) Portrait of cardinal Fernando Niño de Guevara by El Greco. Metropolitan Museum of Art, New York City. [Online]. Available: www.metmuseum.org/art/collection/search/436573.

[404] K. G. Wakefield, *Bennet's Ophthalmic Prescription Work*, 3rd ed. Butterworth-Heinemann, 1994.

[405] (2016, August) Nylon material for cube pro. 3DSystems. [Online]. Available: www.3dsystems.com/shop/cartridges/nylon.

[406] H. J. Lunk, "Discovery, properties and applications of chromium and its compunds," *ChemTexts*, vol. 1, no. 6, pp. 1–17, 2015. [Online]. Available: http://link.springer.com/article/10.1007/s40828-015-0007-z.

[407] *ISO 7998:2005. Optics and Optical Instruments. Spectacle Frames. Vocabulary and List of Equivalent Terms* International Std., 2005.

[408] *ISO 8624:2011 Ophthalmic Optics - Spectacle Frames - Measuring System and Terminology*, International Std., 2011.

[409] *ISO 9456:1991 Optics and Optical Instruments - Ophthalmic Optics - Marking of Spectacle Frames*, International Std., 1991.

[410] E. Krimsky, "The seven primary positions of gaze," *The British Journal of Ophthalmology.*, vol. 51, no. 2, pp. 105–114, 1967.

[411] S. Lipschutz and M. Lipson, *Schaum's Outline of Linear Algebra, Sixth Edition.* McGraw-Hill Education, 2017. [Online]. Available: https://books.google.com/books?id=K6E4DwAAQBAJ.

[412] M. Lipschutz, *Schaum's Outline on Differential Geometry*. McGraw-Hill Education, 1969.

[413] M. P. DoCarmo, *Differential Geometry of Curves and Surfaces*. Prentice Hall, 1976.

[414] J. Alonso, J. A. Gómez-Pedrero, and E. Bernabeu, "Local dioptric power matrix in a progressive addition lens," *Ophthalmic and Physiological Optics*, vol. 17, no. 6, pp. 522–529, 1997. [Online]. Available: www.ncbi.nlm.nih.gov/pubmed/9666927.

[415] V. Mahajan, *Aberration Theory Made Simple:* 2nd ed. SPIE Optical Engineering Press, 2011.

[416] V. N. Mahajan, *Aberration Theory Made Simple*, 1st ed. SPIE Optical Engineering Press, 1991.

[417] R. J. Noll, "Zernike polynomials and atmospheric turbulence," *Journal of the Optical Society of America*, vol. 66, no. 207-211, 1976.

Index